Subcellular Biochemistry

Volume 27

Biology of the Lysosome

SUBCELLULAR BIOCHEMISTRY

SERIES EDITOR

J. ROBIN HARRIS, Institute of Zoology, University of Mainz, Mainz, Germany

ASSISTANT EDITORS

H. J. HILDERSON, University of Antwerp, Antwerp, Belgium
B. B. BISWAS, University of Calcutta, Calcutta, India

Subcellular Biochemistry

Volume 27

Biology of the Lysosome

Edited by

John B. Lloyd and
Robert W. Mason

Thomas Jefferson University
Philadelphia, Pennsylvania, and
Nemours Research Programs
Wilmington, Delaware

SPRINGER SCIENCE+BUSINESS MEDIA, LLC

The Library of Congress cataloged the first volume of this title as follows:

Sub-cellular biochemistry.
   London, New York, Plenum Press.
     v. illus. 23 cm. quarterly.
   Began with Sept. 1971 issue. Cf. New serial titles.
   1. Cytochemistry—Periodicals. 2. Cell organelles—Periodicals.
QH611.S84 574.8'76   73-643479

ISSN 0306-0225

ISBN 978-1-4613-7674-3   ISBN 978-1-4615-5833-0 (eBook)
DOI 10.1007/978-1-4615-5833-0

This series is a continuation of the journal *Sub-Cellular Biochemistry,*
Volumes 1 to 4 of which were published quarterly from 1972 to 1975

© 1996 Springer Science+Business Media New York
Originally published by Plenum Press New York in 1996
Softcover reprint of the hardcover 1st edition 1996

10 9 8 7 6 5 4 3 2 1

Contributors

Thomas Braulke   Institut für Biochemie II, Georg-August-Universität Göttingen, D-37073 Göttingen, Germany

Fergus J. Doherty   Department of Biochemistry, Queens Medical Center, Nottingham NE7 2UH, United Kingdom

William J. Johnson   Department of Biochemistry, MCP Hahnemann School of Medicine, Allegheny University of the Health Sciences, Philadelphia, Pennsylvania 19129

Motoni Kadowaki   Department of Applied Biochemistry, Faculty of Agriculture, Niigata University, Niigata, Japan

John B. Lloyd   Department of Pediatrics, Jefferson Medical College, Philadelphia, Pennsylvania 19107; and Division of Developmental Biology, Nemours Research Programs, Wilmington, Delaware 19899

Robert W. Mason   Division of Developmental Biology, Nemours Research Programs, Wilmington, Delaware 19899; and Department of Pediatrics, Jefferson Medical College, Philadelphia, Pennsylvania 19107

R. John Mayer   Department of Biochemistry, Queens Medical Center, Nottingham NE7 2UH, United Kingdom

Giovanni Miotto   Dipartimento di Chimica Biologica, Universita Degli Studi di Padova, 35121 Padova, Italy

Glenn E. Mortimore   Department of Cellular and Molecular Physiology, Hershey Medical Center, The Pennsylvania State University, Hershey, Pennsylvania 17033

Ronald L. Pisoni   Department of Internal Medicine, Division of Nephrology, The University of Michigan Medical Center, Ann Arbor, Michigan 48109

George H. Rothblat   Department of Biochemistry, MCP Hahnemann School of Medicine, Allegheny University of the Health Sciences, Philadelphia, Pennsylvania 19129

Elizabeth Smythe   Department of Biochemistry, Medical Sciences Institute, University of Dundee, Dundee DD1 4HN, Scotland

Rebecca W. Van Dyke   Department of Internal Medicine, University of Michigan Medical School, Ann Arbor, Michigan 48109

Rina Venerando   Dipartimento di Chimica Biologica, Universita Degli Studi di Padova, 35121 Padova, Italy

Gregory J. Warner   Department of Biochemistry, MCP Hahnemann School of Medicine, Allegheny University of the Health Sciences, Philadelphia, Pennsylvania 19129

Robert Wattiaux   Laboratoire de Chimie Physiologique, Facultés Universitaires Notre-Dame de la Paix, B5000 Namur, Belgium

Simone Wattiaux-De Coninck   Laboratoire de Chimie Physiologique, Facultés Universitaires Notre-Dame de la Paix, B5000 Namur, Belgium

Bryan G. Winchester   Division of Biochemistry and Genetics, Institute of Child Health, London WC1N 1EH, United Kingdom

Patricia G. Yancey   Department of Biochemistry, MCP Hahnemann School of Medicine, Allegheny University of the Health Sciences, Philadelphia, Pennsylvania 19129

Foreword

"Lysosomes are now known to be not just a collection of isolated organelles of interest only to the biochemist, but part of a complex, dynamic, membranous system essential to the cell's economy." So wrote the late Dame Honor Fell and I in the preface to the first volume of *Lysosomes in Biology and Pathology* almost 30 years ago. We went on to say that research on the lysosomal system at the time was in a state of explosive and chaotic growth.

While the chaos has been largely reduced, the growth of research into the biology of the lysosome remains considerable. Biologists worldwide are still fascinated by the diversity of activities and the interaction between the various membranous systems of the cell, both in biological and pathological situations. The present volume, edited by John Lloyd and Robert Mason, who have each made major contributions to research in this field, continues to systematize the growth of information in this important area.

Taken together, the twelve chapters of this volume form an extensive update of our knowledge of the biological and physiological role of the lysosomal system. The book will enhance our knowledge of cell function and help in our understanding of the factors that control cell metabolism in health and disease.

John T. Dingle

Hughes Hall
Cambridge, England

ix

Preface

Lysosomes are still orphan organelles. Even in the 1990s, it is not unknown for members of a scientific or medical audience to reveal, by a question following a lecture, that they confuse *lysosome* with *lysozyme* or *liposome*. And this ignorance has a deeper reason than mere confusion over similar-sounding terms. It arises because lysosomes exist on the periphery of cell function, rather than at its center. Digesting waste macromolecules and recycling the products is certainly an important function, but it does not have the immediate link to cell viability that, for example, mitochondrial energy metabolism has. The remarkably large number of inherited human diseases that result from defects in lysosomal enzymes or lysosome membrane metabolite porters is itself a testimony that lysosome function is of secondary importance to the life of the cell. A strike of garbage collectors can be tolerated for longer than a strike of workers in the electricity supply industry.

However, lysosomes have been fortunate in their protagonists. Their principal discoverer, Christian de Duve, in addition to being a formidable intellect and a fine scientist, is an extraordinarily articulate communicator. His lectures and published review articles over four decades have been models of lucidity, vision, and disciplined speculation. Another major publicist for lysosomes has been John Dingle, who, first with Dame Honor Fell and later with Roger Dean and others, envisioned and edited a series of seven volumes with the general title *Lysosomes in Biology and Pathology*. This series, published over the period from 1969 to 1984, still remains a valuable source of information. We are honored that John Dingle has agreed to contribute a foreword to the present volume.

With the notable and admirable exception of Eric Holtzman's *Lysosomes*, published in 1989, there has been a lack of books in the past decade that describe and discuss the principal facets of lysosome biology. We concluded that the time was ripe for such a book and decided that it should be a multi-author volume with contributions from acknowledged experts in the field. We also decided that the heart of the book should be a series of chapters describing the metabolic activities of lysosomes. We believe that this section of the book (Part II) has no precedent. The section on metabolism is preceded by a description of how the components of the lysosome and the substrates for digestion are delivered (Part I) and is followed by a consideration of how the lysosome interacts with its cytoplasmic environment (Part III). Together, the twelve chapters provide a comprehensive review of current knowledge of the function of lysosomes in mammalian cells.

Having a clear conception of a book's contents is no doubt good, but it is also an anxiety generator for editors. This is because any author who defaults in delivering a promised chapter will leave the book with a painfully obvious gap. We are therefore extremely grateful to the authors who agreed to contribute to this present volume, both for writing their chapters well and for delivering them to us on time.

We thank Robin Harris, the Series Editor of *Subcellular Biochemistry,* for his steady encouragement during the gestation period of this volume, and Joanna Lawrence of Plenum Press for her skill and competence during its final production phase. And we particularly thank Susan-Jean Keenan, whose meticulous attention to detail and sheer hard work have been invaluable to us during many aspects of the editorial process. Also we are grateful to Carole Andrew, who helped us greatly in the final phase of assembling the manuscript for the press.

We believe that *Biology of the Lysosome* will be a useful resource for professional cell biologists and for students who need to know about this most fascinating organelle.

John B. Lloyd
Robert W. Mason

Wilmington, Delaware

Contents

Chapter 3

Endocytosis
Elizabeth Smythe

Chapter 4

Autophagy
Glenn E. Mortimore, Giovanni Miotto, Rina Venerando, and Motoni Kadowaki

Chapter 5

Selective Proteolysis: 70-kDa Heat-Shock Protein and Ubiquitin-Dependent Mechanisms?
R. John Mayer and Fergus J. Doherty

Part II: Metabolism in the Lysosome

Chapter 6

Lysosomal Metabolism of Proteins
Robert W. Mason

Chapter 7

Lysosomal Metabolism of Glycoconjugates
Bryan G. Winchester

Chapter 8

Lysosomal Metabolism of Lipids
William J. Johnson, Gregory J. Warner, Patricia G. Yancey, and
George H. Rothblat

Chapter 9

Lysosomal Nucleic Acid and Phosphate Metabolism and Related Metabolic Reactions
Ronald L. Pisoni

Part III: The Lysosome in Its Cytoplasmic Environment

Chapter 10

Acidification of Lysosomes and Endosomes
Rebecca W. Van Dyke

Chapter 11

Metabolite Efflux and Influx across the Lysosome Membrane
John B. Lloyd

Chapter 12

Lysosome Pharmacology and Toxicology
Robert Wattiaux and Simone Wattiaux-De Coninck

Chapter 1

The Taxonomy of Lysosomes and Related Structures

John B. Lloyd

1. THE VACUOLAR APPARATUS REVISITED

This book on lysosomes is appearing in 1996, 30 years after the publication of the landmark review by Christian de Duve and Robert Wattiaux in the *Annual Review of Physiology*. It so happened that I was visiting Dr. Wattiaux's laboratory in Namur, Belgium, on the day the proofs of this review arrived in the mail. A vivid memory is his amusement that the key word in its title had been misspelled *lusosomes*.

In addition to summarizing most of the recent literature on lysosomes (and most of the literature was recent at that stage), the review of de Duve and Wattiaux (1966) was a significant milestone in setting the lysosome in a broader cell biological framework. The authors presented the lysosome as a component of a family of membrane-bounded vesicles that had been described as associated with and/or fusing with lysosomes. The term *vacuolar apparatus* was proposed for this family of organelles "as a reminder of its historical roots and in order to emphasize its essentially extracellular character." The review also provided the first attempt at a classification of the various organelles comprising the vacuolar apparatus.

It is interesting to reread the de Duve and Wattiaux review as a historical document. The term vacuolar apparatus never gained wide acceptance, nor did

John B. Lloyd   Department of Pediatrics, Jefferson Medical College, Philadelphia, Pennsylvania 19107, and Division of Developmental Biology, Nemours Research Programs, Wilmington, Delaware 19899.

Subcellular Biochemistry, Volume 27: Biology of the Lysosome, edited by Lloyd and Mason. Plenum Press, New York, 1996.

the broader concept of the *vacuome* introduced by de Duve (1969). Neither term was used by de Duve himself in his 1983 retrospective, *Lysosomes Revisited*, but the concept of the lysosome as a pivotal part of the well-defined and rather self-contained intracellular digestive system became the paradigm within which lysosome research has flourished for a generation. In 1996 it is more difficult to define the limits of the vacuolar apparatus as distinct from the wider world of intracellular organelles. Nevertheless, the broad outlines have stood the test of time.

Figure 1 is reproduced from de Duve and Wattiaux (1966) and Fig. 2 from a contemporary article (Jacques, 1966) written by one of their colleagues. The action is in the secondary lysosomes, vacuoles where the digestive enzymes meet their macromolecular substrates. The secondary lysosomes were seen as the only location where both enzymes and substrates are together and thus the only part of the vacuolar system where digestion could take place. The secondary lysosomes are formed by the fusion of a primary lysosome and a phagosome, which could be either a *heterophagosome* or an *autophagosome*. Following a completed digestive event, the secondary lysosome may enter the cycle multiple times by fusing with

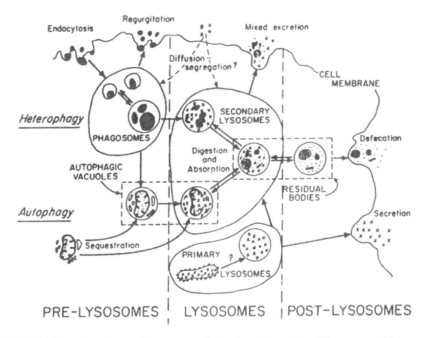

FIGURE 1. The various forms of lysosomes and related particles and the different types of interactions which they may have with each other and with the cell membrane. Each cell type is believed to be the site of one or more of the circuits shown, but not necessarily all. Crosses symbolize acid hydrolases. Reproduced with permission from de Duve and Wattiaux (1966).

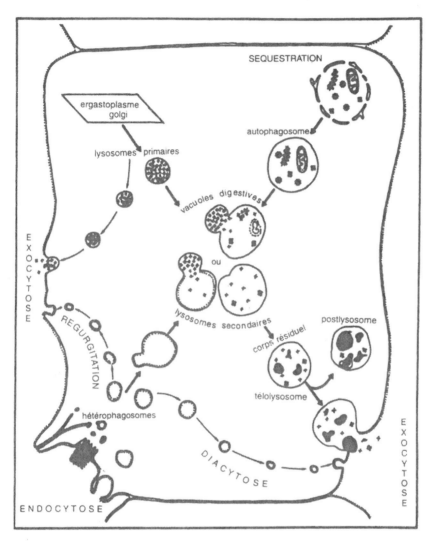

FIGURE 2. The vacuolar apparatus. Reproduced with permission from Jacques (1966). *Reader's note:* Labels on this figure are in French, as in the original text of this historical document.

new, incoming phagosomes. Eventually, it was envisioned, the lysosome might become "constipated" by the accumulation of substances its enzymes could not digest. These residue-filled *telolysosomes* (or residual bodies) could lose their enzymes entirely and become *postlysosomes,* permanent features of the cell from that point on. Alternatively, at least in some cell types, the telolysosomes could release their content by exocytosis. As will be evident from the chapters that follow, this scheme has undergone remarkably little modification.

The origin of the primary lysosome was already a controversial topic in 1966, as indicated by the lack of detail and the question mark in Fig. 1. It was agreed by all parties, however, that the lysosomal enzymes had their origin in the rough endoplasmic reticulum and that the Golgi apparatus was an important way station in their subsequent translocation to the lysosomes. This delivery route and its mechanisms have been the subject of intense investigation over the past decade. The results are discussed in detail in Chapter 2 of this volume. It is noteworthy that Figs. 1 and 2 depict some lysosomal enzymes being released to the extracellular environment.

The process of autophagy was discussed in considerable detail in the 1966 review. More than any other area of lysosome research, this concept relied heavily on morphological evidence, specifically the electron microscope profiles of intracellular vacuoles containing mitochondria and other cytoplasmic organelles. It was proposed that these autophagosomes were formed by the sequestration of some portion of cytoplasm by cellular membranes. The question of whether the sequestration process is random or in some way selective was discussed explicitly. It is only in the past decade that there has been significant progress in understanding the mechanisms and the regulation of autophagy, as summarized in Chapters 4 and 5.

By contrast heterophagy seemed relatively easy to understand in 1966. The heterophagosome was derived from the plasma membrane and brought either solutes (pinocytosis) or particles (phagocytosis) into the cell (see Fig. 2). The possibility of a short circuit from the heterophagosome back to the plasma membrane, now such a prominent feature of current understandings of endosomal function, was already envisioned and termed *regurgitation*.

The fate of the low-molecular-weight products of intralysosomal digestion was not explicitly mentioned in the contributions to the 1963 Ciba symposium or by de Duve (1964). The release of metabolites across the lysosome membrane and their reutilization was presumably considered self-evident and of less interest than the possibilities of cell pathology resulting from incomplete digestion or from damage to the lysosome membrane. By 1966, however, de Duve and Wattiaux were writing that it would be of great interest to find out whether these "micromolecular products of digestion . . . traverse the membrane by simple passive diffusion or are transported by permeases or other carrying mechanisms." Diagrams of the vacuolar apparatus in which metabolite efflux by passive diffusion and perhaps active transport are explicitly featured appear in reviews by de Duve (1966, 1968).

2. THE ASCENDANCY OF THE ENDOSOME

In the early days of lysosome research much of the emphasis was understandably and necessarily on establishing the organelle as a distinct subcellular entity, different from mitochondria and peroxisomes. Although the battles are now long past, it is worth recalling that the lysosome concept was not without its detractors. As late as 1963, an article in the *Biochemical Journal* (Conchie and Hay,

1963) began provocatively, "The lysosome theory, whether it is true or not . . . ," and went on to indicate that the authors clearly thought the theory was not valid. The disagreement was over the data on hydrolase latency that had been the basis for the lysosome concept. Conchie and Hay (1963) maintained that these data had been too simplistically interpreted and that the lysosome concept was either untrue as originally conceived or so loosely defined as to be of little value. Historians and philosophers of science would find this episode interesting, as it illustrates well how a flexibly formulated hypothesis may legitimately be retained in such circumstances because of its unifying and explanatory potential.

A concept from this period that survived too well was of lysosomes as "suicide bags" (de Duve, 1959), prepackaged destruction awaiting a programmed instruction to strike. It is important to remember that in the early decades of lysosome research there was a major emphasis on lysosomes as mediators of cell damage, with the lysosome membrane as a vulnerable *cordon sanitaire* separating the cytoplasm from a seething cauldron of degradative enzymes. Either nature herself or noxious substances in the environment could cause this membrane to break, leading to cell death. There was much interest during this period in lysosome membrane labilizers and stabilizers, reflecting the belief that the membrane was liable to rupture and needed help not to do so. Although de Duve (1969) points out that he never thought of lysosomes as principally "suicide bags," the metaphor he created was so graphic that the notion gained a currency that endured several decades despite the paucity of supporting evidence.

From at least the time of de Duve's 1963 review, the emphasis of lysosome research was on the lysosome as an intracellular digestive tract. The lysosomal enzymes were no longer a sinister enemy within, but were recognized as part of the cellular housekeeping. They were the cell's "recycling plant" (Lloyd, 1990), removing damaged and effete macromolecules from the cell's environment and from within the cell itself, converting them into reusable products. This view of the lysosome has persisted.

The most recent trend, however, has taken the spotlight away from the lysosome entirely. As explained above, heterophagy was the term favored by de Duve and Wattiaux (1966) for the process by which particles or fluid were captured from the cell's environment and delivered to the lysosomes. The heterophagosome was the organelle, bounded by membrane derived from the plasma membrane, that shuttled exogenous material to the lysosomes for digestion. This unitary functional identity is reflected in Table II of de Duve and Wattiaux (1966), where heterophagosomes are classified as prelysosomes.

In the past 30 years heterophagosomes have become *endosomes,* and in the process their functions have been recognized to encompass much more than providing a route to the lysosomes. The endosome has now replaced the lysosome as the center of attention and the true hub of the vacuolar system. Delivery of substances to the lysosomes from the endosomes is now regarded as but one route from the endosome, and perhaps one of the less interesting ones.

3. A TAXONOMY AND A ROUTE MAP

3.1. An Evolving Nomenclature

The word *endocytosis* was proposed by de Duve in a footnote on page 126 of de Reuck and Cameron (1963) as a generic term for pinocytosis and phagocytosis, two processes in which extracellular material is internalized by cells in a vesicle derived by the invagination and pinching-off of plasma membrane. In the same footnote he proposed the term *exocytosis* for the "reverse pinocytosis believed to be involved in secretory mechanisms".

Unfortunately (in the present author's view), *endocytosis* is now widely used in a different and more restricted sense: to denote only receptor-mediated ligand uptake. In current usage the word no longer subsumes the process of phagocytosis. Moreover, receptor-mediated uptake of solutes by so-called endocytosis is often contrasted with nonspecific uptake of liquid and contained solutes, confusingly now called *pinocytosis,* a term that formerly designated both processes. Current usage is a potential source of misunderstanding, and for a more fundamental reason than mere change of meaning with time. An invagination of the plasma membrane that leads to the formation of an intracellular vesicle carries into the cell a volume of exogenous ambient liquid and solutes derived from it. The same vesicle may contain solutes that are captured passively, simply because they are present in the ambient liquid, solutes that are concentrated because they bind nonspecifically to the membrane, and ligands bound to specific receptors that may have been recruited, by the process of binding, into the membrane domain being invaginated. Thus, a cell may be described as capturing sucrose by pinocytosis and transferrin by endocytosis, when at the level of membrane movements a single event is responsible for both phenomena.

It is probably a lost battle to argue that pinocytosis should be reinstated as the term to describe this process. But if endocytosis must win the day, could it not be used to denote what pinocytosis used to denote? Then we could recognize that endocytosis can internalize solutes by fluid-phase, adsorptive, and receptor-mediated mechanisms. The present practice of contrasting endocytosis and pinocytosis is scientifically untenable and a source of unnecessary confusion.

As already explained, the term endocytosis was coined to describe the uptake of particles and the uptake of liquid and solutes. The use of the term endocytosis to denote what was formerly denoted by pinocytosis involves a real semantic loss: We no longer have a term to use when we are not sure which mechanism is operative. There seems no doubt that phagocytosis is the only mechanism by which cells can engulf large particles, but some particulate matter is small enough to be accommodated within pinocytic vesicles. When the uptake of particles in the range of 30–1100 nm by peritoneal macrophages was studied, the results suggested a smooth transition in uptake mechanism from pinocytosis to phagocytosis with increasing particle size (Pratten and Lloyd, 1986). Fortunately for the authors, that paper was written before endocytosis lost its original meaning.

As noted above, the term phagosome no longer embraces all vesicles delivering exogenous materials, but only those derived from phagocytosis. Because the phagocytic process involves the movement of plasma membrane around the particle being engulfed, phagosomes are presumed to contain little if any of the ambient fluid. Thus phagocytosis probably delivers little except the particle itself to the lysosome.

The term endosome is attributed by de Duve (1964) to A. B. Novikoff in a book published the previous year (Novikoff, 1963). In its original usage, endosome was synonymous with phagosome, and indeed Novikoff himself used the latter term in his contribution to the 1963 Ciba symposium. Now, however, endosome and phagosome do not overlap in meaning. Today's sophisticated endosome model is a development and refinement of the earlier pinosome concept, the endosomes being envisioned as highly dynamic structures that "interchange components with other endosomes, the Golgi complex and the plasma membrane through budding and fusion events" (Berón et al., 1995).

3.2. An Updated Vacuolar Apparatus

The aim of this subsection is to summarize current understanding of the dynamics of the endosome–lysosome system. We aim for a simple scheme, which can be modified and amplified by the detailed information contained in subsequent chapters of this book. Simple should not mean simplistic, but it will inevitably mean incomplete.

As already stated, the endosome and not the lysosome now occupies center stage. There is also broad agreement that there are two subsets of endosomes: early (or light) endosomes and late (or dense) endosomes. Whether these are distinct and permanent features of the cell, with ligands and solutes shuttled between them by small vesicles, or whether early endosomes mature into late endosomes by losing certain components while gaining others is a controversial and unresolved issue. It should also be pointed out that the term late endosome is disliked by some authors, who prefer the term prelysosome. Trowbridge et al. (1993) "favor restricting the term endosome to elements that contain rapidly recycling membrane proteins like the transferrin receptor and that can be defined operationally because they are accessible to endocytic tracers at 20 °C." However the term *late endosome* is currently the most widely used, and we shall adopt it in the following discussion.

The early endosome receives membrane, ligands, and dissolved solutes by the constant influx of vesicles derived from the plasma membrane in the process of endocytosis. Most of the membrane components are quickly returned to the plasma membrane, presumably by budding off small vesicles from a surface-rich region of the early endosome (Rome, 1985). Within the early endosome a lower pH causes many ligands to detach from their receptors (in the case of transferrin the ligand remains attached but releases its charge of Fe^{3+}), so that the latter can return to the plasma membrane with the recycled membrane. Volume-rich regions of the early endosome, containing ligands as well as solutes endocytosed in the bulk fluid phase, are transferred to the late endosome.

The late endosome (or prelysosome) is the organelle to which the lysosomal enzymes are delivered from their site of synthesis in the rough endoplasmic reticulum, following their journey through the Golgi apparatus and subsequently in Golgi-derived vesicles. Most lysosomal enzymes are guided to the late endosomes by virtue of their attachment to a receptor that recognizes phosphomannose. In the late endosome a lower pH causes the lysosomal enzymes to detach; the receptor returns to the Golgi apparatus.

Thus, the previously envisioned fusion of the primary lysosome and the heterophagosome is now seen as taking place at the late endosome stage. The late endosome has the key characteristics of de Duve's concept of the secondary lysosome. Here enzymes and substrates are together, both free from the receptors that delivered them. But because the particle's internal pH may not be optimal for digestion and because the enzymes may not yet be in their fully active form, the structure is not considered a lysosome, but a prelysosome. However, digestion of some substrates may begin in the late endosome, and in some cases, such as insulin (Authier *et al.*, 1994; Seabright and Smith, 1995), this may be a major digestion locus. However, most digestion of endocytosed substrates takes place only after transfer of the contents of the late endosome into the lysosome, where the enzymes become fully active.

Some researchers maintain that late endosomes evolve into lysosomes. Berón *et al.* (1995) wrote that late endosomes "progressively acquire lysosomal characteristics." There is, however, experimental evidence that the delivery of ligand from late endosomes involves a fusion event with a preexisting lysosome (Mullock *et al.*, 1994).

The delivery of a phagocytosed particle to the lysosomes is still regarded as fundamentally the inward migration of the phagosome and the fusion with vesicles bearing lysosomal enzymes. This process seems to bypass the early (and perhaps the late) endosome, although there is evidence for exchange of membrane and perhaps soluble constituents with the endosomes, as well as for recycling of some phagosomal membrane to the plasma membrane in a manner reminiscent of recycling from the endosome (Berón *et al.*, 1995).

Turning next to autophagy, it is striking that a recent detailed study of the formation and maturation of autophagic vacuoles (Dunn, 1990*a,b*) confirms the concepts proposed in 1966 and also finds the terminology of that period still acceptable. Thus autophagosomes form when a region of cytoplasm is enveloped by a double membrane comprising "ribosome-depleted rough endoplasmic reticulum." These mature into autolysosomes by acquiring lysosomal enzymes through fusion events that involve Golgi-derived vesicles, late endosomes, or secondary lysosomes.

Although in 1966 the unloading of lysosomal contents to the cell exterior was regarded as possible in certain cell types, the secondary lysosome was considered essentially a dead end to vesicular traffic. Lysosomes have been regarded as vacuoles that digest their contents and do not recycle them to other parts of the vacuolar system. This view remains largely intact, although a recent report (Jahraus et al., 1994) has proposed that an indigestible solute present in the lysosomes may be recycled back to the late endosome compartment by "retrograde fusion." Berthiaume et al. (1995) have recently suggested that a molecule's ability to return from lysosome to endosome may decrease with increasing molecular size.

Figure 3 is an attempt to update the diagram of the vacuolar system published by Jacques in 1966 (Fig. 2). It is surprising how little modification was necessary, much of the change being at the level of nomenclature rather than substance.

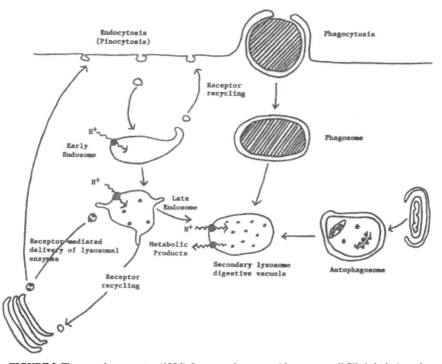

FIGURE 3. The vacuolar apparatus (1996). Lysosomal enzymes (shown as small filled circles) are delivered to the late endosome in Golgi-derived vesicles, which recycle back to the Golgi apparatus together with the receptors that transported the enzymes. Some lysosomal enzymes are delivered to the plasma membrane, where they may be reinternalized by receptor-mediated endocytosis. The latter process captures a range of exogenous ligands, which are delivered to the early endosome compartment, the receptors that delivered them recycling back to and reinserting into the plasma membrane. The early endosome delivers its cargo to the late endosome, where digestion may begin. Digestion of macromolecules takes place principally in the secondary lysosomes, which may also receive substrates for digestion from autophagy and, in a few cell types, from phagocytosis. The products of digestion are released into the cytoplasm, principally on metabolite porters (shown as larger filled circles. *Readers Note:* Some functions are simplified or even omitted if they do not affect lysosome function directly.

Figure 3 is certainly incomplete, but we hope not inaccurately so. Vastly more is now known at the next level of detail, particularly about the protein components of the membranes and their sorting and movements during membrane trafficking. There is also an evident lack of clarity about many issues, particularly about the number of distinct modes of endocytosis/pinocytosis.

4. THE UNKNOWN LYSOSOME

The recent shift of attention from lysosomes to endosomes leaves many questions about the latter without full or adequate answers. We highlight three of these, but there are many more.

4.1. The Internal Composition of Lysosomes

When lysosomes are prepared in the laboratory, they are typically isolated from a tissue homogenate in 250 mM sucrose that has been maintained at 4 °C. When these lysosomes are resuspended in the same medium at a higher temperature, 25 °C or 37 °C, they are osmotically stable, but they break immediately if resuspended in water or hypotonic medium. Clearly the lysosomes contain osmotically active solutes that cannot cross the lysosome membrane rapidly enough to prevent a compensatory influx of water.

The content of lysosomes is a remarkably unexplored field. However, their osmotic pressure is likely to arise principally from inorganic ions rather than from the products of macromolecule digestion. Despite the likely prominence of protein digestion in lysosome, the free amino acid concentration of lysosomes appears to be around 1 mM (Vadgama et al., 1991).

Lysosomes appear to be widely heterogeneous with respect to internal osmolarity. This was first inferred from the shape of the osmotic activation curves of rat liver lysosomes (Appelmans and de Duve, 1955) and has subsequently been confirmed by a range of experiments by other investigators. It can reasonably be conjectured that in vivo, all the lysosomes in a cell are isoosmotic with the surrounding cytoplasm but that lysosomes differ in the proportion of diffusible and nondiffusible solutes contributing to their osmotic pressure. The heterogeneity of osmotic pressure of lysosomes in vitro may reflect differential loss of diffusible solutes during the isolation procedure. This hypothesis could also explain the otherwise surprising observation that endosomes do not demonstrate osmotic activity (Park et al., 1988).

4.2. Lysosomal pH Range

The orthodox view of lysosomes envisions their maintaining an essentially constant pH some 2 units below that of the surrounding cytoplasm. Indeed, this pH differential is often used to define lysosomes operationally, through its effect of causing the accumulation of weak bases. However, Butor *et al.* (1995) have recently proposed that the pH of an individual lysosome may periodically fluctuate towards neutrality, perhaps by fusion events with less acidic vesicles or by variations in the activity of the lysosomal membrane proton pump. The existence of a pH cycle in lysosomes presently remains no more than an intriguing speculation, albeit one with significant implications for the pattern of metabolic activity within the organelle.

4.3. Lysosome Involution

Lysosomes are the constant recipients of both substrates and membrane. The substrates are digested and the products are released. What of the membrane? In a cell whose lysosomal compartment is not increasing in capacity, it seems clear that the superfluous membrane must either be digested or removed. Little is known about the extent or the mechanism of these two phenomena.

Whatever the homeostatic mechanisms that regulate the size of the lysosome compartment are, they can be disturbed when cells accumulate an indigestible solute such as sucrose. The lysosomes become much larger and perhaps more numerous; certainly the volume of the lysosome compartment increases dramatically. The enlarged structures are truly lysosomes, as their membranes contain the lysosome membrane (LAMP) glycoproteins but not the cation-independent mannose 6-phosphate receptor (Jahraus *et al.*, 1994). Often this effect of sucrose is described as lysosomal swelling and regarded as an osmotic phenomenon due to the influx of water into preexisting lysosomes. This is a most implausible explanation, as significant swelling would require a considerable excess of membrane for the previously contained volume, and this is never seen in tissue sections. Moreover, it is known that lysosomes placed in hypoosmotic media do not swell; they break, demonstrating that lysosomes do not possess excess membrane. And finally, some macromolecules with minimal osmotic pressure, such as polyvinylpyrrolidone and dextran, cause the same phenomenon. It seems inescapable that the effect of endocytosed sucrose is to cause lysosomes to fuse with each other.

When two spherical vacuoles coalesce, there is at the moment of fusion an excess of membrane for the contained volume. It has been proposed (Duncan and Pratten, 1977; Lloyd and Williams, 1978) that the loss of turgidity that immediately follows fusion might be the occasion for loss of membrane by an inward or outward budding of membrane to form intravacuolar or intracytosolic vesicles. By this process the new vacuole could itself regain sphericity. Inward budding would presumably lead to membrane digestion and could be a mechanism for the introduction of cytoplasmic constituents into the lysosome. By binding to regions of the membrane destined for internalization, proteins and other cytoplasmic constituents destined for degradation could gain selective entry into the organelle (Lloyd, 1976, 1980). Conversely, outward budding of the lysosome membrane could recycle membrane to endosomal elements or to the plasma membrane. Membrane loss by these same mechanisms could also occur after macromolecule digestion had decreased the solute content of a lysosome; here the flaccidity of the membrane would result from water loss due to a decrease in intralysosomal osmotic pressure.

Although this scheme lacks experimental evidence, it is a plausible explanation for the continuous membrane loss that must be a feature of lysosome physiology. It also provides an explanation for the hypertrophy of the lysosome system induced by agents such as sucrose, for here the loss of lysosome content that normally accompanies the digestive process fails to occur. I propose that lysosome "swelling" is not due to the uptake of water but to the failure of the lysosomal involution that normally accompanies the cycle of fusion and substrate digestion.

ACKNOWLEDGMENTS. The author's work is supported by NIH grant HD29902 and by the Nemours Foundation.

5. REFERENCES

Applemans, F., and de Duve, C., 1955, Tissue fractionation studies. 3. Further observations of the binding of acid phosphatase by rat-liver particles, *Biochem. J.* **59:**426–433.

Authier, F., Rachubinski, R. A., Posner, B. I., and Bergeron, J. J. M., 1994, Endosomal proteolysis of insulin by an acidic thiol metalloprotease unrelated to insulin degrading enzyme, *J. Biol. Chem.* **269:**3010–3016.

Berón, W., Alvarez-Dominguez, C., Mayorga, L., and Stahl, P. D., 1995, Membrane trafficking along the phagocytic pathway, *Trends Cell Biol.* **5:**100–103.

Berthiaume, E. P., Medina, C., and Swanson, J. A., 1995, Molecular size-fractionation during endocytosis in macrophages, *J. Cell Biol.* **129:**989–998.

Butor, C., Griffiths, G., Aronson, N. N., Jr., and Varki, A., 1995, Co-localization of hydrolytic enzymes with widely disparate pH optima: implications for the regulation of lysosomal pH, *J. Cell Sci.* **108:**2213–2219.

Conchie, J., and Hay, A. J., 1963, Mammalian glycosidases. 4. The intracellular localization of β-galactosidase, α-mannosidase, β-*N*-acetylglucosaminidase and α-L-fucosidase in mammalian tissues, *Biochem. J.* **87:**354–361.

de Duve, C., 1959, Lysosomes, a new group of cytoplasmic particles, in *Subcellular Particles* (T. Hayashi, ed.) pp. 128–159, Ronald, New York.

de Duve, C., 1963, The lysosome concept, in *Lysosomes, a Ciba Foundation Symposium* (A. V. S. de Reuck and M. P. Cameron, eds.), pp. 1–31, Churchill, London.

de Duve, C., 1964, From cytases to lysosomes, *Fed. Proc. Fed. Am. Soc. Exp. Biol.* **23:**1045–1049.

de Duve, C., 1966, Introduction à la physiopathologie des lysosomes, *Bruxelles-Méd.* **46:**1087–1094.

de Duve, C., 1968, Lysosomes as targets for drugs, in *The Interaction of Drugs and Subcellular Components in Animal Cells* (P. N. Campbell, ed.), pp. 155–169, Churchill, London.

de Duve, C., 1969, The lysosome in retrospect, in *Lysosomes in Biology and Pathology*, vol. 1 (J. T. Dingle and H. B. Fell, eds.), pp. 3–40, North Holland, Amsterdam.

de Duve, C., 1983, Lysosomes revisited, *Eur. J. Biochem.* **137:**391–397.

de Duve, C., and Wattiaux, R., 1966, Functions of lysosomes, *Annu. Rev. Physiol.* **28:**435–492.

de Reuck, A. V. S., and Cameron, M. P., eds., 1963, *Lysosomes, a Ciba Foundation Symposium*, Churchill, London.

Duncan, R., and Pratten, M. K., 1977, Membrane economics in endocytic systems, *J. Theor. Biol.* **66**:727–735.

Dunn, W. A., 1990a, Studies on the mechanisms of autophagy: Formation of the autophagic vacuole, *J. Cell Biol.* **110**:1923–1933.

Dunn, W. A., 1990b, Studies on the mechanisms of autophagy: Maturation of the autophagic vacuole, *J. Cell Biol.* **110**:1935–1945.

Jacques, P., 1966, Lysosomes et endocytose, *Bruxelles-Méd.* **46**:1053–1059.

Jahraus, A., Storrie, B., Griffiths, G., and Desjardins, M., 1994, Evidence for retrograde traffic between terminal lysosomes and the prelysosomal/late endosome compartment, *J. Cell Sci.* **107**:145–157.

Lloyd, J. B., 1976, Substrate specificity in pinocytosis and intralysosomal protein digestion, in *Proteolysis and Physiological Regulation* (D. W. Ribbons and K. Brew, eds.), pp. 371–386, Academic Press, New York.

Lloyd, J. B., 1980, Insights into mechanisms of intracellular protein turnover from studies on pinocytosis, in *Protein Degradation in Health and Disease,* Ciba Foundation Symposium 75, pp. 151–160, Excerpta Medica, Amsterdam.

Lloyd, J. B., 1990, Lysosomes: The cell's recycling plant, *Biol. Sci. Rev.* **3**:21–24.

Lloyd, J. B., and Williams, K. E., 1978, Lysosomal digestion of endocytosed proteins: Opportunities and problems for the cell, in *Protein Turnover and Lysosomal Function* (H. L. Segal and D. J. Doyle, eds.), pp. 395–416, Academic Press, New York.

Mullock, B. M., Perez, J. H., Kuwana, T., Gray, S. R., and Luzio, J. P., 1994, Lysosomes can fuse with a late endosomal compartment in a cell-free system from rat liver, *J. Cell Biol.* **126**:1173–1182.

Novikoff, A. B., 1963, Lysosomes in the physiology and pathology of cells: Contributions of staining methods, in *Lysosomes, a Ciba Foundation Symposium* (A. V. S. de Reuck and M. P. Cameron, eds.), pp. 36–73, Churchill, London.

Park, R. D., Sullivan, P. C., and Storrie, B., 1988, Hypertonic sucrose inhibition of endocytic transport suggests multiple early endocytic compartments, *J. Cell. Physiol.* **135**:443–450.

Pratten, M. K., and Lloyd, J. B., 1986, Pinocytosis and phagocytosis: The effect of size of a particulate substrate on its mode of capture by rat peritoneal macrophages cultured in vitro, *Biochim. Biophys. Acta* **881**:307–313.

Rome, L. H., 1985, Curling receptors, *Trends Biochem. Sci.* **10**:151.

Seabright, P. J., and Smith, G. D., 1995, The characterisation of insulin degradation products in endosomes, *Biochem. Soc. Trans.* **23**:1S.

Trowbridge, I. S., Collawn, J. F., and Hopkins, C. R., 1993, Signal-dependent membrane protein trafficking in the endocytic pathway, *Annu. Rev. Cell Biol.* **9**:129–161.

Vadgama, J. V., Chang, K., Kopple, J. D., Idriss, J.-M., and Jonas, A. J., 1991, Characteristics of taurine transport in rat liver lysosomes, *J. Cell. Physiol.* **147**:447–454.

Chapter 2

Origin of Lysosomal Proteins

Thomas Braulke

1. INTRODUCTION

Lysosomes are the major intracellular site for the degradation of a variety of macromolecules, including proteins, carbohydrates, nucleic acids, and lipids, that are internalized from the extracellular space by endocytosis, delivered by fusion with phagosomes, or derived from the biosynthetic pathway. Lysosomes are enriched with various acid hydrolases that are separated from other cytoplasmic components by a lysosomal membrane. This membrane has distinct functions, including the formation of a degradation-resistant barrier that protects the cellular structures against destruction, the generation of an acidic intralysosomal milieu, and the selective transport of degradation products to the cytoplasm. The biogenesis of new lysosomes during cell division and the turnover of lysosomal constituents require a continuous substitution with newly synthesized components. The targeting mechanisms and transport pathway of newly synthesized soluble hydrolases to lysosomes are better understood than those involved in targeting of lysosomal membrane proteins. Soluble hydrolases are equipped with mannose 6-phosphate (M6P) residues that are recognized by specific receptors mediating their transport from the Golgi to a prelysosomal compartment. The mechanisms of delivery of lysosomal membrane proteins are different and do not depend on

Thomas Braulke   Institut für Biochemie II, Georg-August-Universitat Göttingen, D-37073 Göttingen, Germany.

Subcellular Biochemistry, Volume 27: Biology of the Lysosome, edited by Lloyd and Mason. Plenum Press, New York, 1996.

M6P signals. Signal structures on the lysosomal proteins and their receptors as well as the identification of additional cytosolic and membrane-bound proteins involved in mediating the specificity of vesicular transport have been described in the last few years, as techniques for mutational cDNA analysis, *in vitro* assays, and analysis of immunocytochemical approaches have allowed the systematic molecular characterization of lysosomal targeting.

This review summarizes our current knowledge of synthesis and transport of lysosomal enzymes and membrane proteins. The main focus will be to describe the identification of structural determinants in lysosomal proteins and M6P receptors and the role of components of the vesicular transport machinery required for the directed intracellular transport to the lysosome.

1.1. Soluble Lysosomal Proteins

Soluble lysosomal enzymes are synthesized as larger precursor polypeptides with an N-terminal signal sequence that ensures the translocation into the lumen of the endoplasmic reticulum (ER). Concomitant to the cleavage of the signal peptide by the signal peptidase, preformed oligosaccharides, (e.g., glucose$_3$ mannose$_9$ N-acetylglucosamine$_2$ [Glc$_3$Man$_9$GlcNac$_2$]) are transferred to selected asparagine residues in the sequence Asn-X-Thr/Ser (where X can be any amino acid except Pro or Asp). The oligosaccharides undergo an extensive processing (see Section 2.1) initiated in the ER before completion of translation by removal of a glucose residue. These first steps in the biosynthetic path are shared with secretory glycoproteins, and both classes of proteins are carried together by vesicular transport to the Golgi via an intermediate (salvage) compartment (for reviews, see von Figura and Hasilik, 1986; Kornfeld and Mellman, 1989). This compartment is defined by the presence of a nonglycosylated homooligomeric transmembrane protein, p53, and as a site where incubation of cultured cells at 15 °C blocks the transport of newly synthesized proteins (Hauri and Schweizer, 1992).

The next major event in the maturation of lysosomal enzymes is the acquisition of phosphomannosyl residues catalyzed by the sequential action of two enzymes. First, a UDP-N-acetylglucosamine: lysosomal enzyme N-acetylglucosamine-1-phosphotransferase (phosphotransferase) transfers N-acetylglucosamine-1-phosphate to selected 6-carbons of mannoses, generating phosphodiester forms. The phosphotransferase recognizes a signal that is common to at least 40 different lysosomal enzymes (see Section 2.1). The specificity of the transferase with respect to the mannose residues is rather weak. Thus, phosphorylation can be detected at one or two of the five outer mannose residues. Furthermore, the number of N-linked oligosaccharides in different lysosomal enzymes that become phosphorylated varies (Table I).

The second enzyme involved in the formation of the phosphomannosyl recognition marker is N-acetylglucosamine-1-phosphodiester α-N-acetylglucosaminidase. This enzyme is located in a later mid-Golgi compartment and

Table I
Properties of Prepro Forms of Human Lysosomal Enzymes

Enzyme	Signal peptide (number of amino acids)	Precursor	Intermediate (in kDa)[a]	Mature form	Secretion (% of total synthesized)	Number of phosphorylated oligosaccharides
Cathepsin B	17	44.5	46	33/27	5–16 (−50)[b]	2 of 2
Cathepsin D	20	53	47	31/14	5–10 (40–66)[b]	2–3 of 2[d]
Cathepsin L	17	39	29	20	11 (94)[b]	2 of 1[d]
β-Hexosaminidase α-chain	22	67	-	54	~20	1 of 3
β-Hexosaminidase β-chain	42	63/61	52	23/21	~20	2 of 4
β-Galactosidase	23	88/84	-	64	10–20	n.a.
G_{M2} activator	23	24	-	21.5–22.5	20–30[c]	1 of 1
Protective protein	28	54	-	32/20	n.a.	1 of 2
Arylsulfatase A	18	62	-	57	< 10	2 of 3
α-Glucosidase	28	110	95	76/70	< 10	2 of 7

[a]Molecular mass calculation based on SDS-PAGE evaluation.
[b]The number in parentheses indicates the percentage of secreted precursor forms in cultured tumor cell lines.
[c]Unpublished results, K. Sandhoff.
[d]Single phosphorylated oligosaccharide with two phosphomonoester moieties.

removes the terminal GlcNac residue, thereby exposing the mannose 6-phosphate residue. The physiological importance of the carbohydrate phosphorylation of lysosomal enzymes is obvious from the I-cell phenotype. In patients suffering from this disease the correct targeting of lysosomal enzymes is affected due to defective phosphotransferase (see Section 6). The M6P residues function as a high-affinity recognition signal for receptors that mediate the segregation of soluble lysosomal enzyme precursors from the secretory route and the targeting to the lysosome. Two M6P-specific receptors are known; they differ in molecular size, ligand binding properties, and transport functions and efficiency (see Section 3). The sorting of lysosomal enzymes is believed to occur in the *trans*-Golgi network (TGN), followed by the budding of clathrin-coated vesicles containing the receptor–ligand complexes. The transport vesicles fuse with an endosomal/prelysosomal compartment that can be subdivided according to the kinetics of endocytosed ligands and the intraluminal pH in early, intermediate, and late endosomes (Trowbridge *et al.*, 1993; this is discussed in greater detail in Chapter 3). It is a matter of debate whether the lysosomal enzyme precursors are transported directly from the TGN to late endosomes or via the early endosomal compartment. Recent studies, detecting both M6P receptors as well as newly synthesized cathepsin D and arylsulfatase A in early endosomes of different cell types (Ludwig *et al.*, 1991; Klumperman *et al.*, 1993; Diesner *et al.*, 1993), indicate that a significant percentage of receptor–ligand complexes are transported intracellularly to lysosomes via early endosomes.

The low pH in the endosomes induces dissociation of lysosomal enzymes, which are then sorted and packed into lysosomes by as yet poorly characterized mechanisms. The M6P receptors recycle to the Golgi complex to mediate further rounds of transport.

Varying amounts of newly synthesized soluble lysosomal enzymes escape the binding to M6P receptors in the Golgi and are secreted (Table I). These M6P-containing precursor forms can be partially internalized and transported to lysosomes through receptor-mediated endocytosis. About 10–20% of total cellular M6P receptors are localized at the plasma membrane or mix with receptors cycling in the biosynthetic pathway.

1.2. Lysosomal Membrane Proteins

In the past 10 years much progress has been made in the identification and characterization of the major protein components of lysosomal membranes (for recent review, see Pisoni and Thoene, 1991; Fukuda, 1994; Peters and von Figura, 1994). About 50% of lysosomal membrane proteins comprise a class of four acidic integral glycoproteins with a molecular mass between 100 and 120 kDa and unknown function. The lysosomal membrane glycoprotein-A (lgp-A, which is identical with rat lgp 107, mouse lysosome-associated membrane protein-1 [LAMP-1], and chicken lysosome-endosome-plasma membrane protein 100 [LEP-100]) is related to lysosomal membrane glycoprotein-B (lgp-B, which is identical with mLAMP-2 or lgp 110). Both of these type I membrane glycoproteins consist of a luminal domain which is separated from a short (10 or 11 amino acids) cytoplasmic domain by a single membrane-spanning region. The luminal domain can be divided into homologous regions separated by a flexible hinge region of approximately 30 amino acids rich in proline and serine/threonine residues. In each luminal subdomain, pairs of cysteine residues form two loops. Furthermore, the luminal domain contains 16–18 N-linked complex carbohydrate chains, some of which are of the poly-N-acetyllactosamine type, and 6–10 O-linked oligosaccharide chains clustered in the hinge region. The increased amount of poly-N-acetyllactosamine and O-glycan chains results in a longer half-life of the molecules and may protect the weakly ordered hinge region against intralysosomal proteolytic degradation.

Lysosomal integral membrane proteins (LIMP I and II) are other members of this class of membrane constituents, with apparent molar masses of 35–55 and 60–85 kDa, respectively. Both are type III membrane glycoproteins with four and two membrane-spanning domains, respectively. Finally, lysosomal acid phosphatase (LAP), another lysosomal protein, is only a transient constituent of lysosomal membranes (see Section 2.3). Lysosomal acid phosphatase is a type I membrane protein with a large luminal domain and a short cytoplasmic domain of 19 amino acids. The luminal domain contains eight N-linked complex carbohydrate chains.

While the first steps in the biosynthetic route of lgp, LIMP, and LAP are shared with the soluble lysosomal proteins, none of the lysosomal membrane proteins acquire the M6P recognition marker. It is proposed that the signal essential for M6P receptor-independent lysosomal targeting is localized in the cytoplasmic domain of the lysosomal membrane proteins, which allows sorting either in the TGN for the direct transport to the endosome and lysosome or at the plasma membrane for the indirect transport from the TGN to the cell surface and subsequently to endosomes and lysosomes. While the majority of lgp-A, lgp-B, and LIMP II are transported via the direct route, LIMP I and LAP are transported via the plasma membrane to lysosomes (Braun *et al.*, 1989; Mathews *et al.*, 1992; Maillet and Shur, 1993). Mutant analysis of lgp-A and LAP demonstrated that the tyrosine residue and its relative position in the cytoplasmic tail is sufficient for efficient lysosomal trafficking and rapid endocytosis from the cell surface. Also, NMR analysis further showed that the tyrosine residue is part of a β-turn structure recognized by cytoplasmic carrier molecules (see Section 5.1). It has been reported that the cytoplasmic tail of LAP binds to such a carrier molecule, called AP-2 adaptor, *in vitro* (Sosa *et al.*, 1993). In the cytoplasmic domain of LIP II, which lacks a tyrosine motif, another lysosomal targeting motif (Leu-Ile-X) in a region with an extended configuration has been identified (Ogata and Fukuda, 1994; Sandoval *et al.*, 1994). This Leu-Ile motif is also found in cytoplasmic domains of both M6P receptors, suggesting that LIMP II uses the same sorting mechanisms as M6P receptors to exit the Golgi (see Section 3.2). A novel lysosome targeting signal sequence has been demonstrated (DGKCPLNPHS, one letter code) in the cytoplasmic domain of P-selectin, a membrane protein facilitating adhesion of leukocytes. This lysosomal targeting sequence is different to sorting signals required for delivery to regulated secretory granules and for rapid internalization from the cell surface (Green *et al.*, 1994).

Despite the detailed characterization of approximately 20 different lysosomal transport systems for amino acids, monosaccharides, nucleosides, cobalamin, phosphate, sulfate, Ca^{2+}, and dipeptides, their molecular identification is still lacking and the transport routes and signal structures for lysosomal targeting is not understood. The V-H$^+$ATPase is a significant component of the lysosomal membrane, and this is discussed in detail in Chapter 10.

2. SYNTHESIS AND POSTTRANSLATIONAL MODIFICATIONS OF LYSOSOMAL ENZYMES

2.1. Carbohydrate Modifications

As the lysosomal enzyme polypeptide chains are translocated across the membrane into the lumen of the ER, core glycosylation with a preformed oligosaccharide, $Glc_3Man_9GlcNac_2$, occurs at selected asparagines. The core units are then processed by glycosidases and mannosidases in the ER in the same way as membrane and soluble glycoproteins (Kornfeld and Kornfeld, 1985). The first outermost glucose residue is rapidly removed (e.g., for cathepsin D, within 1 min of synthesis) by glucosidase I followed by sequential hydrolysis of the two others by glucosidase II in the ER. After removal of the first two glucose residues the membrane-bound molecular chaperone calnexin binds to the monoglucosylated core glycans until the protein is properly folded (Hammond *et al.*, 1994). Alternatively, the monoglucosylated form is achieved by reglucosylation catalyzed by a luminal UDP-glucose:glycoprotein glucosyltransferase of deglucosylated oligosaccharides (Hebert *et al.*, 1995). After completion of folding, the glycoprotein is released from calnexin by removal of the last of three glucoses, catalyzed by glucosidase II. Thus, removal of the three glucose residues functions in the control of protein folding and contributes to the retainment of incompletely folded glycoproteins in the ER. In addition to calnexin, there are other chaperones like BiP that are involved in glycoprotein folding.

After ER mannosidases have removed one to three of the mannoses, the oligosaccharides are of "high-mannose" type. Upon arrival in the *cis*-Golgi, a mannosidase I trims the high-mannose oligosaccharides to $Man_5GlcNac_2$, followed by the transfer of a GlcNac residue by the action of GlcNac transferase I, a marker localized in the mid-Golgi, to yield GlcNac $Man_5GlcNac_2$. A second mannosidase (α-mannosidase II) removes two other mannoses, generating $GlcNacMan_3GlcNac_2$. The transfer of additional GlcNac residues catalyzed by another GlcNac transferase II initiates the formation of complex type oligosaccharides with two or more antennae. The glycosyltransferases responsible for the addition of galactose and sialic acid (and sometimes fucoses) are localized in the *trans*-Golgi and TGN, respectively. Depending on the cell type studied, different types of oligosaccharides are found on lysosomal hydrolases: some chains of high-mannose and complex type and some hybrid oligosaccharides in which one antenna has only mannoses. In many malignant cells, altered glycosylation of lysosomal proteases like cathepsin B, L, or D accompanied with their overexpression are thought to be responsible for the formation of precursor polypeptides with complex type oligosaccharides rather than high-mannose carbohydrates, resulting in increased secretion (for review, see Sloan *et al.*, 1994a; Table I).

The oligosaccharides also contain sulfate groups, some of which are mannose 6-sulfate. Sulfate addition occurs posttranslationally on preexisting lysosomal enzyme precursors. The sulfate groups are lost with half-lives of between 1 and 26 hr depending on the enzyme (Braulke *et al.*, 1987a). The cellular localization of the oligosaccharide sulfotransferase and the possible role of sulfate groups in hydrolase sorting have not been investigated.

A key step in targeting of newly synthesized soluble hydrolases to the lysosome is the formation of phosphomannosyl residues on high-mannose oligosaccharides. The initial transfer of GlcNac-1-phosphate to mannose, forming $\alpha 1,6$-linked branches of the carbohydrate chains, may occur in the intermediate compartment (Hauri and Schweizer, 1992). A second phosphorylation can occur on mannoses, forming $\alpha 1,3$-linked branches, after the removal of terminal mannose residues by an α-mannosidase in the *cis*-Golgi.

To identify critical protein determinants in lysosomal enzymes serving as phosphotransferase recognition markers, Kornfeld and colleagues (Baranski *et al.*, 1990) studied aspartyl proteases consisting of the homologous lysosomal hydrolase cathepsin D (phosphorylated) and the secretory pepsinogen (nonphosphorylated). The aspartyl proteases are bilobed molecules and each lobe contains an active site aspartic acid. Two discontinuous protein determinants (Lys 203 and amino acids 265–292) in the carboxyl lobe of cathepsin D were identified as part of a phototransferase binding site (Baranski *et al.*, 1990). While this site is sufficient to allow phosphorylation of the oligosaccharides of both lobes of the molecule, the C-terminal oligosaccharide was preferentially phosphorylated (Baranski *et al.*, 1992; Cantor *et al.*, 1992). Utilizing a library of chimeric proteins, two regions of the amino lobe of cathepsin D (amino acids 106–116 and 161–167) were identified that may influence the stability and presentation of β-loop 265–292, resulting in increased total and biphosphorylation of the carboxyl lobe oligosaccharide. Moreover, two other regions (the propiece plus 36 amino acids of the mature protein and amino acids 64–81), in combination with the carboxyl lobe region 265–292, directed the phosphorylation of the amino lobe oligosaccharide without reducing phosphorylation in the carboxyl lobe (Dustin *et al.*, 1995). The independence of phosphorylation of either oligosaccharide from that of the other may suggest that the phosphotransferase binds the lysosomal enzyme at multiple sites simultaneously with different binding affinities. Alternatively, the phosphotransferase binds sequentially to the carboxyl lobe, allowing phosphorylation of the nearby oligosaccharide followed by dissociation and rebinding to the amino lobe and phosphorylation of the second oligosaccharide. While it is unclear whether the amino lobe stretches serve as additional phosphotransferase binding sites or contribute to proper exposition of the amino lobe oligosaccharide, these stretches do allow the generation of high-affinity ligands with either single phosphate residues on two oligosaccharides or two phosphate residues on the carboxyl lobe oligosaccharide.

Antibody inhibition experiments affecting the phosphorylation of chimeric arylsulfatase A molecules *in vitro* suggest the localization of topogenic signals for the phosphotransferase in the N-terminal region of the molecule (Sommerlade *et al.*, 1994a). On the other hand, analysis of glycosylation mutants of arylsulfatase A revealed that two of three oligosaccharides (at Asn 158 and Asn 350) are primarily phosphorylated and that these prevent oligosaccharide phosphorylation at Asn 184 (Sommerlade *et al.*, 1994b). However, in the absence of oligosaccharides at Asn 158 or Asn 350 (the latter is a naturally occurring glycosylation site mutant) the oligosaccharide at Asn 184 becomes accessible to phosphotransferase, supporting the idea of a common biological principle ensuring multivalent phosphorylation to guarantee efficient targeting.

During the maturation of precursor forms of lysosomal hydrolases their oligosaccharides become dephosphorylated. The rate and the extent of dephosphorylation of individual acid hydrolases vary and depend on the presence of the 300-kDa M6P receptor. These results led to the suggestion that the mannose 6-phosphatase is controlled by compartmentalization and that hydrolases were not distributed uniformly throughout endosomal/lysosomal subpopulations (Enstein and Gabel, 1991). It has been shown that the lysosomal acid phosphatase is not the mannose 6-phosphatase responsible for dephosphorylation of lysosomal enzymes (Bresciani *et al.*, 1992).

2.2. Other Covalent Modifications of Lysosomal Proteins

In addition to the above-mentioned carbohydrate sulfation, tyrosine sulfation was observed in a few soluble lysosomal proteins. This posttranslational modification occurs in the *trans*-Golgi. The role of protein sulfation in transport, stability, and/or activity of lysosomal proteins is unknown. It has been proposed that after arrival in lysosomes the sulfate is rapidly lost in human fibroblasts (Hille *et al.*, 1990). Furthermore, it was reported that β-glucuronidase and arylsulfatase B can be phosphorylated at serine and threonine residues that may involve cAMP-dependent kinases (Gasa *et al.*, 1987; Ono *et al.*, 1988). Although phosphorylation of arylsulfatase B *in vitro* causes changes in the enzyme activity toward an artificial substrate, it remains to be determined whether protein phosphorylation of lysosomal enzymes is a common modification and related to enzymatic regulation *in vivo*.

Recently, a new cotranslational or posttranslational modification in lysosomal sulfatases has been described that occurs at a conserved cysteine residue embedded in a highly homologous sequence among six human sulfatases (Schmidt *et al.,* 1995a). The cysteine residue is converted into a 2-amino-3-oxopropionic acid residue that is required to generate catalytically active sulfatases, and this modification is lacking in arylsulfatase A and B from cells of patients with multiple sulfatase deficiency.

2.3. Proteolytic Processing

After cleavage of the signal peptide from prepro forms, many inactive lysosomal precursors undergo subsequent proteolytic processing steps, leading to the mature, active enzymes found in lysosomes. The processed polypeptides are 1 to 25 kDa smaller than their precursor forms (Table I). The proteolytic conversion depends on acidic pH and is initiated by some lysosomal enzymes (e.g., cathepsins D and B) within the endosomal compartment and then completed by many others in lysosomes. The sequence and possible mechanisms of the proteolytic processing steps have been studied for the major lysosomal proteinases, cathepsin D, B, and L (for review, see Sloan *et al.,* 1994a). The removal of propeptides results in conversion to active single-chain forms of the enzymes that are further processed to two-chain forms. *In vitro* studies revealed that the cleavage of propeptides of cathepsins D, B, and L may be an autoactivation process (Gal and Gottesman, 1986; Conner, 1989; Mach *et al.,* 1993). However, in a recent report it was shown that mutation at the autocatalytic site in procathepsin D resulted in loss of autoactivation *in vitro* but did not interfere with the transport to lysosomes and its normal processing to the mature form when the mutant cathepsin D was expressed in Ltk⁻ cells (Richo and Conner, 1994). The data suggest that the autocatalytic form of cathepsin D (pseudocathepsin D) found *in vitro* is not a normal proteolytic intermediate of cathepsin D *in vivo.* Inhibitor studies have shown that the endoprotease cleaving propeptides from cathepsins may be cysteine and metalloendopeptidases (Gieselmann *et al.,* 1985; Hanewinkel *et al.,* 1987; Hara *et al.,* 1988).

For cathepsins D, B, and L additional tissue-specific proteolytic steps have been described that cleave the single chain into mature, light and heavy chain doublets. The asymmetric cleavage is accompanied by the loss of two to seven amino acids between the chains. Finally, carboxy-terminal trimming has been reported, resulting in the cleavage of two or six amino acids that may be paralleled by lysosomal aminodipeptidase trimming of the light chain (Faust *et al.,* 1985; Fong *et al.,* 1986; Richo and Conner, 1994). The intracellular conversion of the single chain to the double chain form of cathepsin D appears to be impaired in fibroblasts from patients with I-cell disease (see Section 6), yielding a 33-kDa heavy chain form in contrast to the 31-kDa form seen in normal cells (Hasilik and Neufeld, 1980). A better understanding of the functions of cathepsins require identification of substrates that are targets of the enzymes *in vivo.* Surprisingly, target disruption of the cathepsin D gene in mice does not induce lysosomal storage of undegraded material, but the animals die in the fourth week of life due to progressive atrophy of the intestinal mucosa and a profound destruction of lymphatic organs (Saftig *et al.*, 1995).

The precursor molecule of the human protective protein is cleaved into two chains followed by carboxy-terminal trimming resulting in the reduction of the heavy chain by 2–3 kDa. The 32-kDa heavy chain and 20-kDa light chain are linked by disulfide bridges (Galjart *et al.*, 1988). The protective protein shows characteristics of a lysosomal carboxypeptidase cathepsin A. However, when cathepsin A-like activity was abolished by site-directed mutagenesis the protective protein was still able to restore β-galactosidase and neuraminidase activities in galactosialidosis fibroblasts (Galjart *et al.*, 1991). These results indicate that the proteolytic activity of the protective protein might be important for general protein turnover rather than for proteolytic processing of β-galactosidase and neuraminidase in the complex.

Proteolytic processing of the membrane-spanning lysosomal acid phosphatase involves the removal of the 19-amino acid cytoplasmic tail by a cysteine proteinase and the cleavage of a nonglycosylated carboxy-terminal sequence by an aspartic proteinase, rendering the enzyme soluble (Waheed *et al.*, 1988; Gottschalk *et al.*, 1989). Unlike the proteases, acid phosphatase and glycosidase precursor forms appear to be active (Gottschalk *et al.*, 1989; d'Azzo *et al.*, 1982; Wisselaar *et al.*, 1993; Hasilik *et al.*, 1982). Thus, it has been shown that the acquisition of catalytic activity of β-hexosaminidase requires the correct folding and α and β subunit assembly probably in the Golgi without prior proteolytic processing (Neufeld, 1989). Taken together, synthesis and transport of several lysosomal proteinases as inactive precursor molecules probably prevent undesirable proteolytic processes in the ER and Golgi. Furthermore, the tight binding of precursors to membranes may also facilitate transport and sorting to lysosomes, where proteolytic and oligosaccharide processing converts the enzymes to an active and stable soluble form.

3. MANNOSE 6-PHOSPHATE-DEPENDENT TRANSPORT

3.1. Mannose 6-Phosphate Receptors

Two distinct M6P receptors are involved in the targeting of lysosomal enzymes. Both receptors are type I integral membrane glycoproteins with M_r of 300,000 (MPR 300) and 46,000 (MPR 46). Their subcellular distribution appears to be similar (Bleekemolen *et al.*, 1988). The highest concentration of the two MPRs has been found by immunolocalization techniques in the TGN, followed by endosomal membranes and the plasma membrane, which corresponds to the transport routes of lysosomal enzymes. However, significant differences in the distribution of the two receptors within endosomal membranes (Klumberman *et al.*, 1993) and cell type-specific variations in subcellular localization

(Griffiths *et al.*, 1990) indicate partly different functions of MPR 300 and MPR 46 (for recent reviews, see Kornfeld, 1992; Hille-Rehfeld, 1995).

3.1.1. MPR 300

The cDNAs for MPR 300 have been cloned from various sources, and their deduced amino acid sequences reveal four structural domains: a signal sequence of 40–44 residues, an extracytoplasmic domain of 2264–2269 residues, a 23–residue transmembrane region, and a cytoplasmic tail of 163–164 residues. The extracytoplasmic domain contains 15 repeating segments of approximately 147 amino acids, each sharing 16–38% of identical residues. Ligands containing M6P bind to the repeating units 1–3 and 7–11, independent of divalent cations (Westlund *et al.*, 1991). However, MPR 300 also binds the nonglycosylated insulin-like growth factor II (IGF II). While the IGF II binding site in bovine and human MPR 300 was localized in repeat 11 (Dahms *et al.*, 1994; Garmroudi and MacDonald, 1994; Schmidt *et al.*, 1995b) IGF II does not bind to chicken or frog MPR 300 (Clairmont and Czech, 1989). Domain 13 contains a 43-residue insertion with sequence homology to the collagen-binding domain of fibronectin type II, whose function is unknown. The MPR 300 has been shown to contain different posttranslational modifications: (a) glycosylation of the extracytoplasmic domain which is not required for IGF II binding (Kiess *et al.*, 1991); (b) phosphorylation at four or five sites in the cytoplasmic domain (Meresse *et al.*, 1990; Rosorius *et al.*, 1993a); (c) palmitylation (Westcott and Rome, 1988); and (d) three or four intramolecular disulfide bonds within each repeat (Lobel et al., 1988).

Tong *et al.* report that MPR 300 binds two moles of M6P or one mole of a diphosphorylated oligopeptide per mole receptor, the latter with affinities of approximately 2×10^{-9} M (Tong *et al.*, 1989), and one mole of IGF II per receptor (Tong *et al.*, 1988). Soluble fragments of MPR 300 formed by proteolytic cleavage retain their binding properties for both ligands (Kiess *et al.*, 1987; Causin *et al.*, 1988).

The major function of MPR is sorting newly synthesized lysosomal enzymes in the TGN that are subsequently released in a more distal endosomal compartment. The MPR 300 localized at the plasma membrane mediates the endocytosis of exogenous M6P-containing ligands, which may partially represent missorted newly synthesized enzymes. In addition, cell surface MPR 300 plays a role in IGF II clearance from the circulation and activation of latent TGF β1-precursor molecules (Kiess *et al.*, 1988; Dennis and Rifkin, 1991).

It has been shown that MPR 300 can bind both ligands simultaneously. However, lysosomal enzymes impair IGF II binding, and exogenously added IGF II can inhibit lysosomal enzyme binding and internalization in a cell type-dependent manner (Kiess *et al.*, 1988, 1989; Braulke *et al.*, 1989; Hartmann *et al.*, 1992). In contrast, the sorting of newly synthesized lysosomal enzymes is not affected in either fibroblasts in the presence of exogenous IGF II or in NIH 3T3 cells overexpressing the mature form of IGF II (Braulke *et al.*, 1990a, 1991). When human kidney cells were transfected with the prepro IGF II cDNA to achieve very high expression levels (1 μg IGF II per mg protein in 24 hr) the amount of secreted precursor forms of β-hexosaminidase and cathepsin D was increased, but this was not due to inhibition of reuptake (Hoeflich *et al.*, 1995). These discrepancies may be explained by the different forms and expression levels of IGF II or possibly by variations in the cellular transport capacity. The cellular MPR 300 content is doubled in IGF II overexpressing NIH 3T3 cells compared with control cells (Claussen *et al.*, 1995), while the receptor level is unchanged in kidney cells overexpressing the prepro IGF II form.

3.1.2. MPR 46

The cloning of cDNAs for MPR 46 from various species and the deduced amino acid sequences reveal that the two M6P receptors are related. Like MPR 300, MPR 46 also consists of four structural domains: signal sequence of 21–28 residues, an extracytoplasmic domain of 159–164 residues, a transmembrane region of 20–25 residues, and a 67-residue cytoplasmic tail. The amino acid sequences of the bovine, human, and murine cytoplasmic domain are 100% identical. The extracytoplasmic domain of MPR 46 shares identities with the repeating units of MPR 300 ranging from 14–28% (Dahms *et al.*, 1987). There are no homologies among the transmembrane and cytoplasmic domains of the two receptors. In MPR 46 there are four N-linked oligosaccharide chains. Contrary results have been reported on the effect of removing glycosylation sites in the bovine and human MPR 46 for acquisition of ligand binding conformation (Wendland *et al.*, 1991a; Zhang and Dahms, 1993). Additional modifications of the MPR 46 protein are the formation of three intramolecular disulfide bonds regulating proper folding and assembly of MPR 46 (Wendland *et al.*, 1991b). Human MPR 46 has also been shown to be phosphorylated at one site in its cytoplasmic domain (Hemer *et al.*, 1993).

Chemical cross-linking experiments have shown that MPR 46 exists primarily as a dimer in membranes. In cells that overexpress MPR 46, monomeric, dimeric, and tetrameric forms have been observed (Waheed and von Figura 1990; Ma *et al.*, 1992), and it is proposed that changes in quaternary structure occur during intracellular transport. Furthermore, the presence of divalent cations favored receptor tetramerization, paralleled by an increase in binding of lysosomal β-glucuronidase. The MPR 46 binds 1 mol M6P or 0.5 mol diphosphorylated oligosaccharide per mole of monomeric receptor with a K_D of 2×10^{-7} M but lacks an IGF II binding site (Tong *et al.*, 1989). These data indicate that an MPR 46 dimer can bind in a manner similar to the monomeric MPR 300, an oligosaccharide containing two phosphomannosyl residues.

The lack of increased secretion of newly synthesized lysosomal enzymes after functional immunodepletion suggests that the MPR 46 plays only a minor role in sorting and that the endogenous MPR 300 can compensate for the loss of MPR 46 (Stein *et al.*, 1987). Furthermore, in cells deficient for MPR 300, 30–40%

of newly synthesized lysosomal enzymes remain in the cell, a process that appears to be sorted by MPR 46. This fraction can be increased in these cells to 50% by MPR 46 overexpression (Watanabe *et al.*, 1990). In contrast, recent data on target disruption of the MPR 46 gene in mice showed that, despite a normal phenotype, increased levels of phosphorylated lysosomal enzymes were found in secretions of cultured homozygous mice fibroblasts (Köster *et al.*, 1993; Ludwig *et al.*, 1993). This is paralleled by a 50% reduction of the intracellular enzyme activities compared with control fibroblasts, indicating that MPR 46 is responsible for significant enzyme targeting. The failure of endogenous MPR 300 to compensate for the loss of MPR 46 may be explained by the insufficient endogenous MPR 300 level, differences in the classes of lysosomal enzymes, and their phosphorylated oligosaccharide composition transported by both receptors. Alternatively, an increased secretion of lysosomal enzymes in cells deficient for MPR 46 may result from distinct subcellular organelles that are the targets of both receptors. It was shown that *in vivo* a missorting of lysosomal enzymes in mice deficient in MPR 46 was significantly compensated by other carbohydrate-specific receptors (Köster *et al.*, 1994). In this respect, the physiological significance of the proposed role of MPR 46 in the secretion of lysosomal enzymes (Chao *et al.*, 1990) remains to be further elucidated.

While MPR 46 recycles from intracellular membranes to the plasma membrane, it does not function in endocytosis (Stein *et al.*, 1987). This might be due to the slightly acidic pH required for efficient ligand binding (Tong and Kornfeld, 1989).

3.2. Signal-Dependent Intracellular Routing of Mannose 6-Phosphate Receptors

The directed intracellular transport of the two MPRs along the biosynthetic and endocytic pathway is thought to be mediated by signaling elements, which are proposed to be localized mainly in the cytoplasmic domains of the receptor. Some of these signals have been determined by examination of lysosomal enzyme sorting or endocytosis in receptor-deficient cells transfected with wild-type or mutant receptor cDNA. Thus, an internalization signal representing residues 24–29 (YKYSKV; Tyr 2360 to Val 2365 of the full-length bovine receptor precursor) of the 163-residue MPR 300 cytoplasmic domain has been identified (Canfield *et al.*, 1991). Therefore, Tyr 26 and Val 29 are the most important residues for rapid internalization. The comparison of the MPR 300 internalization signal to those identified in other constitutively recycling receptors indicates common elements with an aromatic residue, especially a tyrosine in the first position separated by two amino acids from a hydrophobic residue (Jadot *et al.*, 1992). Furthermore, it has been shown that internalization sequences of endocytic receptors belonging to either type I or type II membrane proteins are interchangeable (Collawn *et al.*, 1991; Jadot *et al.*, 1992).

For rapid internalization of MPR 46, two signals in the 67-residue cytoplas-mic domain are required: The dominant motif contains Phe 13 and Phe 18 (at po-sitions 225 and 230 of the full-length bovine receptor precursor), while the second includes Tyr 45 (position 257 of the full-length bovine receptor precursor), and the two signals contribute differently to the maximal internalization rate (Johnson et al., 1990). As shown by NMR spectroscopy of some endocytic receptors, the tyrosine-containing internalization sequences form a type I β-turn structure which has been proposed to be required for AP-2 adaptor binding (see Section 5.1 and Chapter 3 of this volume).

For the two MPRs it has been demonstrated that deletion of the carboxy-terminal four or five amino acids of the cytoplasmic domains containing dileucine motifs leads to a decrease in sorting efficiency of lysosomal enzymes (Johnson and Kornfeld, 1992a,b). Since the recycling of these mutant receptors from the plasma membrane to the Golgi was unaffected, the authors suggested that additional signals for efficient sorting may overlap with the internalization sequences. Chen et al. (1993) have reported that there is a casein kinase II phos-phorylation site proximal to the dileucine motif of MPR 300 (Ser 156; at posi-tion 2492) that is part of the sorting signal. This and another site (Ser 85; at position 2421 of the full-length bovine MPR 300 precursor) are phosphorylated in vivo by a casein kinase II-like enzyme that appears to be associated with the 47-kDa subunit of the AP-1 adaptor of the Golgi (Meresse et al., 1990). It was proposed that phosphorylation of both sites occurs in or shortly after the recep-tor exits the TGN (Meresse and Hoflack, 1993). It has been shown that in per-meabilized cells, phosphorylation of MPR 300 at Ser 156 and 85 is important to promote the recruitment of AP-1 to Golgi membranes (LeBorgne et al., 1993). These data suggest that phosphorylation of the MPR 300 cytoplasmic domain is an early event, facilitating the restricted localization of AP-1 adaptor molecules and the proper interaction with components required for efficient lysosomal en-zyme sorting (see Section 5). In contrast, the simultaneous exchange of four po-tential phosphorylation sites, including Ser 156 and 85, to alanine did not impair the efficiency of lysosomal enzyme sorting, suggesting that phosphorylation is not necessary for budding (Johnson and Kornfeld, 1992a). Chen et al. (1993) ex-plain the discrepancies between these and their own data by the different exper-imental approaches and variability depending on the expression level of mutant receptors. Alternatively, different mechanisms may contribute to efficient sort-ing when the phosphorylation sites in the cytoplasmic receptor domain are deleted.

Cross-linking experiments revealed that phosphorylated serine residues in the cytoplasmic domain of the human MPR 300 homologous to those of the bovine receptor are required for interaction with a cytosolic 35-kDa protein (Rosorius et al., 1993b). The identity and function of this protein are unknown. The local-ization of the other three phosphorylation sites in the MPR 300 cytoplasmic domain, the kinases involved and their intracellular location, as well as the role of these phosphorylation sites for MPR 300 function remain to be elucidated. The cytoplasmic domain of MPR 46 is also phosphorylated by a casein kinase II *in vitro* at Ser 56, which is homologous to the C-terminal casein kinase II phosphorylation site in MPR 300 (Körner *et al.,* 1994). Data on the phosphorylation stoichiometry revealed that at steady state about 3% of total MPR 46 are phosphorylated (Breuer *et al.,* manuscript in preparation). Disruption of the phosphorylation site had no detectable effects on sorting or receptor internalization (Johnson and Kornfeld, 1992b; Hemer *et al.,* 1993).

The microinjection of peptide-specific antibody Fab fragments against an epitope corresponding to residues 43–47 of the cytoplasmic domain of MPR 46 (AYRGV) resulted in receptor accumulation in an endosomal subcompartment (Schulze-Garg *et al.,* 1993). The authors speculate that Tyr 44 (position 255 of the full-length human receptor precursor) is not only part of an internalization signal but also serves as a signal for efficient recycling to the TGN.

By constructing a series of chimeric receptors consisting of either EGF receptor extracytoplasmic and transmembrane domain joined with the cytoplasmic tail of MPR 300 (Dintzis and Pfeffer, 1990) or MPR 300 extracytoplasmic and transmembrane domain fused with the cytoplasmic domain of EGF or LDL receptor (Dintzis *et al.,* 1994), it has been shown that the extracytoplasmic domain of MPR 300 contains structural determinants affecting the steady-state distribution of chimeric receptors. The authors proposed an "endosome retention" signal in the extracytoplasmic domain that interferes with rapid receptor recycling by excluding the receptor from tubular recycling endosomes. The origin of the transmembrane domain had no effect on the intracellular distribution of chimeric receptors. In contrast, the expression of a chimeric protein consisting of the transmembrane and cytoplasmic domain of MPR 300 fused to the secretory protein lysozyme was predominantly located in the TGN (Conibear and Pearse, 1994). Furthermore, the fusion products of the transmembrane and cytoplasmic domains of the MPR 300 and MPR 46 to the ectodomain of the hemagglutinin of the influenza virus, a typical plasma membrane protein, codistribute with endogenous MPRs (Mauxion *et al.,* 1995), suggesting that the cytoplasmic domain of MPRs is sufficient to determine the proper intracellular location. The reason for the different localizations of chimeric proteins of various extracellular domains with the same transmembrane and cytosolic domain of MPR 300 is still unclear.

In order to examine the role of MPR 300 in transmembrane signaling in response to IGF II binding, Okamoto *et al.* (1990) synthesized three peptides that correspond to three regions in the cytoplasmic domain of the human MPR 300. These peptides are structurally similar to mastoparan, a wasp venom peptide that activates G_o and G_i proteins. One of the synthetic peptides corresponding to residues 123–136 (RVGLVRGEKARKGK) of the cytoplasmic tail was shown to activate $G_{i2}\alpha$. The G_{i2} proteins are members of a family of trimeric GTP binding proteins which become activated by binding to a family of receptors, resulting in the intracellular formation of second messenger molecules. These receptors are characterized by a common sequence pattern with seven hydrophobic transmembrane helices. Through reconstitution of purified MPR 300 in phospholipid vesicles containing isolated G_{i2} proteins, Nishimoto *et al.* (1989) demonstrated that IGF II induced the activation of G_i proteins and that this could be inhibited by M6P and M6P-containing lysosomal enzymes (Murayama and Nishimoto, 1990). However, reconstitution experiments using purified wild-type and mutant MPR 300 lacking the proposed G protein activation sequence with G_{i2} proteins failed to demonstrate an IGF II-induced interaction of human MPR 300 with G proteins (Körner *et al.*, 1995). These latter results are supported by the observation that in MPR 300-deficient mouse embryos the mitogenic action of IGF II is not altered, while the levels of circulating IGF II was markedly increased (Liu *et al.*, 1993; Baker *et al.*, 1993; Lau *et al.*, 1994; Wang *et al.*, 1994). These data support a role for MPR 300 in the clearance of IGF II from the circulation rather than in signaling.

3.3. Regulation of Mannose 6-Phosphate Receptor Transport

The intracellular distribution of MPRs has been shown to be transiently affected by several agents, such as insulin, IGFs, protein kinase C activators, and G-protein modifying toxins. Treatments of human fibroblasts with IGF I, IGF II, EGF, or phorbol esters increased two- to threefold the number of both MPRs at the plasma membrane due to a redistribution of receptor from intracellular membranes to the cell surface Braulke *et al.*, 1989, 1990b; Damke *et al.*, 1992). Since the internalization rate is unchanged, this receptor redistribution results in an increased uptake of lysosomal enzymes (Braulke *et al.*, 1990a). On the other hand, the sorting of newly synthesized lysosomal enzymes in fibroblasts is not impaired (Braulke *et al.*, 1990a,b), indicating that the redistribution of MPRs does not lead to a functionally relevant deficit of receptors in the Golgi/TGN. In contrast, it is proposed that the platelet-derived growth factor (PDGF)-induced redistribution of MPR 300 to the plasma membrane in NIH 3T3 cells causes the selective increase in cathepsin L secretion by limiting the number of functional receptors in the Golgi (Prence *et al.*, 1990). The intracellular compartment from which the MPRs are recruited is unknown. Since various proteins with different subcellular localization are reported to be translocated to the plasma membrane in response to insulin or IGFs, it appears that these agents generally enhance the fusion between transport vesicles and increase their externalization rate. The underlying mechanisms regulating the distribution of MPRs are still unclear, but it is likely that common components and regulatory processes of vesicular transport are involved (see Section 5 for details). However, several studies that have used the fungal toxin wortmannin provided evidence that the phosphatidylinositol 3-kinase is involved in effector-induced redistribution of membrane proteins including the MPRs to the cell surface (Okada *et al.*, 1994; Körner and Braulke, 1996). In rat adipocytes it has been observed that insulin induces the increase in plasma membrane MPR 300 expression by receptor dephosphorylation resulting in a decrease in its internalization rate (Corvera et al., 1988). In contrast, the redistribution of MPR 300 in response to IGF II is not accompanied by a change in phosphorylation stoichiometry of receptors (Braulke and Mieskes, 1992). However, it cannot be excluded that alterations in the phosphorylation pattern at different sites rather than changes in stoichiometry may contribute to the subcellular localization of MPR 300. Studies using the protein phosphatase 1/2A inhibitor okadaic acid revealed that phosphatases are involved in a general rather than MPR 300-specific, IGF II-modulated recycling (Braulke and Mieskes, 1992). Furthermore, the redistribution of MPR 300 to the cell surface in response to growth factors and activators of protein kinase C is mediated by independent mechanisms whereby the activation of protein kinase C is associated with a stimulation of receptor phosphorylation (Braulke et al., 1990b).

Several immunocytochemical studies have suggested that the recycling of MPR 300 between the binding site of lysosomal enzymes in the Golgi and the delivery site in the endosome is triggered by ligand occupancy (Brown et al., 1984, 1986). Thus, depletion of ligands was reported to result in receptor accumulation in the Golgi, whereas the prevention of ligand dissociation by weak bases led to an accumulation in endosomal structures. However, biochemical studies using receptor ligand-independent approaches (Braulke et al., 1987b; Pfeffer, 1987) and chimeric proteins lacking the extracytoplasmic domain of MPR 300 (Mauxion et al., 1995) support the concept of a constitutive receptor recycling between intracellular and plasma membranes.

4. MANNOSE 6-PHOSPHATE-INDEPENDENT TRANSPORT

Studies of patients with I-cell disease have provided evidence for MPR-independent transport of soluble acid hydrolases to lysosomes. Due to the deficiency of phosphotransferase activity, newly synthesized lysosomal enzymes lack the M6P recognition marker and are secreted by many cell types, including fibroblasts. In other cell types or tissues such as hepatocytes, Kupffer cells, B-lymphocytes, spleen, and kidney, nearly normal levels of lysosomal enzymes have been found (Kornfeld and Sly, 1995; see Section 6). This suggests the existence of a cell type and/or enzyme-specific, M6P-independent sorting pathway. Additionally, several reports have described MPR-independent membrane association of precursor forms of acid hydrolases in other cells like mouse macrophages (Diment et al., 1988), human hepatoma HepG2 cells (Rijnboutt et al., 1991), transformed NIH 3T3 cells (McIntyre and Erickson, 1991), and human breast cancer MCF$_7$ cells (Capony et al., 1994). Procathepsin D is the best studied representative of MPR-independent transported lysosomal enzymes. Thus, 50–90% of the 53-kDa procathepsin D has been shown to be associated with permeabilized membranes in the presence of M6P or acidification and to disappear after proteolytic processing to the intermediate/mature forms (Diment *et al.*, 1988; Rijnboutt *et al.*, 1991; McIntyre and Erickson, 1991). While these studies suggest that membrane association occurs distal to the Golgi primarily in endosomes, it has been shown recently that portions of procathepsin D were complexed with a cosynthesized 68-kDa protein in a soluble form between the rough ER and the Golgi. The M6P-independent membrane association of pro-cathepsin D with a 72-kDa protein has been found in compartments between the late Golgi and dense lysosomal fraction (Zhu and Conner, 1994). The 68- and 72-kDa proteins, which are probably identical to the 60-kDa protein cross-linked to procathepsin D (Grässel and Hasilik, 1992), have been identified as differently glycosylated forms of prosaposin (Zhu and Conner, 1994). It is suggested that M6P-independent targeting mechanisms also are responsible for the routing of procathepsin D to lysosomes in U937 and MCF$_7$ cells even in the presence of NH$_4$Cl (Braulke *et al.*, 1987c; Capony *et al.*, 1994), which prevents the MPR-dependent transport (Gonzalez-Noriega *et al.*, 1980; Braulke *et al.*, 1987b).

The analysis of a number of chimeric proteins between cathepsin D and pepsinogen has determined that residues 188–265 in the carboxyl lobe of cathepsin D are able to ensure M6P-independent lysosomal sorting. This sorting signal partially overlaps with the region of the phosphotransferase recognition motif but is not identical (Glickman and Kornfeld, 1993; see Section 2.1).

Finally, M6P-independent transport of other soluble lysosomal proteins such as procathepsin L (McIntyre and Erickson, 1991), procathepsin B (Sloane *et al.*, 1994b), sphingolipid-activating protein (Rijnboutt *et al.*, 1991), α-glucosidase (Hoefsloot *et al.*, 1990), prosaposin (Zhu and Connor, 1994), as well as proteinase A and carboxypeptidase Y in the lysosome-like vacuole of the yeast *Saccharomyces cerevisiae* (Klionsky *et al.*, 1988; Valls *et al.*, 1990) has been described. While for the latter the receptor protein mediating the membrane association has been identified as *VPS* 10 (vacuolar protein sorting) gene product (Marcusson *et al.*, 1994), the carrier proteins interacting with the M6P-independent sorting sequences in mammalian cells are still unknown.

5. VESICULAR TRANSPORT MACHINERY

Targeting of lysosomal proteins is accomplished by the budding and fusion of transport vesicles. In the past few years genetic and immunolocalization studies as well as reconstituted *in vitro* systems have contributed to the identification of the components and the molecular mechanisms underlying vesicular transport and membrane trafficking in eukaryotic cells. These include polypeptides of the vesicle coat, trimeric and monomeric GTP-binding proteins, and fusion proteins like the N-ethylmaleimide sensitive fusion (NSF) protein, the soluble NSF-associated proteins (SNAPs), and the SNAP receptors (SNAREs).

5.1. Vesicle Coat Components

Many of the transport vesicles have been shown to be coated with proteins recruited from the cytosol onto the appropriate membrane during vesicle formation and then released into the cytosol after budding. Two classes of coated transported vesicles have been characterized so far: clathrin-coated vesicles, derived from the TGN and from the plasma membrane, and nonclathrin-coated or COP-coated vesicles, which are involved in transport from ER to *cis*-Golgi and between Golgi stacks and other steps in the secretory pathway (for recent reviews, see Robinson, 1994; Kreis and Pepperkok, 1994).

Clathrin triskelions are responsible for polygonal lattice formation and are bound to the membrane by a layer of adaptor complexes. Two types of adaptors have been identified in several cells and tissues, one associated with the TGN (AP-1) and one associated with the plasma membrane (AP-2). Both adaptors are composed of homologous, heterotetramer subunits: AP-2 is comprised of α-adaptin, $\beta2$-adaptin (formerly β), $\mu2$ (formerly AP50), and $\sigma2$ (formerly AP17) and AP-1 consists of γ-adaptin, $\beta1$-adaptin (formerly β'), $\mu1$ (formerly AP47), and $\sigma1$ (formerly AP19) (Pearse and Robinson, 1990). A third monomeric clathrin-assembly protein (AP-3; 92 kDa), which is modified by a single O-linked GlcNac residues and phosphorylation, has been identified in brain tissue, suggesting that it plays a role in cycling of synaptic vesicles (Murphy *et al.*, 1994). *In vitro,* adaptins not only bind to clathrin but also to the cytoplasmic domains of membrane proteins, such as the MPR 300, LDL, or EGF receptors, and synaptotagmin I (Glickman *et al.*, 1989; Sorkin and Carpenter, 1993). Neither AP-1 nor AP-2 compete with each other for binding to the MPR 300, indicating that both adaptor complexes recognize distinct and nonoverlapping receptor sites. Analysis of wild-type and mutant MPR 300 cytoplasmic tails suggests that a tyrosine present in a tight β-turn is part of the recognition motif of AP-2. Recently, $\mu1$ and $\mu2$ chains of AP-1 and AP-2 were identified that directly bind tyrosine-based signals of several integral membrane proteins (Ohno *et al.*, 1995).

In non-clathrin-coated vesicles the coats are made from a protein complex, the so-called coatomer, which contains seven coatomer proteins (COPs) (Rothman and Orci, 1992). In addition, the coats contain a small GTP-binding protein, the ADP-ribosylation factor (ARF), which mediates the binding of the coatomer to Golgi membranes (Serafini *et al.*, 1991). Two subunits of the COP complex, β- and ζ-COP, show structural homologies with the $\beta1/\beta2$-adaptin and $\sigma1$ subunits, respectively, suggesting that the coat proteins have related functions in membrane traffic (Duden *et al.*, 1991; Kuge *et al.*, 1993). In addition to their role in the transport through the Golgi, COPs may be involved in recycling proteins from the Golgi back to the ER (Cosson and Letourneur, 1994) as well as in sorting and concentration of cargo in the ER (Balch *et al.*, 1994). In a recent report an alternative model is discussed, in which coatomers are involved only in retrieval of dilysine-tagged proteins back in the ER (Letourneur *et al.*,

1994). Furthermore, a different coat (COP II) has been described as being necessary and sufficient for transport vesicle budding from the ER *in vitro* (Barlowe *et al.,* 1994).

5.2. Heterotrimeric G-Proteins

The large G-proteins composed of α (39–52 kDa), β (35–36 kDa), and γ (7–10 kDa) subunits are classically involved in transmembrane signaling of ligand-activated receptors resulting in either intracellular second messenger formation or regulation of ion channel function (Conklin and Bourne, 1993). In the absence of receptor ligands the α subunit is occupied by GDP while ligand binding stimulates the activation of trimeric G-proteins by catalyzing a GDP/GTP exchange and causes dissociation into α and $\beta\gamma$ subunits. The interaction of the GTP-bound α subunit with effector systems is terminated by the hydrolysis of GTP to GDP by intrinsic GTPase activity followed by a reassociation with $\beta\gamma$ subunits.

Heterotrimeric G-proteins have recently been identified as regulators in multiple vesicular transport steps. Thus, overexpression of $G\alpha_{i3}$, which is localized on Golgi membranes (Leyte *et al.,* 1992), decreased the rate of heparan sulfate proteoglycan secretion (Stow *et al.,* 1991). It is assumed that the overexpression of active (GTP-bound) $G\alpha_{i3}$ decreases vesicle formation, whereas conditions resulting in accumulation of inactive (GDP-bound) G_{i3} protein promote vesicle formation.

A recent report by Huttner and colleagues describes a new type of extra large G-protein (92 kDa $XL\alpha_s$) consisting of a 51-kDa protein linked to a $G\alpha_s$ functional domain which is associated with the TGN (Kehlenback *et al.,* 1994). This might explain why an activation of $G\alpha_s$ led to an accumulation of constitutive secretory vesicles at the TGN (Leyte *et al.,* 1992).

Trimeric G-proteins have also been reported to be involved in processes of endocytosis, endosome–endosome fusion (see also Chapter 3), and sorting into transcytotic vesicles. The precise role of trimeric G-proteins in intracellular membrane transport has not yet been established. However, the prevailing data suggest that trimeric G-proteins regulate membrane traffic by controlling coat assembly.

5.3. Monomeric G-Proteins

Several members of two families of small (20–25 kDa) ras-related GTP-binding proteins have been found to be involved in vesicle targeting and fusion: rabs and ARFs. Both proteins cycle between GDP- and GTP-bound conformations. These proteins are localized predominantly to distinct cytosolic surfaces of intracellular membranes of the biosynthetic and endocytic pathways, reflecting their function in specific membrane transport events (Table II).

Table II
Localization of Rab Proteins and Their
Involvement in Vesicular Transport Steps

Protein	Transport steps
Rab 1	ER-Golgi
Rab 2	Intermediate compartment
Rab 3a/b	Exocytosis of synaptic vesicles
Rab 4	Sorting/recycling through early endosomes
Rab 5	Plasma membrane—early endosomes
Rab 6	Intra-Golgi transport
Rab 7	Late endosomes
Rab 8	TGN—basolateral/dendritic membranes
Rab 9	Late endosome—TGN

More than 30 different rab proteins (called Ypt1/sec4 in yeast) are being identified which are posttranslationally modified by geranylgeranyl or farnesyl moieties (for review, see Magee and Newman, 1992). The prenylation reaction catalyzed by rab geranylgeranyltransferases and rab escort proteins (REPs) is required for functional activity and attachment to membranes. Rab proteins comprise a highly conserved GTP-binding domain and divergent sequences in the C-terminal region which are responsible for specific intracellular membrane association.

Different associated proteins are identified that affect biochemical activities and membrane targeting of rabs. Thus, GDP-bound rab is delivered to the appropriate membrane by a guanine-nucleotide dissociation inhibitor (GDI) and may be accompanied by a GDP/GTP exchange which also involves GDI-displacement factors (GDFs). It has been suggested that different GDFs for each rab protein determine the specificity of membrane attachment. The interaction between membrane-bound rab and guanine-nucleotide exchange factor (GEF) initiates the release of bound GDP and binding of GTP. The GTP-bound form is incorporated into carrier vesicles budding from the donor membrane. Upon fusion of the vesicles with the acceptor membrane, rab-specific GTPase activating proteins (GAPs) promote GTP hydrolysis. The resulting GDP-bound rab form recycles from the target membrane back to the donor compartment (Soldati et al., 1994; Ullrich et al., 1994; for review, see Pfeffer, 1994).

Six distinct ARF proteins have been identified to date, grouped into three classes which are all able to activate cholera toxin-catalyzed ADP-ribosylation. These highly homologous proteins are modified in part at the N-terminus by myristioylation, and this is essential for activity. It is believed that in eukaryotic cells ARFs participate in intra-Golgi vesicular transport, endocytosis, and nuclear membrane assembly (for recent review, see Donaldson and Klausner, 1994; Boman and Kahn, 1995). Like rab-GTPases, the GDP-bound ARF-1 is cytosolic

and associates with Golgi membranes when activated by binding to GTP. The guanine nucleotide status of ARF-1 is presumably regulated by a protease-sensitive
ARF guanine nucleotide exchange protein and GAP activity on Golgi membranes
(Randazzo and Kahn, 1994). It has been suggested that, in addition to the coatomer
(see above), binding of other coat proteins like AP-1 adaptors and a p200 protein
occurs through ARF-1 (Narula *et al.*, 1992; Robinson and Kreis, 1992; Traub *et
al.*, 1993). Whether the recently described activation of phospholipase D by ARF-
1 (Cockcroft *et al.*, 1994) occurs on Golgi membranes independent of coat assembly is not yet known.

5.4. Membrane Fusion Proteins

By using a cell-free system to reconstitute intra-Golgi transport, Rothman
and colleagues have identified a 76-kDa (NSF) protein required for vesicle fusion
(Glick and Rothman, 1987). Thus NSF protein (encoded in yeast by the *SEC*18
gene) forms a soluble tetramer which participates in both constitutive and regulated intracellular fusion processes. It needs three monomeric, soluble NSF attachment proteins (α, β, and γ SNAPs), with mole masses between 35 and 39 kDa,
for binding to membranes (Clary *et al.*, 1990). The SNAPs bind to specific
SNAREs which assemble with NSF protein into a 20S particle (Söllner *et al.*,
1993). Since NSF proteins and SNAPs function in all cell types and in a variety of
intracellular fusion processes it is suggested that multiple SNAREs are responsible for compartment-specific fusion processes. Three SNAREs have been isolated
from nerve terminals: the vesicular SNARE (v-SNARE) synaptobrevin (also
called vesicle-associated membrane protein, VAMP) and the target membrane
receptors (t-SNARE) syntaxin 1 and SNAP-25, a 25-kDa synaptosome-associated
protein. Thus, the first step in the fusion process of synaptic vesicles with the
presynaptic plasma membrane is the docking of the vesicle to the membrane by
binding of synaptobrevin and the vesicular Ca^{2+} phospholipid binding protein
synaptotagmin to the plasma membrane t-SNAREs syntaxin 1 and SNAP-25,
forming a 7S particle *in vitro*. After sequential addition of αSNAP and NSF protein, the hydrolysis of ATP catalyzed by NSF results in the dissociation of the protein complex and initiation of fusion of the membranes (Zhang *et al.*, 1994; for
review, see Rothman and Warren, 1994).

The occurrence of various syntaxin and synaptobrevin isoforms distributed in
neuronal and non-neuronal tissues showing binding selectivity to each other provides a mechanism for the specificity in vesicle transport. The isolation of synaptobrevin, syntaxin, and SNAP-25 homologues in yeast that function in different stages
of the secretory pathway and in vacuolar protein sorting supports the generality of
the SNARE hypothesis (Ferro-Novick and Jahn, 1994; Horazdovsky *et al.*, 1995).

Whether the same v-SNARE or t-SNAREs function in vesicular recycling
processes or whether a second set of SNAREs are needed for vesicle recycling to
the previous compartment is not known. It is likely that the interaction of a par-

ticular SNARE with different rab proteins determines the direction of vesicular transport. Recently it has been found that Ypt1 (rab1 in mammalian cells) selectively activates the v-SNARE of yeast ER-Golgi transport vesicles to form a complex with another membrane protein, which leads to a more efficient interaction with the t-SNARE (Lian *et al.*, 1994).

6. INBORN ERRORS OF LYSOSOMAL ENZYME DELIVERY AND TRANSPORT SYSTEMS

It is clear from the above discussion that the delivery of enzymes and substrates to the lysosome may be impaired due to genetic defects affecting (a) the primary structure of lysosomal proteins resulting in early degradation, (b) the formation of the M6P recognition marker, (c) the function of M6P receptors, and (d) the vesicular and lysosomal transport machinery.

Mutations affecting some of these steps have been observed in humans, leading to the deficiency of one or several lysosomal hydrolases or activator proteins. The resulting lysosomal storage diseases are classified according to the accumulating substrates and defects (for reviews, see Gieselmann, 1995; Pisoni and Thoene, 1991); these classifications describe the clinical variability, age of onset, and progression of the symptoms. The cloning of genes of lysosomal enzymes and the identification of mutations over the last few years has revealed insights into the molecular basis of these disorders. Thus, various mutations resulting in single amino acid substitutions, frameshifts, insertions, or truncations have been described that affect the intracellular transport of lysosomal proteins such as the α- and β-chains of β-hexosaminidase (Gravel *et al.*, 1995), β-D-galactosidase (Oshima *et al.*, 1994), the protective protein (Zhou *et al.*, 1991), sphingolipid activator proteins (Sandhoff *et al.*, 1995), or α-D-galactosidase (Ishii *et al.*, 1993). The introduction of respective mutations into the normal cDNAs and their heterologous expression has led to abnormal processing, folding, subunit assembly, or phosphorylation, resulting in either precursor retainment in the ER or degradation in an early biosynthetic compartment.

In another type of lysosomal disorder (mucolipidosis II and III; I-cell disease and pseudo-Hurler polydystrophy) multiple lysosomal enzymes are secreted instead of being targeted to lysosomes. Mucolipidosis II and III are caused by the inability of UDP-N-acetylglucosamine: lysosomal enzyme N-acetylglucosamine-1-phosphotransferase to catalyze the first step in the M6P recognition marker synthesis. Newly synthesized soluble lysosomal enzymes fail to bind to M6P receptors and become secreted (Kornfeld and Sly, 1995). No patients have been found who lack M6P receptors or show deficiencies in components of the vesicular transport machinery. However, gene targeting of both M6P receptors and IGF II has generated a mouse line resembling the biochemical defects in mucolipidosis II, characterized by a missorting of multiple lysosomal enzymes (A. Efstratiadis, personal communication). The expression of a dominant inhibitory form of rab 9 protein in CHO cells impairs the recycling of M6P receptors from late endosomes to the TGN and causes a decrease in the efficiency of lysosomal enzyme sorting (Riederer *et al.,* 1994). Furthermore, several mutants have been identified in yeast which are defective in vacuolar protein sorting. The analysis of *vps* mutants and their gene products (*VPS*) has revealed that some of these gene products (e.g., *VPS* 1, *VPS* 15, *VPS* 34) are required for sorting of several soluble hydrolases, such as carboxypeptidase Y (CPY) and proteinase A (PrA), which utilize different receptors and/or sorting pathways. Other gene products (*VPS* 10, *VPS* 29, *VPS* 35) are involved in the specific transport of CPY (for review, see Horazdovsky *et al.,* 1995). Cloning, sequencing, and functional studies revealed that the *VPS* 15 and *VPS* 34 gene products encode a 170-kDa Ser/Thr protein kinase and a phosphatidylinositol (PI) 3-kinase. Mutations of *VPS* 15 and *VPS* 34 cause missorting of soluble vacuolar proteins (Herman *et al.,* 1992). Recent studies have demonstrated that the inhibition of PI 3-kinase by wortmannin inhibited the sorting and targeting of newly synthesized cathepsin D in mammalian cells. Kinetic analysis and immunocytochemistry data suggest that PI 3-kinase activity is required for MPR 300-dependent sorting of lysosomal enzymes in the TGN (Brown *et al.,* 1995; Davidson, 1995). The *VPS* 10 gene product was identified as a late-Golgi, transmembrane CPY sorting receptor (Marcusson *et al.,* 1994). It is anticipated that genetic analysis of vacuolar protein sorting in yeast will provide new insights into regulation of vacuole biogenesis, which may be useful in interpreting studies of normal and defective lysosomal protein transport in mammals.

Another group of inherited lysosomal disorders in humans is caused by defects in lysosomal transporters mediating the release of degradation products (for review, see Gahl *et al.,* 1995). Thus, the high lysosomal cystine accumulation observed in patients with nephropathic cystinosis is due to defects in the lysosomal cystine transporter. In two other rare genetic disorders, Salla disease and infantile sialic acid storage disease, the large accumulation of free sialic acid within lysosomes is due to an impairment of a lysosomal transporter responsible for the efflux of sialic acid and glucuronate. Finally, an inborn disorder of lysosomal storage of vitamin B_{12} has been described as being caused by a defect in its release from lysosomes. The molecular structure of the responsible genes and the gene defects for all lysosomal transporter disorders are not known. Lysosomal disorders resulting from defects in transporter activities that mediate the uptake of substrates into lysosomes have not yet been identified.

ACKNOWLEDGMENTS. I wish to thank the colleagues who supplied me data of their recent studies and David Robinson, Victor Armstrong, and Cyrilla Cole-Maelicke for help with the manuscript The experimental work in our laboratory was supported by the Deutsche Forschungsgemeinschaft (SFB 236; B11) and the Fonds der Chemischen Industrie.

7. REFERENCES

Baker, J., Liu, J.-P., Robertson, E., and Efstratiadis, A., 1993, Role of insulin-like growth factors in embryonic and postnatal growth, *Cell* **73**:73–82.

Balch, W. E., McCaffery, J. M., Plutner, H., and Farquhar, M. G., 1994, Vesicular stomatitis virus glycoprotein is sorted and concentrated during export from the endoplasmic reticulum, *Cell* **76**:841–852.

Barnaski, T. J., Faust, P. L., and Kornfeld, S., 1990, Generation of a lysosomal enzyme targeting signal in the secretory protein pepsinogen, *Cell* **63**:281–291.

Baranski, T. J., Cantor, A. B., and Kornfeld, S., 1992, Lysosomal enzyme phosphorylation. I. Protein recognition determinants in both lobes of procathepsin D mediate its interaction with UDP-GlcNac:lysosomal enzyme N-acetylglucosamine-1-phosphotransferase, *J. Biol. Chem.* **267**:23342–23348.

Barlowe, C., Orci, L., Yeung, T., Hosobuchi, M., Hamamoto, S., Salama, N., Rexach, M. F., Ravazzola, M., Amherdt, M., and Schekman, R., 1994, COP II: A membrane coat formed by sec proteins that drive vesicle budding from the endoplasmic reticulum, *Cell* **77**:895–907.

Bleekemolen J. E., Stein, M., von Figura, K., Slot, J. W., and Geuze, H. J., 1988, The two mannose 6-phosphate receptors have almost identical subcellular distributions in U937 monocytes, *Eur. J. Cell Biol.* **47**:366–372.

Boman, A. L., and Kahn, R. A., 1995, Arf proteins: The membrane traffic police?, *Trends Biochem. Sci.* **20**:147–150.

Braulke, T., and Mieskes, G., 1992, Role of protein phosphatases in insulin-like growth factor II (IGF II)-stimulated mannose 6-phosphate/IGF II receptor redistribution, *J. Biol. Chem.* **267**:17347–17353.

Braulke, T., Hille, A., Huttner, W., Hasilik, A., and von Figura, K., 1987a, Sulfated oligosaccharides in human lysosomal enzymes, *Biochem. Biophys. Res. Commun.* **143**:178–185.

Braulke, T., Gartung, C., Hasilik, A., and von Figura, K., 1987b, Is movement of mannose 6-phosphate–specific receptor triggered by binding of lysosomal enzymes? *J. Cell Biol.* **104**:1735–1742.

Braulke, T., Geuze, H. J., Slot, J. W., Hasilik, A., and von Figura, K., 1987c, On the effects of weak bases and monensin on sorting and processing of lysosomal enzymes in human cells, *Eur. J. Cell Biol.* **43**:316–321.

Braulke, T., Tippmer, S., Neher, E., and von Figura, K., 1989, Regulation of the mannose 6-phosphate/IGF II receptor expression at the cell surface by mannose 6-phosphate, insulin-like growth factors and epidermal growth factor, *EMBO J.* **8**:681–686.

Braulke, T., Tippmer, S., Chao, H.-J., and von Figura, K., 1990a, Insulin-like growth factors I and II stimulate endocytosis but do not affect sorting of lysosomal enzymes in human fibroblasts, *J. Biol. Chem.* **265**:6650–6655.

Braulke, T., Tippmer, S., Chao, H.-J., and von Figura, K., 1990b, Regulation of mannose 6-phosphate/insulin-like growth factor II receptor distribution by activators and inhibitors of protein kinase C, *Eur. J. Biochem.* **189**:609–616.

Braulke, T., Bresciani, Buergisser, D. M., and von Figura, K., 1991, Insulin-like growth factor II overexpression does not affect sorting of lysosomal enzymes in NIH-3T3 cells, *Biochem. Biophys. Res. Commun.* **179**:108–115.

Braun, M., Waheed, A., and von Figura, K., 1989, Lysosomal acid phosphatase is transported to lysosomes via the cell surface, *EMBO J.* **8**:3633–3640.

Bresciani, R., Peters, C., and von Figura, K., 1992, Lysosomal acid phosphatase is not involved in the dephosphorylation of mannose 6-phosphate containing lysosomal proteins, *Eur. J. Cell Biol.* **58**:57–61.

Brown, W. J., Constantinescu, E., and Farquhar, M. G., 1984, Redistribution of mannose 6-phosphate specific receptors induced by tunicamycin and chloroquine, *J. Cell Biol.* **99**:320–326.

Brown, W. J., Goodhouse, J., and Farquhar, M. G., 1986, Mannose 6-phosphate receptors for lysosomal enzymes cycle between the Golgi-complex and endosomes, *J. Cell Biol.* **103**:1235–1247.

Brown, W. J., DeWald, D. B., Emr, S. D., Plutner, H., and Balch, W. E., 1995, Role of phosphatidylinositol 3-kinase in the sorting and transport of newly synthesized lysosomal enzymes in mammalian cells, *J. Cell Biol.* **130**:781–796.

Canfield, W. M., Johnson, K. F., Ye, R. D., Gregory, W., and Kornfeld, S., 1991, Localization of the signal for rapid internalization of the bovine cation-independent mannose 6-phosphate/insulin-like growth factor-II receptor to amino acids 24–29 of the cytoplasmic tail, *J. Biol. Chem.* **266**:5682–5688.

Cantor, A. B., Baranski, T. J., and Kornfeld, S., 1992, Lysosomal enzyme phosphorylation. II. Protein recognition determinants in either lobe of procathepsin D are sufficient for phosphorylation of both the amino and carboxyl lobe oligosaccharides, *J. Biol. Chem.* **267**:23349–23356.

Capony, F., Braulke, T., Rougeot, C., Roux, S., Montcourrier, P., and Rochefort, H., 1994, Specific mannose 6-phosphate receptor-independent sorting of pro-cathepsin D in breast cancer cells, *Exp. Cell Res.* **215**:154–163.

Causin, C., Waheed, A., Braulke, T., Junghans, U., Maly, P., Humbel, R. E., and von Figura, K., 1988, Mannose 6-phosphate/insulin-like growth factor II-binding proteins in human serum and urine, *Biochem. J.* **252**:795–799.

Chao, H.-J., Waheed, A., Pohlmann, R., Hille, A., and von Figura, K., 1990, Mannose 6-phosphate receptor dependent secretion of lysosomal enzymes, *EMBO J.* **9**:3507–3513.

Chen, H. J., Remmler, J., Delaney, J. C., Messner, D. J., and Lobel, P., 1993, Mutational analysis of the cation-independent mannose 6-phosphate/insulin-like growth factor II receptor, *J. Biol. Chem.* **268**:22338–22346.

Clairmont, K. B., and Czech, M. P., 1989, Chicken and *Xenopus* mannose 6-phosphate receptors fail to bind insulin-like growth factor II, *J. Biol. Chem.* **264**:16390–16392.

Clary, D. O., Griff, I. C., and Rothman, J. E., 1990, SNAPs, a family of NSF attachment proteins involved in intracellular membrane fusion in animals and yeast, *Cell* **61**:709–721.

Claussen, M., Buergisser, D., Schuller, A. G. P., Matzner, U., and Braulke, T., 1995, Regulation of insulin-like growth factor (IGF)-binding protein-6 and mannose 6-phosphate/IGF II receptor expression in IGF II overexpressing NIH 3T3 cells, *Mol. Endocrinol.* **9**:902–912.

Cockcroft, S., Thomas, G. M. H., Fensome, A., Geny, B., Cunningham, E., Gout, I., Hiles, I., Totty, N. F., Truong, O., and Hsuan, J. J., 1994, Phospholipase D: A downstream effector of ARF in granulocytes, *Science* **263**:523–526.

Collawn, J., Kuhn, L. A., Liu, L.-F. S., Tainer, J. A., and Trowbridge, I. S., 1991, Transplanted LDL and mannose-6-phosphate receptor internalization signals promote high-efficiency endocytosis of the transferrin receptor, *EMBO J.* **10**:3247–3253.

Conibear, E., and Pearse, B. M. F., 1994, A chimera of the cytoplasmic tail of the mannose 6-phosphate/IGF-II receptor and lysozyme localizes to the TGN rather than prelysosomes where the bulk of the endogenous receptor is found, *J. Cell Sci.* **107**:923–932.

Conklin, B. R., and Bourne, H. R., 1993, Structural elements of $G\alpha$ subunits that interact with $G\beta\gamma$, receptors, and effectors, *Cell* **73**:631–641.

Conner, G. E., 1989, Isolation of procathepsin D from mature cathepsin D by pepstatin affinity chromatography. Autocatalytic proteolysis of the zymogen form of the enzyme, *Biochem. J.* **263**:601–604.

Corvera, S., Folander, K., Clairmont, K. B., and Czech, M. P., 1988, A highly phosphorylated subpopulation of insulin-like growth factor II/mannose 6-phosphate receptors is concentrated in a clathrin-enriched plasma membrane fraction, *Proc. Natl. Acad. Sci. USA* **85**:7567–7571.

Cosson, P., and Letourneur, F., 1994, Coatomer interaction with di-lysine endoplasmic retention motifs, *Science* **263**:1629–1631.

Dahms, N. M., Lobel, P., Breitmeyer, J., Chirgwin, J. M., and Kornfeld, S., 1987, 46kD mannose 6-phosphate receptor: Cloning, expression, and homology to the 215 kD mannose 6-phosphate receptor, *Cell* **50**:181–192.

Dahms, N. M., Wick, D. A., and Brzycki-Wessell, M. A., 1994, The mannose 6-phosphate/insulin-like growth factor II receptor. Localization of the insulin-like growth factor II binding site to domains 5–11, *J. Biol. Chem.* **269**:3802–3809.

Damke, H., von Figura, K., and Braulke, T., 1992, Simultaneous redistribution of mannose 6-phosphate and transferrin receptors by insulin-like growth factors and phorbol ester, *Biochem. J.* **281**:225–229.

Davidson, H. W., 1995, Wortmannin causes mistargeting of procathepsin D. Evidence for the involvement of a phosphatidylinositol 3-kinase in vesicular transport to lysosomes, *J. Cell Biol.* **130**:797–805.

d'Azzo, A., Hoogeven, A., Reuser, A. J., Robinson, D., and Galjaard, H., 1982, Molecular defect in combined β-galactosidase and neuraminidase deficiency in man, *Proc. Natl. Acad. Sci. USA* **79**:4535–4539.

Dennis, P. A., and Rifkin, D. B., 1991, Cellular activation of latent transforming growth factor β requires binding to the cation-independent mannose 6-phosphate/insulin-like growth factor type II receptor, *Proc. Natl. Acad. Sci. USA* **88**:580–584.

Diesner, F., Sommerlade, H.-J., and Braulke, T., 1993, Transport of newly synthesized arylsulfatase A to the lysosome via transferrin receptor-positive compartments, *Biochem. Biophys. Res. Commun.* **197**:1–7.

Diment, S., Leech, M. S., and Stahl, P. D., 1988, Cathepsin D is membrane associated in macrophage endosomes, *J. Biol. Chem.* **263**:6901–6907.

Dintzis, S. M., and Pfeffer, S. R., 1990, The mannose 6-phosphate receptor cytoplasmic domain is not sufficient to alter the cellular distribution of a chimeric EGF receptor, *EMBO J.* **9**:77–84.

Dintzis, S. M., Velculescu, V. E., and Pfeffer, S. R., 1994, Receptor extracellular domains may contain trafficking information, *J. Biol. Chem.* **269**:12159–12166.

Donaldson, J. G., and Klausner, R. D., 1994, ARF: A key regulatory switch in membrane traffic and organelle structure, *Curr. Opin. Cell Biol.* **6**:527–532.

Duden, R., Griffiths, G., Frank, R., Argos, P., and Kreis, T. E., 1991, β-COP, a 100 kD protein associated with non-clathrin coated vesicles and the Golgi complex, shows homology with β-adaptin, *Cell* **64**:649–665.

Dustin, M. L., Baranski, T. J., Sampath, D., and Kornfeld, S., 1995, A novel mutagenesis strategy identifies distantly spaced amino acid sequences that are required for the phosphorylation of both the oligosaccharides of procathepsin D by N-acetylglucosamine 1-phosphotransferase, *J. Biol. Chem.* **270**:170–179.

Einstein, R., and Gabel, C. A., 1991, Cell- and ligand-specific dephosphorylation of acid hydrolases: Evidence that the mannose 6-phosphatase is controlled by compartmentalization, *J. Cell Biol.* **112**:81–94.

Faust, P. L., Kornfeld, S., and Chirgwin, J. M., 1985, Cloning and sequence analysis of cDNA for human cathepsin D, *Proc. Natl. Acad. Sci. USA* **82**:4910–4914.

Ferro-Novick, S., and Jahn, R., 1994, Vesicle fusion from yeast to man, *Nature* **370**:191–193.

Fong, D., Calhoun, D. H., Hsieh, W. T., Lee, B., and Wells, R. D., 1986, Isolation of a cDNA clone for the human lysosomal proteinase cathepsin B, *Proc. Natl. Acad. Sci. USA* **83**:2909–2913.

Fukuda, M., 1994, Biogenesis of the lysosomal membrane, in *Subcellular Biochemistry*, Vol. 22: Membrane Biogenesis (A. H. Maddy and J. R. Harris, eds.), pp. 199–230, Plenum Press, New York.

Gahl, W. A., Schneider, J. A., and Aula, P. P., 1995, Lysosomal transport disorders: Cystinosis and sialic acid storage disorders, in *The Metabolic and Molecular Bases of Inherited Disease*, Vol. II (C. R. Scriver, A. L. Beaudet, W. S. Sly, and D. Valle, eds.), pp. 3763–3797, McGraw-Hill, Inc., New York.

Gal, S., and Gottesman, M. M., 1986, The major excreted protein of transformed fibroblasts is an acid activable protease, *J. Biol. Chem.* **262**:1760–1765.

Galjart, N. J., Gillemans, N., Harris, A., von der Horst, G. T. J., Verheijen, F. W., Galjaard, H., and d'Azzo, A., 1988, Expression of cDNA encoding the human "protective protein" associated with lysosomal β-galactosidase and neuraminidase: Homology to yeast proteases, *Cell* **54**:755–764.

Galjart, N. J., Morreau, H., Willemsen, R., Gillemans, N., Bonten, E. J., and d'Azzo, A., 1991, Human lysosomal protective protein has cathepsin A-like activity distinct from its protective function, *J. Biol. Chem.* **266**:14754–14762.

Garmroudi, F., and MacDonald, R. G., 1994, Localization of the insulin-like growth factor (IGF II) binding/cross-linking site of the IGF II/mannose 6-phosphate receptor to extracellular repeats 10–11, *J. Biol. Chem.* **269**:26944–26952.

Gasa, S., Balbaa, M., Nakamura, M., Yonemori, H., and Makita, A., 1987, Phosphorylation of human lysosomal arylsulfatase B by cAMP-dependent protein kinase, *J. Biol. Chem.* **262**:1230–1238.

Gieselmann, V., 1995, Lysosomal storage diseases, *Biochim. Biophys. Acta* **1270**:103–136.

Gieselmann, V., Hasilik, A., and von Figura, K., 1985, Processing of human cathepsin D in lysosomes *in vitro, J. Biol. Chem.* **260**:3215–3220.

Glick, B. S., and Rothman, J. E., 1987, Possible role for fatty acyl-coenzyme A in intracellular protein transport, *Nature* **326**:309–312.

Glickman, J. N., and Kornfeld, S., 1993, Mannose 6-phosphate–independent targeting of lysosomal enzymes in I-cell disease B lymphosblasts, *J. Cell Biol.* **123**:99–108.

Glickman, J. N., Conibear, E., and Pearse, B. M. F., 1989, Specificity of binding of clathrin adaptors to signals on the mannose 6-phosphate/insulin-like growth factor, *EMBO J.* **8**:1041–1047.

Gonzalez-Noriega, A., Grubb, J. H., Talkad, V., and Sly, W. S., 1980, Chloroquine inhibits lysosomal enzyme pinocytosis and enhances lysosomal enzyme secretion by impairing receptor recycling, *J. Cell Biol.* **85**:839–852.

Gottschalk, S., Waheed, A., Schmidt, B., Laidler, P., and von Figura, K., 1989, Sequential processing of lysosomal acid phosphatase by a cytoplasmic thiol proteinase and a lysosomal aspartyl proteinase, *EMBO J.* **8**:3215–3219.

Grässel, S., and Hasilik, A., 1992, Human cathepsin D precursor is associated with a 60 kDa glycosylated polypeptide, *Biochem. Biophys. Res. Commun.* **182**:276–282.

Gravel, R. A., Clarke, J. T. R., Kaback, M. M., Mahuran, D., Sandhoff, K., and Suzuki, K., 1995, The GM2 gangliosidoses, in *The Metabolic and Molecular Bases of Inherited Disease,* Vol. II (C. R. Scriver, A. L. Beaudet, W. S. Sly, and D. Valle, eds.), pp. 2839–2879, McGraw-Hill, Inc., New York.

Green, S. A., Setiadi, H., McEver, R. P., and Kelly, R. B., 1994, The cytoplasmic domain of P-selectin contains a sorting determinant that mediates rapid degradation in lysosomes, *J. Cell Biol.* **124**:435–448.

Griffiths, G., Matteoni, R., Back, R., and Hoflack, B., 1990, Characterization of the cation-independent mannose 6-phosphate receptor-enriched prelysosomal compartment in NRK cells, *J. Cell Sci.* **95**:441–461.

Hammond, C., Braakman, I., and Helenius, A., 1994, Role of N-linked oligosaccharides, glucose trimming and calnexin during glycoprotein folding in the endoplasmic reticulum, *Proc. Natl. Acad. Sci. USA* **91**:913–917.

Hanewinkel, H., Glössl, J., and Kresse, H., 1987, Biosynthesis of cathepsin B in cultured normal and I-cell fibroblasts, *J. Biol. Chem.* **262**:12351–12355.

Hara, K., Kominami, E., and Katunuma, N., 1988, Effect of proteinase inhibitors on intracellular processing of cathepsin B, H, and L in rat macrophages, *FEBS Lett.* **231**:229–231.

Hartmann, H., Meyer-Alber, A., and Braulke, T., 1992, Metabolic actions of insulin-like growth factor II in cultured adult rat hepatocytes are not mediated through the insulin-like growth factor II receptor, *Diabetologia* **35**:216–223.

Hasilik, A., and Neufeld, E. F., 1980, Biosynthesis of lysosomal enzymes in fibroblasts. Synthesis as precursors of higher molecular weight, *J. Biol. Chem.* **255**:4937–4945.

Hasilik, A., von Figura, K., Conzelmann, E., Nehrkorn, H., and Sandhoff, K., 1982, Lysosomal enzyme precursors in human fibroblasts, *Eur. J. Biochem.* **125:**317–321.

Hauri, H.-P., and Schweizer, A., 1992, The endoplasmic reticulum–Golgi intermediate compartment, *Curr. Opin. Cell Biol.* **4:**600–608.

Hebert, D. N., Foellmer, B., and Helenius, A., 1995, Glucose trimming and reglucosylation determine glycoprotein association with calnexin in the endoplasmic reticulum, *Cell* **81:**425–433.

Hemer, F., Körner, C., and Braulke, T., 1993, Phosphorylation of the human 46-kDa mannose 6-phosphate receptor in the cytoplasmic domain at serine 56, *J. Biol. Chem.* **268:**17108–17113.

Herman, P. K., Stack, J. H., and Emr, S. D., 1992, An essential role for a protein kinase and lipid kinase complex in secretory protein sorting, *Trends Cell Biol.* **2:**363–368.

Hille, A., Braulke, T., von Figura, K., and Huttner, W. B., 1990, Occurrence of tyrosine sulfate in proteins—a balance sheet. 1. Secretory and lysosomal proteins, *Eur. J. Biochem.* **188:**577–586.

Hille-Rehfeld, A., 1995, Mannose 6-phosphate receptors in sorting and transport of lysosomal enzymes, *Biochim. Biophys. Acta* **1241:**177–194.

Hoeflich, A., Wolf, E., Braulke, T., Koepf, G., Kessler, U., Brem, G., Rascher, W., Blum, W., and Kiess, W., 1995, Does the overexpression of prepro-IGF II in transfected human embryonic kidney fibroblasts increase the secretion of lysosomal enzymes?, *Eur. J. Biochem.* **232:**172–178.

Hoefsloot, L. H., Willemsen, R., Kroos, M. A., Hoogeveen-Westerveld, M., Hermans, M. M. P., van der Ploeg, A. T., Oostra, B., and Reuser, A. J. J., 1990, Expression and routing of human lysosomal α-glucosidase in transiently transfected mammalian cells, *Biochem. J.* **272:**485–492.

Horazdovsky, B. F., DeWald, D. B., and Emr, S. D., 1995, Protein transport to the yeast vacuole, *Curr. Opin. Cell Biol.* **7:**544–551.

Ishii, S., Kase, R., Sakuraba, H., and Suzuki, Y., 1993, Characterization of a mutant α-galactosidase gene product for the late-onset cardiac form of Fabry disease, *Biochem. Biophys. Res. Commun.* **197:**1585–1589.

Jadot, M., Canfield, W. M., Gregory, W., and Kornfeld, S., 1992, Characterization of the signal for rapid internalization of the bovine mannose 6-phosphate/insulin-like growth factor-II receptor, *J. Biol. Chem.* **267:**11069–11077.

Johnson, K. F., and Kornfeld, S., 1992a, The cytoplasmic tail of the mannose 6-phosphate/insulin-like growth factor-II receptor has two signals for lysosomal enzyme sorting in the Golgi, *J. Cell Biol.* **119:**249–257.

Johnson, K. F., and Kornfeld, S., 1992b, A His-Leu-Leu sequence near the carboxyl terminus of the cytoplasmic domain of the cation-dependent mannose 6-phosphate receptor is necessary for the lysosomal enzyme sorting function, *J. Biol. Chem.* **267:**17110–17115.

Johnson, K. F., Chan, W., and Kornfeld, S., 1990, Cation-dependent mannose 6-phosphate receptor contains two internalization signals in its cytoplasmic domain, *Proc. Natl. Acad. Sci. USA* **87:**10010–10014.

Kehlenbach, R. H., Matthey, J., and Huttner, W. B., 1994, XLαs is a new type of G protein, *Nature* **372:**804–809.

Kiess, W., Greenstein, L. A., White, R. M., Lee, L., Rechler, M. M., and Nissley, S. P., 1987, Type II insulin-like growth factor receptor is present in rat serum, *Proc. Natl. Acad. Sci. USA* **84:**7720–7724.

Kiess, W., Blickenstaff, G. D., Sklar, M. M., Thomas, C. L., Nissley, S. P., and Sahagian, G. G., 1988, Biochemical evidence that the type II insulin-like growth factor receptor is identical to the cation-independent mannose 6-phosphate receptor, *J. Biol. Chem.* **263:**9339–9344.

Kiess, W., Thomas, C., Greenstein, L. A., Lee, L., Sklar, M. M., Rechler, M. M., Sahagian, G. G., and Nissley, S. P., 1989, Insulin-like growth factor II (IGF II) inhibits both the cellular uptake of β galactosidase and the binding of β galactosidase to purified IGF II/mannose 6-phosphate receptor, *J. Biol. Chem.* **264**:4710–4714.

Kiess, W., Greenstein, L. A., Lee, L., Thomas, C., and Nissley, S. P., 1991, Biosynthesis of the insulin-like growth factor-II (IGF-II)/mannose 6-phosphate receptor in rat C6 glial cells. The role of N-linked glycosylation in binding IGF-II to the receptor, *Mol. Endocrinol.* **5**:282–291.

Klionsky, D. J., Banta, L. M., and Emr, S. D., 1988, Intracellular sorting and processing of a yeast vacuolar hydrolase: Proteinase A propeptide contains vacuolar targeting information, *Mol. Cell Biol.* **8**:2105–2116.

Klumperman, J., Hille, A., Veenendaal, T., Oorschot, V., Stoorvogel, W., von Figura, K., and Geuze, H. J., 1993, Differences in the endosomal distributions of the two mannose 6-phosphate receptors, *J. Cell Biol.* **121**:997–1010.

Körner, C., Herzog, A., Weber, B., Rosorius, O., Hemer, F., Schmidt, B., and Braulke, T., 1994, *In vitro* phosphorylation of the 46-kDa mannose 6-phosphate receptor by casein kinase II, *J. Biol. Chem.* **269**:16529–16532.

Körner, C., Nürnberg, B., Uhde, M., and Braulke, T., 1995, Mannose 6-phosphate/insulin-like growth factor II receptor fails to interact with G-proteins, *J. Biol. Chem.* **270**:287–295.

Körner, C., and Braulke, T., 1996, Inhibition of IGF II-induced redistribution of mannose 6-phosphate receptors by the phosphatidylinositol 3-kinase inhibitor, wortmannin, *Mol. Cell Endocrinol* **118**:201–205.

Kornfeld, R., and Kornfeld, S., 1985, Assembly of asparagine-linked oligosacchardies, *Annu. Rev. Biochem.* **54**:631–664.

Kornfeld, S., 1992, Structure and function of the mannose 6-phosphate/insulin-like growth factor II receptors, *Annu. Rev. Biochem.* **61**:307–330.

Kornfeld, S., and Mellman, I., 1989, The biogenesis of lysosomes, *Annu. Rev. Cell Biol.* **5**:483–525.

Kornfled, S., and Sly, W. S., 1995, I-cell disease and pseudo-Hurler polydystrophy: Disorders of lysosomal enzyme phosphorylation and localization, in *The Metabolic and Molecular Bases of Inherited Disease*, Vol. II (C. R. Scriver, A. L. Beaudet, W. S. Sly, and D. Valle, eds.), pp. 2495–2508, McGraw-Hill, Inc., New York.

Köster, A., Saftig, P., Matzner, U., von Figura, K., Peters, C., and Pohlmann, R., 1993, Target disruption of the Mr 46000 mannose 6-phosphate receptor gene in mice results in misrouting of lysosomal proteins, *EMBO J.* **12**:5219–5223.

Köster, V., von Figura, K., and Pohlmann, R., 1994, Mistargeting of lysosomal enzymes in Mr 46000 mannose 6-phosphate receptor-deficient mice is compensated by carbohydrate-specific endocytotic receptors, *Eur. J. Biochem.* **224**:685–689.

Kreis, T. E., and Pepperkok, R., 1994, Coat proteins in intracellular membrane transport, *Curr. Opin. Cell Biol.* **6**:533–537.

Kuge, O., Hara-Kuge, S., Orci, L., Ravazzola, M., Amherdt, M., Tanigawa, G., Wieland, F. T., and Rothman, J. E., 1993, ζ-COP, a subunit of coatomer, is required for COP-coated vesicle assembly, *J. Cell Biol.* **123**:1727–1734.

Lau, M. M. H., Stewart, C. E. H., Liu, Z., Bhatt, H., Rotwein, P., and Stewart, C. L., 1994, Loss of the imprinted IGF2/cation-independent mannose 6-phosphate receptor results in fetal overgrowth and perinatal lethality, *Genes Dev.* **8**:2953–2963.

LeBorgne, R., Schmidt, A., Mauxion, F., Griffiths, G., and Hoflack, B., 1993, Binding of AP-1 Golgi adaptors to membranes requires phosphorylated cytoplasmic domains of mannose 6-phosphate/insulin-like growth factor II receptor, *J. Biol. Chem.* **268**:22552–22556.

Letourneur, F., Gaynor, E. C., Hennecke, S., Demolliere, C., Duden, R., Emr, S. D., Riezman, H., and Cosson, P., 1994, Coatomer is essential for retrieval of dilysine-tagged proteins to the endoplasmic reticulum, *Cell* **79**:1199–1207.

Leyte, A., Barr, F. A., Kehlenbach, R. H., and Huttner, W., 1992, Multiple trimeric G-proteins on the trans-Golgi network exert stimulatory and inhibitory effects on secretory vesicle formation *EMBO J.* **11**:4795–4804.

Lian, J. P., Stone, S., Jiang, Y., Lyons, P., and Ferro-Novick, S., 1994, Ypt1p implicated in v-SNARE activation, *Nature* **372**:698–701.

Liu, J. P., Baker, J., Perkins, A. S., Robertson, E. J., Efstratiadis, A., 1993, Mice carrying null mutations of the genes encoding insulin-like growth factor I and type I IGF receptor, *Cell* **75**:59–72.

Lobel, P., Dahms, N. M., and Kornfeld, S., 1988, Cloning and sequence analysis of the cation-independent mannose 6-phosphate receptor, *J. Biol. Chem.* **263**:2563–2570.

Ludwig, T., Griffiths, G., and Hoflack, B., 1991, Distribution of newly synthesized lysosomal enzymes in the endocytic pathway of normal rat kidney cells, *J. Cell Biol.* **115**:1561–1572.

Ludwig, T., Ovitt, C. E., Bauer, U., Hollinshead, M., Remmler, J., Lobel, P., Rüther, U., and Hoflack, B., 1993, Targeted disruption of the mouse cation-dependent mannose 6-phosphate receptor results in partial missorting of multiple lysosomal enzymes, *EMBO J.* **12**:5225–5235.

Ma, Z., Grubb, J. H., and Sly, W. S., 1992, Divalent cation-dependent stimulation of ligand binding to the 46-kDa mannose 6-phosphate receptor correlates with divalent cation-dependent tetramerization, *J. Biol. Chem.* **267**:19017–19022.

Mach, L., Schwihla, H., Stüwe, K., Rowan, A. D., Mort, J. S., and Glössl, J., 1993, Activation of procathepsin B in human hepatoma cells: The conversion into the mature enzyme relies on the action of cathepsin B itself, *Biochem. J.* **293**:437–442.

Magee, T., and Newman, C., 1992, The role of lipid anchors for small G proteins in membrane trafficking, *Trends Cell Biol.* **2**:318–323.

Maillet, C., and Shur, B. D., 1993, Uvomorulin, LAMP-1 and laminin are substrates for cell surface β-1,4-galactosyltransferase on F9 embryonal carcinoma cells: Comparisons between wild-type and mutant 5.51 att cells, *Exp. Cell Res.* **208**:282–295.

Marcusson, E. G., Horazdovsky, B. F., Cereghino, J. L., Gharakhanian, E., and Emr, S. D., 1994, The sorting receptor for yeast vacuolar carboxypeptidase Y is encoded by the *VPS10* gene, *Cell* **77**:579–586.

Matthews, P. M., Martinie, J. B., and Fambrough, D. M., 1992, The pathway and targeting signal for delivery of the integral membrane glycoprotein LEP100 to lysosomes, *J. Cell Biol.* **118**:1027–1040.

Mauxion, F., Schmidt, A., LeBorgne, R., and Hoflack, B., 1995, Chimeric proteins containing the cytoplasmic domains of the mannose 6-phosphate receptors codistribute with the endogenous receptors, *Eur. J. Cell Biol.* **66**:119–126.

McIntyre, G. F., and Erickson, A. H., 1991, Procathepsins L and D are membrane-bound in acidic microsomal vesicles, *J. Biol. Chem.* **266**:15438–15445.

Meresse, S., and Hoflack, B., 1993, Phosphorylation of the cation-independent mannose 6-phosphate receptor is closely associated with its exit from the trans-Golgi network, *J. Cell Biol.* **120**:67–75.

Meresse, S., Ludwig, T., Frank, R., and Hoflack, B., 1990, Phosphorylation of the cytoplasmic domain of the bovine cation-independent mannose 6-phosphate receptor, *J. Biol. Chem.* **265**:18833–18842.

Murayama, Y., and Nishimoto, I., 1990, Distinctive regulation of the functional linkage between the human cation-independent mannose 6-phosphate receptor and GTP-binding proteins by insulin-like growth factor II and mannose 6-phosphate, *J. Biol. Chem.* **265**:17456–17462.

Murphy, J.-E., Hanover, J. A., Froehlich, M., DuBois, G., and Keen, J. H., 1994, Clathrin assembly protein AP-3 is phosphorylated and glycosylated on the 50-kDa structural domain, *J. Biol. Chem.* **269**:21346–21352.

Narula, N., McMorrow, I., Plopper, G., Doherty, J., Matlin, K. S., Burke, B., and Stow, J. L., 1992, Identification of the 200 kDa, brefeldin A-sensitive protein on Golgi membranes, *J. Cell Biol.* **117**:27–38.

Neufeld, E. F., 1989, Natural history and inherited disorders of a lysosomal enzyme, β-hexosaminidase, *J. Biol. Chem.* **264**:10927–10930.

Nishimoto, I., Murayama, Y., Katada, T., Ui, M., and Ogata, E., 1989, Possible direct linkage of insulin-like growth factor-II receptor with guanine nucleotide-binding proteins, *J. Biol. Chem.* **264:**14029–14038.

Ogata, S., and Fukuda, M., 1994, Lysosomal targeting of Limp II membrane glycoprotein requires a novel Leu-Ile motif at a particular position in its cytoplasmic tail, *J. Biol. Chem.* **269:**5210–5217.

Ohno, H., Stewart, J., Fournier, M.-C., Bosshart, H., Rhee, I., Miyatake, S., Saito, T., Gallusser, A., Kirchhausen, T., and Bonifacino, J. S., 1995, Interaction of tyrosine-based sorting signals with clathrin-associated proteins, *Science* **269:**1872–1875.

Okada, T., Kawano, Y., Sakakibara, T., Hazeki, O., and Ui, M., 1994, Essential role of phosphatidylinositol 3-kinase in insulin-induced glucose transport and antilipolysis in rat adipocytes, *J. Biol. Chem.* **269:**3568–3573.

Okamoto T., Katada, T., Murayama, Y., Ui, M., Ogata, E., and Nishimoto, I., 1990, A simple structure encodes G protein-activating function of the IGF-II/mannose 6-phosphate receptor, *Cell* **62:**709–717.

Ono, M., Taniguchi, N., Makita, A., Fujita, M., Sekiya, C., and Namiki, M., 1988, Phosphorylation of β-glucuronidase from human normal liver and hepatoma by cAMP-dependent protein kinase, *J. Biol. Chem.* **263:**5884–5889.

Oshima, A., Yoshida, K., Ithoh, K., Kase, R., Sakuraba, H., and Suzuki, Y., 1994, Intracellular processing and maturation of mutant gene products in hereditary β-galactosidase deficiency (β-galactosidosis), *Hum. Genet.* **93:**109–114.

Pearse, B. M. F., and Robinson, M. S., 1990, Clathrin, adaptors and sorting, *Annu. Rev. Cell Biol.* **6:**151–171.

Peters, C., and von Figura, K., 1994, Biogenesis of lysosomal membranes, *FEBS Lett.* **346:**108–114.

Pfeffer, S. R., 1987, The endosomal concentration of a mannose 6-phosphate receptor is unchanged in the absence of ligand synthesis, *J. Cell Biol.* **105:**229–234.

Pfeffer, S. R., 1994, Rab GTPases: Master regulators of membrane trafficking, *Curr. Opin. Cell Biol.* **6:**522–526.

Pisoni, R. L., and Thoene, J. G., 1991, The transport systems of mammalian lysosomes, *Biochim. Biophys. Acta* **1071:**351–373.

Prence, E. M., Dong, J., and Sahagian, G. G., 1990, Modulation of the transport of a lysosomal enzyme by PDGF, *J. Cell Biol.* **110:**319–326.

Randazzo, P. A., and Kahn, R. A., 1994, GTP hydrolysis by ADP-ribosylation factor is dependent on both an ADP-ribosylation factor GTPase-activating protein and acid phospholipids, *J. Biol. Chem.* **269:**10758–10763.

Richo, G. R., and Conner, G. E., 1994, Structural requirements of procathepsin D activation and maturation, *J. Biol. Chem.* **269:**14806–14812.

Riederer, M. A., Soldati, T., Shapiro, A. D., Lin, J., and Pfeffer, S. R., 1994, Lysosome biogenesis requires rab 9 function and receptor recycling from endosomes to the trans-Golgi network, *J. Cell Biol.* **125:**573–582.

Rijnboutt, S., Aerts, H. M. F. G., Geuze, H. J., Tager, J. M., and Strous, G. J., 1991, Mannose 6-phosphate–independent membrane association of cathepsin D, glucocerebrosidase, and sphingolipid-activating protein in HepG2 cells, *J. Biol. Chem.* **266:**4862–4868.

Robinson, M. S., 1994, The role of clathrin, adaptors and dynamin in endocytosis, *Curr. Opin. Cell Biol.* **6:**538–544.

Robinson, M. S., and Kreis, T. E., 1992, Recruitment of coat proteins onto Golgi membranes in intact and permeabilized cells: Effects of brefeldin A and G protein activators, *Cell* **69:**129–138.

Rosorius, O., Mieskes, G., Issinger, O.-G., Körner, C., Schmidt, B., von Figura, K., and Braulke, T., 1993a, Characterization of phosphorylation sites in the cytoplasmic domain of the 300 kDa mannose 6-phosphate receptor, *Biochem. J.* **292:**833–838.

Rosorius, O., Issinger, O.-G., and Braulke, T., 1993b, Phosphorylation of the cytoplasmic tail of the 300 kDa mannose 6-phosphate receptor is required for the interaction with a cytosolic protein, *J. Biol. Chem.* **268**:21470–21473.

Rothman, J. E., and Orci, L., 1992, Molecular dissection of the secretory pathway, *Nature* **355**:409–416.

Rothman, J. E., and Warren, G., 1994, Implications of the SNARE hypothesis for intracellular membrane topology and dynamics, *Curr. Biol.* **4**:220–233.

Saftig, P., Hetmann, M., Schmahl, W., Weber, K., Heine, L., Mossmann, H., Köster, A., Hess, B., Evers, M., von Figura, K., and Peters, C., 1995, Mice deficient for the lysosomal proteinase cathepsin D exhibit progressive atrophy of the intestinal mucosa and profound destruction of lymphoid cells, *EMBO J.* **14**:3599–3608.

Sandhoff, K., Harzer, K., and Fürst, W., 1995, Sphingolipid activator proteins, in *The Metabolic and Molecular Bases of Inherited Disease*, Vol. II (C. R. Scriver, A. L. Beaudet, W. S. Sly, and D. Valle, eds.), pp. 2427–2441, McGraw-Hill, Inc., New York.

Sandoval, I. V., Arredondo, J. J., Alcalde, J., Gonzalez Noriega, A., Vandekerckhove, J., Jimenez, M. A., and Rico, M., 1994, The residues Leu (Ile)[475]-Ile(Leu, Val, Ala)[476], contained in the extended carboxyl cytoplasmic tail, are critical for targeting of the resident lysosomal membrane protein LIMP II to lysosomes, *J. Biol. Chem.* **269**:6622–6631.

Schmidt, B., Selmer, T., Ingendoh, A., and von Figura, K., 1995a, A novel amino acid modification in sulfatases that is defective in multiple sulfatase deficiency, *Cell* **82**:271–278.

Schmidt, B., Kiecke-Siemsen, C., Waheed, A., Braulke, T., and von Figura, K., 1995b, Localization of the insulin-like growth factor II binding site to amino acid 1508–1566 in repeat 11 of the mannose 6-phosphate/insulin-like growth factor II receptor, *J. Biol. Chem.* **270**:14975–14982.

Schulze-Garg, C., Böker, C., Nadimpalli, S. K., von Figura, K., and Hille-Rehfeld, A., 1993, Tail-specific antibodies that block return of 46,000 M_r mannose 6-phosphate receptor to the *trans*-Golgi network, *J. Cell Biol.* **122**:541–551.

Serafini, T., Orci, L., Amherdt, M., Brunner, M., Kahn, R. A., and Rothman, J. E., 1991, ADP-ribosylation factor (ARF) is a subunit of the coat of Golgi-derived COP-coated vesicles: A novel role for a GTP-binding protein, *Cell* **67**:239–253.

Sloane, B. F., Moin, K., and Lah, T. T., 1994a, Regulation of lysosomal endopeptidases in malignant neoplasia, in *Biochemical and Molecular Aspects of Selected Cancers*, Vol. 2 (T. G. Pretlow and T. P. Pretlow, eds.), pp. 411–466, Academic Press, San Diego.

Sloane, B. F., Moin, K., Sameni, M., Tait, L. R., Rozhin, J., and Ziegler, G., 1994b, Membrane association of cathepsin B can be induced by transfection of human breast epithelial cells with c-Ha-*ras* oncogene, *J. Cell Sci.* **107**:373–384.

Soldati, T., Shapiro, A. D., Dirac-Svejstrup, A. B., and Pfeffer, S. R., 1994, Membrane targeting of the small GTPase rab9 is accompanied by nucleotide exchange, *Nature* **369**:76–78.

Söllner, T., Whiteheart, S. W., Brunner, M., Erdjument-Bromage, H., Geromanos, S., Tempst, P., and Rothman, J. E., 1993, SNAP receptors implicated in vesicle targeting and fusion, *Nature* **362**:318–324.

Sommerlade, H. J., Hille-Rehfeld, A., von Figura, K., and Gieselmann, V., 1994a, Four monoclonal antibodies inhibit the recognition of arylsulfatase A by the lysosomal enzyme phosphotransferase, *Biochem. J.* **297**:123–130.

Sommerlade, H.-J., Selmer, T., Ingendoh, A., Gieselmann, V., von Figura, K., Neifer, K., and Schmidt, B., 1994b, Glycosylation and phosphorylation of arylsulfatase A, *J. Biol. Chem.* **269**:20977–20981.

Sorkin, A., and Carpenter, G., 1993, Interaction of activated EGF receptors with coated pit adaptins, *Science* **261**:612–615.

Sosa, M. A., Schmidt, B., von Figura, K., and Hille-Rehfeld, A., 1993, *In vitro* binding of plasma membrane-coated vesicle adaptors to the cytoplasmic domain of lysosomal acid phosphatase, *J. Biol. Chem.* **268**:12537–12543.

Stein, M., Zijderhand-Bleekemolen, J. E., Geuze, H., Hasilik, A., and von Figura, K., 1987, M_r 46,000 mannose 6-phosphate specific receptor: Its role in targeting of lysosomal enzymes, *EMBO J.* **6:**2677–2681.

Stow, J. L., DeAmeida, J. B., Narula, N., Holtzman, K. J., Ercolani, L., and Ausiello, S. A., 1991, A heterotrimeric G protein, $G\alpha i3$, on Golgi membranes regulates the secretion of a heparan sulfate proteoglycan in LLC-PK$_1$ epithelial cells, *J. Cell Biol.* **114:**1113–1124.

Tong, P. Y., and Kornfeld, S., 1989, Ligand interactions of the cation-dependent mannose 6-phosphate receptor, *J. Biol. Chem.* **264:**7970–7975.

Tong, P. Y., Tollefsen, S. E., and Kornfeld, S., 1988, The cation-independent mannose 6-phosphate receptor binds insulin-like growth factor II, *J. Biol. Chem.* **263:**2585–2588.

Tong, P. Y., Gregory, W., and Kornfeld, S., 1989, Ligand interactions of the cation-independent mannose 6-phosphate receptor, *J. Biol. Chem.* **264:**7962–7969.

Traub, L. M., Ostrom, J. A., and Kornfeld, S., 1993, Biochemical dissection of AP-1 recruitment onto Golgi membranes, *J. Cell Biol.* **123:**561–573.

Trowbridge, I. S., Collawn, J. F., and Hopkins, C. R., 1993, Signal-dependent membrane protein trafficking in the endocytic pathway, *Annu. Rev. Cell Biol.* **9:**129–161.

Ullrich, O., Horiuchi, H., Bucci, C., and Zerial, M., 1994, Membrane association of rab5 mediated by GDP-dissociation inhibitor and accompanied by GDP/GTP exchange, *Nature* **368:**157–160.

Valls, L. A., Winther, J. R., and Stevens, T. H., 1990, Yeast carboxypeptidase Y vacuolar targeting signal is defined by four propeptide amino acids, *J. Cell Biol.* **111:**361–368.

von Figura, K., and Hasilik, A., 1986, Lysosomal enzymes and their receptors, *Annu. Rev. Biochem.* **55:**167–193.

Waheed, A., and von Figura, K., 1990, Rapid equilibrium between monomeric, dimeric and tetrameric forms of the 46-kDa mannose 6-phosphate receptor at 37 °C, *Eur. J. Biochem.* **193:**47–54.

Waheed, A., Gottschalk, S., Hille, A., Krentler, C., Pohlmann, R., Braulke, T., Heuser, H., Geuze, H., and von Figura, K., 1988, Human lysosomal acid phosphatase is transported as a transmembrane protein to lysosomes in transfected baby hamster kidney cells, *EMBO J.* **7:**2351–2358.

Wang, Z.-Q., Fung, M. R., Barlow, D. P., and Wagner, E. F., 1994, Regulation of embryonic growth and lysosomal targeting by the imprinted Igf2/Mpr gene, *Nature* **372:**464–467.

Watanabe, H., Grubb, J. H., and Sly, W. S., 1990, The overexpressed human 46-kDa mannose 6-phosphate receptor mediates endocytosis and sorting of β-glucuronidase, *Proc. Natl. Acad. Sci. USA* **87:**8036–8040.

Wendland, M., Waheed, A., Schmidt, B., Hille, A., Nagel, G., and von Figura, K., 1991a, Glycosylation of the M_r 46,000 mannose 6-phosphate receptor. Effect on ligand binding, stability, and conformation, *J. Biol. Chem.* **266:**4598–4604.

Wendland, M., von Figura, K., and Pohlmann, R., 1991b, Mutational analysis of disulfide bridges in the Mr 46,000 mannose 6-phosphate receptor, *J. Biol. Chem.* **266:**7132–7136.

Westcott, K. R., and Rome, L. H., 1988, Cation-independent mannose 6-phosphate receptor contains covalently bound fatty acid, *J. Cell. Biochem.* **38:**23–33.

Westlund, B., Dahms, N. M., and Kornfeld, S., 1991, The bovine mannose 6-phosphate/insulin-like growth factor II receptor. Localization of mannose 6-phosphate binding sites to domains 1–3 and 7–11 of the extracytoplasmic region, *J. Biol. Chem.* **266:**23233–23239.

Wisselaar, H. A., Kroos, M. A., Hermans, M. M., van Beeumen, J., and Reuser, A. J., 1993, Structural and functional changes of lysosomal acid alpha-glucosidase during intracellular transport and maturation, *J. Biol. Chem.* **268:**2223–2231.

Zhang, Y., and Dahms, N. M., 1993, Site-directed removal of N-glycosylation sites in the bovine cation-dependent mannose 6-phosphate receptor: Effects on ligand binding, intracellular targeting and association with binding immunoglobulin protein, *Biochem. J.* **295:**841–848.

Zhang, J. Z., Davletov, B. A., Südhoff, T. C., and Anderson, R. G. W., 1994, Synaptotagmin is a high affinity receptor for clathrin AP-2: Implications for membrane recycling, *Cell* **78:**751–760.

Zhou, X. Y., Galjart, N. J., Willemsen, R., Gillemans, N., Galjaard, H., and d'Azzo, A., 1991, A mutation in a mild form of galactosialidosis impairs dimerization of the protective protein and renders it unstable, *EMBO J.* **10**:4041–4048.

Zhu, Y., and Conner, G. E., 1994, Intermolecular association of lysosomal protein precursors during biosynthesis, *J. Biol. Chem.* **269**:3846–3851.

Chapter 3

Endocytosis

Elizabeth Smythe

1. INTRODUCTION

The plasma membrane of cells provides a critical barrier between the intracellular and extracellular environment that protects the cell from deleterious factors in the surrounding medium and also provides the means of transmission of signals from the outside to the inside of the cell. Endocytosis is an essential physiological function for most cells. It allows modulation of the expression of cell surface receptors, internalization of a wide variety of biologically important macromolecules, essential functions of antigen processing and presentation, and transport of material across the cell. Much of the material which is found in lysosomes enters the cell by this process.

This chapter describes the various routes utilized by cells for the uptake of material from the extracellular environment. These routes are determined by the nature of the material to be taken up and also by the intracellular and extracellular environment. The biochemical mechanism of these uptake processes will be emphasized.

Elizabeth Smythe   Department of Biochemistry, Medical Sciences Institute, University of Dundee, Dundee DD1 4HN, Scotland.

Subcellular Biochemistry, Volume 27: Biology of the Lysosome, edited by Lloyd and Mason. Plenum Press, New York, 1996.

2. ROUTES OF ENTRY INTO THE CELL

2.1. Phagocytosis

2.1.1. Physiological Role

Phagocytosis has traditionally been defined as cellular "eating" and distinguished from pinocytosis (cell drinking) on the basis of the size of the ingested particle (>0.5 μm diameter). However, size is not the only determinant of phagocytosis; a transmission of signals within the cell is also required. This is the route of entry into the cell for a variety of pathogenic microorganisms and cellular debris. Phagocytosis plays a major role in host defense mechanisms because of its roles in the ingestion and degradation of microorganisms and also in tissue remodeling and inflammation (Greenberg and Silverstein, 1993). In contrast to pinocytosis, which occurs constitutively in almost all cell types, phagocytosis is generally carried out in higher eukaryotes in response to a signal by professional phagocytes, such as polymorphonuclear granulocytes, monocytes, or macrophages. So-called nonprofessional phagocytes may also be induced to phagocytose a limited range of particles. In addition, there are examples of cells that phagocytose specific particles under particular conditions (Rabinovitch, 1995), so-called paraprofessional phagocytes. These include the uptake of collagen by fibroblasts in tissue remodeling (McCulloch and Knowles, 1993) and the uptake of certain bacteria by dendritic cells (Inaba *et al.*, 1993).

2.1.2. Ingestion

Particles that are to be phagocytosed interact with receptors, often of several different types, on the surface of the phagocyte. Receptors involved in phagocytosis include the immunoglobulin Fc receptors (FcRs), which interact with the Fc domain of immunoglobulins, and the CR3 complement receptor, an integrin involved both in phagocytosis and cell adhesion (Greenberg and Silverstein, 1993). The abundance of receptors for immunoglobulins and complement displayed by professional phagocytes enhances their abilities to phagocytose. Opsonization of the particle, where opsonins such as complement or immunoglobulins coat the surface of the particle, promotes interactions with these cell surface receptors. The two receptor types can bind independently, but can also act in a synergistic fashion. Binding of the opsonized particle to the receptors is independent of energy and temperature and results in the formation of a pseudopod from the phagocyte, the length of which is determined by the receptor–ligand interactions (Fig. 1). Pseudopods extend from the phagocytic cell around the particle as more receptors and ligands become engaged. Ingestion occurs when the extending pseudopods meet and become closely apposed.

The mechanism by which the phagosome pinches off is as yet undefined. Attempts by two macrophages to engulf the same particle results in stalemate, so-

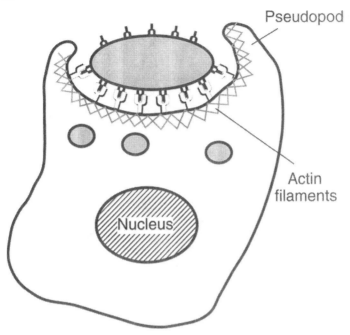

FIGURE 1. Binding of a particle to a phagocyte. Binding of a particle to a phagocyte and the induction of phagocytosis is dependent on sequential engagement of receptors on the surface of the phagocyte with ligands on the surface of the particle. Engulfment proceeds by a zipper mechanism and involves the recruitment of actin filaments.

called frustrated phagocytosis, where pseudopods from competing cells become closely apposed (Fig. 2). However, intracellular fusion does not occur, demonstrating that phagocytosis is an autonomous event requiring complex regulation within individual phagocytic cells. The size of the phagosome formed is generally dependent on the size of the particle being ingested. There are exceptions to this rule in that certain bacteria, for example, *Salmonella typhimurium*, are phagocytosed into "spacious" phagosomes that are much larger than the particle (Francis *et al.*, 1993). However, uptake of *Salmonella* is really more similar to macropinocytosis (see below).

Early experiments by Silverstein and colleagues (Griffin *et al.*, 1975, 1976) demonstrated that ingestion appears to proceed via a "zipper" mechanism. The phagocytosis of lymphocytes by macrophages was used as a model system. Membrane IgG is randomly distributed along the cell surface of lymphocytes. For ingestion to occur the lymphocyte needed to be uniformly coated with anti-IgG immunoglobulin at 4°C. Warming of cells to which anti-IgG has been bound results in the process of cell capping, the relocation of the membrane IgG to one area of the cell surface. Macrophages are not capable of ingesting lymphocytes that are

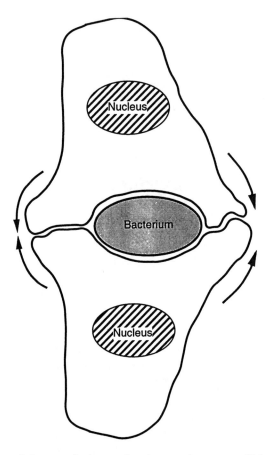

FIGURE 2. Frustrated phagocytosis. Attempts by two macrophages to engulf the same bacterium result in frustrated phagocytosis, with the pseudopods from the two macrophages becoming closely apposed. The bacterium cannot be engulfed by either cell because partial binding of the bacterium is insufficient to trigger the intracellular signals required for ingestion within the macrophages.

capped or not coated (Griffin *et al.*, 1976). These experiments indicate that it is not simply a threshold level of anti-IgG that triggers phagocytosis but, rather, complete coating of the particle which allows ingestion to proceed. It should be noted, however, that there are phagocytic processes where ingestion appears to occur via a triggered mechanism (see Section 2.3.1 on macropinocytosis and review by Swanson and Baer, 1995).

Formation of the pseudopod requires a recruitment of actin filaments (Fig. 1) as shown by sensitivity of phagocytosis to drugs such as cytochalasin B (Greenberg and Silverstein, 1993). Actin assembly appears to be initiated by binding of the particle to the surface of the phagocyte, although the assembly mechanism has not yet been elucidated. What is clear is that ligation of phagocytic receptors such as FcγRs activates a variety of signal transduction cascades which ultimately result in actin assembly. Tyrosine phosphorylation of several proteins was demonstrated to occur early after particle binding, and the appearance of phosphoproteins occurred with the same time course as the appearance of foci of F-actin beneath phagocytic particles (Greenberg *et al.,* 1993). Furthermore, genistein, a specific inhibitor of tyrosine phosphorylation, blocked the appearance of phosphoproteins and F-actin and also inhibited ingestion. Further analysis of the proteins phosphorylated after receptor ligation demonstrated that the γ subunit of the FcγR is one of the targets of the activated tyrosine kinase (Greenberg and Chang, 1994). Many FcγRs contain a consensus sequence for tyrosine phosphorylation, YxxLx 5-12Yx2-3L/l, which is termed a tyrosine activation motif (TAM; Samelson and Klausner, 1992). Although the tyrosine kinase responsible for their phosphorylation has not yet been identified it may be a member of the Src family of kinases, members of which can be coprecipitated with FcγR (Ghazizadeh *et al.,* 1994; Sarmay *et al.,* 1994; Hamada *et al.,* 1993). Tyrosine phosphorylation of the TAMs results in the recruitment of another kinase, Syk kinase, to the membrane and its subsequent activation (Benhamou *et al.,* 1993; Kiener *et al.,* 1993; Law *et al.,* 1993). Activation of the Syk kinase leads to actin polymerization and phagocytosis (Greenberg *et al.,* 1991; Phatak and Packman, 1994), which suggests that it directly links those receptors involved in phagocytosis with the actin cytoskeleton. One hypothesis is that Syk phosphorylation of itself or its target substrate(s) result in the formation of actin nucleation sites at the plasma membrane (Greenberg, 1995). Consistent with this, transfection of chimeric molecules containing the Syk kinase domain in COS cells was sufficient to trigger phagocytosis (Greenberg, *et al.,* 1996).

Other signal transducing processes that occur in response to receptor ligation include activation of PI3-kinase (Kapeller and Cantley, 1994), release of arachidonic acid, and activation of a variety of serine and threonine protein kinases (Greenberg and Silverstein, 1993). However, the order and interrelationships of these pathways have not yet been fully determined.

2.1.3.   Particle Degradation

After internalization of pathogens, the phagocyte generates toxic oxidants such as hydrogen peroxide and hypochlorous acid from superoxide anion, O_2^-, within the lumen of the phagosome to kill and degrade the internalized organism (Bokoch, 1995). The respiratory burst which results in the production of O_2^- is effected by NADPH oxidase and the activity of this oxidase is regulated by a small ras-related GTP-binding protein, rac (see also Section 2.3.1). NADPH oxidase is present in the cell in an inactive state and becomes active in response to signals that induce phagocytosis. *In vitro* studies have demonstrated that the NADPH oxidase activity may be reconstituted using four recombinant proteins, cytochrome b_{558}, which is normally associated with the plasma membrane, and three cytosolic proteins, p67*phox,* p47*phox,* and rac, which are recruited to the membrane in response to phagocytic signals (Heyworth *et al.,* 1994; Rotrosen *et al.,* 1993; Abo *et al.,* 1992). The translocation of rac to the membrane involves its dissociation from rho guanine dissociation inhibitor (rhoGDI), with which it is complexed in the cytosol, and this dissociation is regulated by the nucleotide status of rac. Conversion of rac from the GDP-bound form to the GTP-bound form is necessary for membrane association, and this may be catalyzed by a membrane-associated exchange factor (GEF) which is also activated in response to phagocytosis (Bokoch *et al.,* 1994). Translocation of the other components of the NADPH oxidase is separately regulated, probably by reversible phosphorylation (Rotrosen and Leto, 1990; Dusi and Rossi, 1993). Rac has been demonstrated to interact directly with p67*phox in vitro.* In addition, in cells where cytochrome b_{558} is missing, rac translocation to the membrane is substantially reduced (Heyworth *et al.,* 1994). This indicates that rac interacts directly with components of the oxidase system. It has been suggested that activation of phagocytosis and NADPH oxidase activation are under the coordinate control of rac (Bokoch, 1995). As with other small GTP-binding proteins, the activation of rac itself would be dependent on its nucleotide status, which is governed by a variety of exchange and GTPase activating proteins (see Chapter 2 for a more detailed description of these proteins).

As is often the case with cellular uptake pathways, the phagocytic pathway may be exploited by pathogenic organisms. Bacteria such as *Yersinia pseudotuberculosis* are internalized by phagocytosis using the integrin family of cell adhesion molecules as phagocytic receptors. Again, their internalization requires a series of signal transduction events which results in cytoskeletal rearrangements that promote phagocytosis (for review, see Isberg and Tran Van Nhieu, 1995).

2.2. Clathrin-Mediated Pinocytosis

Receptor-mediated endocytosis via clathrin-coated pits results in the cellular uptake of a variety of biologically important macromolecules. Ligands which enter the cell by this route include hormones such as insulin, growth factors such as epidermal growth factor, and essential nutrients such as low density lipoprotein (LDL), and the iron carrying molecule transferrin. These ligands, often present in low amounts in the extracellular environment, bind with high affinity to cell surface transmembrane receptors. Opportunistic viruses can also utilize this pathway to gain access to the cell. Receptors are already clustered or become clustered as a result of ligand binding into specialized areas of the cell surface, termed coated pits. Coated pits originate as planar structures which form beneath the surface of the plasma membrane.

They become increasingly invaginated until they eventually pinch off, forming coated vesicles.

2.2.1.   Coated Pits and Coated Vesicles

Coated pits were originally identified morphologically (Anderson *et al.,* 1977) because of their very characteristic bristle-like appearance when viewed by electron microscopy. Freeze fracture deep-etch studies (Heuser and Evans, 1980) revealed that the coat has a basket-like structure. The major structural component of the coat is clathrin, which was first isolated and purified in the 1970s (Pearse, 1975). Clathrin exists in solution as a triskelion, composed of three molecules of clathrin heavy chain and three molecules of clathrin light chain. The clathrin heavy chain, which is highly conserved between mammalian species (99%), appears to be the product of a single gene (Kirchhausen *et al.,* 1987a). The carboxy termini of three clathrin heavy chains meet at the hub of the triskelion (Kirchhausen and Harrison, 1984). There are three defined regions in each heavy chain, the proximal end containing the carboxy terminus. The chain then kinks at the knee and the next linear array of amino acids constitutes the distal region and the terminal domains are globular regions at the N-terminus connected by a protease-sensitive link region (Kirchhausen and Harrison, 1984). *In vitro,* clathrin triskelia can assemble under conditions of low pH into cage-like structures which very much resemble isolated coated vesicles.

Extensive studies using monoclonal antibodies that disrupt clathrin function have identified the proximal region as being important for trimerization, light chain binding, and assembly (Blank and Brodsky, 1987; Nathke *et al.,* 1992). These studies have been confirmed and extended by the expression of the C-terminal third of the clathrin heavy chain in bacteria (Liu *et al.,* 1995). The recombinant clathrin fragment was capable of trimerization, light chain binding, and assembly. However, although recombinant clathrin hubs can assemble into planar lattices, the structures formed appeared by ultrastructural analysis to be both considerably more variable and also flatter than assembled triskelia, suggesting that the globular domains are responsible for the direction in which clathrin assembly occurs (Liu *et al.,* 1995).

In mammalian and avian cells, there are two types of light chain, LCa and LCb (Kirchhausen *et al.,* 1987b). By contrast there is only one type of light chain in yeast (Silveira *et al.,* 1990). LCa and LCb are encoded by two separate genes and are 60% identical in protein sequence. Alternative RNA splicing gives rise to three different subtypes of LCa and two subtypes of LCb. These subtypes have a molecular weight in the range of 30–40 kDa when analyzed by SDS-PAGE. The higher molecular weight subtypes appear to be expressed mainly in neuronal tissues (Wong *et al.,* 1990; Stamm *et al.,* 1992) with the smaller forms expressed in non-neuronal tissue (Jackson *et al.,* 1987; Kirchhausen *et al.,* 1987b). Although both light chains bind to the same sites on the clathrin heavy chain and appear to act interchangeably in assembly (Ungewickell, 1983; Winkler and Stanley, 1983), it is thought that they may also have independent functions. The LCb is phosphorylated by a casein kinase-like activity associated with coated vesicles (Hill *et al.*, 1988), but the physiological significance of this observation has yet to be determined. Assembly at low pH is independent of the presence of light chains. Studies with the bacterially expressed clathrin hub have implicated the light chains as having a regulatory role in preventing the assembly of clathrin at physiological pH, thus ensuring that the assembly of clathrin is dependent on the presence of the adaptor molecules (Liu *et al.*, 1995).

The other major structural components of coated pits are the adaptor complexes. They were first identified because of their ability to promote the formation of clathrin cages at physiological pH (Zaremba and Keen, 1983). They are a family of heterotetrameric proteins of which several classes have been identified. Immunofluorescence studies have demonstrated that AP-1 is specifically localized to clathrin-coated vesicles derived from the Golgi complex, while AP-2 is localized to the plasma membrane (Robinson, 1987). The protein AP-3 is a neuron-specific adaptor which very efficiently promotes adaptor assembly (Lindner and Ungewickell, 1992) but whose precise role in membrane trafficking remains unclear. The AP-2 protein has a subunit composition of two large "adaptin" polypeptides, α-adaptin (105–112 kDa) and β-adaptin (104 kDa), a medium size subunit, AP50, and a small subunit, AP17. Limited proteolysis resulted in the cleavage of a 30 kDa fragment from each of the adaptin molecule (Zaremba and Keen, 1985). An analysis of the sequence of the cloned adaptor molecules indicated an enrichment of proline/glycine residues, characteristic of flexible regions in other proteins adjacent to the protease cleavage sites (Kirchhausen *et al.*, 1989; Matsui and Kirchhausen, 1990). This suggested that the adaptin molecules are composed of an N-terminal domain of approximately 60 kDa, connected to a 30–40 kDa "ear" by a protease-sensitive flexible hinge. Rotary shadowing electron microscopy of purified adaptors supported this analysis, demonstrating that the complex is composed of a head group and two "ears" (Heuser and Keen, 1988). The smaller subunits appear to be associated with the N-terminal portion of the molecule. The N-terminal regions of α and β appear to interact with each other to form the core molecule. Bacterially expressed β or β' (the AP-1 homologue) subunits are effective in binding clathrin and in promoting cage formation, suggesting that the clathrin binding site is contained within this subunit (Gallusser and Kirchhausen, 1993).

It is of some significance that there are very specific differences between neuronal and non-neuronal tissue with respect to components of the endocytic apparatus (Morris and Schmid, 1995). All of the coat proteins are present in neuronal tissue at levels 10- to 50-fold higher than their non-neuronal counterparts. Many of the coat proteins of brain have brain-specific inserts that arise from differential RNA splicing (Jackson *et al.*, 1987; Kirchhausen *et al.*, 1987b). In addition, there are proteins associated with neuronal coated vesicles which are neuron-specific. These include AP-3, mentioned above, and auxilin which binds to the clathrin heavy chain and supports its assembly into very regular-sized cages (Ahle and

Ungewickell, 1990) and which has recently been shown to have a critical role in uncoating (Ungewickell, et al., 1995). The significance of the differences observed in neuronal tissue has been ascribed to the requirement for very rapid endocytosis of membrane following release of synaptic vesicles after stimulation of the nerve terminal.

2.2.2. Mechanism of Coated Pit Assembly

The clathrin lattice provides the structural basis of the coated pit, acting as a mechanical scaffold pulling the membrane into a bud. Clathrin assembles initially onto the membrane as a planar lattice composed of a hexagonal array of triskelia. Increased invagination of the lattice appears to arise from a conversion of hexagons to pentagons, although the mechanism of this conversion remains a mystery. One proposal is that the conversion occurs via a heptagonal intermediate and such intermediates have been visualized by freeze fracture deep-etch studies (Heuser and Evans, 1980). An alternative proposal is that curvature is coupled to the growth of the pit (Pearse and Crowther, 1987).

The adaptors have been visualized using computer reconstructions of tilt series of hydrated coated vesicles frozen in vitreous ice and appear to form an inner layer within the clathrin lattice (Vigers et al., 1986). Other biochemical evidence supporting this location came from early studies where clathrin was stripped from coated vesicles. Clathrin could rebind these stripped vesicles but not if the vesicles were treated with the protease elastase to which the adaptins are exquisitely sensitive (Unanue et al., 1981). Furthermore, the adaptor proteins were demonstrated to bind to an integral membrane protein in alkaline extracted membranes from bovine brain, and clathrin binding to these membranes was dependent on prebinding of adaptors (Virshup and Bennett, 1988). These lines of evidence all point to the adaptors acting, as their name suggests, to link clathrin to the membrane.

Coated pits act as molecular filters, selectively enriching some components, while actively excluding others (Bretscher et al., 1980). This selectivity was first demonstrated when it was shown that Thy 1, a resident plasma membrane protein, is excluded from coated pits (Bretscher et al., 1980), while the low density lipoprotein receptor is clustered in coated pits (Anderson et al., 1977). The signals which result in clustering into coated pits reside in the cytoplasmic tails of receptor molecules. Early evidence that this was the case came from work on the LDL receptor. Several mutant receptors had been identified in patients suffering from familial hypercholesterolemia. Three of these mutations caused alterations in the cytoplasmic tails of the receptors, resulting in impaired ability to be internalized (Lehrman et al., 1985). Two of these receptors had truncated cytoplasmic tails but the third mutation was caused by a single base substitution resulting in the conversion of a tyrosine to a cysteine at residue 807. Further work from a number of laboratories implicated tyrosine as a critical residue in the internalization motif of several receptors (Prywes *et al.*, 1986; Livneh *et al.*, 1986; Mostov *et al.*, 1986; Rothenberger *et al.*, 1987; Jing *et al.*, 1990). However, apart from the tyrosine residue there was no apparent homology between the cytoplasmic tails of those receptors analyzed. Lazarovits and Roth (1988) demonstrated that it was not simply the presence of tyrosine alone that was sufficient to induce endocytosis of a transmembrane protein. The influenza virus hemagglutinin protein is largely excluded from coated pits. Substitution of a tyrosine residue for a cysteine residue in the cytoplasmic tail of the hemagglutinin molecule resulted in efficient internalization through coated pits. However, replacement of only one of the three cysteine residues in the cytoplasmic tail, residue 543, was effective, and it was the appearance of the tyrosine rather than the removal of the cysteine which was required for uptake (Lazarovits and Roth, 1988). This strongly suggested that it is the context in which the tyrosine is presented that is important for the generation of an internalization motif.

Extensive analysis using site-directed mutagenesis of the transferrin receptor demonstrated that a sequence of four amino acids YXRF, was necessary and sufficient for receptor internalization (Collawn *et al.*, 1990). The contribution made by individual amino acids to this internalization motif was analyzed by sequentially mutating each residue and analyzing the resulting mutant receptor for its ability to be internalized. From this study it was shown that an amino-terminal aromatic residue and either a carboxy-terminal aromatic or largely hydrophobic residue are required for activity. While a positively charged residue is optimal at position 3, neutral (glycine) or negatively charged (glutamic acid) residues in this position only reduce the internalization efficiency by 50%. The only other structural requirement was that this tetrapeptide needed to be located at least seven amino acids distal to the membrane. The composition of these seven amino acids does not appear to be critical.

Using a similar experimental approach, the internalization motif of the LDL receptor was also identified and found to be the tetrapeptide NPXY (Chen *et al.*, 1990). The lack of sequence homology between these sequences and those of other receptors, as mentioned above, led to the concept that internalization motifs encode sequences which adopt specific three-dimensional structures and it is these structural determinants that allow inclusion into a coated pit. To test this hypothesis, Trowbridge and co-workers analyzed proteins of known three-dimensional structure for the presence of this sequence or analogues of this sequence which permit up to 50% internalization. Of the 28 structures analyzed, most were demonstrated to adopt a β1-type tight turn and were found either at the end of helices or between secondary structures. Interestingly, a similar analysis of the NPXY sequence of the LDL receptor also revealed that 80% of the analogous sequences formed a tight turn structure (Collawn *et al.*, 1990). This provided evidence that receptors rely on a structural motif for internalization.

Support for this hypothesis also came from the fact that substitution of the internalization motif from either the mannose 6-phosphate receptor or the LDL receptor into the transferrin receptor was effective in promoting efficient internalization. Also internalization motifs could be successfully substituted into both type I or type II integral membrane proteins (Collawn et al., 1991). Analysis by NMR of peptides containing the LDL internalization motif (Bansal and Gierasch, 1991) and that of lysosomal acid phosphatase (Eberle et al., 1991) also provides evidence that the internalization motifs adopt a tight turn in solution.

In similar experimental approaches, a dileucine motif has also been implicated as an internalization motif for several plasma membrane proteins. These include IgG FcRs (Hunziker and Fumey, 1994), the insulin receptor (Haft et al., 1994), and CD3 γ and δ chains (Letourneur and Klausner, 1992). It has been postulated that this dileucine motif may also act to stabilize a tight turn (Hunziker and Fumey, 1994). Obviously definitive evidence that internalization motifs adopt a tight turn must await the crystallization of a transmembrane protein which is internalized through coated pits.

The transferrin and LDL receptors are constitutively recycling receptors, that is, their internalization rate is independent of ligand binding. Other receptors such as the epidermal growth factor (EGF) receptor are not normally localized to coated pits and only become clustered as a result of ligand binding. This suggests that ligand binding induces a conformational change in the receptor such that an internalization motif is revealed. Within the cytoplasmic tail of the EGF receptor are three internalization motifs which are analogous to those found in constitutively endocytosed receptors. Substitution of motifs from the transferrin or LDL receptor for these motifs results in efficient ligand dependent uptake, but only in the presence of the receptor tyrosine kinase (Chang et al., 1993). In vitro studies have demonstrated that although mutant EGF receptors whose autophosphorylation sites have been deleted are efficiently sequestered into coated pits, there is still a requirement for an active kinase for uptake. This suggests that, in addition to the requirements for structural determinants in the cytoplasmic tail of the receptor, there is a requirement for the phosphorylation of some other cellular component (Lamaze and Schmid, 1995a).

The lymphoid differentiation antigen, CD4, is an interesting example of the regulation of the selective uptake of some ligands by specific cells. In nonlymphoid cells, CD4 is internalized at a rate of 2–3% per minute. Although this is somewhat less than the rate of transferrin or LDL uptake, it still represents a significantly selective uptake compared to the internalization of bulk membrane. By contrast, in lymphoid cells, CD4 internalization is very slow and this reduced endocytosis has been shown to result from association of the CD4 molecule with the tyrosine kinase, p56[lck]. Transfection of p56[lck] into nonlymphoid cells reduces CD4 internalization to rates comparable to those observed in lymphoid cells (Pelchen-Matthews et al., 1991, 1992). Also, CD4 is used as a receptor by HIV. Nef, an early protein of HIV, has been shown to induce downregulation of CD4 and the basis of this downregulation has been demonstrated to be an increase in endocytosis (Aiken et al., 1994). Endocytosis induced by Nef is absolutely dependent on the presence of a dileucine motif in the cytoplasmic tail of CD4. Interestingly, this determinant is independent of the amino acid sequence required for p56[lck] binding (Aiken *et al.,* 1994).

A number of studies have indicated that the adaptor molecules interact with the cytoplasmic tails of receptors *in vitro.* AP-2 was shown to interact with an affinity matrix composed of the cytoplasmic tail of the LDL receptor, and binding was competitively inhibited by the cytoplasmic domains of other receptors known to cluster in coated pits (Pearse, 1988). The mannose 6-phosphate receptor is involved both in transport of newly synthesized lysosomal enzymes from the Golgi to the lysosomes and also in retrieval of secreted lysosomal enzymes from the extracellular medium. Therefore, it is localized both in clathrin-coated vesicles in the *trans*-Golgi network (TGN) and also at the plasma membrane. Consistent with these two localizations, AP-1 and AP-2 were both demonstrated to bind to the cytoplasmic tail of the mannose 6-phosphate receptor at different and nonoverlapping sites (Glickman *et al.,* 1989). This indicates that the binding of the relevant adaptor complex at each location must be regulated by other components. In these experiments, substitution of the tyrosine residues shown to be essential for internalization resulted in decreased binding of AP-2. Similarly, AP-2 was shown to bind to the cytoplasmic tail of the lysosomal enzyme lysosomal acid phosphatase (Sosa *et al.,* 1993). Again, binding of the AP-2 molecules was dependent on the presence of either tyrosine or phenylalanine. Similarly, binding of AP-2 molecules to dileucine motifs has been demonstrated using surface plasmon resonance (Heilker *et al.,* 1996). More direct evidence for the interaction of receptor tails and adaptor molecules has come from the work of Sorkin and Carpenter (1993), who demonstrated that the AP-2 complex could be co-immunoprecipitated from intact cells with the EGF receptor, but only in the presence of EGF.

2.2.3. Regulation of Coated Pit Assembly

The preceding data provide good evidence for direct interactions between the cytoplasmic tails of receptors and the adaptor molecules. Indeed, it was originally thought that coated pits form by the assembly of adaptor molecules around the receptor cytoplasmic tails. However, as described above, the adaptor complexes demonstrate a very precise subcellular localization (Robinson, 1987). This localization occurs even though these two membrane compartments show overlapping receptor compositions with, for example, the mannose 6-phosphate receptor being present in both locations. This lends support to the notion of nucleation sites on the membrane to which adaptors are targeted (Morris *et al.,* 1989). These nucleation sites are hypothesized to be transmembrane proteins, which would bind adaptors with a higher affinity than that observed between receptor tails and adaptors. A high-affinity binding site for adaptors was observed in alkaline stripped membranes (Virshup and Bennett, 1988). In addition, in an *in vitro* system using a plasma membrane preparation from sonicated fibroblasts, it was demonstrated that adaptors bind to a limited number of sites on the membrane (Moore *et al.*, 1987) and that binding is mediated by an integral plasma membrane protein (Mahaffey *et al.*, 1990). It is possible that after adaptors bind to the membrane, receptors may diffuse into the growing coated pit and remain there as a result of lower affinity interactions between receptor tails and adaptors.

After the scission of coated pits, clathrin and the adaptor molecules are very rapidly removed from the newly formed coated vesicle and are recycled back through the soluble pool to participate in further rounds of uptake. There are, therefore, large soluble pools of the coat proteins (Goud *et al.*, 1985; Robinson and Kreis, 1992). It is from these soluble pools that new coated pits are assembled. A key question is how adaptors and clathrin are targeted to membranes. It is clear that it is a highly regulated process because conditions that lead to a disruption of the intracellular environment, such as hypertonic shock, result in the misassembly of clathrin into "microcages," small aberrant structures too small to contain a vesicle (Heuser, 1989; Heuser and Anderson, 1989). Furthermore, *in vitro* studies have demonstrated that both the formation of new coated pits and scission to form coated vesicles are absolutely dependent on ATP and on cytosolic factors (Smythe *et al.*, 1989), and, as mentioned above, adaptors bind to a saturable number of sites on the membrane (Moore *et al.*, 1987).

Several experimental approaches have been used to examine the molecular mechanism of targeting. Chimeric bovine adaptin molecules were generated where the ear domains of alpha and gamma adaptin were swapped and the chimeric proteins transfected into rat cells (Robinson, 1993). Using a species-specific antibody which only recognizes the chimeric adaptors, it was shown that the major targeting signal appears to be contained within the head region (Robinson, 1993). Further analysis of the sequences required for targeting to the correct membrane revealed a 200 amino acid stretch in the head domain to be essential. Interestingly, this sequence was also required for interaction with the correct medium and small chains implicating these subunits in targeting (Page and Robinson, 1995). These studies also revealed that there is a weaker targeting signal contained within the ear domains of the adaptins which is revealed when the major targeting signal is removed (Page and Robinson, 1995).

Work from a number of laboratories has shown that there are multiple GTP-dependent steps involved in clathrin-coated pit assembly and coated vesicle formation (Carter *et al.*, 1993; Robinson and Kreis, 1992). Dynamin is a GTP binding protein involved in the late steps of coated vesicle assembly (see Section 2.2.4), but there may be others. There is a requirement for ADP-ribosylation factor (ARF1), a small GTP-binding protein, for the formation of both COP-coated vesicles in the Golgi (Serafini *et al.*, 1991; See also Chapter 2) and clathrin-coated vesicles at the *trans*-Golgi network (Stamnes and Rothman, 1993; Traub *et al.*, 1993). ADP-ribosylation factors are a family of small GTP-binding proteins of which there are 17 structurally related gene products, 6 ARFs and 11 ARF-related proteins (Clark *et al.*, 1993; Schurmann *et al.*, 1994). ARF1, which has been studied to the greatest extent, cycles between the membrane and cytosol, and its localization is dependent upon its guanine nucleotide status (Donaldson and Klausner, 1994). *In vitro* studies have implicated ARF1 in fusion reactions as diverse as intra-Golgi trafficking (Taylor *et al.*, 1992), nuclear envelope fusion (Boman *et al.*, 1992), and endosomal fusion (Lenhard *et al.*, 1992). However, the existence of a family of related proteins suggested that ARFs may be able to substitute for each other *in vitro* while *in vivo* a given member of the family would be functional only at one location. Consistent with this hypothesis, transient transfection of ARF1 and ARF6 revealed that the latter was localized to the plasma membrane and early endosomes while the former, as expected, was localized to the Golgi complex (D'Souza-Schorey *et al.*, 1995; Peters *et al.*, 1995). Unlike ARF1, the transfected form of ARF6 does not appear to cycle between the membrane and cytosol, but remains membrane-associated along the endocytic pathway (Peters *et al.*, 1995). Overexpression of both the wild-type ARF6 and a mutant deficient in GTP hydrolysis resulted in an accumulation of transfected protein at the cell surface and a reduction in the uptake of transferrin (D'Souza-Schorey *et al.*, 1995). It might be expected that there would be strong mechanistic similarities between the formation of both clathrin-coated vesicles and also of COP vesicles. However, to date no direct involvement of ARF6 in plasma-membrane clathrin-coated vesicle formation has been demonstrated.

In contrast to clathrin-coated pit formation in the TGN, where adaptor recruitment is significantly enhanced by GTPγS (Robinson and Kreis, 1992), treatment of permeabilized cells with GTPγS results in their mislocalization to an endosomal compartment (Seaman *et al.*, 1993). This mislocalization can also be effected by cationic amphiphilic drugs (Wang *et al.*, 1993). These results have been interpreted to mean that there is a storage compartment where excess nucleation proteins reside and which may be mobilized depending on the endocytic requirements of the cell. It is proposed that GTPγS can act as a regulator switch activating these proteins in the endosomal compartment (Seaman *et al.*, 1993). The effect of GTPγS on the mislocalization of AP-2 complexes may be blocked by pretreatment of cells with brefeldin A (Seaman *et al.*, 1993), the target of which is believed to be the ARF receptor (Donaldson *et al.*, 1992; Helms and Rothman, 1992). One interpretation of these data is that if an ARF is involved in endocytosis it has an essentially negative regulatory role (Robinson, 1994).

Other *in vitro* studies which measure the sequestration of transferrin into functional coated pits in permeabilized cells (Schmid and Smythe, 1991) have revealed that clathrin-coated pit assembly requires the recruitment of cytosolic clathrin (Smythe *et al.*, 1992). Surprisingly, clathrin isolated from coated vesicles is incapable of supporting coated pit assembly, demonstrating that there is a functional difference between the membrane and soluble pools of clathrin. Elucidation of the molecular basis for the difference between the membrane and soluble pools of clathrin should reveal the mechanism by which clathrin is converted into an assembly-competent form.

2.2.4. Dynamin

A protein which has recently been shown to have a major role in clathrin-coated vesicle formation is dynamin. Dynamin was originally identified as the product of an essential gene, shibire, in *Drosophila* (Guigliatti *et al.,* 1973). Flies with a temperature-sensitive mutation in this gene show a reversible paralysis when exposed to the nonpermissive temperature. Electron microscopy revealed that the defect occurred as a result of inhibition of scission of coated pits at the neuromuscular junction (Kosaka and Ikeda, 1983). Cloning of the mammalian homologue of the shibire mutant revealed it to be dynamin (Van der Bliek and Meyerowitz, 1991), a protein previously shown to have microtubule bundling activity *in vitro* (Shpetner and Vallee, 1989). Dynamin is a member of a family of GTP-binding proteins of high molecular weight. Transfection of GTP-binding domain mutant of dynamin into mammalian cells results in the inhibition of early stages of transferrin uptake (Herskovits *et al.,* 1993; Van der Bliek *et al.,* 1993). Further studies in permeabilized neuronal cells demonstrated that the presence of GTPγS resulted in the formation of coated pits with very elongated necks. These elongated necks could be decorated with anti-dynamin antibodies (Takei *et al.,* 1995). These data, in conjunction with the observed phenotype of the shibire mutant, led to the proposal that dynamin forms a ring or collar-like structure around the neck of the deeply invaginated coated pit, constricting it so that the final scission event may occur. This hypothesis is supported by the observation that recombinant dynamin, isolated from insect cells, can assemble into collar-like structures under certain conditions of salt and buffer (Fig. 3; Hinshaw and Schmid, 1995). Dynamin is found to be associated with coated vesicles (Takei *et al.,* 1995), supporting its role in the final stages of coated vesicle budding. In addition, it appears to act after the clathrin lattice has assembled, because in the presence of GTPγS where coated pits with elongated necks were formed, clathrin is localized to the coated bud while dynamin is localized to the neck (Takei *et al.,* 1995).

2.2.5. Uncoating

After scission of the deeply invaginated pit the resultant coated vesicle has a very short life. The coat is rapidly lost and studies on the uncoating of coated vesicles *in vitro* demonstrate that this is mediated by an "uncoating ATPase" (Schlossman *et al.,* 1984; Braell *et al.,* 1984). Uncoating ATPase was subsequently identified as hsp70c, a member of the cytosolic family of heat-shock proteins (Chappell *et al.,* 1986; Ungewickell, 1985). Uncoating ATPase remains bound to clathrin triskelia after uncoating and is only released in the presence of other factors such as intact cages for which it has a higher affinity (Schmid and Rothman, 1985). The binding of uncoating ATPase to clathrin triskelia has been postulated as a mechanism by which the cell prevents futile cycles of assembly and disassembly (Schmid and

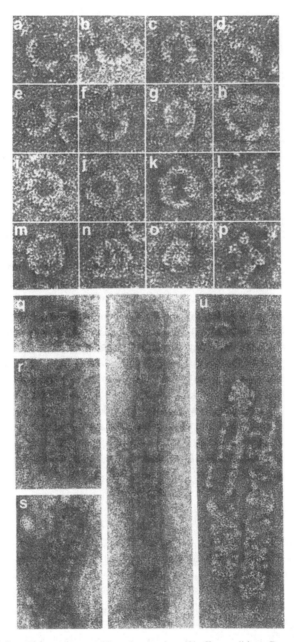

FIGURE 3. Gallery of dynamin assembly under a variety of buffer conditions. Dynamin expressed in insect cells was purified and analyzed under a range of buffer conditions. The micrograph illustrates that the purified molecule can self-assemble into a variety of ring-like structures consistent with its proposed function in forming a collar at the neck of a deeply invaginated coated pit. Courtesy of Sandra Schmid, The Scripps Research Institute, La Jolla, California.

Rothman, 1985). In this model, dissociation of clathrin and uncoating ATPase would occur when soluble clathrin became "primed" for assembly into a new lattice.

One of the difficulties in establishing an *in vivo* role for uncoating ATPase has been its high concentration in the cell relative to clathrin. It is present in brain in a 3:1 molar ratio with soluble clathrin (Schmid and Rothman, 1985). Recent studies have shed light on the way in which the interaction of uncoating ATPase with clathrin cages is regulated. The chaperone function of the Hsp70 family had been shown in a number of studies to be modulated by interactions of Hsp 70 with DnaJ-like proteins (reviewed in Cyr *et al.*, 1994). The interactions between DnaJ proteins and members of the Hsp 70 family are believed to confer specificity of targeting on the diverse functions of the latter. In the case of uncoating it has been shown that, in addition to uncoating ATPase, a 100kDa cofactor, auxilin, is required (Ungewickell *et al.*, 1995). This cofactor acts substoichiometrically and appears to facilitate the association of uncoating ATPase with clathrin cages. Auxilin has a DnaJ domain which, if deleted, results in the loss of cofactor ability (Ungewickell *et al.*, 1995). Auxilin is a brain-specific protein and its function may be related to the requirement for rapid endocytosis following synaptic vesicle release (see Section 2.2.1.). It seems likely, however, that a functionally related protein would confer the same specificity in other cell types. The ability of uncoating ATPase and auxilin to disassemble clathrin lattices is independent of the presence of the light chains or the N-terminal domains (Ungewickell *et al.*, 1995).

A further question in relation to the specificity of uncoating is the ability of uncoating ATPase (or its DnaJ-like protein partner) to distinguish between coated pits and coated vesicles such that it only uncoats the latter. Various mechanisms have been proposed, including loss of an inhibitory subunit after coated vesicle budding (Rothman and Schmid, 1986), changes in ionic fluxes within the coated vesicle which regulate binding of uncoating ATPase (Deluca Flaherty *et al.*, 1990), and differences in the energy state of the lattice at different stages in invagination (Smythe and Warren, 1991). Resolution of this issue will require the complete reconstitution of the coated pit/coated vesicle cycle *in vitro*.

2.3. Non-Clathrin-Mediated Pinocytosis

Studies on fluid-phase uptake of material into cells have revealed other non-clathrin-mediated pathways (Lamaze and Schmid, 1995b). Fluid-phase pinocytosis arises from engulfment of solutes by entrapment in vesicles that bud from the cell surface. It is essentially nonspecific uptake since there is no active inclusion or exclusion mechanism. The amount of material taken up is proportional to its concentration in the extracellular environment. Early studies on the uptake of fluid-phase markers suggest that their uptake into the cell could be completely accounted for by clathrin-coated vesicle formation (Marsh and Helenius, 1980). However, more recent studies using conditions that inhibit clathrin-mediated endocytosis indicate the existence of other non-clathrin-mediated endocytic pathways. Under conditions of potassium depletion (Larkin *et al.,* 1986), cytosol acidification (Sandvig *et al.,* 1988), and hypertonic shock (Daukas and Zigmond, 1985), the clathrin-coated pit pathway, as assayed by the uptake of markers such as transferrin and LDL, is inhibited, but substantial uptake of fluid-phase markers persists, indicating that the two pathways coexist in cells. Morphological studies indicate that the vesicles involved in this internalization event are of similar diameter to clathrin-coated vesicles (Hansen *et al.,* 1991). Receptor bound ligands have also been demonstrated to enter cells via a non-clathrin-mediated pathway (Raposo *et al.,* 1989; Subtil *et al.,* 1994).

Because this non-clathrin-mediated pathway was detected only when the normal intracellular environment was disrupted, it was unclear whether it became switched on in response to cellular stress or whether it was constitutively active. A very careful analysis of the uptake of fluid-phase markers in rat fetal fibroblasts subjected to K^+ depletion or hypertonic shock revealed that the initial rates of internalization were identical in control and treated cells (Cupers *et al.,* 1994), suggesting that non-clathrin-mediated pathways are either constitutively active or are able to be rapidly turned on. These results suggest that clathrin may only be required for the concentration of receptors and that the mechanisms of vesicle budding in the presence and absence of clathrin may be identical. However, it has been demonstrated that in cells expressing a temperature-sensitive dynamin mutant, fluid-phase uptake is partially inhibited initially but then is restored to the original level. This indicates that, at least in this cell line, non-clathrin-mediated uptake proceeds via a clathrin- and dynamin-independent pathway (Damke *et al.,* 1995). However, this issue will not be fully resolved until all of the components necessary for both clathrin- and non-clathrin-mediated uptake have been identified. Although internalized through a non-clathrin-mediated pathway, material taken into the cell by this route is delivered to the early endosome, the next stage on the endocytic pathway, in a manner analogous to that internalized through clathrin-coated pits (Hansen *et al.,* 1993; Tran *et al.,* 1987).

2.3.1. Macropinocytosis

The process of macropinocytosis (reviewed in Swanson and Watts, 1995) occurs as a result of ruffling at the leading edge of spread cells, often in response to growth factors or phorbol esters (West *et al.,* 1989; Hewlett *et al.,* 1994). Treatment of macrophages with monocyte-macrophage colony stimulating factor or treatment of A431 cells with epidermal growth factor results in the formation of ruffles or unguided pseudopodia (Swanson and Baer, 1995). These fold back on themselves, trapping extracellular fluid (Fig. 4). The macropinosomes thus formed are large vacuoles between 0.5 and 2 μm in diameter which result in the internalization of large amounts of soluble fluid phase material. There is no apparent selectivity in either the membrane or the fluid taken up. Recent studies have indicated a role for this apparently disadvantageous process in the immune

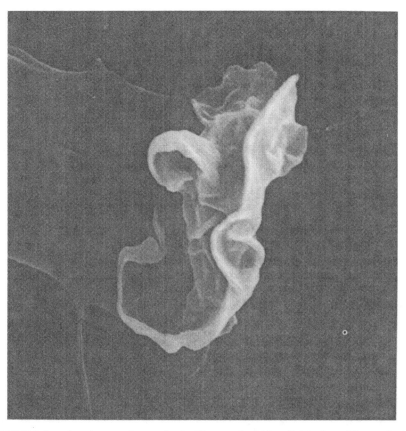

FIGURE 4. Scanning electron micrograph of ruffles on a dendritic cell. Circular ruffling results in parts of the plasma membrane folding back onto itself and trapping large volumes of extracellular fluid in the process of macropinocytosis. Magnification: ×22220. Courtesy of Chris Norbury and Alan Prescott, University of Dundee.

response. Macropinocytosis of soluble material results in the capture of sufficient antigen by dendritic cells to generate a T-cell response (Sallusto *et al.*, 1995). Similarly, macropinocytosis may provide a route for the entry of exogenous antigen into the class I major histocompatibility complex (MHC) pathway (Norbury *et al.*, 1995).

Macropinocytosis is dependent on the actin cytoskeleton being inhibited in the presence of cytochalasins which disrupt actin function (Allison *et al.*, 1971). In this regard it is akin to phagocytosis, but appears to proceed by a different mechanism. As described above, phagocytosed particles are generally taken up via a zippering mechanism, whereas macropinocytosis appears to be more of a triggered response. For example, invasive bacteria such as *Salmonella* use this pathway for entry into the cell, stimulating the internalization of noninvasive bacteria and inert particles such as latex beads into macropinosomes (Francis *et al.,* 1993). Thus macropinocytosis may be subverted by pathogenic organisms such as *Salmonella typhimurium* for entry into the cell (Francis *et al.,* 1993; Alpuche-Aranda *et al.,* 1994).

Membrane ruffling in response to growth factors appears to occur as a result of activation of rac. (See also Section 2.1.3). Microinjection of an activated form of this protein, that is, one which has reduced GTPase activity and which is insensitive to GTPase activating proteins, results in membrane ruffling and pinocytosis in the absence of an extracellular signal (Ridley *et al.,* 1992). In addition, a dominant inhibitory mutant rac protein blocks the ruffling response to growth factors such as epidermal growth factor and platelet-derived growth factor. Rac proteins are part of the rho subfamily of GTP-binding proteins, and the latter are known to be involved in regulation of the actin cytoskeleton and in the formation of focal adhesion complexes and stress fibers (Ridley, 1994). In the formation of membrane ruffles it is postulated that rac acts to nucleate actin at sites on the plasma membrane and this serves to direct membrane ruffles (Hall, 1994). Other regulators of macropinocytosis include phorbol esters such as PMA. In at least some cell types, the ability of phorbol esters to promote ruffling requires endogenous rac (Ridley *et al.,* 1992). Phorbol esters are potent activators of protein kinase C, but it is still unclear whether activation of protein kinase C is required for macropinocytosis (Swanson and Watts, 1995).

Macropinocytosis is exquisitely sensitive to changes in intracellular pH. Very minor reduction in pH caused by inhibition of the plasma membrane sodium/proton exchange protein inhibits membrane ruffling and macropinocytosis, whereas an increase in cytosolic pH leads to enhanced uptake by this pathway (West *et al.,* 1989; Heuser, 1989). The fate of macropinosomes after they have been formed differs between cell types. In A431 cells the macropinosomes do not appear to fuse with conventional components of the endocytic pathway, although they are capable of fusing with each other (Hewlett *et al.,* 1994). This contrasts with macrophages, where macropinosomes have been demonstrated to fuse with the endosomal network (Racoosin and Swanson, 1993), and so their cargo would ultimately be delivered to lysosomes.

2.3.2. Caveolae

Caveolae are flask-shaped invaginations of the plasma membrane of 50–80 nm diameter. When viewed by conventional electron microscopy no coat is visible, but quick freeze deep-etch images reveal that the caveolae have a coat composed of a granular spiraling material (Peters *et al.,* 1985; Rothberg *et al.,* 1992), a component of which is caveolin (Rothberg *et al.,* 1992). Analysis of the lipid composition of caveolae has revealed that they consist of specialized domains on the cell surface enriched in glycosphingolipids and cholesterol. This enrichment is critical for caveolae structure and function (Rothberg *et al.*, 1992; Schnitzer *et al.*, 1994a). Caveolae have been implicated in the uptake of small molecules such as folate directly into the cytoplasm (Anderson *et al.*, 1992) via the GPI-linked folate receptor. It was also thought that GPI-linked proteins were clustered in caveolae, but this has recently been shown to result artifactually from cross-linking of the GPI-anchored proteins with antibodies (Mayor *et al.*, 1994). Caveolae have also been implicated in transcytosis (Ghitescu *et al.*, 1986; Schnitzer, 1993) and in the clustering of signal transduction molecules (Lisanti *et al.*, 1995). However, their precise function has remained somewhat controversial. Until recently, it was unclear whether these structures could be considered as truly endocytic organelles because of doubts as to whether caveolae actually pinch off from the plasma membrane (Anderson, 1993; Van Deurs *et al.*, 1993).

Recent experiments have provided strong evidence that caveolae do indeed undergo endocytosis and deliver their cargo into endocytic vesicles (Parton *et al.*, 1994). Cholera toxin binds to the ganglioside GM1, which is enriched in caveolae (Parton, 1994). Alkaline phosphatase is a GPI-linked cell surface protein that becomes clustered in caveolae after cross-linking with antibodies. The internalization of both cholera and alkaline phosphatase through caveolae was shown to be inhibited in the presence of cytochalasin D (a drug which disrupts the function of the actin cytoskeleton) and stimulated by the protein phosphatase inhibitor okadaic acid (Parton *et al.*, 1994). The effects of these drugs on caveolae internalization contrast sharply with their effects on clathrin-mediated endocytosis, where cytochalasin D has no effect and okadaic acid inhibits internalization (Beauchamp and Woodman, 1994). In addition, treatment of cells with filipin, which sequesters cholesterol, disrupts the internalization of modified albumin in endothelial cells, whereas the uptake of α_2-macroglobulin which enters cells by the clathrin-mediated pathway is unaffected (Schnitzer *et al.*, 1994b). Modified albumins internalized via caveolae appear within the endosomal system, although their point of entry has yet to be determined (Schnitzer *et al.*, 1994b). Again, in contrast to receptor-mediated uptake via clathrin-coated pits, uptake via caveolae appears to be slow (Parton *et al.*, 1994).

3. DELIVERY TO ENDOSOMES

Irrespective of the mode of entry of material into the endocytic pathway, apart from the exceptions noted above, endocytic vesicles deliver their contents to some component of the endosomal network. The endocytic vesicle must therefore be targeted to and dock with endosomes in order to deliver its cargo. Docking and fusion appear to be mediated by members of a group of protein factors which are involved in other vesicle fusion reactions within the cell and have been dealt with in greater detail elsewhere (for review, see Rothman, 1994, and also Chapter 2). Briefly,

N-ethylmaleimide (NEM) sensitive factor (NSF) and soluble NSF-associated proteins (SNAPs) have been implicated in endosome–endosome fusion (Lenhard *et al.,* 1992). In addition to NSF, other NEM-sensitive components appear to be involved in these fusion events (Rodriguez *et al.,* 1994). The specificity of docking and fusion appears to be mediated at least in part by individual vesicular and target SNAP receptors (v- and t-SNAREs; Rothman and Warren, 1994). Studies *in vivo* and *in vitro* have demonstrated that vectorial transport occurs along the endocytic pathway. According to the SNARE hypothesis, this would be effected by specific receptors on target membranes, t-SNAREs, which recognize components of incoming vesicles (v-SNAREs), permitting docking and fusion. Such components have been identified on the secretory pathway. After interaction of the cognate v- and t-SNAREs, fusion can occur in the presence of NSF and SNAPs (Rothman and Warren, 1994; Chapter 2). To date, the only v-SNARE that has been identified in the endocytic pathway is cellubrevin, the non-neuronal homologue of synaptobrevin (or vesicle-associated membrane protein, VAMP). Cellubrevin has been implicated as the v-SNARE involved in formation of vesicles containing material to be recycled back to the cell surface (Galli *et al.,* 1994). This was demonstrated by treatment of permeabilized cells with tetanus toxin, a metallo-endoprotease that is very specific for synaptobrevin and related proteins (McMahon *et al.,* 1993). Although proteolytic degradation of all detectable cellubrevin resulted in inhibition of transferrin recycling, there was a significant residual extent of recycling that was toxin independent (Galli *et al.,* 1994). This suggested that other exocytic mechanisms exist which are independent of cellubrevin but that may be mediated by a related protein(s) that is insensitive to tetanus toxin.

Recently the amino acid sequence required for targeting of VAMP to synaptic vesicles was identified. The same sequence is present in cellubrevin, and indeed transfection of cellubrevin into PC12 cells, a neuronal cell line, results in its targeting to synaptic vesicles (Grote *et al.,* 1995). This indicates that cellubrevin and VAMP both have a general role in the regulation of exocytic traffic, be it constitutive or regulated (Grote *et al.,* 1995). Other components must therefore mediate specificity of intracellular fusion events, and the family of small ras-related GTP-binding proteins, the rab proteins, and proteins which interact with them are other key regulators of docking and fusion.

Individual rab proteins have been localized to specific subcellular compartments including those found along the endocytic pathway (Chavrier *et al.,* 1991). Rab 4 and rab 5 have been localized to the plasma membrane and early endosomes, while rab 7 and rab 9 appear to be localized to late endosomes. The implication of rab 5 in early endosome dynamics first came from *in vitro* studies which reconstitute these steps using isolated populations of endosomes (Gruenberg and Howell, 1989). These *in vitro* assays are dependent on cytosol. Fusion was markedly stimulated by cytosol prepared from cells overexpressing rab 5, while cytosol prepared from cells overexpressing mutant forms of rab 5 deficient in GTP binding was unable to support fusion. Purified rab 5 could be used to rescue the mutant cytosol

(Gorvel *et al.*, 1991). In intact cells, overexpression of rab 5 resulted in an enhanced rate of internalization of the transferrin receptor, while the rate of transferrin recycling was unaffected (Bucci *et al.*, 1992). By contrast, over-expression of protein defective in the ability to bind GTP led to a reduced rate of endocytosis. Morphological analysis of the cells in both cases demonstrated that overexpression of the wild-type protein led to an accumulation of many large endosomal structures, whereas in the cells overexpressing the mutant protein there was an accumulation of small vesicular structures (Bucci *et al.*, 1992). These experiments demonstrated that rab 5 acts at the level of fusion between coated vesicles and endosomes and is rate-limiting in the cell since its overexpression leads to enhanced rates of internalization. Overexpression of wild-type rab 4 resulted in a redistribution of transferrin receptors to the cell surface, while the initial rate of transferrin uptake was unaffected. In addition transferrin did not appear to reach a compartment which was sufficiently acidic to allow discharge of iron (Van der Sluijs *et al.*, 1992). These results implicate rab 4 as a regulatory component of the recycling pathway.

Rab proteins can associate reversibly with membranes as a result of a geranylgeranyl modification at their C-terminus (reviewed in Magee and Newman, 1992; Armstrong, 1993). The active form of rab proteins is thought to be the GTP form, which is present on transport vesicles. RabGDI transports rabGDP from the cytosol to its target membrane. RabGDI dissociates from the membrane and GDP/GTP exchange is mediated by a guanine-nucleotide-exchange factor (GEF). RabGTP is then in a form capable of participating in docking and fusion. After GTP hydrolysis, rabGDP is retrieved from the membrane by rabGDI and delivered back to the cytosol (Novick and Brennwald; 1993, Zerial and Stenmark, 1993; Pfeffer, 1994). Although there is much evidence to support a role for rab proteins in intracellular transport, the precise mechanism by which they act has not yet been fully elucidated. A role in proofreading, that is, in ensuring that the appropriate v-SNARE interacts with the appropriate t-SNARE, has been assigned to these proteins. However, there appears to be an extra level of specificity built into the targeting mechanism. For example, in polarized cells, although both apical and basolateral cells exhibit rab5-dependent fusion, these endosomes are not capable of fusion with each other *in vitro* (Bomsel *et al.*, 1990; Parton *et al.*, 1991), suggesting that rab proteins do not interact directly with v- and t-SNARES but mediate their effects via other targets. Recently, a direct effector of rab 5, rabaptin-5, has been identified using the yeast two-hybrid system (Stenmark *et al.*, 1995). Rabaptin-5 binds preferentially to the GTP-bound form of rab5. The bulk of the protein is present in the cytosol, but it is recruited to the endosomal membrane in a GTP dependent manner. Endosomal fusion *in vitro* is also dependent on the presence of rabaptin (Stenmark *et al.*, 1995). A number of possibilities exist as to the role of rabaptin-5 in vesicle docking and fusion. It may interact directly with SNAREs or it may act upstream of SNAREs via other proteins. Interestingly, although rab proteins have been implicated in the formation of SNARE complexes, they cannot be detected in the SNARE complex (Lian *et al.*, 1994; Sogaard *et al.*, 1994).

4.   PROPERTIES OF ENDOSOMAL COMPARTMENTS

4.1.   Structure of the Endosomal Network

The endosomal network consists of a complex collection of vacuolar and tubular reticular structures (Fig. 5). The heterogeneity of the endosomal network and, until recently, the lack of specific markers has resulted in the endosomal network being defined operationally in terms of times of appearance of internalized ligands in a given compartment (Schmid, 1993). Early endosomes are reached first ($t_{1/2}$ ~5 min) by internalized ligands. They are located in peripheral areas of the cell and are tubulovesicular in structure with large vacuolar components and tubular extensions. Their content is positive for recycling receptors such as the transferrin and LDL receptors. Endosomal carrier vesicles

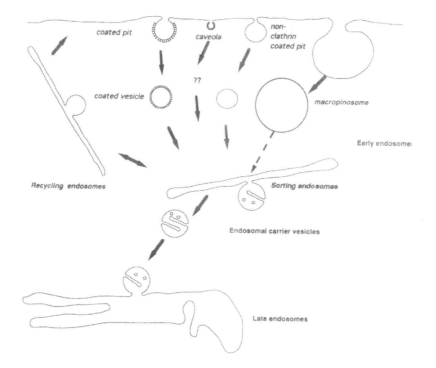

FIGURE 5. Overview of the endocytic pathway. This cartoon presents a schematic view of the endocytic pathway illustrating the consensus view of the dynamic interrelationships that occur between different compartments on the pathway.

are distributed throughout the cytoplasm; they are spherical structures with internal membranes and are believed to mediate transport from early endosomes to late endosomes. Late endosomes, again originally defined in terms of arrival of ligands from the cell surface ($t_{1/2} \sim 15$–20 min), are heterogeneous structures with many internal membranes and are often termed multivesicular bodies. These are found in the perinuclear region of the cell. Late endosomes and/or prelysosomes are characterized by the presence of both lysosomal glycoproteins and also the mannose 6-phosphate receptor, which is involved in the delivery of newly synthesized lysosomal enzymes from the biosynthetic pathway to lysosomes (see Chapter 2). Late endosomes, being the site of delivery of these proteins, represent the intersection of the endocytic and biosynthetic pathways. It should be noted that the heterogeneity of the endosomal compartment and its dynamic nature and different manifestations in various cell types has led to some confusion in the literature. However, this chapter has tried to describe a consensus view of the endosomal network, although it should be borne in mind that there is overlap in how compartments are defined. The issue has been greatly clarified by the identification of specific rab proteins that localize to particular endocytic compartments (Chavrier *et al.*, 1991).

As discussed above, material that has entered the cell by pinocytosis is generally delivered to the early endosome. In the case of phagocytosed particles it has been known for many years that the phagosome ultimately becomes a phagolysosome as its luminal pH drops to below 5 and it acquires a full complement of lysosomal enzymes (Geisow *et al.*, 1981). However, it was not clear whether the phagolysosome arose as a result of fusion of the phagosome with existing lysosomes or entered the endocytic pathway at an earlier stage. Recent experiments using a combination of biochemical and morphological techniques have shed light on this question (Desjardins *et al.*, 1994a,b). Highly purified phagosomes were prepared at various times after formation and their protein composition analyzed by two-dimensional gel electrophoresis. With increasing time, phagosomes were demonstrated to gain and lose various endocytic markers. These include rab5, rab7, and members of the LAMP family of proteins. When viewed by video microscopy, a series of transient interactions between phagosomes and components of the endosomal network was observed. Furthermore, it was observed that there was a difference in the time course of transfer of soluble versus membrane content between phagosomes and the endocytic pathway, with transfer of membrane components being considerably slower (Desjardins *et al.*, 1994a,b). This has led to the "kiss-and-run" hypothesis for phagolysosome biogenesis (Desjardins, 1995), which argues that phagosomes make several rapid fusion–fission interactions with components of the endosomal pathway. This would explain the more rapid mixing of content compared to membrane, before the fully functional phagolysosome is formed. Results from *in vitro* systems demonstrating the fusion of phagosomes with early endosomes (Mayorga *et al.*, 1991) and with late endosomes (Rabinowitz *et al.*, 1992) are consistent with this hypothesis and illustrate the dynamic nature of the endosomal network.

4.2. Membrane Recycling

A key feature of endocytic uptake is that membrane which is internalized into the cell is continuously being recycled to the cell surface. This is borne out by the magnitude of membrane flow, measured using fluid-phase markers along the endocytic pathway. Markers such as the enzyme horseradish peroxidase, radio-labeled sucrose, fluorescein isothiocyanate (FITC) dextrans do not bind specifically to the cell, and uptake is proportional to their concentration in the extracellular medium. Markers of this type have been used to carry out a stereological analysis of membrane flow along the endocytic pathway in baby hamster kidney (BHK) cells (Griffiths *et al.*, 1989). In this study the amount of membrane in each subcellular organelle in BHK cells was measured as well as the rate of fluid-phase uptake of horseradish peroxidase. It was clear that the rate of *de novo* membrane synthesis was insufficient to account for maintenance of the plasma membrane given the high rate of internalization from the cell surface. Therefore, membrane must be recycled. In addition, receptor-mediated uptake of some ligands can occur for many hours even when protein synthesis is inhibited, indicating that internalized receptors can participate in multiple rounds of endocytic uptake (Goldstein *et al.*, 1985). Recycling of internalized soluble components to the cell surface also occurs to a large extent.

More direct evidence of recycling came from a close look at the kinetics of internalization of fluid-phase markers. There is a biphasic curve of uptake with a faster initial rate of uptake, followed by a slower steady-state rate (Besterman *et al.*, 1981). This indicates that at later time points fluid-phase markers were being returned to the cell surface. In addition, in experiments where cells were pulsed with a fluid-phase marker for increasing lengths of time and then chased in medium lacking the marker, the greatest amount of material was returned to the medium in cells pulsed for the shortest times. This indicated that recycling occurs predominantly from a compartment which is labeled early ($t_{1/2} \sim 5$ min). Material delivered to later compartments therefore becomes increasingly refractory to recycling.

4.3. Sorting

Early endosomes act as a major sorting station, sending receptors and ligands to a variety of destinations within the cell. It is from this compartment that some molecules are targeted to lysosomes, some receptors and ligands recycled to the cell surface, and other molecules transcytosed across the cell. Recent biochemical and morphological experiments have begun to elucidate how these complex sorting and targeting processes are effected.

It has long been established that ligands encounter an increasingly acidic pathway as they traverse the endocytic pathway. This has been determined using fluid-phase markers coupled to pH-sensitive fluorophores (Yamashiro et al., 1989). The internal pH of early endosomes lies between 6 and 6.5, whereas that of endosomal carrier vesicles and late endosomes is approximately 5.5. Endosome acidification results from the action of an energy-dependent proton pump (Galloway et al., 1983; Merion et al., 1983). The extent of acidification of early endosomes is regulated by Na^+, K^+-ATPase (Cain et al., 1989; Fuchs et al., 1989). The increasing acidity of the endocytic pathway plays a role in the functions of different populations of vesicles. The pH of the early endosome is sufficiently low that it can cause dissociation of a number of ligands from their receptors. One such ligand is LDL, whose receptor is recycled to the cell surface while LDL is targeted to lysosomes.

Peptide determinants in recycling molecules do not appear to be required for entry into the recycling pathway in contrast to delivery to late endosomes. This pathway appears to be the default pathway, as was illustrated in experiments using the lipid analogue C6-NBD-sphingomyelin as a marker of bulk membrane. These experiments showed that greater than 95% of the lipid was returned to the cell surface after internalization (Koval and Pagano, 1989, 1991). Further studies comparing the kinetics of recycling of transferrin versus C6-NBD-sphingomyelin showed that the rates of recycling were identical (Mayor et al., 1993).

The acidity of the early endosome does not only appear to be required for dissociation of ligands from receptors, but also seems to be important for sorting and transport to the next destination. Treatment of cells with lysosomotropic agents—e.g., chloroquine, which raises the pH of intracellular organelles—results in an inhibition of recycling and an accumulation of plasma-membrane proteins in the endosome (Ohkuma and Poole, 1981; Schwartz et al., 1984; Zijderhand-Bleekemolen et al., 1987; Braun et al., 1989). Interestingly, two resident trans-Golgi network proteins, TGN 38 and the protease furin, also accumulate in the endosome in the presence of chloroquine (Chapman and Munro, 1994; Reeves and Banting, 1994). These proteins have been shown to recycle from the Golgi to the plasma membrane and back to the Golgi using the endocytic pathway. Furthermore, the cytoplasmic tail of furin alone is capable of conferring chloroquine sensitivity on a heterologous protein (Chapman and Munro, 1994). It has also been shown recently that the formation of the endosomal carrier vesicle (ECV) requires both a functional vacuolar ATPase which is responsible for maintaining acid conditions in the endosome (Clague et al., 1994) and also the pH-dependent recruitment of a subset of the COP proteins to the cytoplasmic face of the early endosome (Aniento et al., 1996). Intriguingly, therefore, the lumenal pH is important in the regulation of events occurring on the cytoplasmic face of the endosome. Taken together, these data suggest that reduced pH is required for the exit of components from this compartment.

The question then arises how receptors and ligands are segregated within the early endosome. Recent studies have shown that early endosomes may be subdivided into two classes, namely, sorting and recycling endosomes, which are both functionally and physically distinct even though they undergo dynamic interactions (Gruenberg and Maxfield, 1995). Sorting endosomes are the first endosomes encountered by incoming ligands. They have an internal pH of approximately 6.2 and a tubulovesicular structure as described above. At early stages of internalization, both transferrin (a recycling ligand) and LDL (a ligand destined for lysosomes) are found within the same compartment (Ghosh et al., 1994). Interestingly, LDL accumulates in this compartment while transferrin is sorted from LDL with a half-time of 2.5 min and is recycled to the cell surface with a half-time of 7 min, indicating that transferrin is rapidly being segregated to another compartment. At early time points transferrin and LDL were viewed by microscopy in the same compartment—the tubulovesicular component of the early endosomes. After sorting, transferrin appears to enter recycling endosomes, which are tubular structures often located close to the nucleus (Hopkins et al., 1994). It should be noted that recycling endosomes also appear capable of performing sorting functions. This was demonstrated by analysis of the trafficking of oligomerized transferrin receptors (Marsh et al., 1995). Transferrin receptors were cross-linked using multivalent transferrin complexes of up to ten transferrin molecules. Receptors thus cross-linked showed a significant reduction in their rate of recycling, which was shown to result from retention in the recycling compartment. Retention was independent of the cytoplasmic tail of the receptors and was thus mediated by lumenal sorting signals.

The mechanism of sorting has not yet been fully elucidated. One hypothesis is that after dissociation of ligands from receptors, recycling components are segregated into tubular areas of the endosome and the high surface area-to-volume ratio prevents trapping of excess soluble material (Rome, 1985). This was supported by morphological analysis which indicted that after dissociation of asialoglycoprotein from its receptor, the ligand was located in the vacuolar components of the endosome while the receptor was localized to the tubular extensions (Geuze et al., 1983). Another model that might explain how high-efficiency sorting could occur involves repeated sorting, whereby material delivered to endosomes is sorted with a relatively low efficiency and rapidly returned to the plasma membrane and then reinternalized. This process is repeated until receptor–ligand sorting is complete. Support for this model came from the observation that transferrin recycles through a compartment where LDL accumulates (Dunn et al., 1989; Ghosh et al., 1994). It appears that the recycling compartment is physically distinct from the sorting endosome. Evidence that this is the case comes from the fact that the two types of endosome are morphologically distinct. The recycling compartment appears to be a long-lived compartment located in the perinuclear area of the cell (Hopkins et al., 1994; Ghosh et al., 1994). It is composed of a network of tubules of approximately 50 nm diameter. It is defined as an early endosome by the presence of recycling receptors and transferrin but differs significantly from sorting endosomes in that LDL is absent. In addition, the pH of the recycling endosome in Chinese hamster ovary (CHO) cells is significantly higher (~0.3–0.4 pH units) than that of the sorting endosome (Yamashiro et al., 1984; Sipe and Murphy, 1991). Although distinct entities, they are likely involved in dynamic interactions with each other. This is supported by the demonstration of extensive homotypic fusion between endosomal compartments in vitro (Gruenberg and Howell, 1989) and also by examination of internalized transferrin by confocal microscopy in living cells, which illustrates that there is retrograde movement of transferrin from recycling endosomes back to sorting endosomes (Ghosh and Maxfield, 1995).

Transport from early to late endosomes appears to be mediated by ECVs or multivesicular bodies as they are variously referred to in the literature. These carrier vesicle structures, whose formation is microtubule-dependent, appear to arise from invaginations of the limiting membrane of the early endosome which results in the presence of internal membranes within large spherical structures. Their existence as discrete organelles has been supported by in vitro fusion assays (Gruenberg et al., 1989). These assays have shown that despite extensive homotypic fusion in vitro between endosomal populations, early and late endosomes are not capable of fusing with each other. By contrast, ECVs are capable of fusing with late endosomes, but not with each other (Bomsel et al., 1990). Interestingly, analysis of the fate of internalized fluorescently-labeled transferrin and EGF receptors using video-enhanced microscopy showed that transferrin receptor was present in an extensive reticular network while after addition of EGF the EGF receptor was found in boluses which moved along the network. Electron microscopy revealed these boluses to be multivesicular bodies (Hopkins et al., 1990). Possibly ECVs arise as a result of these boluses budding from the reticular network.

In contrast to the recycling pathway, positive sorting signals appear to be required for incorporation into ECVs and subsequent delivery of membrane components to late endosomes, the next stage on the pathway to lysosomes. The EGF receptor after internalization is generally transported to lysosomes for degradation in the process of receptor downregulation. However, receptors deficient in the cytoplasmic domain tyrosine kinase recycle from the early endosome back to the cell surface (Felder et al., 1990). The major substrate for the endosomal kinase has been identified as annexin I (Futter et al., 1993), a member of a family of calcium and phospholipid binding proteins. Some members of this family have been implicated in a variety of intracellular trafficking events (Gruenberg and Emans, 1993). Other examples of proteins with defined signals required for transit to late endosomes are P-selectin (Green et al., 1994) and the mannose 6-phosphate receptor (Johnson and Kornfeld, 1992a,b).

A key question in terms of trafficking of material along the endocytic pathway is the mechanism by which material moves through the endosomal compartments. This issue has been the subject of much debate and in the early 1980s two different hypotheses were proposed (Helenius et al., 1983; van Deurs et al., 1989).

The vesicle shuttle model proposed that early and late endosomes represent stable, long-lived organelles, and material destined for lysosomes would be packaged into vesicles that bud from one endosomal compartment and fuse with the next. The maturation model predicts that material taken up by endocytosis is contained within an organelle composed initially of internalized plasma membrane and that with time this organelle develops into an early endosome and then a late endosome. According to the maturation model, components specific for a given organelle, such as lysosomal enzymes, would be inserted by vesicle fusion while material to be recycled would be removed by vesicles budding off.

According to the vesicle shuttle model of endocytic trafficking, the t-SNAREs on early and late endosomes would be resident components of a stable compartment recognizing v-SNAREs on incoming vesicles. After fusion, the v-SNAREs would be removed by recycling vesicles. According to the maturation model, t-SNAREs would have to be inserted into the relevant membrane during the process of maturation and then either inactivated or removed once they had completed their role. A fundamental premise of the maturation model is that internalized plasma membrane ultimately forms the late endosomal membrane, and one prediction of this model is that a common population of protein would be found in both compartments. To test this possibility the proteins of plasma membrane and early and late endosomes were radiolabeled with iodine and analyzed by two-dimensional gel electrophoresis (Schmid *et al.*, 1988). Each endosomal compartment was found to have a distinct protein composition; proteins common to plasma membrane and late endosomes were not observed. Furthermore, during the purification of the endosomal compartments used in this study, no electrophoretic intermediates were observed between early and late endosome, suggesting that transfer of material between these two compartments occurs as a distinct step. Electron microscopy studies of intact cells supported the premise that the endosomal carrier vesicles represent a transport intermediate from early to late endosomes (Gruenberg *et al.*, 1989). In the presence of the microtubule disrupting drug nocodazole, an accumulation of ECVs was observed, consistent with ECVs being a discrete transport intermediate. Further evidence that transport of material occurs through distinct organelles comes from the observation that an *in vitro* system which reconstitutes transport from late endosomes to lysosomes requires the presence of preexisting lysosomes (Mullock *et al.*, 1994). Late endosomes incubated under conditions which support transport but in the absence of lysosomes did not mature into lysosomal fractions. Additionally, the small GTP-binding rab proteins show a very specific localization to either early endosomes (rab 5) or late endosomes (rab 7) (Chavrier *et al.*, 1991), and significantly ECVs formed *in vitro* do not contain rab 5 (Aniento *et al.*, 1996). One prediction based on the existence of a stable endosomal compartment is that fluid-phase markers will be delivered to a compartment that does not change volume. However, the concentration of material within the compartment should increase with time. This was borne out by quantitative immunocytochemical analysis of the uptake of horseradish peroxidase in baby-hamster-kidney cells (Griffiths *et al.*, 1990).

The maturation of early endosomes into late endosomes and lysosomes has been postulated to occur by several variations on a common theme (Murphy, 1991). The basic premise of this model is that endocytosed material is delivered to an organelle to which new components are continuously being added and from which other material is removed and recycled. One possibility by which this might occur would be by iterative sorting as mentioned above (Dunn *et al.*, 1989). Experimental evidence that supports the maturation model is based on the ability of a recycling ligand, transferrin conjugated to horseradish peroxidase, to access all parts of the degradative endocytic pathway labeled with asialoorosomucoid, suggesting that the loss of early endosomal markers is a gradual process (Stoorvogel *et al.*, 1991). In addition, a compartment which is filled at early times with LDL becomes refractory to the addition of more LDL at later times (Ghosh *et al.*, 1995). These workers also demonstrated the non-vectorial transport of material from recycling endosomes back to sorting endosomes and proposed that such non-vectorial transport could be a mechanism by which proteins stored in a long-lived recycling compartment could be delivered to a transient sorting endosome (Ghosh *et al.*, 1995).

The issue of a vesicle shuttle model versus a maturation model will be fully resolved only with a complete molecular analysis of all of the compartments en route to lysosomes. It may well be that material is delivered to lysosomes by a combination of these two mechanisms.

5. CONCLUSIONS

A great many unanswered questions remain about the regulation and mechanism of the endocytic pathway but this chapter has tried to present an overview of our current knowledge of the way in which internalized material is transported to lysosomes. We are beginning to understand the molecular level of the regulation and mechanism of this complex process and look forward to further insights in the coming years.

ACKNOWLEDGMENTS. E.S. is a Medical Research Council Senior Fellow.

6. REFERENCES

Abo, A., Boyhan, A., West, I., Thrasher, A. J., and Segal, A. W., 1992, Reconstitution of neutrophil nadph oxidase activity in the cell-free system by 4 components—p67-*phox*, p47-*phox*, p21rac1, and cytochrome b_{554}, *J. Biol. Chem.* **267**:16767–16770.
Ahle, S., and Ungewickell, E., 1990, Auxilin, a newly identified clathrin-associated protein in coated vesicles from bovine brain, *J. Cell Biol.* **111**:19–29.

Aiken, C., Konner, J., Landau, N. R., Lenburg, M. E., and Trono, D., 1994, Nef induces CD4 endocytosis—requirement for a critical dileucine motif in the membrane-proximal CD4 cytoplasmic domain, *Cell* **76**:853–864.

Allison, A. C., Davies, P., and de Petris, S., 1971, Role of contractile microfilaments in macrophage movement and endocytosis, *Nat., New Biol.* **232**:153–155.

Alpuche-Aranda, C. M., Racoosin, E. L., Swanson, J. A., and Miller, S. I., 1994, Salmonella stimulate macrophage macropinocytosis and persist within spacious phagosomes, *J. Exp. Med.* **179**:601–608.

Anderson, R. G. W., 1993, Caveolae—where incoming and outgoing messengers meet, *Proc. Natl. Acad. Sci. USA* **90**:10909–10913.

Anderson, R. G. W., Brown, M. S., and Goldstein, J. L., 1977, Role of the coated endocytic vesicle in the uptake of the low density lipoprotein receptor bound to low density lipoprotein in human fibroblasts, *Cell* **10**:351–364.

Anderson, R. G. W., Kamen, B. A., Rothberg, K. G., and Lacey, S. W., 1992, Potocytosis: Sequestration and transport of small molecules by caveolae, *Science* **255**:410–411.

Aniento, F., Gu, F., Parton, R. G., and Gruenberg, J., 1996, An endosomal beta COP is involved in the pH-dependent formation of transport vesicles destined for late endosomes, *J. Cell Biol.* **133**:29–41.

Armstrong, J., 1993, Two fingers for membrane traffic, *Curr. Biol.* **3**:33–35.

Bansal, A., and Gierasch, L. M., 1991, The NPXY internalization signal of the LDL receptor adopts a reverse-turn conformation, *Cell* **67**:1195–1201.

Beauchamp, J. R., and Woodman, P. G., 1994, Regulation of transferrin receptor recycling by protein-phosphorylation, *Biochem. J.* **303**:647–655.

Benhamou, M., Ryba, N. J. P., Kihara, H., Nishikata, H., and Siraganian, R. P., 1993, Protein-tyrosine kinase p72syk in high-affinity IgE receptor signaling: Identification as a component of pp72 and association with the receptor γ chain after receptor aggregation, *J. Biol. Chem.* **268**:23318–23324.

Besterman, J. M., Airhart, J. A., Woodworth, R. C., and Low, R. B., 1981, Exocytosis of pinocytosed fluid in cultured cells: Kinetic evidence for rapid turnover and compartmentation, *J. Cell Biol.* **91**:716–727.

Blank, G. S., and Brodsky, F. M., 1987, Clathrin assembly involves a light chain binding region, *J. Cell Biol.* **105**:2011–2019.

Bokoch, G. M., 1995, Regulation of the phagocyte respiratory burst by small GTP-binding proteins, *Trends Cell Biol.* **5**:109–113.

Bokoch, G. M., Bohl, B. P., and Chuang, T. H., 1994, Guanine-nucleotide exchange regulates membrane translocation of rac/rho GTP-binding proteins, *J. Biol. Chem.* **269**:31674–31679.

Boman, A. L., Taylor, T. C., Melancon, P., and Wilson, K. L., 1992, A role for ARF (ADP-ribosylation factor), a small GTP-binding protein in nuclear vesicle dynamics, *Nature* **358**:512–514.

Bomsel, M., Parton, R., Kuznetsov, S. A., Schroer, T. A., and Gruenberg, J., 1990, Microtubule-dependent and motor-dependent fusion *in vitro* between apical and basolateral endocytic vesicles from MDCK cells, *Cell* **62**:719–731.

Braell, W. A., Schlossman, D. M., Schmid, S. L., and Rothman, J. E., 1984, Dissociation of clathrin coats coupled to the hydrolysis of ATP. Role of an uncoating ATPase, *J. Cell Biol.* **99**:734–741.

Braun, M., Waheed, A. and von Figura, K., 1989, Lysosomal Acid Phosphatase is transported to lysosomes via the cell surface, *EMBO J.* **8**:3633–3640.

Bretscher, M. S., Thomson, J. N., and Pearse, B. M. F., 1980, Coated pits act as molecular filters, *Proc. Natl. Acad. Sci. USA.* **77**:4156–4159.

Bucci, C., Parton, R. G., Mather, I. H., Stunnenberg, H., Simons, K., Hoflack, B., and Zerial, M., 1992, The small GTPase rab5 functions as a regulatory factor in the early endocytic pathway, *Cell* **70**:715–728.

Cain, C. C., Sipe, D. M., and Murphy, R. F., 1989, Regulation of endocytic pH by the Na+, K+-ATPase in living cells, *Proc. Natl. Acad. Sci. USA.* **86**:544–548.

Carter, L. L., Redelmeier, T. E., Woollenweber, L. A., and Schmid, S. L., 1993, Multiple GTP-binding proteins participate in clathrin-coated vesicle-mediated endocytosis, *J. Cell Biol.* **120**:37–45.

Chang, C. P., Lazar, C. S., Walsh, B. J., Komuro, M., Collawn, J. F., Kuhn, L. A., Tainer, J. A., Trowbridge, I. S., Farquhar, M. G., Rosenfeld, M. G., Wiley, H. S., and Gill, G. N., 1993, Ligand-induced internalization of the epidermal growth-factor receptor is mediated by multiple endocytic codes analogous to the tyrosine motif found in constitutively internalized receptors, *J. Biol. Chem.* **268**:19312–19320.

Chapman, R. E., and Munro, S., 1994, Retrieval of TGN proteins from the cell-surface requires endosomal acidification, *EMBO J.* **13**:2305–2312.

Chappell, T. G., Welch, W. J., Schlossman, D. M., Palter, K. B., Schlesinger, M. J., and Rothman, J. E., 1986, Uncoating ATPase is a member of the 70 kilodalton family of stress proteins, *Cell* **45**:3–13.

Chavrier, P., Gorvel, J. P., Stelzer, E., Simons, K., Gruenberg, J., and Zerial, M., 1991, Hypervariable C-terminal domain of rab proteins acts as a targeting signal, *Nature* **353**:769–772.

Chen, W. J., Goldstein, J. L., and Brown, M. S., 1990, NPXY, a sequence often found in cytoplasmic tails, is required for coated pit-mediated internalization of the low-density-lipoprotein receptor, *J. Biol. Chem.* **265**:3116–3123.

Clague, M. J., Urbe, S., Aniento, F., and Gruenberg, J., 1994, Vacuolar ATPase activity is required for endosomal carrier vesicle formation, *J. Biol. Chem.* **269**:21–24.

Clark, J., Moore, L., Krasinskas, A., Way, J., Battey, J., Tamkun, J., and Kahn, R. A., 1993, Selective amplification of additional members of the ADP-ribosylation factor (ARF) family: Cloning of additional human and Drosophila ARF-like genes, *Proc. Natl. Acad. Sci. USA* **90**:8952–8956.

Collawn, J. F., Stangel, M., Kuhn, L. A., Esekogwu, V., Jing, S. Q., Trowbridge, I. S., and Tainer, J. A., 1990, Transferrin receptor internalization sequence YXRF implicates a tight turn as the structural recognition motif for endocytosis, *Cell* **63**:1061–1072.

Collawn, J. F., Kuhn, L. A., Liu, L. F. S., Tainer, J. A., and Trowbridge, I. S., 1991, Transplanted LDL and mannose-6-phosphate receptor internalization signals promote high-efficiency endocytosis of the transferrin receptor, *EMBO J.* **10**:3247–3253.

Cupers, P., Veithen, A., Kiss, A., Baudhuin, P., and Courtoy, P. J., 1994, Clathrin polymerization is not required for bulk-phase endocytosis in rat fetal fibroblasts, *J. Cell Biol.* **127**:725–735.

Cyr, D. M., Langer, T., and Douglas, M. G., 1994, DnaJ-like proteins: molecular chaperones and specific regulators of Hsp70, *Trends Biochem. Sci.* **19**:176–181.

Damke, H., Baba, T., van der Bliek, A. M., and Schmid, S. L., 1995, Clathrin-independent pinocytosis is induced in cells overexpressing a temperature-sensitive mutant of dynamin, *J. Cell Biol.* **131**:69–80.

Daukas, G., and Zigmond, S. H., 1985, Inhibition of receptor-mediated but not fluid phase endocytosis in polymorphonuclear leukocytes, *J. Cell Biol.* **101**:1673–1679.

Deluca-Flaherty, C., McKay, D. B., Parham, P., and Hill, B. L., 1990, Uncoating protein (hsc70) binds a conformationally labile domain of clathrin light chain LCa to stimulate ATP hydrolysis, *Cell* **62**:875–887.

Desjardins, M., 1995, Biogenesis of phagolysosomes: The kiss and run hypothesis, *Trends Cell Biol.* **5**:183–186.

Desjardins, M., Celis, J. E., Vanmeer, G., Dieplinger, H., Jahraus, A., Griffiths, G., and Huber, L. A., 1994a, Molecular characterization of phagosomes, *J. Biol. Chem.* **269**:32194–32200.

Desjardins, M., Huber, L. A., Parton, R. G., and Griffiths, G., 1994b, Biogenesis of phagolysosomes proceeds through a sequential series of interactions with the endocytic apparatus, *J. Cell Biol.* **124**:677–688.

Donaldson, J. G., and Klausner, R. D., 1994, Arf: A key regulatory switch in membrane traffic and organelle structure, *Curr. Opin. Cell Biol.* **6**:527–532.

Donaldson, J. G., Finazzi, D., and Klausner, R. D., 1992, Brefeldin-A inhibits Golgi membrane-catalyzed exchange of guanine nucleotide onto ARF protein, *Nature* **360**:350–352.

D'Souza-Schorey, C., Li, G. P., Colombo, M. I., and Stahl, P. D., 1995, A regulatory role for ARF6 in receptor-mediated endocytosis, *Science* **267**:1175–1178.

Dunn, K. W., McGraw, T. E., and Maxfield, F. R., 1989, Iterative fractionation of recycling receptors from lysosomally destined ligands in an early sorting endosome, *J. Cell Biol.* **109**:3303–3314.

Dusi, S., and Rossi, F., 1993, Activation of NADPH oxidase of human neutrophils involves the phosphorylation and the translocation of cytosolic p67phox, *Biochem. J.* **296:**367–371.

Eberle, W., Sander, C., Klaus, W., Schmidt, B., Vonfigura, K., and Peters, C., 1991, The essential tyrosine of the internalization signal in lysosomal acid-phosphatase is part of a beta turn, *Cell* **67:**1203–1209.

Felder, S., Miller, K., Moehren, A., Ullrich, A., Schlessinger, N. J., and Hopkins, C. R., 1990, Kinase activity controls the sorting of the epidermal growth factor receptor within the multivesicular body, *Cell* **61:**623–634.

Francis, C. L., Ryan, T. A., Jones, B. D., Smith, S. J., and Falkow, S., 1993, Ruffles induced by salmonella and other stimuli direct macropinocytosis of bacteria, *Nature* **364:**639–642.

Fuchs, R., Schmid, S., and Mellman, I., 1989, A possible role for Na$^+$, K$^+$-ATPase in regulating ATP-dependent endosome acidification, *Proc. Natl. Acad. Sci. USA* **86:**539–543.

Futter, C. E., Felder, S., Schlessinger, J., Ullrich, A., and Hopkins, C. R., 1993, Annexin I is phosphorylated in the multivesicular body during the processing of the epidermal growth factor receptor, *J. Cell Biol.* **120:**77–83.

Galli, T., Chilcote, T., Mundigl, O., Binz, T., Niemann, H., and DeCamilli, P., 1994, Tetanus toxin-mediated cleavage of cellubrevin impairs exocytosis of transferrin receptor-containing vesicles in CHO cells, *J. Cell Biol.* **125:**1015–1024.

Galloway, C. J., Dean, G. E., Marsh, M., Rudnick, G., and Mellman, I., 1983, Acidification of macrophage and fibroblast endocytic vesicles *in vitro, Proc. Natl. Acad. Sci. USA* **80:**3334–3338.

Gallusser, A., and Kirchhausen, T., 1993, The β subunit and β'subunit of the AP complexes are the clathrin coat assembly components, *EMBO J.* **12:**5237–5244.

Geison, M. J., D'Arcy Hart, P., and Young, M. R., 1981, Temporal changes of lysosome and phagosome pH during phagolysosome formation in macrophages: studies by fluorescence spectroscopy, *J. Cell. Biol.* **89:**645–652.

Geuze, H. J., Slot, J. W., Strous, G., Lodish, H. F., and Schwartz, A. L., 1983, Intracellular site of asialoglycoprotein receptor ligand uncoupling: Double-label immunoelectron microscopy during receptor-mediated endocytosis, *Cell* **32:**277–287.

Ghazizadeh, S., Bolen, J. B., and Fleit, H. B., 1994, Physical and functional association of src-related protein-tyrosine kinases with FcγRII in monocytic THP-1 cells, *J. Biol. Chem.* **269:**8878–8884.

Ghitescu, L., Fixman, A., Simionescu, M., and Simionescu, N., 1986, Differentiated uptake and transcytosis of albumin by the continuous endothelium of successive vascular segments, *J. Cell Biol.* **103:**A449.

Ghosh, R. N., and Maxfield, F. R., 1995, Evidence for nonvectorial, retrograde transferrin trafficking in the early endosomes of Hep2 cells, *J. Cell Biol.* **128:**549–561.

Ghosh, R. N., Gelman, D. L., and Maxfield, F. R., 1994, Quantification of low-density-lipoprotein and transferrin endocytic sorting in Hep2 cells using confocal microscopy, *J. Cell Biol.* **107:**2177–2189.

Glickman, J. N., Conibear, E., and Pearse, B. M. F., 1989, Specificity of binding of clathrin adaptors to signals on the mannose-6-phosphate insulin-like growth factor II receptor, *EMBO J.* **8:**1041–1047.

Goldstein, J. L., Brown, M. S., Anderson, R. G. W., Russell, D. W., and Schneider, W. J., 1985, Receptor-mediated endocytosis: Concepts emerging from the LDL receptor system, *Annu. Rev. Cell Biol.* **1:**1–39.

Gorvel, J. P., Chavrier, P., Zerial, M., and Gruenberg, J., 1991, Rab5 controls early endosome fusion *in vitro,* Cell **64:**915–925.

Goud, B., Huet, C., and Louvard, D., 1985, Assembled and unassembled pools of clathrin: A quantitative study using an enzyme-immunoassay, *J. Cell Biol.* **100:**521–527.

Green, S. A., Setiadi, H., McEver, R. P., and Kelly, R. B., 1994, The cytoplasmic domain of P-selectin contains a sorting determinant that mediates rapid degradation in lysosomes, *J. Cell Biol.* **124:**435–448.

Greenberg, S., 1995, Signal-transduction of phagocytosis, *Trends Cell Biol.* **5:**93–99.

Greenberg, S., and Chang, P., 1994, Aggregation of FcγRIII induces f-actin-rich membrane ruffles and association of p72^{syk} with the γ subunit of FcγRIII, *Clin. Res.* **42**:A133.

Greenberg, S., and Silverstein, S. C., 1993, Phagocytosis, *Fundamental Immunology,* 3d Ed. (W. E. Paul, ed.), pp. 941–964, Raven Press, New York.

Greenberg, S., Chang, P., and Silverstein, S. C., 1993, Tyrosine phosphorylation is required for Fc-receptor mediated phagocytosis in mouse macrophages, *J. Exp. Med.* **177**:529–534.

Greenberg, S., Chang, P., Wang, D. C., Xavier, R., and Seed, B., 1996, Clustered syk tyrosine kinase domains trigger phagocytosis, *Proc. Natl. Acad. Sci.* **93**:1103–1107.

Greenberg, S., Elkhoury, J., Divirgilio, F., Kaplan, E. M., and Silverstein, S. C., 1991, Ca^{2+}-independent F-actin assembly and disassembly during Fc receptor mediated phagocytosis in mouse macrophages, *J. Cell Biol.* **113**:757–767.

Griffin, F. M., Griffin, J. A., Leider, J. E., and Silverstein, S. C., 1975, Requirements for circumferential attachment of particle-bound ligands to specific receptors on the macrophage plasma membrane, *J. Exp. Med.* **142**:1263–1282.

Griffin, F. M., Griffin, J. A., and Silverstein, S. C., 1976. The interaction of macrophages with anti-immunoglobulin IgG-coated bone marrow–derived macrophages, *J. Exp. Med.* **144**:788–809.

Griffiths, G., and Gruenberg, J., 1991, The arguments for pre-existing early and late endosomes, *Trends Cell Biol.* **1**:5–9.

Griffiths, G., Back, R., and Marsh, M., 1989, A quantitative analysis of the endocytic pathway in baby hamster kidney cells, *J. Cell Biol.* **109**:2703–2720.

Grote, E., Hao, J. C., Bennett, M. K., and Kelly, R. B., 1995, A targeting signal in VAMP regulating transport to synaptic vesicles, *Cell* **81**:581–589.

Gruenberg, J., and Emans, N., 1993, Annexins in membrane traffic, *Trends Cell Biol.* **3**:224–227.

Gruenberg, J., and Howell, K. E., 1989, Membrane traffic in endocytosis: Insights from cell-free assays, *Ann. Rev. Cell Biol.* **5**:453–481.

Gruenberg, J., and Maxfield, F. R., 1995, Membrane transport in the endocytic pathway, *Curr. Opin. Cell Biol.* **7**:552–563.

Gruenberg, J., Griffiths, G., and Howell, K. E., 1989, Characterization of the early endosome and putative endocytic carrier vesicles *in vivo* and with an assay of vesicle fusion *in vitro, J. Cell Biol.* **108**:1301–1316.

Guigliatti, T. A., Hall, L., Rosenbluth, R., and Suzuki, D. T., 1973, Temperature sensitive mutations in Drosophila melanogaster. XIV. A selection of immobile adults, *Mol. Gen. Genet.* **120**:107–114.

Haft, C. R., Klausner, R. D., and Taylor, S. I., 1994, Involvement of dileucine motifs in the internalization and degradation of the insulin-receptor, *J. Biol. Chem.* **269**:26286–26294.

Hall, A., 1994, Small GTP-binding proteins and the regulation of the actin cytoskeleton, *Annu. Rev. Cell Biol.* **10**:31–54.

Hamada, F., Aoki, M., Akiyama, T., and Toyoshima, K., 1993, Association of immunoglobulin-g FcγRII with src-like protein tyrosine kinase fgr in neutrophils, *Proc. Natl. Acad. Sci. USA* **90**:6305–6309.

Hansen, S. H., Sandvig, K., and Van deurs, B., 1991, The preendosomal compartment comprises distinct coated and noncoated endocytic vesicle populations, *J. Cell Biol.* **113**:731–741.

Hansen, S. H., Sandvig, K., and Van Deurs, B., 1993, Molecules internalized by clathrin-independent endocytosis are delivered to endosomes containing transferrin receptors, *J. Cell Biol.* **123**:89–97.

Heilker, R., Manning-Kreig, U., Zuber, J. F., and Spiess, M., 1996, In vitro binding of clathrin adaptors to sorting signals correlates with endocytosis and basolateral sorting, *EMBO J.* **15**:2893–2899.

Helenius, A., Mellman, I., Wall, D., and Hubbard, A., 1983, Endosomes, *Trends Biochem. Sci.* **8**:245–250.

Helms, J. B., and Rothman, J. E., 1992, Inhibition by brefeldin A of a Golgi membrane enzyme that catalyzes exchange of guanine-nucleotide bound to ARF, *Nature* **360**:352–354.

Herskovits, J. S., Burgess, C. C., Obar, R. A., and Vallee, R. B., 1993, Effects of mutant rat dynamin on endocytosis, *J. Cell Biol.* **122**:565–578.

Heuser, J., 1989, Effects of cytoplasmic acidification on clathrin lattice morphology, *J. Cell Biol.* **108**:401–411.

Heuser, J. E., and Anderson, R. G. W., 1989, Hypertonic media inhibit receptor-mediated endocytosis by blocking clathrin-coated pit formation, *J. Cell Biol.* **108**:389–400.

Heuser, J. E., and Evans, L., 1980, Three-dimensional visualization of coated vesicle formation in fibroblasts, *J. Cell Biol.* **84**:560–583.

Heuser, J. E., and Keen, J., 1988, Deep-etch visualization of proteins involved in clathrin assembly, *J. Cell Biol.* **107**:877–886.

Hewlett, L. J., Prescott, A. R., and Watts, C., 1994, The coated pit and macropinocytic pathways serve distinct endosome populations, *J. Cell Biol.* **124**:689–703.

Heyworth, P. G., Bohl, B. P., Bokoch, G. M., and Curnutte, J. T., 1994, Rac translocates independently of the neutrophil NADPH oxidase components p47*phox* and p67*phox:* Evidence for its interaction with flavocytochrome b_{558}, *J. Biol. Chem.* **269**:30749–30752.

Hill, B. L., Drickamer, K., Brodsky, F. M., and Parham, P., 1988, Identification of the phosphorylation sites of clathrin light chain LCb, *J. Biol. Chem.* **263**:5499–5501.

Hinshaw, J. E., and Schmid, S. L., 1995, Dynamin self-assembles into rings suggesting a mechanism for coated vesicle budding, *Nature* **374**:190–192.

Hopkins, C. R., Gibson, A., Shipman, M., and Miller, K., 1990, Movement of internalized ligand-receptor complexes along a continuous endosomal reticulum, *Nature* **346**:335–339.

Hopkins, C. R., Gibson, A., Shipman, M., Strickland, D. K., and Trowbridge, I. S., 1994, In migrating fibroblasts, recycling receptors are concentrated in narrow tubules in the pericentriolar area, and then routed to the plasma-membrane of the leading lamella, *J. Cell Biol.* **125**:1265–1274.

Hunziker, W., and Fumey, C., 1994, A di-leucine motif mediates endocytosis and basolateral sorting of macrophage IgG Fc-receptors in MDCK cells, *EMBO J.* **13**:2963–2969.

Inaba, K., Inaba, M., Naito, M., and Steinman, R. M., 1993, Dendritic cell progenitors phagocytose particulates, including bacillus-calmette-guerin organisms, and sensitize mice to mycobacterial antigens *in vivo*, *J. Exp. Med.* **178**:479–488.

Isberg, R. R., and Tran Van Nhieu, G., 1995, The mechanism of phagocytic uptake promoted by invasin integrin interaction, *Trends Cell Biol.* **5**:120–124.

Jackson, A. P., Seow, H. F., Holmes, N., Drickamer, K., and Parham, P., 1987, Clathrin light-chains contain brain-specific insertion sequences and a region of homology with intermediate filaments, *Nature* **326**:154–159.

Jing, S. Q., Spencer, T., Miller, K., Hopkins, C., and Trowbridge, I. S., 1990, Role of the human transferrin receptor cytoplasmic domain in endocytosis: Localization of a specific signal sequence for internalization, *J. Chem. Biol.* **110**:283–294.

Johnson, K. F., and Kornfeld, S., 1992a, The cytoplasmic tail of the mannose 6-phosphate insulin-like growth factor II receptor has two signals for lysosomal-enzyme sorting in the Golgi, *J. Cell Biol.* **119**:249–257.

Johnson, K. F., and Kornfeld, S., 1992b, A his-leu-leu sequence near the carboxyl terminus of the cytoplasmic domain of the cation-dependent mannose 6-phosphate receptor is necessary for the lysosomal enzyme sorting function, *J. Biol. Chem.* **267**:17110–17115.

Kapeller, R., and Cantley, L. C., 1994, Phosphatidylinositol 3-kinase, *Bioessays* **16**:565–576.

Kiener, P. A., Rankin, B. M., Burkhardt, A. L., Schieven, G. L., Gilliland, L. K., Rowley, R. B., Bolen, J. B., and Ledbetter, J. A., 1993, Cross-linking of FcγRI and FcγRII on monocytic cells activates a signal-transduction pathway common to both Fc-receptors that involves the stimulation of p72 syk protein-tyrosine kinase, *J. Biol. Chem.* **268**:24442–24448.

Kirchhausen, T., and Harrison, S. C., 1984, Structural domains of clathrin heavy-chains, *J. Cell Biol.* **99**:1725–1734.

Kirchhausen, T., Harrison, S. C., Chow, E. P., Mattaliano, R. J., Ramachandran, K. L., Smart, J., and Brosius, J., 1987a, Clathrin heavy-chain: molecular cloning and complete primary structure. *Proc. Natl. Acad. Sci. USA* **84**:8805–8809.

Kirchhausen, T., Scarmato, P., Harrison, S. C., Monroe, J. J., Chow, E. P., Mattaliano, R. J., Ramachandran, K. L., Smart, J. E., Ahn, A. H., and Brosius, J., 1987b, Clathrin light-chains LCa and LCb are similar, polymorphic, and share repeated heptad motifs, *Science* **236**:320–324.

Kirchhausen, T., Nathanson, K. L., Matsui, W., Vaisberg, A., Chow, E. P., Burne, C., Keen, J. H., and Davis, A. E., 1989, Structural and functional division into two domains of the large (100-kDa to 115-kDa) chains of the clathrin-associated protein complex AP 2, *Proc. Natl. Acad. Sci. USA* **86**:2612–2616.

Kosaka, T., and Ikeda, K., 1983, Reversible blockage of membrane retrieval and endocytosis in the Garland cell of the temperature-sensitive mutant of Drosophila melanogaster, shibire ts1, *J. Cell Biol.* **97**:499–507.

Koval, M., and Pagano, R. E., 1989, Lipid recycling between the plasma membrane and intracellular compartments: Transport and metabolism of fluorescent sphingomyelin analogs in cultured fibroblasts, *J. Cell Biol.* **108**:2169–2181.

Koval, M., and Pagano, R. E., 1991, Intracellular transport and metabolism of sphingomyelin, *Biochim. Biophys. Acta* **1082**:113–125.

Lamaze, C., and Schmid, S. L., 1995a, Recruitment of epidermal growth factor receptors into coated pits requires their activated tyrosine kinase, *J. Cell Biol.* **129**:47–54.

Lamaze, C., and Schmid, S. L., 1995b, The emergence of clathrin-independent pinocytic pathways, *Curr. Opin. Cell Biol.* **7**:573–580

Larkin, J. M., Donzell, W. C., and Anderson, R. G. W., 1986, Potassium-dependent assembly of coated pits: New coated pits form as planar clathrin lattices, *J. Cell Biol.* **103**:2619–2627.

Law, D. A., Chan, V. W. F., Datta, S. K., and Defranco, A. L., 1993, B-cell antigen receptor motifs have redundant signaling capabilities and bind the tyrosine kinases ptk72, lyn and fyn, *Curr. Biol.* **3**:645–657.

Lazarovits, J., and Roth, M., 1988, A single amino acid change in the cytoplasmic domain allows the influenza-virus hemagglutinin to be endocytosed through coated pits, *Cell* **53**:743–752.

Lehrman, M. A., Goldstein, J. L., Brown, M. S., Russell, D. W., and Schneider, W. J., 1985, Internalization-defective LDL receptors produced by genes with nonsense and frameshift mutations that truncate the cytoplasmic domain, *Cell* **41**:735–743.

Lenhard, J. M., Kahn, R. A., and Stahl, P. D., 1992, Evidence for ADP-ribosylation factor (ARF) as a regulator of in vitro endosome-endosome fusion, *J. Biol. Chem.* **267**:13047–13052.

Letourneur, F., and Klausner, R. D., 1992, A novel di-leucine motif and a tyrosine-based motif independently mediate lysosomal targeting and endocytosis of CD3 chains, *Cell* **69**:1143–1157.

Lian, J. P., Stone, S., Jiang, Y., Lyons, P., and Ferro, N. S., 1994, Ypt1p implicated in v-SNARE activation, *Nature* **372**:698–701.

Lindner, R., and Ungewickell, E., 1992, Clathrin-associated proteins of bovine brain coated vesicles: An analysis of their number and assembly promoting activity, *J. Biol. Chem.* **267**:16567–16573.

Lisanti, M. P., Tang, Z. L., Scherer, P. E., Kubler, E., Koleske, A. J., and Sargiacomo, M., 1995, Caveolae, transmembrane signaling and cellular transformation, *Mol. Membr. Biol.* **12**:121–124.

Liu, S. H., Wong, M. L., Craik, C. S., and Brodsky, F. M., 1995, Regulation of clathrin assembly and trimerization defined using recombinant triskelion hubs, *Cell* **83**:257–267.

Livneh, E., Prywes, R., Kashles, O., Reiss, N., Sasson, I., Mory, Y., Ullrich, A., and Schlessinger, J., 1986, Reconstitution of human epidermal growth-factor receptors and its deletion mutants in cultured hamster cells, *J. Biol. Chem.* **261**:2490–2497.

Magee, T., and Newman, C., 1992, The role of lipid anchors for small G proteins in membrane trafficking, *Trends Cell Biol.* **2**:318–323.

Mahaffey, D. T., Peeler, J. S., Brodsky, F. M., and Anderson, R. G. W., 1990, Clathrin-coated pits contain an integral membrane protein that binds the AP2 subunit with high affinity, *J. Biol. Chem.* **265**:16514–26520

Marsh, M, and Helenius, A., 1980, Adsorptive endocytosis of semliki forest virus, *J. Mol. Biol.* **142**:439–454.

March, E. W., Leopold, P. L., Jones, N.l., and Maxfield, F. R., 1995, Oligomerized transferrin receptors are selectively retained by a lumenal sorting signal in a long-lived endocytic recycling compartment, *J. Cell Biol.* **129:**1509–1522.

Matsui, W., and Kirchhausen, T., 1990, Stabilization of clathrin coats by the core of the clathrin-associated protein complex AP2, *Biochemistry* **29:**10791–10798.

Mayor, S., Presley, J. F., and Maxfield, F. R., 1993, Sorting of membrane components from endosomes and subsequent recycling to the cell surface occurs by a bulk flow process, *J. Cell Biol.* **121:**1257–1269.

Mayor, S., Rothberg, K. G., and Maxfield, F. R., 1994, Sequestration of GPI-anchored proteins in caveolae triggered by cross-linking, *Science* **264:**1948–1951.

Mayorga, L. S., Bertini, F., and Stahl, P. D., 1991, Fusion of newly formed phagosomes with endosomes in intact cells and in a cell-free system, *J. Biol. Chem.* **266:**6511–6517.

McCulloch, C. A. G., and Knowles, G. C., 1993, Deficiencies in collagen phagocytosis by human fibroblasts *in vitro:* A mechanism for fibrosis, *J. Cell. Physiol.* **155:**461–471.

McMahon, H. T., Ushkaryov, Y. A., Edelmann, L., Link, E., Binz, T., Niemann, H., Jahn, R., and Sudhof, T. C., 1993, Cellubrevin is a ubiquitous tentanus-toxin substrate homologous to a putative synaptic vesicle fusion protein, *Nature* **364:** 346–349.

Merion, M., Schlesinger, P., Brooks, R. M., Moehring, J. M., Moehring, T. J., and Sly, W. S., 1983, Defective acidification of endosomes in chinese-hamster ovary cell mutants cross-resistant to toxins and viruses, *Proc. Natl. Acad. Sci. USA* **80:**5315–5319.

Moore, M. S., Mahaffey, D. T., Brodsky, F. M., and Anderson, R. G. W., 1987, Assembly of clathrin-coated pits onto purified plasma-membranes, *Science* **236:**558–563.

Morris, S. A., and Schmid, S. L., 1995, Synaptic vesicle recycling—the Ferrari of endocytosis, *Curr. Biol.* **5:**113–115.

Morris, S. A., Ahle, S., and Ungewickell, E., 1989, Clathrin-coated vesicles, *Curr. Opin. Cell Biol.* **1:**684–690.

Mostov, K. E., Kops, A. D., and Deitcher, D. L., 1986, Deletion of the cytoplasmic domain of the polymeric immunoglobulin receptor prevents basolateral localization and endocytosis, *Cell* **47:**359–364.

Mullock, B. M., Perez, J. H., Kuwana, T., Gray, S. R., and Luzio, J. P., 1994, Lysosomes can fuse with a late endosomal compartment in a cell-free system from rat-liver, *J. Cell Biol.* **126:**1173–1182.

Murphy, R. F., 1991, Maturation models for endosome and lysosome biogenesis, *Trends Cell Biol.* **1:**77–82.

Nathke, I. S., Heuser, J., Stock, J., Turck, C. W., and Brodsky, F. M., 1992, Folding and trimerisation of clathrin triskelia at the hub, *Cell* **68:**899–910.

Norbury, C. C., Hewlett, L. J., Prescott, A. R., Shastri, N., and Watts, C., 1995, Class I MHC presentation of exogenous antigen via macropinocytosis in bone marrow macrophages, *Immunity,* **3:**783–791.

Novick, P., and Brennwald, P., 1993, Friends and family: the role of the Rab GTPases in vesicular traffic, *Cell* **75:**597–601.

Ohkuma, S., and Poole, B., 1981, Cytoplasmic vacuolation of mouse peritoneal-macrophages and the uptake into lysosomes of weakly basic substances, *J. Cell Biol.* **90:**656–664.

Page, L. J., and Robinson, M. S., 1995, Targeting signals and subunit interactions in coated vesicle adaptor complexes, *J. Cell Biol.* **131:**619–630.

Parton, R. G., 1994, Ultrastructural-localization of gangliosides—GM1 is concentrated in caveolae, *J. Histochem. Cytochem.* **42:**155–166.

Parton, R. G., Bomsel, M., Griffiths, G., Gruenberg, J., and Simons, K., 1991, The polarized organization of the endocytic pathways in MDCK epithelial-cells, *J. Histochem. Cytochem.* **39:**717.

Parton, R. G., Joggerst, B., and Simons, K., 1994, Regulated internalization of caveolae, *J. Cell Biol.* **127:**1199–1215.

Pearse, B. M. F., 1975, Coated vesicles from pig brain: Purification and biochemical characteristics, *J. Mol. Biol.* **97:**93–98.

Pearse, B. M. F., 1988, Receptors compete for adaptors found in plasma-membrane coated pits, *EMBO J.* **7**:3331–3336.

Pearse, B. M. F., and Crowther, R. A., 1987, Structure and assembly of coated vesicles, *Annu. Rev. Biophys. Biophys. Chem.* **16**:49–68.

Pelchen-Matthews, A., Armes, J. E., Griffiths, G., and Marsh, M., 1991, Differential endocytosis of CD4 in lymphocytic and nonlymphocytic cells, *J. Exp. Med.* **173**:575–587.

Pelchen-Matthews, A., Boulet, I., Littman, D. R., Fagard, R., and Marsh M., 1992, The protein tyrosine kinase p56[lck] inhibits CD4 endocytosis by preventing entry of CD4 into coated pits, *J. Cell Biol.* **117**:279–290.

Peters, K. R., Carley, W. W., and Palade, G. E., 1985, Endothelial plasmalemmal vesicles have a characteristic striped bipolar surface-structure, *J. Cell Biol.* **101**:2233–2238.

Peters, P. J., Hsu, V. W., Ooi, C. E., Finazzi, D., Teal, S. B., Oorschot, V., Donaldson, J. G., and Klausner, R. D., 1995, Overexpression of wild-type and mutant ARF1 and ARF6: Distinct perturbations of nonoverlapping membrane compartments, *J. Cell Biol.* **128**:1003–1017.

Pfeffer, S. R., 1994, Rab GTPases: master regulators of membrane trafficking, *Curr. Opin. Cell Biol.* **6**:522–526.

Phatak, P. D., and Packman, C. H., 1994, Engagement of the T-cell antigen receptor by anti-CD3 monoclonal antibody causes a rapid increase in lymphocyte F-actin, *J. Cell. Physiol.* **159**:365–370.

Prywes, R., Livneh, E., Ullrich, A., and Schlessinger, J., 1986, Mutations in the cytoplasmic domain of EGF receptor affect EGF binding and receptor internalization, *EMBO J.* **5**:2179–2190.

Rabinowitz, M., 1995, Professional and nonprofessional phagocytes—an introduction, *Trends Cell Biol.* **5**:85–87.

Rabinovitch, S., Horstmann, H., Gordon, S., and Griffiths, G., Immunocytochemical characterization of the endocytic and phagolysomal compartments in peritoneal macrophages, *J. Cell Biol.* **116**:95–112.

Racoosin, E. L., and Swanson, J. A., 1993, Macropinosome maturation and fusion with tubular lysosomes in macrophages, *J. Cell Biol.* **121**:1011–1020.

Raposo, G., Dunia, I., Delavierklutchko, C., Kaveri, S., Strosberg, A. D., and Benedetti, E. L., 1989, Internalization of β-adrenergic receptor in A431 cells involves non-coated vesicles, *Eur. J. Cell Biol.* **50**:340–352.

Reaves, B., and Banting, G., 1994, Vacuolar ATPase inactivation blocks recycling to the *trans*-Golgi network from the plasma membrane, *FEBS Lett.* **345**:61–66.

Ridley, A. J., 1994, Membrane ruffling and signal transduction, *Bioessays* **16**:321–327.

Ridley, A. J., Paterson, H. F., Johnston, C. L., Diekmann, D., and Hall, A., 1992, The small GTP-binding protein rac regulates growth-factor induced membrane ruffling, *Cell* **70**:401–410.

Robinson, M. S., 1987, 100-kd coated vesicle proteins: Molecular heterogeneity and intracellular-distribution studied with monoclonal antibodies, *J. Cell Biol.* **104**:887–895.

Robinson, M. S., 1993, Assembly and targeting of adaptin chimeras in transfected cells, *J. Cell Biol.* **123**:67–77.

Robinson, M. S., 1994. The role of clathrin, adapters and dynamin in endocytosis, *Curr. Opin. Cell Biol.* **6**:538–544.

Robinson, M. S., and Kreis, T. E., 1992, Recruitment of coat proteins onto Golgi membranes in intact and permeabilized cells: Effects of brefeldin A and G-protein activators, *Cell* **69**:129–138.

Rodriguez, L., Stirling, C. J., and Woodman, P. G., 1994, Multiple N-ethylmaleimide-sensitive components are required for endosomal vesicle fusion, *Mol. Biol. Cell* **5**:773–783.

Rome, L. H., 1985, Curling receptors, *Trends Biochem. Sci.* **10**:151.

Rothberg, K. G., Heuser, J. E., Donzell, W. C., Ying, Y. S., Glenney, J. R., and Anderson, R. G. W., 1992, Caveolin, a protein-component of caveolae membrane coats, *Cell* **68**:673–682.

Rothenberger, S., Iacopetta, B. J., and Kuhn, L. C., 1987, Endocytosis of the transferrin receptor requires the cytoplasmic domain but not its phosphorylation site, *Cell* **49**:423–431.

Rothman, J. E., 1994, Mechanism of intracellular protein-transport, *Nature* **372**:55–63.

Rothman, J. E., and Schmid, S. L., 1986, Enzymatic recycling of clathrin from coated vesicles, *Cell* **46**:5–9.

Rothman, J. E., and Warren, G., 1994, Implications of the SNARE hypothesis for intracellular membrane topology and dynamics, *Curr. Biol.* **4**:220–233.

Rotrosen, D., and Leto, T. L., 1990, Phosphorylation of neutrophil 47-kDa cytosolic oxidase factor: Translocation to membrane is associated with distinct phosphorylation events, *J. Biol. Chem.* **265**:19910–19915.

Rotrosen, D., Yeung, C. L., and Katkin, J. P., 1993, Production of recombinant cytochrome b_{558} allows reconstitution of the phagocyte NADPH oxidase solely from recombinant proteins, *J. Biol. Chem.* **268**:14256–14260.

Sallusto, F., Cella, M., Danieli, C., and Lanzavecchia, A., 1995, Dendritic cells use macropinocytosis and the mannose receptor to concentrate macromolecules in the major histocompatibility complex class II compartment: Down-regulation by cytokines and bacterial products, *J. Exp. Med.* **182**:389–400.

Samelson, L. E., and Klausner, R. D., 1992, Tyrosine kinases and tyrosine-based activation motifs: Current research on activation via the T-cell antigen receptor, *J. Biol. Chem.* **267**:24913–24916.

Sandvig, K., Olsnes, S., Petersen, O. W., and Vandeurs, B., 1988, Inhibition of endocytosis from coated pits by acidification of the cytosol, *J. Cell. Biochem.* **36**:73–81.

Sarmay, G., Pecht, I., and Gergely, J., 1994, Protein-tyrosine kinase activity tightly associated with human FcγRII receptors, *Proc. Natl. Acad. Sci. USA* **91**:4140–4144.

Schlossman, D. M., Schmid, S. L., Braell, W. A., and Rothman, J. E., 1984, An enzyme that removes clathrin coats: Purification of an uncoating ATPase, *J. Cell Biol.* **99**:723–733.

Schmid, S. L., 1993, Towards a biochemical definition of the endosome compartment. Studies using free flow electrophoresis, in *Subcellular Biochemistry*, Vol. 19 (J. J. M. Bergeron and J. R. Harris, eds.) pp. 1–23 Plenum Press, New York.

Schmid, S. L., Fuchs, R., Male, P., and Mellman, I., 1988, Two distinct subpopulation of endosomes involved in membrane recycling and transport to lysosomes, *Cell* **52**:73–83.

Schmid, S. L., and Rothman, J. E., 1985, Enzymatic dissociation of clathrin cages in a two-stage process, *J. Biol. Chem.* **260**:44–49.

Schmid, S. L., and Smythe, E., 1991, Stage-specific assays for coated pit formation and coated vesicle budding *in vitro*, *J. Cell Biol.* **114**:869–880.

Schnitzer, J. E., 1993, Update on the cellular and molecular basis of capillary permeability, *Trends Cardiovasc. Med.* **3**:124–130.

Schnitzer, J. E., Allard, J., and Oh, P., 1994a, Ablation of caveolae inhibits endothelial endocytosis, transcytosis and capillary permeability, *FASEB J.* **8**:A1057.

Schnitzer, J. E., Oh, P., Pinney, E., and Allard, J., 1994b, Filipin-sensitive caveolae-mediated transport in endothelium: Reduced transcytosis, scavenger endocytosis, and capillary permeability of select macromolecules, *J. Cell Biol.* **127**:1217–1232.

Schurmann, A., Breiner, M., Becker, W., Huppertz, C., Kainulainen, H., Kentrup, H., and Joost, H. G., 1994, Cloning of two novel ADP-ribosylation factor–like proteins and characterization of their differential expression in 3T3-I1 cells, *J. Biol. Chem.* **269**:15683–15688.

Schwartz, A. L., Bolognesi, A., and Fridovitch, S. E., 1984, Recycling of the asialoglycoprotein receptor and the effect of lysosomotropic amines in hepatoma cells, *J. Cell. Biol.* **98**:732–738.

Seaman, M. N. J., Ball, C. L., and Robinson, M. S., 1993, Targeting and mistargeting of plasma-membrane adapters in vitro, *J. Cell Biol.* **123**:1093–1105.

Serafini, T., Orci, L., Amherdt, M., Brunner, M., Kahn, R. A., and Rothman, J. E., 1991, ADP-ribosylation factor is a subunit of the coat of Golgi-derived COP-coated vesicles: A novel role for a GTP-binding protein, *Cell* **67**:239–253.

Shpetner, H. S., and Vallee, R. B., 1989, Identification of dynamin, a novel mechanochemical enzyme that mediates interactions between microtubules, *Cell* **59**:421–432.

Silveira, L. A., Wong, D. H., Masiarz, F. R., and Schekman, R., 1990, Yeast clathrin has a distinctive light chain that is important for cell growth, *J. Cell Biol.* **111**:1437–1449.

Sipe, D. M., and Murphy, R. F., 1991, Binding to cellular receptors results in increased iron release from transferrin at mildly acidic pH, *J. Biol. Chem.* **266**:8002–8007.

Smythe, E., and Warren, G., 1991, The mechanism of receptor-mediated endocytosis, *Eur. J. Biochem.* **202**:689–699.

Smythe, E., Carter, L. L., and Schmid, S. L., 1992, Cytosol-dependent and clathrin-dependent stimulation of endocytosis in vitro by purified adapters, *J. Cell Biol.* **119**:1163–1171.

Smythe, E., Pypaert, M., Lucocq, J., and Warren, G., 1989, Formation of coated vesicles from coated pits in broken A431 cells, *J. Cell Biol.* **108**:843–853.

Sogaard, M., Tani, K., Ye, R. R., Geromanos, S., Tempst, P., Kirchhausen, T., Rothman, J. E., and Sollner, T., 1994, A rab protein is required for the assembly of SNARE complexes in the docking of transport vesicles, *Cell* **78**:937–948.

Sorkin, A., and Carpenter, G., 1993, Interaction of activated EGF receptors with coated pit adaptins, *Science* **261**:612–615.

Sosa, M. A., Schmidt, B., Von Figura, K., and Hillerehfeld, A., 1993, *In vitro* binding of plasma membrane coated vesicle adapters to the cytoplasmic domain of lysosomal acid-phosphatase, *J. Biol. Chem.* **268**:12537–12543.

Stamm, S., Casper, D., Dinsmore, J., Kaufmann, C. A., Brosius, J., and Helfman, D. M., 1992, Clathrin light chain B: gene structure and neuron-specific splicing, *Nucleic Acids Res.* **20**:5097–5103.

Stamnes, M. A., and Rothman, J. E., 1993, The binding of AP1 clathrin adapter particles to Golgi membranes requires ADP-ribosylation factor, a small GTP-binding protein, *Cell* **73**:999–1005.

Stenmark, H., Parton, R. G., Steelemortimer, O., Lutcke, A., Gruenberg, J., and Zerial, M., 1994, Inhibition of rab5 GTPase activity stimulates membrane-fusion in endocytosis, *EMBO J.* **13**:1287–1296.

Stenmark, H., Vitale, G., Ullrich, O., and Zerial, M., 1995, Rabaptin-5 is a direct effector of the small GTPase Rab5 in endocytic membrane fusion, *Cell* **83**:423–432.

Stoorvogel, W., Strous, G. J., Geuze, H. J., Oorschot, V., and Schwartz, A. L., 1991, Late endosomes derive from early endosomes by maturation, *Cell* **65**:417–427.

Subtil, A., Hemar, A., and Dautryvarsat, A., 1994, Rapid endocytosis of interleukin-2 receptors when clathrin-coated pit endocytosis is inhibited, *J. Cell Sci.* **107**:3461–3468.

Swanson, J. A., and Baer, S. C., 1995, Phagocytosis by zippers and triggers, *Trends Cell Biol.* **5**:89–93.

Swanson, J. A., and Watts, C., 1995, Macropinocytosis, *Trends Cell Biol.* **5**:424–428.

Takei, K., McPherson, P. S., Schmid, S. L., and Decamilli, P., 1995, Tubular membrane invaginations coated by dynamin rings are induced by GTPγS in nerve-terminals. *Nature* **374**:186–190.

Taylor, T. C., Kahn, R. A., and Melancon, P., 1992, Two distinct members of the ADP-ribosylation factor family of GTP-binding proteins regulate cell-free intro-Golgi transport, *Cell* **70**:69–79.

Tran, D., Carpentier, J. L., Sawano, F., Gorden, P., and Orci, L., 1987, Ligands internalized through coated or noncoated invaginations follow a common intracellular pathway, *Proc. Natl. Acad. Sci. USA* **84**:7957–7961.

Traub, L. M., Ostrom, J. A., and Kornfeld, S., 1993, Biochemical dissection of AP-1 recruitment onto Golgi membranes, *J. Cell Biol.* **123**:561–73.

Unanue, E. R., Ungewickell, E., and Branton, D., 1981, The binding of clathrin triskelions to membranes from coated vesicles, *Cell* **26**:439–446.

Ungewickell, E., 1983, Biochemical and immunological studies on clathrin light chains and their binding sites on clathrin triskelions, *EMBO J.* **2**:1401–1408.

Ungewickell, E., 1985, The 70-kd mammalian heat-shock proteins are structurally and functionally related to the uncoating protein that releases clathrin triskelia from coated vesicles, *EMBO J.* **4**:3385–3391.

Ungewickell, E., Ungewickell, H., Holstein, S. E., Lindner, R., Prasad, K., Barouch, W., Martin, B., Greene, L. E., and Eisenberg, E., 1995, Role of auxilin in uncoating clathrin-coated vesicles, *Nature* **378**:632–635.

Van der Bliek, A. M., and Meyerowitz, E. M., 1991, Dynamin-like protein encoded by the Drosophila shibire gene associated with vesicular traffic, *Nature* **351**:411–414.

Van der Bliek, A. M., Redelmeier, T. E., Damke, H., Tisdale, E. J., Meyerowitz, E. M., and Schmid, S. L., 1993, Mutations in human dynamin block an intermediate stage in coated vesicle formation, *J. Cell Biol.* **122**:553–563.

Van der Sluijs, P., Hull, M., Webster, P., Male, P., Goud, B., and Mellman, I., 1992, The small GTP-binding protein rab4 controls an early sorting event on the endocytic pathway, *Cell* **70**:729–740.

van Deurs, B., Peterson, O. W., Olsnes, S., and Sandvig, K., 1989, The ways of endocytosis, *Int. Rev. Cytol.* **117**:131–177.

van Deurs, B., Holm, P. K., Sandvig, K., and Hansen, S. H., 1993, Are caveolae involved in endocytosis? *Trends Cell Biol.* **3**:69–72.

van Deurs, B., Holm, P. K., Kayser, L., Sandvig, K., and Hansen, S. H., 1993, Multivesicular bodies in HEp-2 cells are maturing endosomes, *Eur. J. Cell Biol.* **61**:208–224.

Vigers, G. P. A., Crowther, R. A., and Pearse, B. M. F., 1986, Location of the 100 kD–50 kD accessory proteins in clathrin coats, *EMBO J.* **5**:2079–2085.

Virshup, D. M., and Bennett, V., 1988, Clathrin-coated vesicle assembly polypeptides: Physical properties and reconstitution studies with brain membranes, *J. Cell Biol.* **106**:39–50.

Wang, L. H., Rothberg, K. G., and Anderson, R. G. W., 1993, Mis-assembly of clathrin lattices on endosomes reveals a regulatory switch for coated pit formation, *J. Cell Biol.* **123**:1107–1117.

West, M. A., Bretscher, M. S., and Watts, C., 1989, Distinct endocytotic pathways in epidermal growth factor-stimulated human carcinoma A431 cells, *J. Cell Biol.* **109**:2731–2739.

Winkler, F. K., and Stanley, K. K., 1983, Clathrin heavy-chain, light chain interactions, *EMBO J.* **2**:1393–1400.

Wong, D. H., Ignatius, M. J., Parosky, G., Parham, P., Trojanowski, J. Q., and Brodsky, F. M., 1990, Neuron-specific expression of high-molecular-weight clathrin light chain, *J. Neurosci.* **10**:3025–3031.

Yamashiro, D. J., Tycko, B., Fluss, S. R., and Maxfield, F. R., 1984, Segregation of transferrin to a mildly acidic (pH 6.5) para-Golgi compartment in the recycling pathway, *Cell* **37**:789–800.

Yamashiro, D. J., Borden, L. A., and Maxfield, F. R., 1989, Kinetics of α-2-macroglobulin endocytosis and degradation in mutant and wild-type chinese-hamster ovary cells, *J. Cell. Physiol.* **139**:377–382.

Zaremba, S., and Keen, J. H., 1983, Assembly polypeptides from coated vesicles mediate reassembly of unique clathrin coats, *J. Cell Biol.* **97**:1339–1347.

Zaremba, S., and Keen, J. H., 1985, Limited proteolytic digestion of coated vesicle assembly polypeptides abolishes reassembly activity, *J. Cell. Biochem.* **28**:47–58.

Zerial, M., and Stenmark, H., 1993, Rab GTPases in vesicular transport. *Curr. Opin. Cell Biol.* **5**:613–620.

Zijderhand-Bleekemolen, J. E., Schwartz, A. L., Slot, J. W., Srous, G. J., and Geuze, H. J., 1987, Ligand-induced and weak base induced redistribution of asialoglycoprotein receptors in hepatoma cells, *J. Cell Biol.* **104**:1647–1654.

Chapter 4

Autophagy

Glenn E. Mortimore, Giovanni Miotto, Rina Venerando, and Motoni Kadowaki

1. INTRODUCTION

Autophagy and *autophagocytosis* are terms given to a membrane-mediated process in eukaryotic cells in which portions of cytoplasm are sequestered within vacuoles and degraded by acid hydrolases acquired by fusion with lysosomes. Although vacuoles of this type may be formed under pathologic conditions, autophagy is fundamentally a physiological process that plays indispensable roles in cell restructuring and in the ongoing turnover of cellular protein and other macromolecules. Cytoplasmic sequestration is the one step in this pathway that distinguishes autophagy from other degradative processes in the cell. It is relatively nonselective as a volume uptake mechanism (one apparent exception is discussed in Section 3.1.2). Accordingly, most organelles and macromolecules are subject to vacuolar isolation in proportion to their cytoplasmic abundance. Moreover, non-selective uptake permits the simultaneous handling of more than one class of macromolecule. This is illustrated in the

Glenn E. Mortimore   Department of Cellular and Molecular Physiology, Hershey Medical Center, The Pennsylvania State University, Hershey, Pennsylvania 17033.   **Giovanni Miotto and Rina Venerando**   Dipartimento di Chimica Biologica, Università Degli Studi de Padova, 35121 Padova, Italy.   **Motoni Kadowaki**   Department of Applied Biochemistry, Faculty of Agriculture, Niigata University, Niigata, Japan.
Subcellular Biochemistry, Volume 27: Biology of the Lysosome, edited by Lloyd and Mason. Plenum Press, New York, 1996.

Table I

Effects of Amino Acids on the Fractional Turnover of Protein and RNA in Livers of Fed Rats

| | Fractional turnover percent hr^{-1} | | | |
| | Protein | | RNA | |
Amino acids (x)	Total	Minus 10×	Total	Minus 10×
0	4.57 ± 0.13	3.16	3.48 ± 0.23	3.19
0.5	3.25 ± 0.09	1.84	2.24 ± 0.25	1.95
4	1.64 ± 0.10	0.23	0.45 ± 0.06	0.16
10	1.41 ± 0.08	0	0.29 ± 0.04	0

[a]Livers were perfused for 40 min in the single-pass mode with fractions/multiples (x) of a normal plasma amino acid mixture. Rates of protein and RNA degradation were determined as described elsewhere (Pösö *et al.*, 1982b; Lardeux *et al.*, 1987). The columns headed "Minus 10×" represent accelerated turnover, corrected for basal or 10× rates. Results are means ± S.E. of six to eight experiments.

perfused rat liver by the striking similarity in the accelerated responses of protein and RNA degradation during amino acid deprivation (see Table I).

Although the appearance of the vacuoles varies widely among cells, the process is highly conserved and found in nearly all lower plants and animals as well as in higher species. In yeast, for example, a vacuole expressing autophagic function plays a major role in the supply of endogenous amino acids (Takeshige *et al.*, 1992; reviewed by Jones and Murdock, 1994). A similar role for the vacuole is found in germinating seeds (Nishimura and Beevers, 1979) and in the turnover of intracellular proteins in protoplasts of cultured plant cells (Canut *et al.*, 1985). Interesting variants of autophagy are found in the plant rootlet (Marty, 1978) and in mammalian corpus luteum in association with irreversible cell remodeling (Paavola, 1978a,b). Another form of autophagy in the amoeba *Tetrahymena pyriformis* sequesters and degrades intracellular membranes to provide lipid substrate for gluconeogenesis by the glyoxalate pathway (May *et al.*, 1982). In the mammalian hepatocyte, where both protein turnover and the need for endogenous amino acids are large, autophagy is highly expressed and closely regulated by complex amino acid feedback and hormonal mechanisms (reviewed by Mortimore and Pösö, 1987). Taken together, these findings attest to a fundamental role of autophagy in cellular homeostasis.

This chapter covers the main features of general intracellular protein and RNA degradation and the major classes of autophagy, as well as autophagic regulation and its mechanism, focusing primarily on the mammalian hepatocyte, which has been extensively studied as a model for the pathway.

2. GENERAL TURNOVER OF CELLULAR PROTEIN AND RNA

2.1 Protein Degradation

At the present time the only feasible way of determining general rates of protein degradation in isolated cells or tissues is based on amino acid release. Changes in total cellular protein are usually too small for the accuracy needed in most experimental situations, and estimates derived from posttranscriptionally modified amino acids, such as 3-methylhistidine in muscle (Young and Munro, 1978), would be useful only for specific proteins or protein groups. It may be noted, however, that since modified amino acids are not reused for protein synthesis, the latter approach could be employed for estimating breakdown *in vivo,* provided the distribution of the probe is well characterized (Young and Munro, 1978).

Regardless of which method is used in cell experiments, all share two strict requirements: The first is that protein synthesis or, in the case of prelabeled cells, the reutilization by synthesis of newly released labeled amino acids must be effectively suppressed either by inhibiting synthesis with an agent such as cycloheximide (Khairallah and Mortimore, 1976) or by flooding the cells (and precursor sites) with a massive dose of unlabeled amino acid (Mortimore *et al.,* 1972, 1988b). Secondly, the mode of suppression must not perturb the underlying proteolytic process. The choice of amino acid depends on the type of cell studied. Ideally, it should be metabolically stable and widely distributed in proteins. Valine fulfills these requirements in liver and the isolated hepatocyte (Mortimore and Mondon, 1970; Seglen and Solheim, 1978) by its relative abundance, the virtual lack of biosynthesis, and almost negligible oxidization (Mortimore and Mondon, 1970). For similar reasons, tyrosine and phenylalanine have been used in muscle (Rannels *et al.,* 1975; Fulks *et al.,* 1975). Leucine has also been widely employed as a marker, but because it is an effective inhibitor of autophagy (Seglen *et al.,* 1980; Pösö *et al.,* 1982b), its application in experiments involving regulation should be carefully evaluated before use.

2.2. Classes of General Protein Turnover

Despite the fact that the turnover of individual cellular proteins varies widely, only two classes of degradation have been documented under conditions where degradation rates are determined from amino acid release (reviewed in Mortimore and Pösö, 1987). The first is a rapidly turning over or *short-lived* class of proteins. The second is *long-lived* and consists largely of the breakdown of resident cellular proteins.

Experiments that clearly illustrate the difference between these two classes have been carried out with the perfused mouse liver (Hutson and Mortimore, 1982). [^{14}C]Valine that was previously incorporated into liver protein during a

10-min pulse was released in two readily separable components: a rapidly turning over fraction with a half-life of ~10 min and a slow release derived from the degradation of resident proteins. The half-life of the slow component was ≥ 100-fold greater than that of the rapidly released component, and no intermediate fractions were found (Hutson and Mortimore, 1982). Similar results have been reported in cultured cells (Poole and Wibo, 1973; Epstein *et al.*, 1975; Auteri *et al.*, 1983).

2.2.1. Degradation of Short-Lived Proteins

The nature of the short-lived protein pool is not clear. While one cannot exclude the possibility that the group contains some functional proteins, it is probable that most of them consist of nascent polypeptides such as signal peptides or errors in synthesis destined for rapid removal (Wheatley, 1984). Their rapid degradation to acid-soluble oligopeptides (Solheim and Seglen, 1980; Hutson and Mortimore, 1982) is very likely mediated through the ubiquitin-ATP-dependent proteolytic system (Ciechanover *et al.*, 1984) in conjunction with proteosomes (Goldberg, 1992; Rechsteiner *et al.*, 1993).

It is important to point out that although the turnover of the short-lived pool may constitute as much as one third of total protein synthesis in liver (Hutson and Mortimore, 1982; Wheatley, 1984), the probable size of the pool is very small, representing less than 0.3% of total cell protein (G. E. Mortimore, unpublished). Owing to its rapid turnover, the pool can be effectively depleted of label during a 60-min chase (Hutson and Mortimore, 1982). Of relevance to autophagic function (see discussion later in this chapter) is the fact that only the breakdown of long-lived proteins, which constitute more than 99% of total cell protein in liver (Schworer *et al.*, 1981), is physiologically regulated by serum, amino acids, and hormones (Poole and Wibo, 1973; Epstein *et al.*, 1975; Knowles and Ballard, 1976; Vandenburgh and Kaufman, 1980; Hutson and Mortimore, 1982; Auteri *et al.*, 1983); short-lived protein degradation is unaffected by these agents.

2.3. Cytoplasmic RNA Degradation

Lardeux and co-workers have devised a method for measuring cytoplasmic RNA degradation in the perfused rat liver (Lardeux *et al.*, 1987; Lardeux and Mortimore, 1987) and isolated rat hepatocyte (Balavoine *et al.*, 1990) that is similar in principle to the use of labeled valine in proteolytic determinations. The procedure takes advantage of the absence of cytidine deaminase in rat liver, which is normally required in the oxidation of cytidine or cytosine, to measure RNA breakdown from cytidine release. Thus in rat livers labeled with [¹⁴C]orotic acid, [¹⁴C]cytidine will accumulate in cell water and the medium in direct proportion to 3'-CMP release when its reutilization is prevented by flooding with un-labeled cytidine (Lardeux *et al.*, 1987). It has been possible in this way to mea-sure rates of RNA breakdown instantaneously, an important asset in following rapid responses to regulatory agents (Lardeux and Mortimore, 1987; Lardeux *et al.*, 1988).

Virtually the entire unidirectional flux of labeled cytidine is believed to originate from the degradation of structural RNA (rRNA and tRNA) in auto-phagic vacuoles. Any 5'-CMP arising from RNA processing in the nucleus or breakdown of mRNA in the cytosol will not form cytidine but will be rephospho-rylated to CTP and immediately reutilized. Recent studies employing isotope dilution kinetics have shown that 5'-CMP generated in the cytosol closely matches that of mRNA breakdown based on poly A turnover (Heydrick *et al.*, manuscript submitted), supporting the argument that mature mRNA is broken down by extralysosomal mechanisms.

3. MECHANISM OF AUTOPHAGY

Apart from its principal role in sequestering and degrading cytoplasmic macromolecules, autophagy is structurally diverse and difficult to categorize from a mechanistic point of view. As the reader may appreciate from references cited in Section 1, major differences exist between cells in the nature of membranes that form the vacuoles and the manner by which acid hydrolases are acquired. In ad-dition, rates of vacuole formation and their mode of regulation within a given cell may vary depending on the particular needs of the cell or organism. This is clearly illustrated by differences that exist between micro- and macroautophagy in the he-patocyte (see Sections 3.1 and 3.2) and by the presence of two distinct forms of macroautophagy in the involuting corpus luteum of the guinea pig (Paavola, 1978a,b). Rather than attempt simply to list this information, we have chosen to use the mammalian animal cell as a model for integrating autophagic structure and function. Since considerable morphological and biochemical information that is relevant to the process has been obtained from these cells, we hope this choice will be of value to the reader in suggesting how autophagy might fit into an overall scheme of regulation at the cellular level.

The remainder of this chapter focuses largely on the mammalian hepatocyte where protein turnover is especially large, making autophagy comparatively easy to define in morphologic terms. As depicted in Fig. 1, two major classes of au-tophagy, termed macro- and microautophagy, have been recognized. The first is the classic, overt variety, while the second consists of small, less conspicuous vac-uoles (Mortimore *et al.*, 1988a). Although both types occur spontaneously in the normal, untreated cell, the macro form is induced by amino acid deprivation (re-viewed in Mortimore and Pösö, 1987) while microautophagy predominates under

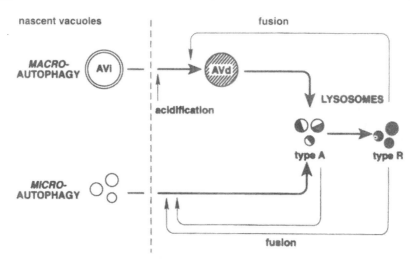

FIGURE 1. Major lysosomal-vacuolar compartments and routes of acid hydrolase flow in macro- and microautophagocytosis in the rat hepatocyte. AVi, autophagosome; AVd, autolysosome (de Duve and Wattiaux, 1966); type A and R secondary lysosomes (Mortimore *et al.,* 1988a). *Note:* For representative electron micrographs (EM) of AVi and AVd, see Schworer *et al.* (1981) and Mortimore and Kadowaki (1994); for EM examples of type A and type R secondary lysosomes, see Mortimore *et al.* (1988a). In the formation of nascent vacuoles (left of the broken line), AVi utilize rough ER (Dunn, 1990a); vesicles that form type A lysosomes are probably derived from smooth ER (Mortimore *et al.,* 1988a). Newly synthesized enzymes enter via Golgi-derived primary lysosomes (Lawrence and Brown, 1992); acidification of AVi occurs before fusion (Dunn, 1990b). The point or points where endosomes join the lysosomal-vacuolar pathway has not been settled and, for this reason, was omitted (see text for discussion).

basal conditions (Mortimore *et al.,* 1988a). Together, these two autophagic systems are capable of explaining virtually all cytoplasmic sequestration in the hepatocyte (Hutson and Mortimore, 1982; Mortimore *et al.,* 1983).

3.1. Macroautophagy

3.1.1. Vacuole Formation

Macroautophagy is an intrinsic, ongoing process that is restrained by amino acids, serum, or other growth-promoting agents in cell media (Ballard and Gunn, 1982). Removal of one or more of these factors will accelerate autophagic responses in the perfused rat liver (Neely *et al.,* 1974; Mortimore and Schworer, 1977; Neely *et al.,* 1977; Ward *et al.,* 1977), perfused rat heart (Jefferson *et al.,* 1974), and cultured cells (Mitchener *et al.,* 1976; Amenta and Brocher, 1981). Vacuoles are first seen as double-walled membranes that envelop and isolate portions of cytoplasm. As observed after stringent amino acid deprivation in liver, for example, their formation is rapid and widely distributed throughout the cytoplasmic space (Mortimore and Schworer, 1977; Schworer et al., 1981; reviewed by Hirsimäki et al., 1983). The results of immunocytochemical studies using antibodies to specific proteins of the rough endoplasmic reticulum (RER) have demonstrated that the smooth-surfaced, double enveloping membranes of the initial or nascent vacuole (termed AVi; see Fig. 1 for definition of terms) are derived from RER denuded of ribosomes (Dunn, 1990a; Yokata, 1993). It is not known whether the vacuoles are formed by a "wrapping" movement of preexisting RER or possibly assembled in situ from smaller membrane units.

The AVi formation is clearly ATP-dependent, although the steps involved have not been defined (Plomp et al., 1989; Kadowaki et al., 1994). It is possible that the requirement is related to ER membrane movement. In this regard it should be mentioned that the microfilament inhibitors cytochalasins B and D have been reported to decrease vacuole formation in cultured rat kidney cells (Aplin et al., 1992). The effect could be cell-specific, since results from an earlier study in Ehrlich tumor cells showed that cytochalasin B did not inhibit AVi formation induced by vinblastin, despite evidence of microfilament disorganization (Hirsimäki and Hirsimäki, 1984). Cycloheximide and other inhibitors of protein synthesis strongly suppress macroautophagy (Khairallah and Mortimore, 1976; Kovács and Seglen, 1981). It should be pointed out, however, that the inhibition is consistently delayed by 15–20 min in the cyclically perfused liver (Khairallah and Mortimore, 1976), an observation that has been used advantageously to determine proteolysis during single-pass liver perfusions (Schworer et al., 1981). It is of interest that 3-methyladenine, which has little effect on protein synthesis, strongly inhibits AVi formation (Seglen and Gordon, 1982) although its mechanism of action is not known.

Evidence has recently been obtained implicating one or more GTP-binding proteins in the initiation of autophagy. Studies by Kadowaki et al. (1994) using the isolated rat hepatocyte permeabilized by α-toxin from Staphylococcus aureus have revealed that AVi formation, monitored by both electron microscopy and the sequestration of ^{125}I-tyramine-cellobiitol in vacuoles separated on colloidal silica density gradients, is strongly inhibited by GTPγS and other nonhydrolyzable GTP analogs. More recent findings by Ogier-Denis et al. (1995) have specifically implicated the heterotrimeric G_{i3} protein in autophagic expression in the human colon cancer cell line HT-29.

Finally, amino acids are the prime inhibitors of macroautophagy in the perfused liver and isolated hepatocyte (Mortimore and Schworer, 1977; Schworer and Mortimore, 1979; Schworer et al., 1981; Kovács et al., 1981; Venerando et al., 1994). Since they selectively inhibit AVi formation, they could act at a site similar to that of GTPγS. It is of added interest that heat stress augments the deprivation response in temperature-sensitive CHO mutant cells (Schwartz et al., 1992). The proposed mechanism of the amino acid inhibition and its hormonal modulation is discussed in Section 4.

3.1.2. Selectivity of Autophagic Sequestration

Because macroautophagic vacuoles are widely distributed in the cell, the content of AVi, based on EM studies, tends to reflect the composition of cytoplasm. In the hepatocyte, for example, rough and smooth ER are abundant while mitochrondria, peroxisomes, glycogen, and free ribosomes occupy smaller fractions (Pfeifer, 1973, 1978; Schworer *et al.,* 1981). Since space between membranes is also sequestered, one may presume that free cytosol is taken up as well. Indeed, this prediction is supported by the nonselective uptake of such cytosolic enzymes as ornithine decarboxylase, tyrosine aminotransferase, and lactate dehydrogenase into autophagic vacuoles observed after proteolytic suppression by the proteinase inhibitor leupeptin (Kominami *et al.,* 1983; Henell and Glaumann, 1984; Kopitz *et al.,* 1990).

One apparent exception to the nonselective uptake by macroautophagy should be noted. It is known that the removal of alanine from 1x amino acid mixtures in the perfused rat liver induces maximal rates of sequestration and proteolysis comparable to those seen with complete amino acid omission (see Section 4). Despite the similarity in the aggregate volume of autophagic vacuoles, vacuolar composition differs strikingly in the two conditions. The deletion of alanine from a 1x mixture, for example, results in a 4.5-fold greater proportion of smooth endoplasmic reticulum (SER) to rough endoplasmic reticulum (RER) and 65% fewer mitochondria and peroxisomes than seen after stringent amino acid omission (Mortimore *et al.,* 1987). Interestingly, the same effect is obtained when macroautophagy is maximally induced by glucagon at 1x amino acid levels (Lardeux and Mortimore, 1987). There is no obvious explanation for this change in composition although past observations may be relevant (Schworer and Mortimore, 1979). An EM study of the contents of vacuoles formed at 0x and 1x plasma levels of amino acids indicates that the latter resemble those induced by alanine deletion and by the addition of glucagon at 1x, except that the amount is less. The possibility exists then that macroautophagic stimulation at 1x is achieved largely through an increase in the existing vacuolar population.

3.1.3. Vacuole Maturation and Kinetics of Turnover

As shown in Fig. 2, the autophagic response to stringent plasma amino acid deprivation in the perfused liver is rapid (Schworer *et al.,* 1981; Dunn, 1990a). Initially, only AVi are seen, but after a lag of 7–8 min degradative forms or AVd appear (see Fig. 1 for definition; Schworer *et al.,* 1981). Typically, AVd are characterized by (a) a single limiting membrane that is somewhat thicker than the outer membranes of AVi, (b) cytochemical evidence of acid hydrolase activity within the vacuoles, and (c) varying degrees of disruption of the vacuolar contents (Mortimore and Schworer, 1977; Schworer *et al.,* 1981; Dunn, 1990a). While AVd obtain most of their acid hydrolases from the fusion of AVi with mature type R

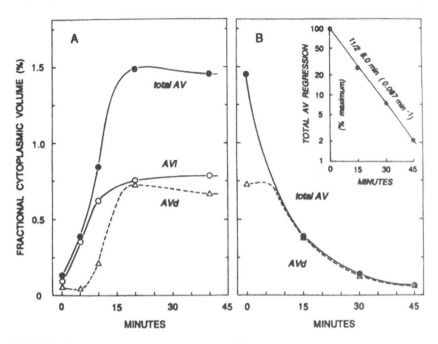

FIGURE 2. Time course of macroautophagic vacuole (AV) formation and regression. (A) Livers from normal fed rats were perfused in parallel without added amino acids and, at intervals, fixed for electron microscopy. Volume densities of AVi, AVd, and total AV were determined stereologically (modified from Schworer et al., 1981). (B) Livers were perfused in parallel for 20 min without amino acids; at time zero 10× plasma amino acids were added. Volume densities were determined as in A. (Inset) Semilogarithmic plot of the regression of total AV volume. (Modified from Schworer et al., 1981, with permission of the publisher.)

lysosomes (Fig. 1; Lawrence and Brown, 1992), the transformation of AVi to AVd does not involve just a single event, but rather multiple steps with acidification appearing before lysosomal fusion (Dunn, 1990b).

The possibility that fluid-phase endosomes fuse with AVi before lysosomal fusion was raised by Gordon and Seglen (1988) and Gordon et al. (1992) from indirect biochemical evidence in the electropermeabilized hepatocyte. Although such a pathway appears to exist in the pancreatic exocrine cell (Tooze et al., 1990) and fibroblast (Punnonen et al., 1993), the question remains controversial in liver. Dunn (1990a), for example, failed to observe a transfer of plasma membrane markers to AVi. But more unsettling evidence was reported by Lawrence and Brown (1992). In a detailed study of fluid-phase flow in cultured hepatocytes using colloidal gold markers, these investigators observed that AVi fuse rapidly with preexisting lysosomes, but rarely with prelysosomal vacuoles. Cell differences certainly occur (Punnonen et al., 1993), but the opposing views on the hepatocyte question have more likely arisen from differences in technical approaches. While

Gordon *et al.* (1992) are probably correct in concluding that AVi fuse with the prelysosomal compartment, the morphometric findings of Lawrence and Brown (1992) and autophagic uptake measurements of Kadowaki *et al.* (1994) show that flow through the pathway is small relative to autophagic fluxes.

The lysosomal/vacuolar circuit is completed with the digestion of sequestered cytoplasm and reappearance of mature (type R) lysosomes as end products (see definition in Fig. 1). With ongoing autophagic vacuole formation, a large part of the acid hydrolase activity acquired by AVi arises from the recycling of pre-existing lysosomes (Lawrence and Brown, 1992). At the same time, newly processed enzymes may enter the lysosomal/vacuolar pathway via primary lysosomes ultimately derived from Golgi vesicles.

While the formation of lysosomal end products was thought to be a straight-forward process of intravacuolar digestion and disposal of excess membrane, new findings indicate that the ubiquitin system could play an important role in autophagic/lysosomal function (Gropper *et al.*, 1991; Lenk *et al.*, 1992; Schwartz *et al.*, 1992; Löw *et al.*, 1993; see Chapter 5). Of particular relevance to autophagy are studies with temperature-sensitive CHO mutant cells that restrict ubiquitin conjugate formation at the initial step (E1) when the temperature is elevated above permissive levels (Gropper *et al.*, 1991). The results showed that the conversion of AVd to lysosomes is inhibited at elevated temperatures, although AVi formation and their maturation to AVd are not affected (Lenk *et al.*, 1992; Schwartz *et al.*, 1992). It is of interest that the effects are limited to macroautophagy; basal protein turnover (microautophagy) is apparently not involved (Gropper *et al.*, 1991).

Based on the time course of autophagic vacuole formation in the perfused liver (Fig. 2), cytoplasmic sequestration is an ongoing process in which a steady state between the formation of AVi and their maturation to AVd and lysosomes is attained within 20 min (Schworer *et al.*, 1981). In the same experiments, the addition of 10x amino acids after maximal levels of macroautophagy were attained immediately stopped new AVi formation. The existing population of AVi was rapidly transformed to AVd, and these regressed exponentially with a half-life of 8 min (Fig. 2). Half-life measurements from other studies, where vacuolar half-life in the hepatocyte was determined from the regression of autophagic volume and sensitivity to osmotic shock, have averaged ~8 min (Neely *et al.*, 1974; Pfeifer, 1978; Papadopoulos and Pfeifer, 1986; Kadowaki *et al.*, 1994). Similar half-lives have been obtained in mouse pancreatic acinar and seminal vesicle cells (Kovács *et al.*, 1987).

3.1.4. Correlation Between Macroautophagy and Deprivation-Induced Degradation of Protein and RNA

In livers perfused in the single-pass (nonrecirculating) mode with graded levels of plasma amino acids, the aggregate volumes of AVi and AVd, determined stereologically from electron micrographs, decreased in direct proportion to

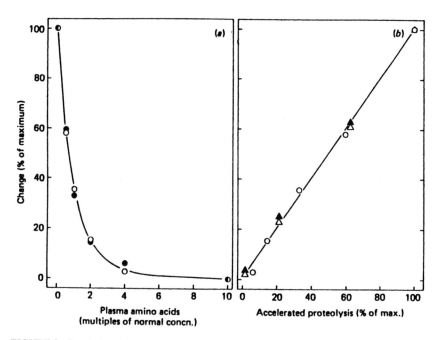

FIGURE 3. Correlations between proteolysis, aggregate volumes of AVi and AVd, and the shift of β-hexosaminidase from lysosomes to AVi during deprivation-induced macroautophagy in perfused livers of fed rats. (a) Relationship between accelerated long-lived proteolysis (●) and the above shift of β-hexosaminidase (O) at various levels of plasma amino acids. The shift, which was determined in lysosomal-vacuolar fractions separated in colloidal silica gradients, is a measure of lysosomal-AVi fusion. (b) Correlation of AVi (△), AVd (▲), and the β-hexosaminidase shift (O) *versus* rates of deprivation-induced proteolysis. (From Surmacz *et al.,* 1987, with permission of the publisher.)

decreases in rates of degradation of long-lived proteins over the full range of macroautophagy (Schworer *et al.,* 1981). The results are depicted in Fig. 3, together with measurements of lysosomal fusion based on the proportion of β-hexosaminidase in colloidal silica density gradients that had shifted from lysosomes to macroautophagic vacuoles as a result of fusion (Surmacz *et al.,* 1987). It is evident that deprivation-induced changes in lysosomal fusion as well as in the aggregates volume of AVi and AVd are directly proportional to induced rates of proteolysis.

The direct relationship between fusion and AVi induction in Fig. 3 suggests that the proportion of type R lysosomes that fuse with vacuoles is strictly maintained despite wide differences in vacuole size and shape (Fig. 1; Surmacz *et al.,* 1987). Because the size of AVi vacuoles rises sharply with increasing deprivation (Mortimore and Schworer, 1977; Schworer *et al.,* 1981), it is probable that the direct correlation shown in Fig. 3 would not exist without a mechanism that regulates the number of vacuolar fusions according to their individual volumes

(Surmacz *et al.*, 1987). In other words, the fusion dose (roughly equivalent to the number of fusion "hits") for each AVi vacuole must remain constant per unit of vacuolar volume. How this is achieved is not known.

Since autophagy induced by plasma amino acid deprivation is relatively non-selective, one would expect comparable increases above basal turnover in the fractional rates of protein and cytoplasmic RNA degradation. Results in Table I show that this is so. The increases, which are directly attributable to macroautophagy, were the same within experimental error over the full range of deprivation. Because cytoplasmic RNA breakdown, unlike that of protein, lacks a short-lived component, these results also support the tacit assumption that the proteolytic measurement reflects only the breakdown of long-lived (resident) proteins (see Section 2). However, one exception to the correspondence between turnover of RNA and long-lived protein must be mentioned. As was described earlier in Section 3.1.2, the specific deletion of alanine from a normal plasma amino acid mixture or the coadministration of glucagon along with a normal mixture evokes an autophagic response that selectively sequesters SER over RER (Mortimore *et al.*, 1987; Lardeux and Mortimore, 1987). Measurements of RNA degradation appeared to reflect this differential effect in failing to show an enhancement of RNA breakdown by glucagon at normal plasma concentrations of amino acids (Lardeux and Mortimore, 1987).

Although the accelerated fractional rates of RNA and protein breakdown were similar (Table I), basal rates differed strikingly. With RNA, the value was approximately 20% of that found for protein. The reason for the difference is not entirely understood, but it is probable that membrane-bound ribosomes, which make up at least two-thirds of rRNA, are not sequestered in the basal state. On the other hand, free ribosomes have been seen frequently in microautophagic vacuoles under basal conditions (Mortimore *et al.*, 1983; 1988a). In addition, because basal RNA turnover is decreased by short-term starvation in parallel with that of protein, it is likely that both are degraded in the same vacuolar compartment (Mortimore *et al.*, manuscript in preparation).

3.1.5. Intralysosomal Pools of Degradable Protein and RNA

Apart from vacuolar volumes and their turnover, the quantity of substrate undergoing hydrolysis and its relationship to rates of degradation in the intact cell represent additional elements in the development of a unified mechanism of autophagic function. These elements were first explored in experiments in which the content of degradable protein within lysosomal particles, including autophagic vacuoles, was correlated with rates of degradation of long-lived proteins (Mortimore and Ward, 1981). Protein pool size was ascertained from the total release of valine in isotonic liver homogenates at pH7.0, during incubations lasting 90–120 min. All but 4–5% of the released valine was derived from pro-

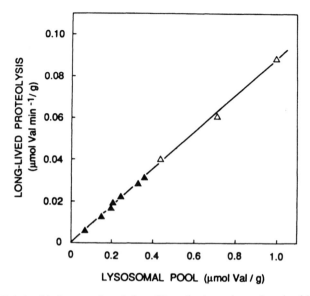

FIGURE 4. Relationship between degradation of long-lived proteins and pools of intralysosomal degradable protein in rat liver over the full range of protein turnover. The quantity of valine residues in the protein pools, expressed per gram of liver, from livers in various accelerated (open triangles) and basal (filled triangles) states of protein degradation was correlated with corresponding rates of proteolysis monitored from valine release. The above regression equation was $y = -0.0002 + 0.0882x$ ($r = 0.999$), based on data from three studies (Mortimore and Ward, 1981; Hutson and Mortimore, 1982; Mortimore *et al.*, 1988a), all corrected for an intralysosomal pH of 4.5 as described by Mortimore and Pösö (1984).

teins previously internalized within lysosomes, and none could be attributed to interactions between cytosolic proteins and lysosomes during the incubations; lysosomal latency was maintained over the course of the experiments. The quantity of sequestered protein was determined from the total quantity of free valine released, corrected for an intralysosomal pH of 4.5 as described by Mortimore and Pösö (1984) and an average valine content of rat liver protein of 0.591 μmol/mg (Schworer *et al.*, 1981).

Degradable intralysosomal protein in three studies, expressed as μmol valine released per g liver (Mortimore and Ward, 1981; Hutson and Mortimore, 1982; Mortimore *et al.*, 1988a), was found to correlate directly with breakdown of long-lived proteins ($r = 0.999$) over the full range of turnover (Fig. 4). The regression exhibits three important features. First, the slope of 0.088 min^{-1} ($t_{1/2} = 7.9$ min) is nearly identical to the apparent rate constant of macroautophagic turnover shown in Fig. 2, thus independently confirming the initial estimate. Second, inasmuch as the released valine is derived mostly from two major classes of autophagic vacuoles, macro and micro, it follows that the slope must be about the same in each class. This is demonstrated by the fact that the lower seven points depict basal conditions in which macroautophagic vacuoles

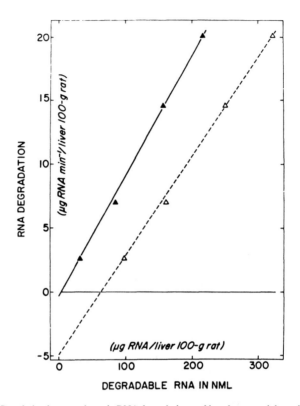

FIGURE 5. Correlation between hepatic RNA degradation and intralysosomal degradable RNA. Livers from nonstarved rats, previously labeled with [^{14}C]orotic acid, were perfused with graded levels of plasma amino acids over the full range of RNA turnover (compare with Fig. 4). RNA breakdown in liver and liver lysosomal particles was determined from [^{14}C]cytidine release as detailed in Heydrick *et al.* (1991). The broken line displays total homogenate RNA breakdown (lysosome + cytosol) while the solid line gives total minus cytosolic breakdown rates. (From Heydrick *et al.,* 1991, with permission of the publisher.)

are absent (Fig. 2; Hutson and Mortimore, 1982; Mortimore *et al.,* 1988a). It is probable that the foregoing rate constant of vacuole turnover is a general feature of mammalian autophagy that is independent of the size and type of vacuole, a point also raised by Papadopoulos and Pfeifer (1986). Third, the absence of a significant zero intercept suggests that, within the limits of experimental error, degradation of long-lived proteins in the adult hepatocyte is a function of autophagic sequestration. A similar direct correlation has been demonstrated by Heydrick *et al.* (1991) between RNA breakdown in the perfused rat liver and lysosomal pools of degradable RNA (Fig. 5).

The third point is not inconsistent with the coexistence of important nonautophagic degradative mechanisms. It should be emphasized that their significance is not related to their impact on protein turnover generally, but rather to their effect on specific proteins (frequently regulatory) whose pool sizes may be small in relation to the total quantity of resident cellular proteins.

3.2. Microautophagy

Owing to the large variation in size of autophagic vacuoles, de Duve and Wattiaux (1966) coined the term *microautophagy* to advance the notion that the cytoplasmic "bite" could extend below the limit of macroautophagy into the molecular range. In the hepatocyte, the term has been used to denote the process of sequestration in the basal state, the existence of which is shown by the lower seven points in Fig. 4 and the lowest point in Fig. 5. It is probable that most cells possess similar mechanisms although most of the evidence has come from the liver parenchymal cell. By the above criterion of de Duve and Wattiaux, the lysosomal uptake of specific proteins should fall into this category. Such a mechanism does in fact exist. Dice and co-workers (reviewed by Dice and Terlecky, 1994; see Chapter 5) have demonstrated that cytosolic proteins containing sequences related to Lys-Phe-Glu-Arg-Gln (KFERQ) are selectively taken up. The uptake is saturable and ATP-dependent and appears to require the association of a 73-kDa heat-shock protein. Although it has been categorized as autophagic in nature, the uptake actually bears a closer relationship to mitochondrial protein import and differs sharply from autophagic processes is not involving vacuole formation. Thus it would not contribute to the pools of degradable protein and RNA in hepatic lysosomes that constitute sequestered substrate (Figs. 4 and 5).

The possibility that cytosolic proteins are sequestered by lysosomes under basal conditions was first recognized from the breakdown of cellular protein internalized in isolated liver lysosomes (Mortimore *et al.*, 1973); similar findings were obtained by Ahlberg *et al.* (1985). The results of later studies with the starved–refed mouse pointed to the possible involvement of a secondary lysosome, termed type A (for autophagic), in basal sequestration (Hutson and Mortimore, 1982; Mortimore *et al.*, 1983). A direct quantitative correlation between microautophagic vacuoles and degradation of long-lived proteins was established in the perfused rat liver in which proteolytic interference from macroautophagy was eliminated by amino acid suppression (Mortimore *et al.*, 1988a). In this study, basal proteolytic rates decreased > 50% during 48 hr of starvation in association with a fall in the volume density of type A lysosomes. By contrast, a few lysosomes ($\sim$10% volume fraction) comprising various types of membrane invagination (Marzella *et al.*, 1980; Pfeifer, 1981; de Waal *et al.*, 1986) remained constant.

A close quantitative correlation was found between the basal degradation of long-lived proteins and the combined volumes of all microautophagic elements (90% type A lysosomes + 10% various invagination forms) during 48 hours of starvation (Fig. 6; see also Fig. 1; Mortimore *et al.*, 1988a). It is

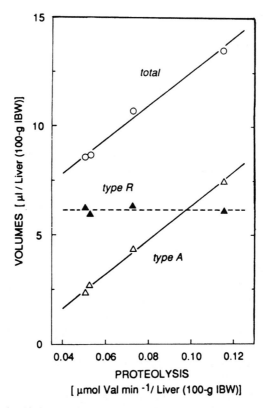

FIGURE 6. Relationship between basal protein turnover and volume densities of type A and type R lysosomes during short-term starvation in rat liver. Livers from fed and 48 hr starved rats were perfused with 10× plasma amino acids for the measurement of basal hydrolysis of long-lived protein and stereological analysis of secondary lysosomes (Mortimore *et al.,* 1988a). Under- and overestimation of type A and type R elements, respectively, were corrected by multiplying the observed type A volumes in Mortimore *et al.* (1988a) by 1.285, which equalized its slope to that of the total (see discussion in text); R was then determined by subtracting type A from the total. Type A lysosomes were also assumed to be underestimated ~2% because of random sections through electron-lucent zones; this correction increased the total ~1% but did not affect type R. The type A volumes shown include closed invagination forms illustrated in Mortimore *et al.* (1988a).

important to point out that the sharply demarcated, electron-lucent and electron-dense zones that characterize type A lysosomes will unavoidably generate some morphometric error. For example, a section through a dense zone, which missed an electron-lucent area, will be viewed as a type R lysosome. Thus for reasons of particle geometry alone, type A lysosomes will be consistently underestimated (Mortimore *et al.,* 1988a). In practice though, the underestimation has proved to be predictable and was corrected (in Fig. 6). The correction was validated and is shown in Table II.

Table II

Prediction of Hydrolysis of Long-Lived Proteins from Cytoplasmic Protein Concentration and the Volume Flux through Autophagic Vacuoles (AV): A Unified Compartmental Mechanism of Autophagy (AV)

AV class/AV flux method	AV volume µl A	AV flux µl/hr B	Protein concentration nmol Val/µl Cytoplasm C	Protein concentration nmol Val/µl AV D/B	Long-lived protein degradation mol Val/hr Predicted [BC]	Long-lived protein degradation mol Val/hr Observed D
Perfused livers						
Macro AV/ EM stereology[a]	16.802	88.71 (3.27)[f]	171.6	170.4	15.22	15.12 (3.25)[f]
Micro AV/ EM stereology[b]	7.444	39.30 (1.40)[f]	177.5	175.6	6.98	6.90 (1.37)[f]
Micro AV/ EM stereology[b]	3.128	16.52 (1.00)[f]	209.2	213.1	3.46	3.52 (1.01)[f]
α-Toxin permeabilized hepatocytes						
Total AV/ ITC uptake[c]	—	12.18 (1.88)[f]	163.7	164.2	2.00	2.00 (1.89)[f]
Macro AV/ ITC uptake[c,d]	—	6.60 (1.02)	163.7	164.4	1.08	1.09 (1.03)
Macro AV/ EM stereology[e]	1.262	6.66 (1.03)	163.7	161.2	1.09	1.07 (1.01)

[a]A was determined from the maximal deprivation-induced increase in fractional cytoplasmic volumes of AVd in perfused rat livers (Schworer and Mortimore, 1979), multiplied by the total cytoplasmic space of 2.71 ml per 100-g rat as calculated in Mortimore *et al.* (1979); B is the product of A and the first-order rate constant of AV turnover 0.088 min^{-1}; C was computed from the valine content of liver protein, 465 µmol per 100-g rat, divided by the cytoplasmic volume; D is the maximal deprivation-induced rate of proteolysis per 100-g rat (Schworer *et al.*, 1979). Note: Data from Schworer *et al.* (1981) were not used in this summary as AVd volumes were available only in the graphs.

[b]Values corresponding to those above were taken from rat liver perfusion studies of Mortimore *et al.* (1988a) with type A fractional cytoplasmic volumes corrected for underestimation as detailed in Fig. 6 and the text; values from the three 48-hr starved rats were averaged. Valine in liver protein was computed from the following proportion: 0.591 µmol Val per mg protein (Schworer *et al.*, 1981); the same AV rate constant (0.088 min^{-1}) was used, and all hepatic volume and rates were expressed per 100-g initial body weight.

[c]Data were obtained from experiments with isolated rat hepatocytes permeabilized with *S. aureus* α-toxin (Kadowaki *et al.*, 1994); B was determined from the total AV uptake of ^{125}I-tyramine-cellobiitol (ITC). The cytoplasmic volume and valine content of protein were 646.5 µl and 105.8 µmol; except for concentration, all values are expressed per 10^8 cells.

[d]B represents the decrease in ^{125}I-tyramine-cellobiitol uptake over total uptake elicited by a maximal dose of GTPγS; the effect parallels the inhibition of macroautophagy shown in the following row.

[e]The experiments were carried out as explained in Note c except that the macroautophagic volumes in B were determined stereologically from electron micrographs (EM).

[f]The numbers in parentheses are fractional rates of the corresponding absolute values, expressed as percentages of the respective total cytoplasmic volumes and quantities of valine in liver protein given above.

How type A lysosomes are formed in not fully understood. While a few may be end products of macroautophagy, most are generated *de novo* since they persist in the presence of amino acid concentrations high enough to inhibit macroautophagy by 95–97% (Schworer and Mortimore, 1979; Schworer *et al.*, 1981; Mortimore *et al.*, 1988a). Their close association with the SER and the observation that they frequently contain glycogen (Amherdt *et al.*, 1974) suggest that they could arise from the fusion of dense bodies and small vesicles derived from the SER (Mortimore *et al.*, 1988a). If so, distinctive early vacuoles containing newly sequestered material might be difficult to identify, although they are readily differentiated from AVi by the presence of only a single delimiting membrane. Novikoff and Shin (1978) have suggested that membranes of the SER are in constant motion and could isolate bits of cytoplasm. Such a notion is attractive because it fits with the above theory and could explain the decline in autophagic activity and proteolysis during starvation; ER is rapidly lost early in starvation (Pfeifer, 1973), and microsequestration could decrease in parallel with the loss.

3.3. A Unified Compartmental Mechanism of Autophagic Proteolysis

Macro- and microqutophagy are volume uptake processes that have the same apparent rate constants as AV turnover. Hence, if sequestration is the rate-determining step in proteolysis, it should be possible to compute rates of degradation of long-lived proteins for the two autophagic processes as a product of (1) the AV rate constant, (2) the aggregate volume of the degradative vacuoles, and (3) the concentration of cytoplasmic protein sequestered within them. The product of (1) and (2) is termed the *autophagic flux,* and in the case of macroauthphagy it can be computed from the aggregate volume of AVd and the observed turnover constant of 0.087-0.088 min^{-1} (see Fig. 2 and Section 3.1.5). The corresponding value for microautophagy is less certain, as it is necessary to assume that the proteolytic compartment is limited to type A lysosomes augmented by a few invaginated forms (see Section 3.2). The assumption is reasonable, though, since volumes based on type A, but not type R (shown in Fig. 6), correlate directly with hydrolysis of long-lived proteins during short-term starvation.

An evaluation of the apparent concentration of protein at degradative sites emerged unexpectedly from studies in the isolated rat hepatocyte permeabilized with α-toxin from *Staphylococcus aureus* (Kadowaki *et al.*, 1994). Kadowaki *et al.* (1994) showed that the above-mentioned autophagic flux can be determined from the steady-state uptake of externally added ^{125}I-iodotyramine-cellobiitol into macro- and microautophagic vacuoles separated from other organelles in colloidal silica gradients (Surmacz *et al.*, 1983a,b). No lysosomotropic uptake of iodotyramine-cellobiitol into lysosomes was observed in control experiments (M. Kadowaki and G. E. Mortimore, unpublished results). Of interest was the finding that the ratio of proteolytic rate to autophagic flux, which provides an indirect measure of the concentration of sequestered protein within autophagic vacuoles, was higher than average intracellular values but equivalent to the apparent cytoplasmic protein concentration computed from total cell protein divided by the cytoplasmic volume (measured stereologically). As summarized in Table II, remarkably close agreement between the ratio and the computed cytoplasmic protein concentration was found in all previous studies for which complete data were available. Moreover, equally close agreement was obtained between predicted and observed rates of proteolysis.

Several important conclusions may be drawn from these findings. The close agreements between predicted and observed cytoplasmic protein concentrations and proteolytic rates (Table II) obtained through two independent measurements of autophagic flux, validate previous assessments of flux based on stereologic measurements alone. The first was computed as the product of the stereologic AV volumes and the rate constant of vacuole turnover ($\sim 0.088^{-1}$); the second represented the direct vacuolar uptake of iodotyramine-cellobiitol (Kadowaki et al., 1994). These agreements also confirm the method used to correct the underestimation of type A lysosomes (Fig. 6). It can readily be seen from Fig. 6 that the correlations between predicted and observed volumes of total and type A lysosomes are optimal only when constancy of type R volume is assumed over 48 hr of starvation; any upward or downward deviation in the slope of type R during the period would adversely affect them.

Why the apparent protein concentrations in AVd are higher than in AVi and the intracellular space is not entirely clear, although it has been attributed to vacuolar acidification and the associated net loss of K^+, Na^+, and water (Schworer et al., 1981). Since the aggregate volume of AVd is $\sim 18\%$ less than that of AVi (Schworer et al., 1981), the volume loss is enough to explain the increase in protein concentration. It should be emphasized that these cellular adjustments in AVd volume and protein level result in almost exact correspondence between the fractional turnover of cytoplasm by autophagic sequestration and the turnover of total cell protein (Table II). Considering the uniformity of the kinetics of sequestration and vacuolar turnover that has been observed over the full range of deprivation (Figs. 3, 4 and 5 and Table II), it is reasonable to extend conclusions based on macroautophagy to microautophagy as well. From this standpoint, the two forms appear to be indistinguishable, differing mainly in the source of membrane for sequestration and in their mode of regulation.

4. REGULATION OF AUTOPHAGY

4.1 Amino Acid Control of Macroautophagy

Amino acids are prime regulators of hepatic macroautophagy and, as shown earlier in Fig. 2a, can control AVi formation over its full range without hormonal assistance (Mortimore and Schworer, 1977; Schworer et al., 1981; Surmacz et al., 1987). Since structural RNA and long-lived proteins are both degraded by this pathway, their responses are similar (Schworer *et al.,* 1981; Lardeux and Mortimore, 1987). Although the experiments in Fig. 6 were carried out with multiples (×) of a complete plasma amino acid mixture, only a few amino acids are required for regulation, the number and kind varying among the different cells and tissues. In skeletal and cardiac muscle, for example, only leucine is active (Buse and Reid, 1975; Fulks *et al.,* 1975; Chua *et al.,* 1979), while proteolytic inhibition in liver (Woodside and Mortimore, 1972; Seglen *et al.,* 1980, Pösö *et al.,* 1982b) and cultured kidney cells (Rabkin *et al.,* 1991) requires the concerted action of more than one amino acid.

4.1.1. Regulatory and Nonregulatory Amino Acids

It is now generally agreed that at least eight amino acids (Leu, Gln, Tyr, Phe, Pro, Met, His, and Trp) contribute to the direct suppression of autophagic proteolysis in the adult rat hepatocyte (Woodside and Mortimore, 1972; Hopgood *et al.,* 1980; Seglen *et al.,* 1980; Sommercorn and Swick, 1981; Pösö *et al.,* 1982b). Inhibition has been observed with asparagine alone (Seglen *et al.,* 1980) although it is not effective as a direct suppressor of vacuole formation (Grinde and Seglen, 1981). Whether or not all eight amino acids are "true" inhibitors is not easily determined. In many instances, individual effects are too small to measure or are obscured by differences between test systems. In other cases, high doses of amino acids evoke effects that cannot be observed at physiological plasma concentrations.

In elucidating regulatory mechanisms, a strategy that we have found useful has been to determine the smallest number of regulatory amino acids that can evoke maximal inhibition of autophagic proteolysis (Pösö *et al.,* 1982b). In perfused livers of synchronously fed rats, for example, a combination of leucine and glutamine consistently evokes maximal inhibition (Mortimore *et al.,* 1991), while in livers from rats fed *ad libitum,* leucine and glutamine, together with tyrosine or phenylalanine produce comparable effects (Kadowaki *et al.,* 1992). Distinct differences between regulatory effects of tyrosine and phenylalanine are discussed in Section 4.1.4. Although the original number of regulatory amino acids may have been overestimated, it is unlikely that an important regulator was overlooked inasmuch as the remaining 12 complementary amino acids are totally ineffective as a group at plasma levels as high as 10× normal concentrations (Pösö *et al.,* 1982a).

4.1.2. Multiphasic Control by Regulatory Amino Acids: L and H Modes

Although proteolytic dose responses to leucine in skeletal and cardiac muscle (Tischler *et al.,* 1982) and cultured kidney cells (Rabkin *et al.,* 1991) reveal a single phase of inhibition, responses to regulatory amino acids in the perfused rat liver (Mortimore *et al.,* 1987) as well as in the isolated rat hepatocyte (Venerando *et al.,* 1994) are more complex and are mediated by two alternate, amino acid concentration-dependent mechanisms. One, termed L mode, conveys direct inhibition at low and high physiological levels of regulatory amino acids but specifically requires the coaddition of alanine (see Section 4.1.3) for the expression of inhibition at normal plasma levels. The second, or H mode, expresses no inhibition at normal levels and below, although it is fully effective at higher physiologic levels; alanine exhibits no coregulatory activity in this mode. Detailed experiments with synchronously fed rats have shown that the two modes evolve in a predetermined manner during each feeding cycle (Mortimore *et al.*, 1991): H appears first after food intake; shortly thereafter, H spontaneously begins to alternate randomly with L; with omission of the next feeding, H disappears and L becomes a constant feature during starvation.

4.1.3. Multiphasic Control: L and H Modes and Coregulation by Alanine

The L mode of amino acid regulation represents the norm in rats fed *ad libitum*. But, as observed by Venerando *et al.* (1994), there are exceptions to this pattern of expression, in which alternating L and H modes may spontaneously appear during *ad libitum* feeding (Fig. 7). It is possible that the manner of L/H expression is affected seasonally and by the timing of food intake in addition to environmental factors. Typical multiphasic (L mode) proteolytic dose responses to leucine, glutamine, and tyrosine and the complete regulatory mixture (except Phe) are depicted in Fig. 8 for the perfused rat liver; similar responses to leucine are shown in panel B of Fig. 7 for the isolated rat hepatocyte. It is important to emphasize that the regulatory amino acids (except Phe), individually and as a group, inhibit proteolysis at the same low (0.5×) and high (4×) plasma amino acid levels and express the same zonal peak (1×) of accelerated proteolysis. The latter is directly related to the aforementioned *coregulatory* action of alanine and can be completely abolished by the addition of 0.5 mM (1×) alanine (Fig. 7B, Pösö and Mortimore, 1984; Mortimore *et al.*, 1991). Alanine is not a direct inhibitor and is ineffective in cells expressing the H mode (Fig. 7A).

The coregulatory effect is highly specific for alanine. Although 10- to 20-fold higher quantities of pyruvate or lactate, or equimolar additions of glutamate or aspartate can substitute for alanine (probably through their conversion to alanine), alanine is the only amino acid capable of increasing proteolysis to near-maximal rates when it is deleted from a 1× normal plasma mixture (Pösö and Mortimore, 1984; Mortimore *et al.*, 1991). Coregulation by alanine is not affected when its transamination is blocked by the aminotransferase inhibitor aminooxyacetate at a dose that virtually abolishes its conversion to glucose (Mortimore *et al.*, 1991). This indicates that the site of alanine's action lies upstream from its transamination, possibly at or near plasma membrane loci that have been proposed for amino acid recognition (Miotto *et al.*, 1992, 1994).

Meijer and co-workers have shown, using rat hepatocytes perfused in the single-pass mode (termed perifusion), that leucine acts synergistically with alanine, glutamate, and aspartate at high physiological concentrations (Leverve *et al.*, 1987; Caro *et al.*, 1989). This interesting effect resembles alanine coregulation but differs

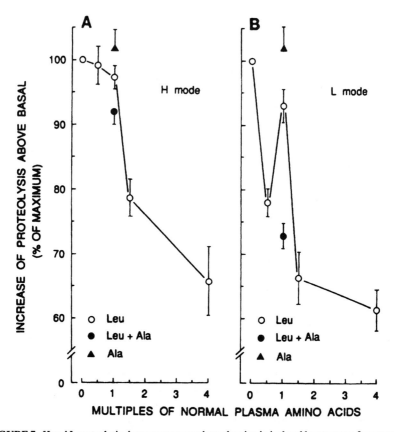

FIGURE 7. H and L proteolytic dose-response modes to leucine in isolated hepatocytes from rats fed *ad libitum.* The responses to leucine and alanine shown above were carried out in 13 experiments, each on a different day. Responses to leucine (1× Leu + 1× Ala) on different days spontaneously fell into one of two randomly alternating modes, termed H and L. (From Venerando *et al.,* 1994, with permission of the publisher.)

from it in exhibiting broader amino acid concentration and specificity requirements and by the fact that it is observed only in starvation (Pösö and Mortimore, 1984; Mortimore *et al.,* 1991; Venerando *et al.,* 1994). Whether the cells used in this study possessed coregulatory effectiveness as well as synergism was not reported.

4.1.4. Divergent Regulation by Tyrosine and Phenylalanine

The fact that the points of inflection in the dose–response curves to leucine, glutamine, and tyrosine individually and the regulatory amino acids as a group are virtually identical when concentrations are related to their respective normal plasma values (Fig. 8) suggests that they act through similar mechanisms (Pösö *et al.,* 1982b; Mortimore *et al.,* 1987). Phenylalanine, however, is a notable exception. In contrast to the effects of tyrosine (Fig. 8), phenylalanine inhibits proteolysis sharply to near-basal values at concentrations between 0.5× and 1× and thus there is no point of inflection in the dose-response curve for this amino acid (Kadowaki *et al.*, 1992). Maximal effects of the two aromatic amino acids are the same (Kadowaki *et al.*, 1992). These findings, especially the extraordinary sensitivity to low levels of phenylalanine and lack of multiphasic features, indicate that phenylalanine inhibits autophagy and hydrolysis of long-lived proteins by a mechanism different from that of tyrosine. Why this anomaly exists is not known. Its sensitivity would probably prevent it from enhancing autophagy during starvation. It might play a evolutionarily conserved role, along with the rapid conversion of phenylalanine to tyrosine (Shiman *et al.*, 1982), in protecting adult or fetal cells against excessive levels of phenylalanine and phenylpyruvate.

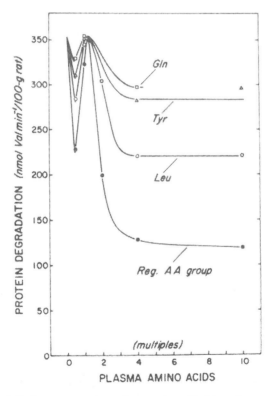

FIGURE 8. Proteolytic dose responses to regulatory amino acids. Livers from normal fed rats were perfused in the single-pass mode with various regulatory (Reg) amino acids at multiples of their concentrations in a 1× reference plasma mixture. The 1× Reg mixture is composed of the following (μM): Leu, 204; Gln, 716; Tyr, 98; Pro, 437; Met, 60; His, 92; Trp, 93 (Mortimore *et al.*, 1991; Venerando *et al.*, 1994). (From Mortimore *et al.*, 1987, with permission of the publisher.)

4.1.5. Concerted Multiphasic Control by Regulatory Amino Acids

The similarity of dose responses to leucine, glutamine, and tyrosine shown in Fig. 8 provides grounds for supposing that the regulatory amino acids act in a concerted manner at sites close to each other. Indeed, in livers from rats fed *ad libitum*, inhibitory effects of leucine, glutamine, and tyrosine are fully additive (Kadowaki *et al.*, 1992) and, when given as a group, yield a combined effect equal to the sum of the individual effects; in synchronously fed rats (Fig. 9), a combination of only leucine and glutamine will evoke multiphasic responses equivalent to the complete regulatory group (Mortimore *et al.*, 1991; Miotto *et al.*, 1992).

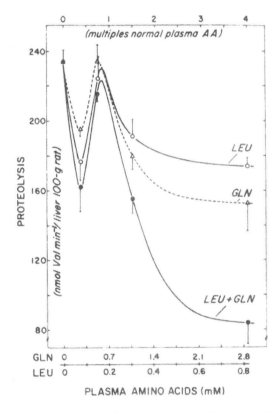

FIGURE 9. Proteolytic dose responses to leucine, glutamine, and leucine + glutamine. The experiments were carried out as in Fig. 8 except that the livers were obtained from synchronously fed, 24 hr starved rats. The molar values for leucine and glutamine on the abscissa (bottom) correspond to fractions/multiples of their concentrations in the reference plasma mixture (top) given in Fig. 8. The values are means ± S.E. of 3 to 33 experiments, normalized to 100 g of initial body weight. (From Miotto *et al.*, 1992, with permission of the publisher.)

As described in Section 4.1.3, alanine serves as a coregulator for the expression of autophagic/proteolytic inhibition by regulatory amino acids at normal plasma levels. In its presence, the zonal peak is lost and responses are transformed to continuous inhibitory functions that are proportional over the range of regulation (Miotto *et al.*, 1992). These findings suggest that recognition sites for leucine and glutamine are directly accessible to plasma amino acids and that proteolytic inhibition is determined in some way by concentration-dependent reactions between the sites and the regulatory amino acids.

Support for this contention was obtained by the use of conventional Michaelis–Menten kinetics in plotting the relationship between proteolytic inhibition (V) and fraction/multiples of normal amino acid concentrations (S). In Fig. 10, linear V/S *versus* V plots were obtained from which values of about $0.5\times$ were computed for the *relative* apparent K_m values of the plasma amino acid mixture and of leucine (+ alanine). In molar terms this indicates that the apparent K_m values for regulatory amino acids are one half of their normal plasma concentrations. Thus the agreement in *relative* apparent K_m values between leucine and the complete amino acid mixture is probably not fortuitous but an essential feature of a complex regulatory mechanism that responds in a concerted way to more than one amino acid.

4.1.6. Site of Regulation by Leucine

Leucine's inhibitory effectiveness is widely distributed among cells, perhaps more so than any other amino acid, and it is the dominant amino acid regulator in liver (Seglen *et al.*, 1980; Pösö *et al.*, 1982b). Past studies have suggested that leucine mediates its proteolytic inhibition in muscle through metabolic products of oxidation (Tischler *et al.*, 1982; Mitch and Clark, 1984). This notion, however, was never seriously considered in the hepatocyte because of the cell's extremely low transaminating activity for branched-chain amino acids (Pösö *et al.*, 1982b). This opened the possibility that the site of leucine recognition lies upstream from its locus of transamination. Impetus for this notion was fueled by the finding that the hydroxyl analog of L-leucine, L-α-hydroxyisocaproate, closely mimics leucine's multiphasic dose response (Mortimore *et al.*, 1987). While this finding currently excludes leucyl-tRNA as a mediator (Tischler *et al.*, 1982), it does not rule out a site of recognition at or near the plasma membrane.

The discovery by Miotto *et al.* (1989) that isovaleryl-L-carnitine (IVC) also mimics leucine's action in perfused rat liver prompted inquiry into the structural nature of leucine recognition (Miotto *et al.*, 1992). Bearing structural similarities to leucine, IVC was found to be as inhibitory as leucine at the same concentrations and to interact with alanine in the expression of coregulation (Miotto *et al.*, 1992). The biological effectiveness of IVC, which is stereospecific (Miotto *et al.*, 1992), was not achieved with either isovalerate or L-carnitine alone (Miotto *et al.*, 1989). From an examination of other leucine derivatives, the following requirements were proposed

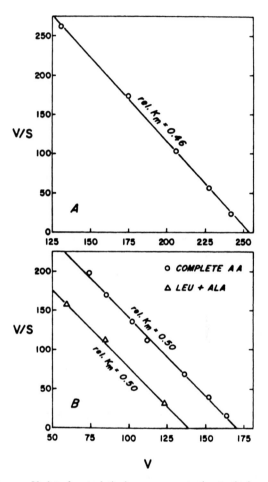

FIGURE 10. V/S *versus* V plot of proteolytic dose responses to the standard complete amino acid mixture and to leucine + alanine. V represents proteolytic inhibition, expressed as nmol valine min^{-1} per liver (100 g of initial body weight); S denotes fractions/multiples of amino acids in the medium. In this plot V_{max} is conveniently shown as the V-intercept. The apparent K_m values are also relative (rel.) in that they are based on fractions/multiples of the molar values in the standard plasma mixture. (A) Responses of livers from normal fed rats to the complete amino acid mixture. (B) Responses of livers from synchronously fed, 24 hr starved rats to the same mixture and to leucine + alanine. (From Miotto *et al.*, 1992, with permission of the publisher.)

(Miotto *et al.*, 1992). First, a side chain closely matching the isovaleryl moiety is needed. Second, the presence of a group, at or near the α-carbon that participates in hydrogen bonding or can be protonated appears to be required. This second requirement was proposed since α-keto and α-chloro derivatives, which have little or no ability to form hydrogen bonds, are ineffective at plasma concentrations lower than leucine's normal value (Miotto *et al.*, 1989; 1992), while derivatives with potentially reactive substituents such as an OH, NH_2, or a quaternary amine (as in IVC) are as biologically active as leucine (Mortimore *et al.*, 1987; Miotto *et al.*, 1992). Finally, there is no specific requirement for a carboxyl group (Wert *et al.*, 1992).

With regard to the cellular locus of recognition, accumulating evidence now points to the plasma membrane. Much of it is based on the strong correlation between autophagic responses and external concentrations where little relationship to internal concentrations may exist. As illustrated in Figs. 8 and 9, the identical, sharp inflections of the dose-response curves for leucine, glutamine, and tyrosine correspond closely with their respective normal plasma concentrations. By contrast, intracellular glutamine is known to remain comparatively stable despite wide plasma changes. A second example brings the question into clearer focus. Since IVC is rapidly metabolized in the hepatocyte, its intracellular levels equilibrate at about 10% of external values (Miotto *et al.*, 1992). By contrast, leucine is rapidly transported by system L and is metabolically stable, and its internal and external concentrations are approximately equal. Since leucine and IVC share a common recognition site, the plasma membrane must be considered the likely location (Miotto *et al.*, 1992).

This possibility was substantiated by using the multiple antigen peptide derivative Leu_8-Lys_4-Lys_2-Lys-βAla (Leu_8-MAP) to determine whether leucine can elicit autophagic inhibition from a site on the cell exterior (Miotto *et al.*, 1994). Synthesis of the derivative was prompted by the above-mentioned finding that the carboxyl of leucine is not specifically required for biological activity (Wert *et al.*, 1992). Thus the attachment of leucine residues through their carboxyl groups to a branched molecule of high molecular weight such as MAP (Tam, 1989) might yield a nontransportable structure that retains regulatory activity. On a molar basis (independent of the number of leucine residues) Leu_8-MAP proved to be metabolically stable and as effective as leucine in suppressing macroautophagy and proteolysis; moreover, it exhibited the same apparent K_m, about 0.1 mM. The intracellular:extracellular distribution ratio of [³H]Leu_8-MAP was 100:1 or greater, indicating that plasma membrane transport of Leu_8-MAP into the cytosolic compartment was insignificant. Because cytosolic levels of Leu_8-MAP were at least 100-fold smaller than those of leucine giving equal inhibition, these results offered the first compelling evidence that the inhibition of autophagy from leucine recognition could not be explained by signals arising from within the cell.

This, of course, leaves the cell surface as the logical alternative. Indeed, the results of recent photoaffinity experiments have suggested that specific leucine binding is demonstrable in hepatocyte plasma membrane fractions (Mortimore *et al.*, 1994). A ~340,000 M_r protein complex has been demonstrated in membrane fractions obtained from intact rat hepatocytes after UV irradiation with photoreactive Leu-MAP (Fig. 11). The latter derivative, which was biologically active, was synthesized from Leu_8-MAP by replacing one of its leucine residues with an ¹²⁵I-labeled azido group. The 340,000 M_r protein was readily reduced by

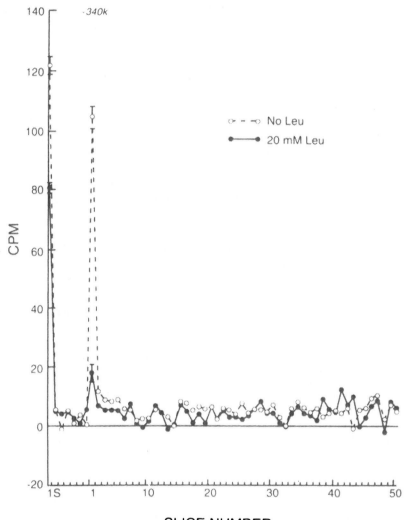

SLICE NUMBER

FIGURE 11. Nonreducing polyacrylamide gel electrophoresis showing a photolabeled protein of M_r ~340,000 that had entered a 7:5–20% gradient gel. Note the strong protection afforded by 20 mM leucine in parallel runs. Fresh rat hepatocytes were photolyzed with (^{125}I-ASA)Leu$_7$-MAP ± 20 mM leucine and the plasma membrane fraction extracted with 2% SDS for electrophoresis. No photolabeling was obtained with similar photoprobes constructed with valine or isoleucine. (From Mortimore *et al.*, 1994, with permission of the publisher.)

dithiothreitol to smaller fractions, possibly subunits (not shown). Unlabeled leucine (20 mM) exerted a strong protective effect on binding of the photoligand to the 340k protein (Fig. 11) and the smaller fractions, creating maximal signal-to-noise ratios in the vicinity of 6:1.

The binding (^{125}I-ASA)Leu$_7$-MAP to the protein was saturable, specific for leucine, and had an apparent affinity of 10^{-4} M, equaling the K_m of leucine for autophagic inhibition (Fig. 10). The possibility that the site of binding was the system L transporter can be excluded by the lack of photoreactivity with other branched-chain amino acids. Similar MAP probes, for example, synthesized with isoleucine or valine in place of leucine, failed to bind (Mortimore et al., 1994). An answer to the question of whether the M_r 340k protein is actually the mediator of leucine's effect on macroautophagy must await its isolation and characterization.

4.2. Amino Acids, Cell Swelling, and Proteolytic Regulation

The concentration of protein in most cells is relatively stable despite alterations in total cell protein. Because this relationship prevails widely among cells, the notion that protein and volume are closely regulated by a common factor or factors has long been considered. Insulin, for example, increases cell volume and inhibits protein degradation in muscle and liver cells. Autophagy is established as the major pathway for hydrolysis of long-lived proteins in the hepatocyte (see Section 4.3.1), so the question of a link between autophagy and cell swelling arises.

In the last five years a number of investigators have examined the possible relationship between liver proteolysis and induced alterations in cell volume (Häussinger et al., 1990; Hallbrucker et al., 1991; Meijer et al., 1993). In experiments with the isolated rat liver previously labeled with [^{3}H]leucine and perfused in the nonrecirculating mode, Häussinger and co-workers have shown that glutamine, glycine, serine, alanine, and some other amino acids associated with Na$^+$ transport elicit changes in labeled leucine release and that these changes decrease in proportion to gain in liver weight. It should be noted that glycine and serine, which are active in the cell volume phenomenon, are noninhibitory amino acids in the perfused rat liver (Pösö et al., 1982a,b) and isolated rat hepatocyte (Seglen et al., 1980), while leucine, the dominant inhibitor of proteolysis, does not cause cell swelling. Although Häussinger and coworkers did not conclude that the autophagic pathway was responsible for the decrease in proteolysis, they did suggest that it could be involved in some instances. However, no assessment of autophagy by EM or other means was made. Moreover, their proteolytic measurements, based on [^{3}H]leucine release, very likely underestimated actual rates of protein degradation, as the amount of unlabeled leucine (0.1 mM) added to prevent label reutilization by protein synthesis was probably too small, about 50% of leucine's normal plasma level. Thus to some degree, the changes in released leucine in these studies might have reflected effects on protein synthesis.

Although cell volume changes appear to be relevant to many important aspects of cellular regulation (Häussinger *et al.*, 1994), it is difficult to imagine how amino acid-induced swelling could explain glutamine's multiphasic dose response (Figs. 8 and 9). For example, the sharp change in direction from proteolytic acceleration to inhibition as plasma glutamine moves from half-normal to above-normal concentrations is not consistent with effects arising from transport-mediated uptake. The problem is made even more complex by the fact that the proteolytic effects of glutamine are additive to those of leucine (Fig. 9). Because the transport of leucine is not Na^+-dependent and is unrelated to changes in cell volume, one may conclude that the concerted effects of glutamine and leucine are not manifestations of cell swelling, but are more likely mediated by signaling from specific cell surface recognition site(s).

4.3.   Hormonal Regulation of Autophagy

4.3.1.   Insulin

Growth promoting hormones or factors in mammalian cells, such as insulin and fetal calf serum, consistently inhibit proteolysis (reviewed by Ballard and Gunn, 1982; Mortimore and Pösö, 1987) and, in many cases, autophagy (Jefferson *et al.*, 1974; Neeley *et al.*, 1977; Mortimore *et al.*, 1987). Yet, despite the fact that the importance of factors such as insulin is well established, the molecular details of its action are unclear. In experiments with cultured kidney cells, Tsao *et al.* (1990) tested the possibility that the observed autophagic and proteolytic inhibition by insulin was the result of Na^+-H^+ antiporter stimulation. Although enhancement of this activity is generally associated with cytoplasmic growth, no stimulation was found.

Since amino acids alone are capable of regulating macroautophagy over its full range, it is instructive to evaluate responses to insulin in relation to the underlying amino acid dose response. The results of one such study are shown in Fig. 12, where inhibitory effects of insulin on multiphasic dose responses to regulatory amino acids were obtained from rat livers perfused in the single-pass mode (Mortimore *et al.*, 1987). Two distinct effects on degradation of long-lived proteins were seen. The first is dependent on plasma amino acid concentrations and results in complete loss of the zonal peak of inhibitory effectiveness at $1\times$ plasma levels. It thus mimics the coregulatory effect of alanine. In a possibly related function, insulin can switch the basic mode of regulation from H to L (Mortimore *et al.*, 1991). The second effect of insulin shown in Fig. 12 is independent of amino acid concentration and is about the same in magnitude of $0\times$ and $10\times$ amino acid levels. A comparable response to insulin was observed with cytoplasmic RNA breakdown (Lardeux and Mortimore, 1987). These effects on protein and RNA degradation can be obtained in the virtual absence of macroautophagy (e.g., at $10\times$ plasma amino acids) and are thus presumed to re-

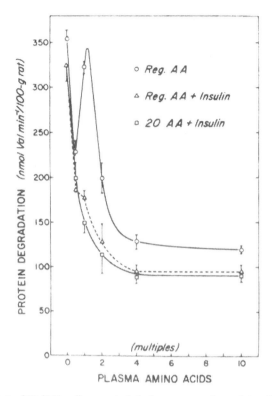

FIGURE 12. Effects of 10^{-9} M insulin on proteolytic dose responses to regulatory (Reg AA) and complete plasma amino acid mixtures. In other respects the experiments were the same as those in Fig. 8. (From Mortimore *et al.*, 1987, with permission of the publisher.)

flect decreases in the rate of microautophagic sequestration. Although small in comparison with the first effect, which represents a near-maximal macroautophagic response, it is sufficient in size to account for approximately one half of the net loss of resident proteins in liver during starvation (Mortimore *et al.*, 1983). Since microautophagy rather than macroautophagy is the dominant sequestrational process in basal turnover, the former would play a more important role in cellular regulation under conditions closer to the physiological steady state than is apparent from the effects shown in Fig. 12.

4.3.2.   Glucagon and β-Agonists

Following Ashford and Porter's initial report (1962), glucagon has been widely employed as an inducer of hepatic macroautophagy (Deter *et al.*, 1967; Arstila and Trump, 1968; Rosa, 1971; Schworer and Mortimore, 1979; Bleiberg-Daniel *et al.*, 1994) and degradation of long-lived proteins (Woodside *et al.*, 1974;

Schworer and Mortimore, 1979; Mortimore *et al.*, 1987; Lardeux and Mortimore, 1987; Bleiberg-Daniel *et al.*, 1994). Although responses to glucagon in the perfused liver are similar to those observed after stringent amino acid deletion (Schworer and Mortimore, 1979), effects in the isolated rat hepatocyte are comparatively small in magnitude (Hopgood *et al.*, 1980; Poli *et al.*, 1981). The reason for this difference is not readily apparent. Results in the perfused liver indicate that glucagon acts by blocking the inhibitory effectiveness of amino acids at low concentrations (~0.5×) thereby increasing macroautophagy and proteolysis maximally at normal levels (Fig. 13). Suppression, though, is ultimately achieved at higher concentrations through a second mechanism that overrides the glucagon stimulation. The resulting dose response resembles the H mode seen in livers of synchronously fed rats perfused 18 hr after food intake (Mortimore *et al.*, 1991).

Microautophagy, in contrast to its macro counterpart, is a vacuolar process that has been regarded as unresponsive to physiological regulation except for moderate decreases after short-term starvation (Mortimore *et al.*, 1988a) and exposure

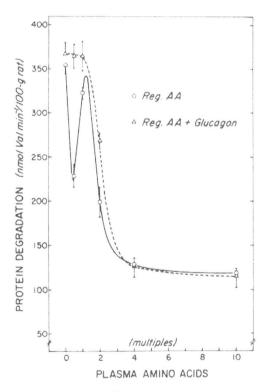

FIGURE 13. Effect of 8×10^{-9} M glucagon on the proteolytic dose response to regulatory (Reg AA) amino acids. The conditions otherwise were the same as those in Fig. 8. (From Mortimore *et al.*, 1987, with permission of the publisher.)

to insulin (Fig. 12); it is not inhibited by amino acids (Mortimore *et al.,* 1988a). Surprisingly, though, we have recently found (G. E. Mortimore, M. Kadowaki, K. K. Khurana, and J. J. Wert, Jr., manuscript in preparation) that infusing glucagon or cAMP to livers of synchronously fed rats perfused with 10× plasma amino acids 18 hr after last food intake will induce the appearance of large numbers of unique, single-walled vacuoles and particles reminiscent of type A lysosomes (Mortimore *et al.,* 1988a) and glycogenosomes, glycogen-filled lysosomes seen in livers of streptozotocin-induced diabetic rats (Amherdt *et al.,* 1974). In these experiments, AVi were strongly induced by amino acid depletion and nearly abolished by 10× plasma amino acids both in the presence and absence of glucagon enabling us to exclude macroautophagy as an active participant in this autophagic induction at 10× plasma concentrations. Based on proteolytic measurements, the magnitude of glucagon stimulation in 10× plasma amino acids was comparable to the normal deprivation response and was blocked by the lysosomotropic inhibitor chloroquine; stimulation by glucagon of a similar magnitude was observed in the absence of amino acids. Both effects disappeared after short-term starvation.

Both the above induction of microautophagic-like elements under basal conditions and the previously documented macroautophagic response to glucagon described in connection with Fig. 13 appear to be mediated by cAMP. With regard to the latter, effects have been elicited with epinephrine (Rosa, 1971; Woodside *et al.,* 1974; Hopgood *et al.,* 1980) and with cAMP alone (Rosa, 1971; Hopgood *et al.,* 1980). The α-adrenergic pathway does not seem to be involved, since vasopressin and phenylephrine fail to stimulate proteolysis in rat liver perfusion experiments (G. E. Mortimore, unpublished results). By contrast, in muscle, cAMP and β-agonists suppress proteolysis. In rat heart, for example, glucagon and isoproterenol inhibit macroautophagy and proteolysis (Chua *et al.,* 1978; Dämmrich and Pfeifer, 1981); similar results have been reported in skeletal muscle (Garber *et al.,* 1976; Li and Jefferson, 1977).

How cAMP enhances autophagy is not well understood although it is clear that protein synthesis is not required (Woodside *et al.,* 1974; Hopgood *et al.,* 1980) and it is not mediated by the depletion of glucogenic amino acids as was once thought (Schworer and Mortimore, 1979). Current evidence implicates two mechanisms. The first is related to the observation that cAMP switches the mode of amino acid control from L to H, possibly at plasma membrane sites of amino acid recognition. The second derives from the finding that glucagon and cAMP can stimulate autophagy independently of amino acids in livers of 18 hr synchronously fed rats (G. E. Mortimore, M. Kadowaki, K. K. Khurana, and J. J. Wert, Jr., manuscript in preparation). Both mechanisms are potentially important because they demonstrate that micro- and macroautophagy are subject to direct hormonal control. In addition, they point to the existence of phosphorylation/dephosphorylation mechanisms that very likely operate through the basic amino acid control pathway. In this connection, it is of interest that Blommaart *et al.* (1995) have recently found that amino acids increase the phosphorylation state of the 31-kDa ribosomal protein S6. The extent to which insulin is capable of counterbalancing the stimulation of autophagy by cAMP has not been determined although there is little doubt that it is a potential player in the control of proteolysis at normal ($1\times$) plasma amino acid concentrations (Fig. 12).

Finally, it should be mentioned that because of the relatively slow changes in plasma amino acids, it is probable that most of the active, moment-to-moment physiological regulation of macroautophagy is achieved hormonally (e.g., insulin and glucagon), rather than by direct amino acid inhibition. In this sense, intracellular signaling generated from amino acid recognition at plasma membrane sites would mediate the primary control which in turn would be modulated through superimposed effects of the hormones.

4.4. Control of Macro- and Microautophagy during Starvation and Refeeding

Virtually nothing is known of how microautophagy, or macoratophagy for that matter, is regulated *in vivo* or how the two forms of autophagy interrelate in complex cellular functions such as cytoplasmic involution and growth. Both are ongoing sequestrational functions that are controlled by amino acids and hormones in distinctly different ways. For example, findings from rat liver perfusion studies have shown that plasma amino acids regulate macroautophagy on a moment-to-moment basis over a wide range without appreciably affecting microautophagy (Mortimore and Schworer, 1977; Schworer *et al.*, 1981; Mortimore *et al.*, 1988a), while short-term starvation can bring about significant decreases in microautophagy (Mortimore *et al.*, 1988a). Both types of autophagy are modulated by insulin and glucagon (see Figs. 12 and 13; Sections 4.3.1 and 4.3.2).

The induction of cytoplasmic growth in mouse liver by starvation and refeeding is associated with sharp decreases in rates of protein breakdown to levels that could only be explained by the suppression of both macro- and microautophagy (Conde and Scornik, 1976; Hutson and Mortimore, 1982). Using stereologic techniques and an approach similar to that described in Table II, Mortimore *et al.* (1983) determined the aggregate volumes of the two types of autophagy in livers of fed, 48 hr starved, and starved-refed mice in an attempt to assess their relative contributions to total protein degradation. Volumes were expressed in absolute terms per liver of a standard body weight and thus were independent of variations in liver weight. The results revealed no change in macroautophagy during 48 hr of starvation although it decreased 93% after 12 hr of refeeding. By contrast, microautophagy (type A secondary lysosomes) fell 53% during starvation and decreased even further during refeeding.

Predicted rates of degradation of long-lived proteins in this study were computed for the course of starvation and refeeding (Mortimore *et al.*, 1983) by an approach similar to that shown in Table II. The predictions were based on (1) measurements of cytoplasmic protein, (2) the relevant vacuolar volumes, and (3) the first-order rate constant of autophagic turnover. The values were then compared with observed rates computed from differences between net rates of loss or gain of liver protein and rates of resident protein synthesis in parallel experiments (Hutson and Mortimore, 1982; Mortimore *et al.,* 1983). In accord with the findings and conclusions reported in Table II, close agreement between predicted and observed rates was found at all points in the course of starvation–refeeding.

What emerged unexpectedly from this study (Mortimore *et al.* 1983) was the pronounced decline in microautophagy that occurred during starvation in the absence of any fall in macroautophagy. Although the lack of change in macroautophagy agrees with earlier results of Pfeifer (1973) and the decline in microautophagy matches recent evidence of our own (Mortimore *et al.,* 1988a), we still have no clear explanation for the microautophagic decrease. Since amino acid levels were probably normal or low and the hormonal milieu favorable to an acceleration of autophagy, one can only speculate on a possible cause. As noted earlier, Novikoff and Shin (1978) have called attention to the probable role of the smooth endoplasmic reticulum (SER) in isolating bits of cytoplasm, and we have encountered type A lysosomes most frequently in areas filled with dilated SER. Because the quantity of SER per cell falls appreciably during starvation (Cardell, 1977), one might expect to see a progressive decline in microsequestration during starvation if type A nascent vacuoles are in fact derived from the SER.

The almost total suppression of macroautophagic activity that was found during refeeding (Mortimore *et al.,* 1983) can be attributed to the rise of plasma amino acids during absorption and the almost certain increase in the ratio of insulin to glucagon in plasma. Although type A lysosomes were also decreased, the effect was moderate in degree. Hence, most of the very small pool of sequestered, degradable protein found within lysosomal elements under these conditions (Hutson and Mortimore, 1982) would probably exist within micro- rather than macroautophagic vacuoles (Mortimore *et al.,* 1983).

5. OVERVIEW

Autophagy is a highly conserved, energy-dependent vacuolar process in plant and animal cells that sequesters and degrades organelles and other macromolecular constituents of cytoplasm for cellular restructuring and as a source of free amino acids for metabolic use in early starvation. The sequestering vacuoles, which are typically formed from endoplasmic reticulum, are converted to digestive vacuoles after acidification an fusion with lysosomes. As a volume uptake process, autophagy is able to capture most classes of macromolecules in a relatively nonselective manner and, hence, to balance the overall synthesis of structural elements in a way that does not disturb their cytoplasmic proportions.

Owing to its inherently high rate of protein turnover, the mammalian hepatocyte has proved useful as a model for correlating degradation of long-lived proteins with autophagy. Two distinct classes of vacuoles in this cell have been noted: (1) an overt, double-walled form (macroautophagy) that is induced by deprivation and strongly suppressed by amino acids; and (2) a less conspicuous, single-walled type (microautophagy) that is not regulated by amino acids and accounts for basal protein turnover. Both processes sequester protein and RNA in intralysosomal pools whose sizes are directly proportional to rates of RNA and protein breakdown in the intact cell. From an analysis of these correlations and a comparison of vacuole turnover by quantitative electron microscopy and by the direct uptake of ^{125}I-tyramine-cellobiitol in permeabilized cells, we conclude that both classes of vacuoles consistently turn over with a first-order rate constant of ~0.888 min^{-1} and that most cytoplasmic proteins and RNA are candidates for autophagic uptake.

Vacuole formation and the sequestration of cytoplasm together act as the rate-determining step in macroautophagy and are rapidly regulated by the concerted actions of seven regulatory amino acids (Leu, Gln, Tyr, Pro, Met, His, Trp) and the coregulator alanine. Inhibitory responses to complete mixtures and to regulatory amino acids, singly or in combination, yield apparent K_m values that are 0.5 times their normal concentration (1×) in plasma. In the fed state, alanine alone is not inhibitory but is specifically required for expression of inhibition by the regulatory amino acids at 1× concentrations. Hence, the omission of alanine leads to a complex *multiphasic* dose response that features a nearly total loss of inhibition between 0.5× and 2×. In the isolated liver these amino acid responses are strongly modulated by insulin, glucagon, and β-agonists. Insulin acts like alanine in permitting the full expression of autophagic inhibition by amino acids at normal concentrations, whereas glucagon blocks amino acid inhibition at normal amino acid levels and lower. Thus glucagon evokes a maximal macroautophagic effect at 1× while insulin produces the opposite effect. Since these hormonal responses are amino acid-dependent, it is likely that they are achieved through interactions of the hormones with amino acid recognition proteins.

The notion that the locus of amino acid recognition is at the plasma membrane was prompted from mounting indirect evidence suggesting that the complex responses to regulatory amino acids are predictable from external but not internal amino acid pools. The hypothesis was confirmed by use of Leu$_8$-MAP, a derivative that retains leucine's effectiveness as a macroautophagic inhibitor but fails to enter the cytosolic compartment. The conclusion that the recognition site for leucine is at the plasma membrane was given strong additional support by the recent demonstration of the specific binding of a photoreactive derivative of Leu-MAP to a high-molecular-weight protein on the plasma membrane having an affinity matching that of leucine. The isolation and characterization of the binding protein or proteins could help to unlock the complex molecular mechanisms mediating the amino acid and hormonal control of autophagy.

6. REFERENCES

Ahlberg, J., Berkenstam, A., Henell, F., and Glaumann, H., 1985, Degradation of short- and long-lived proteins in isolated rat liver lysosomes. Effects of pH, temperature, and proteolytic inhibitors, *J. Biol. Chem.* **260:**5847–5854.

Amenta, J. S., and Brocher, S. C., 1981, Mechanisms of protein turnover in cultured cells, *Life Sci.* **28:**1195–1208.

Amherdt, J. S., Harris, V., Renold, A. E., Orci, L., and Unger, G. H., 1974, Hepatic autophagy in uncontrolled experimental diabetes and its relationship to insulin and glucagon, *J. Clin. Invest.* **54:**188–193.

Aplin, A., Jasionowski, T., Tuttle, D. L., Lenk, S. E., and Dunn, W. A., Jr., 1992, Cytoskeletal elements are required for the formation and maturation of autophagic vacuoles, *J. Cell. Physiol.* **152:** 458–466.

Arstila, A. U., and Trump, B. F., 1968, Studies on autophagocytosis: The formation of autophagic vacuoles in the liver following glucagon administration, *Am. J. Pathol.* **53:**687–733.

Ashford, T. P., and Porter, K. R., 1962, Cytoplasmic components in hepatic cell lysosomes, *J. Cell Biol.* **12:**198–202.

Auteri, J. S., Okada, A., Bochaki, V., and Dice, J. F., 1983, Regulation of intracellular protein degradation in IMR-90 human diploid fibroblasts, *J. Cell Physiol.* **115:**167–174.

Balavoine, S., Feldmann, G., and Lardeux, B., 1990, Rates of RNA degradation in isolated rat hepatocytes. Effects of amino acids and inhibitors of lysosomal function, *Eur. J. Biochem.* **189:**617–623.

Ballard, F. J., and Gunn, J. M., 1982, Nutritional and hormonal effect on intracellular protein catabolism, *Nutr. Rev.* **40:**33–42.

Bleiberg-Daniel, F., Lamri, Y., Feldmann, G., and Lardeux, B., 1994, Glucagon administration *in vivo* stimulates hepatic RNA and protein breakdown in fed and fasted rats, *Biochem. J.* **299:**645–649.

Blommaart, E. F. C., Luiken, J. J., Blommaart, P. J., Van Woerkom, G. M., and Meijer, A. J., 1995, Phosphorylation of ribosomal protein S6 is inhibitory for autophagy in isolated rat hepatocytes, *J. Biol. Chem.* **270:**2320–2326.

Buse, M. G., and Reid, S. S., 1975, Leucine: A possible regulator of protein turnover in muscle, *J. Clin. Invest.* **56:**1250–1261.

Canut, H., Alibert, G., and Boudet, A. M., 1985, Hydrolysis of intracellular proteins in vacuoles isolated from *Acer Pseudoplatanus L.* cells, *Plant Physiol.* **79:**1090–1093.

Cardell Jr., R. R., 1977, Smooth endoplasmic reticulum in rat hepatocytes during glycogen deposition and depletion, *Int. Rev. Cytol.* **48:**221–279.

Caro, L. H. P., Plomp, P. J. A. M., Leverve, X. M., and Meijer, A. J., 1989, A combination of intracellular leucine with glutamate or aspartate inhibits autophagic proteolysis in isolated rat hepatocytes, *Eur. J. Biochem.* **181:**717–720.

Chua, B. H. L., Kao, R. L., Rannels, D. E., and Morgan, H. E., 1978, Effects of epinephrine and glucagon on protein turnover in perfused rat heart, *Fed. Proc. Fed. Am. Soc. Exp. Biol.* **37:**1333A.

Chua, B. H. L., Watkins, C. A., Siehl, D. L., and Morgan, H. E., 1978, Inhibition of protein degradation by anoxia and ischemia in perfused rat hearts, *J. Biol.Chem.* **254:**6617–6623.

Ciechanover, A., Finley, D., and Varshavsky, A., 1984, The ubiquitin-mediated proteolytic pathway and mechanisms of energy-dependent intracellular protein degradation, *J. Cell. Biochem.* **24:**27–53.

Conde, R. D., and Scornik, O. A., 1976, Role of protein degradation in the growth of livers after a nutritional shift, *Biochem. J.* **158:**385–390.

Dämmrich, J., and Pfeifer, U., 1981, Acute effects of isoproterenol on cellular autophagy. Inhibition in myocardium but stimulation in liver parenchyma, *Virchows Arch B* **38:**209–218.

de Duve, C., and Wattiaux, R., 1966, Functions of lysosomes, *Annu. Rev. Physiol.* **28:**435–492.

Deter, R. L., Baudhuin, P., and de Duve, C., 1967, Participation of lysosomes in cellular autophagy induced in rat liver by glucagon, *J. Cell Biol.* **35**:C11–C16.

de Waal, E. J., Vreeling-Sindelárová, H., Schellens, J. P., Houtkooper, J. M., and James, J., 1986, Quantitative changes in the lysosomal vacuolar system of rat hepatocytes during short-term starvation. A morphometric analysis with special reference to macro- and microautophagy, *Cell Tissue Res.* **243**:641–648.

Dice, F. J., and Terlecky, S. R., 1994, Selective degradation of cytosolic proteins by lysosomes, in *Cellular Proteolytic Systems* (A. J. Ciechanover and A. L. Schwartz, eds.), pp. 55–64, Wiley-Liss, New York.

Dunn Jr., W. A., 1990a, Studies on the mechanism of autophagy: Formation of the autophagic vacuole, *J. Cell Biol.* **110**:1923–1933.

Dunn Jr., W. A., 1990b, Studies on the mechanism of autophagy: Maturation of the autophagic vacuole, *J. Cell Biol.* **110**:1935–1945.

Epstein, D., Elias-Bishko, S., and Hershko, A., 1975, Requirement for protein synthesis in the regulation of protein breakdown in cultured hepatoma cells, *Biochemistry* **14**:5199–5204.

Fulks, R. M., Li, J. B., and Goldberg, A. L., 1975, Effects of insulin, glucose, and amino acids on protein turnover in rat diaphragm, *J. Biol. Chem.* **250**:290–298.

Garber, A. J., Karl, I. E., and Kipnis, D. M., 1976, Alanine and glutamine synthesis and release from skeletal muscle. IV. β-Adrenergic inhibition of amino acid release, *J. Biol. Chem.* **251**:851–857.

Goldberg, A. L., 1992, The mechanisms and functions of ATP-dependent proteases in bacterial and animal cells, *Eur. J. Biochem.* **203**:9–23.

Gordon, P. B., and Seglen, P. O., 1988, Prelysosomal convergence of autophagic and endocytic pathways, *Biochem. Biophys. Res. Commun.* **151**:40–47.

Gordon, P. B., Høyvik, H., and Seglen, P. O., 1992, Prelysosomal and lysosomal connections between autophagy and endocytosis, *Biochem. J.* **283**:361–369.

Grinde, B., and Seglen, P. O., 1981, Leucine inhibition of autophagic vacuole formation in isolated rat hepatocytes, *Exp. Cell Res.* **134**:33–39.

Gropper, R., Brandt, R. A., Elias, S., Bearer, C. F., Mayer, A., Schwartz, A. L., and Ciechanover, A., 1991, The ubiquitin-activating enzyme, E1, is required for stress-induced lysosomal degradation of cellular proteins, *J. Biol. Chem.* **266**:3602–3610.

Hallbrucker, C., vom Dahl, S., Lang, F., and Häussinger, D., 1991, Control of hepatic proteolysis by amino acids. The role of cell volume, *Eur. J. Biochem.* **197**:717–724.

Häussinger, D., Hallbrucker, C., vom Dahl, S., Lang, F., and Gerok, W., 1990, Cell swelling inhibits proteolysis in perfused rat liver, *Biochem. J.* **272**:239–242.

Häussinger, D., Newsome, W., vom Dahl, S., Stoll, B., Noe, B., Schreiber, R., Wettstein, M., and Lang, F., 1994, The role of cell volume changes in metabolic regulation, *Biochem. Soc. Trans.* **22**:497–502.

Henell, F., and Glaumann, H., 1984, Effect of leupeptin on the autophagic vacuolar system of rat hepatocytes. Correlation between ultrastructure and degradation of membrane and cytosolic proteins, *Lab. Invest.* **51**:46–56.

Heydrick, S. J., Lardeux, B. R., and Mortimore, G. E., 1991, Uptake and degradation of cytoplasmic RNA by hepatic lysosomes: Quantitative relationship to RNA turnover, *Biol. Chem.* **266**:8790–8796.

Hirsimäki, P., Arstila, A. U., Trump, B. F., and Marzella, L., 1983, Autophagocytosis, in *Pathobiology of Cell Membranes* (A. U. Arstila and B. F. Trump, eds.), pp. 201–235, Academic Press, New York/London.

Hirsimäki, Y., and Hirsimäki, P., 1984, Vinblastine-induced autophagocytosis: The effect of disorganization of microfilaments by cytochalasin B, *Exp. Mol. Pathol.* **40**:61–69.

Hopgood, M. F., Clark, M. G., and Ballard, F. J., 1980, Protein degradation in hepatocyte monolayers. Effects of glucagon, adenosine 3′:5′-cyclic monophosphate and insulin, *Biochem. J.* **186**:71–79.

Hutson, N. J., and Mortimore, G. E., 1982, Suppression of cytoplasmic protein uptake by lysosomes as the mechanism of protein regain in livers of starved-refed mice, *J. Biol. Chem.* **257**:9548–9554.

Jefferson, L. S., Rannels, D. E., Munger, B. L., and Morgan, H. E., 1974, Insulin in the regulation of protein turnover in heart and skeletal muscle, *Fed. Proc. Fed. Am. Soc. Exp. Biol.* **33**:1098–1104.

Jones, E. W., and Murdock, D. G., 1994, Proteolysis in the yeast vacuole, in *Cellular Proteolytic Systems* (A. J. Ciechanover and A. L. Schwartz, eds.) pp. 55–64, Wiley-Liss, New York.

Kadowaki, M., Pösö, A. R., and Mortimore, G. E., 1992, Parallel control of hepatic proteolysis by phenylalanine and phenylpyruvate through independent sites at the plasma membrane, *J. Biol. Chem.* **267**:22060–22065.

Kadowaki, M., Venerando, R., Miotto, G., and Mortimore, G. E., 1994, De novo autophagic vacuole formation in hepatocytes permeabilized by *Staphylococcus aureus* α-toxin: Inhibition by nonhydrolyzable GTP analogs, *J. Biol. Chem.* **269**:3703–3710.

Khairallah, E. A., and Mortimore, G. E., 1976, Assessment of protein turnover in perfused rat liver: Evidence for amino acid compartmentation from differential labeling of free and tRNA-bound valine, *J. Biol. Chem.* **251**:1375–1384.

Knowles, S. E., and Ballard, F. J., 1976, Selective control of the degradation of normal and aberrant proteins in Reuber H35 hepatoma cells, *Biochem. J.* **156**:609–617.

Kominami, E., Hashida, S., Khairallah, E. A., and Katunuma, N., 1983, Sequestration of cytoplasmic enzymes in autophagic vacuole-lysosomal system induced by injection of leupeptin, *J. Biol. Chem.* **258**:6093–6100.

Kopitz, J., Kisen, G. Ø., Gordon, P. B., Bohley, P., and Seglen, P. O., 1990, Nonselective autophagy of cytosolic enzyme by isolated rat hepatocytes, *J. Cell Biol.* **111**:941–953.

Kovács, A. L., and Seglen, P. O., 1981, Inhibition of hepatocytic protein degradation by methylaminopurines and inhibitors of protein synthesis, *Biochim. Biophys. Acta* **676**:213–220.

Kovács, A. L., Grinde, B., and Seglen, P. O., 1981, Inhibition of autophagic vacuole formation and protein degradation by amino acids in isolated hepatocytes, *Exp. Cell Res.* **133**:431–436.

Kovács, J., Fellinger, E., Kárpáti, A. P., Kovács, A. L., László, L., and Réz, G., 1987, Morphometric evaluation of the turnover of autophagic vacuoles after treatment with Triton X-100 and vinblastine in murine pancreatic acinar and seminal vesicle epithelial cells, *Virchows Arch B* **53**:183–190.

Lardeux, B. R., and Mortimore, G. E., 1987, Amino acid and hormonal control of macromolecular turnover in perfused rat liver: Evidence for selective autophagy, *J. Biol. Chem.* **262**:14514–14519.

Lardeux, B. R., Heydrick, S. J., and Mortimore, G. E., 1987, RNA degradation in perfused rat liver as determined from the release of [^{14}C]cytidine, *J. Biol. Chem.* **262**:14507–14513.

Lardeux, B. R., Heydrick, S. J., and Mortimore, G. E., 1988, Rates of rat liver RNA degradation *in vivo* as determined from cytidine release during brief cyclic perfusion *in situ*, *Biochem. J.* **252**: 363–367.

Lawrence, B. P., and Brown, W. J., 1992, Autophagic vacuoles rapidly fuse with pre-existing lysosomes in cultured hepatocytes, *J. Cell Sci.* **102**:515–526.

Lenk, S. E., Dunn Jr., W. A., Trausch, J. S., Ciechanover, A., and Schwartz, A. L., 1992, Ubiquitin-activating enzyme is associated with maturation of autophagic vacuoles, *J. Cell. Biol.* **118**: 301–308.

Leverve, X. M., Caro, L. H., Plomp, P. J., and Meijer, A. J., 1987, Control of proteolysis in perfused rat hepatocytes, *FEBS Lett.* **219**:455–458.

Li, J. B., and Jefferson, L. S., 1977, Effect of isoproterenol on amino acid levels and protein turnover in skeletal muscle, *Am. J. Physiol.* **232**:E243–E249.

Löw, P., Doherty, F. J., Sass, M., Kovács, J., Mayer, R. J., and Lászó, L., 1993, Immunogold localisation of ubiquitin-protein conjugates in Sf9 insect cells. Implications for the biogenesis of lysosome-related organelles, *FEBS Lett.* **316**:152–156.

Marty, F., 1978, Cytochemical studies on GERL, provacuoles, and vacuoles in root meristematic cells of *Euphorbia, Proc. Natl. Acad. Sci. USA* **75**:852–856.

Marzella, L., Ahlberg, J., and Glaumann, H., 1980, *In vitro* uptake of particles by lysosomes, *Exp. Cell Res.* **129**:460–466.

May, L. T., Anderson, O. R., and Hogg, J. F., 1982, Changes of cellular structure and subcellular enzymatic patterns during the activation of glyconeogenesis in Tetrahymena pyriformis, *J. Ultrastruct. Res.* **81**:271–289.

Meijer, A. J., Gustafson, L. A., Luiken, J. J., Blommaart, P. J., Caro, L. H., Van Woerkom, G. M., Spronk, C., and Boon, L., 1993, Cell swelling and the sensitivity of autophagic proteolysis to inhibition by amino acids in isolated rat hepatocytes, *Eur. J. Biochem.* **15**:449–454.

Miotto, G., Venerando, R., and Siliprandi, N., 1989, Inhibitory action of isovaleryl-L-carnitine on proteolysis in perfused rat liver, *Biochem. Biophys. Res. Commun.* **158**:797–802.

Miotto, G., Venerando, R., Khurana, K. K., Siliprandi, N., and Mortimore, G., 1992, Control of hepatic proteolysis by leucine and isovaleryl-L-carnitine through a common locus: Evidence for a possible mechanism of recognition at the plasma membrane, *J. Biol. Chem.* **267**:22066–22072.

Miotto, G., Venerando, R., Marin, O., Siliprandi, N., and Mortimore, G. E., 1994, Inhibition of macroautophagy and proteolysis in the isolated rat hepatocyte by a nontransportable derivative of the multiple antigen peptide Leu$_8$-Lys$_4$-Lys$_2$-Lys-βAla, *J. Biol. Chem.* **269**:25348–25353.

Mitch, W. E., and Clark, A. S., 1984, Specificity of the effects of leucine and its metabolites on protein degradation in skeletal muscle, *Biochem. J.* **222**:579–586.

Mitchener, J. S., Shelburne, J. D., Bradford, W. D., and Hawkins, H. K., 1976, Cellular autophagocytosis induced by deprivation of serum and amino acids in HeLa cells, *Am. J. Pathol.* **83**:485–491.

Mortimore, G. E., and Kadowaki, M., 1994, Autophagy: Its mechanism and regulation, in *Cellular Proteolytic Systems* (A. J. Ciechanover and A. L. Schwartz, eds.), pp. 65–87, Wiley-Liss, New York.

Mortimore, G. E., and Mondon, C. E., 1970, Inhibition by insulin of valine turnover in liver: Evidence for a general control of proteolysis, *J. Biol. Chem.* **245**:2375–2383.

Mortimore, G. E., and Pösö, A. R., 1984, Lysosomal pathways in hepatic protein degradation: Regulatory role of amino acids, *Fed. Proc. Fed. Am. Soc. Exp. Biol.* **43**:1289–1294.

Mortimore, G. E., and Pösö, A. R., 1987, Intracellular protein catabolism and its control during nutrient deprivation and supply, *Annu. Rev. Nutr.* **7**:539–564.

Mortimore, G. E., and Schworer, C. M., 1977, Induction of autophagy by amino acid deprivation in perfused rat liver, *Nature* **270**:174–176.

Mortimore, G. E., and Ward, W. F., 1981, Internalization of cytoplasmic protein by hepatic lysosomes in basal and deprivation-induced proteolytic states, *J. Biol. Chem.* **256**:7659–7665.

Mortimore, G. E., Woodside, K. H., and Henry, J. E., 1972, Compartmentation of free valine and its relation to protein turnover in perfused rat liver, *J. Biol. Chem.* **247**:2776–2784.

Mortimore, G. E., Neely, A. N., Cox, J. R., and Guinivan, R. A., 1973, Proteolysis in homogenates of perfused rat liver: Responses to insulin, glucagon and amino acids, *Biochem. Biophys. Res. Commun.* **54**:89–95.

Mortimore, G. E., Hutson, N. J., and Surmacz, C. A., 1983, Quantitative correlation between proteolysis and macro- and microautophagy in mouse hepatocytes during starvation and refeeding, *Proc. Natl. Acad. Sci. USA* **80**:2179–2183.

Mortimore, G. E., Pösö, A. R., Kadowaki, M., and Wert Jr., J. J., 1987, Multiphasic control of hepatic protein degradation by regulatory amino acids: General features and hormonal modulation, *J. Biol. Chem.* **262**:16322–16327.

Mortimore, G. E., Lardeux, B. R., and Adams, C. E., 1988a, Regulation of microautophagy and basal protein turnover in rat liver: Effects of short-term starvation, *J. Biol. Chem.* **263**:2506–2512.

Mortimore, G. E., Wert Jr., J. J., and Adams, C. E., 1988b, Modulation of the amino acid control of hepatic proteolysis by caloric deprivation: Two modes of alanine co-regulation, *J. Biol. Chem.* **263**:19545–19551.

Mortimore, G. E., Khurana, K. K., and Miotto, G., 1991, Amino acid control of proteolysis in perfused livers of synchronously fed rats, *J. Biol. Chem.* **266**:1021–1028.

Mortimore, G. E., Wert Jr., J. J., Miotto, G., Venerando, R., and Kadowaki, M., 1994, Leucine-specific binding of photoreactive Leu₇-MAP to a high molecular weight protein on the plasma membrane of the isolated rat hepatocyte, *Biochem. Biophys. Res. Commun.* **203**:200–208.

Neely, A. N., Nelson, P. B., and Mortimore, G. E., 1974, Osmotic alterations of the lysosomal system during rat liver perfusion: Reversible suppression by insulin and amino acids, *Biochim. Biophys. Acta* **338**:458–472.

Neely, A. N., Cox, J. R., Fortney, J. A., Schworer, C. M., and Mortimore, G. E., 1977, Alterations of lysosomal size and density during rat liver perfusion. Suppression by insulin and amino acids, *J. Biol. Chem.* **252**:6948–6954.

Nishimura, M., and Beevers, H., 1979, Hydrolysis of protein in vacuoles isolated from higher plant tissue, *Nature* **277**:412–413.

Novikoff, A. B., and Shin, W. Y., 1978, Endoplasmic reticulum and autophagy in rat hepatocytes, *Proc. Natl. Acad. Sci. USA* **75**:5039–5042.

Ogier-Denis, E., Couvineau, A., Maoret, J. J., Houri, J. J., Bauvy, C., De Stefanis, D., Isidoro, C., Laburthe, M., and Codogno, P., 1995, A heterotrimeric G₁₃-protein controls autophagic sequestration in the human colon cancer cell line HT-29, *J. Biol. Chem.* **270**:13–16.

Paavola, L. G., 1978a, The corpus luteum of the guinea pig. II. Cytochemical studies on the Golgi complex, GERL, and lysosomes in luteal cells during maximal progesterone secretion, *J. Cell Biol.* **79**:45–58.

Paavola, L. G., 1978b, The corpus luteum of the guinea pig. III. Cytochemical studies on the Golgi complex and GERL during normal postpartum regression of luteal cells, emphasizing the origin of lysosomes and autophagic vacuoles, *J. Cell Biol.* **79**:59–73.

Papadopoulos, T., and Pfeifer, U., 1986, Regression of rat liver autophagic vacuoles by locally applied cycloheximide, *Lab. Invest.* **54**:100–107.

Pfeifer, U., 1973, Cellular autophagy and cell atrophy in rat liver during long-term starvation, *Virchows Arch. B* **12**:195–211.

Pfeifer, U., 1978, Inhibition by insulin of the formation of autophagic vacuoles in rat liver. A morphometric approach to the kinetics of intracellular degradation by autophagy, *J. Cell Biol.* **78**:152–167.

Pfeifer, U., 1981, Morphological aspects of intracellular protein degradation: Autophagy, *Acta Biol. Med. Ger.* **40**:1619–1624.

Plomp, P. J., Gordon, P. B., Meijer, A. J., Høyvik, H., and Seglen, P. O., 1989, Energy dependence of different steps in the autophagic-lysosomal pathway, *J. Biol. Chem.* **264**:6699–6704.

Poli, A., Gordon, P. B., Schwarze, P. E., Grinde, B., and Seglen, P. O., 1981, Effect of insulin and anchorage on hepatocytic protein metabolism and amino acid transport, *J. Cell Sci.* **48**:1–18.

Poole, B., and Wibo, M., 1973, Protein degradation in cultured cells, *J. Biol. Chem.* **248**:6221–6226.

Pösö, A. R., and Mortimore, G. E., 1984, Requirement for alanine in the amino acid control of deprivation-induced protein degradation in liver, *Proc. Natl. Acad. Sci. USA* **81**:4270–4274.

Pösö, A. R., Schworer, C. M., and Mortimore, G. M., 1982a, Acceleration of proteolysis in perfused rat liver by deletion of glucogenic amino acids: Regulatory role of glutamine, *Biochem. Biophys. Res. Commun.* **107**:1433–1439.

Pösö, A. R., Wert Jr.,. J. J., and Mortimore, G. E., 1982b, Multi-functional control by amino acids of deprivation-induced proteolysis in liver: Role of leucine, *J. Biol. Chem.* **257**:12114–12120.

Punnonen, E. L., Autio, S., Kaija, H., and Reunanen, H., 1993, Autophagic vacuoles fuse with the prelysosomal compartment in cultured rat fibroblasts, *Eur. J. Cell Biol.* **6**:54–66.

Rabkin, R., Tsao, T., Shi, J. D., and Mortimore, G., 1991, Amino acids regulate kidney cell protein breakdown, *J. Lab. Clin. Med.* **117**:505–513.

Rannels, D. E., Kao, R., and Morgan, H. E., 1975, Effect of insulin on protein turnover in heart muscle, *J. Biol. Chem.* **250:**1694–1701.

Rechsteiner, M., Hoffman, L., and Dubiel, W., 1993, The multicatalytic and 26S proteases, *J. Biol. Chem.* **268:**6065–6068.

Rosa, F., 1971, Ultrastructural changes produced by glucagon, cyclic 3'5'AMP and epinephrine on perfused rat livers, *J. Ultrastruct. Res.* **34:**205–213.

Schwartz, A. L., Brandt, R. A., Geuze, H. J., and Ciechanover, A., 1992, Stress induced alteration in the autophagic pathway: Relationship to ubiquitin system, *Am. J. Physiol.* **262:**C1031–C1038.

Schworer, C. M., and Mortimore, G. M., 1979, Glucagon-induced autophagy and proteolysis in rat liver: Mediation by selective deprivation of intracellular amino acids, *Proc. Natl. Acad. Sci. USA* **76:**3169–3173.

Schworer, C. M., Shiffer, K. A., and Mortimore, G. E., 1981, Quantitative relationship between autophagy and proteolysis during graded amino acid deprivation in perfused rat liver, *J. Biol. Chem.* **256:**7652–7658.

Seglen, P. O., and Gordon, P. B., 1982, 3-Methyladenine: Specific inhibitor of autophagic/lysosomal protein degradation in isolated rat hepatocytes, *Proc. Natl. Acad. Sci. USA* **79:**1889–1892.

Seglen, P. O., and Solheim, A. E., 1978, Valine uptake and incorporation into protein in isolated rat hepatocytes. Nature of the precursor pool for protein synthesis, *Eur. J. Biochem.* **85:**15–25.

Seglen, P. O., Gordon, P. B., and Poli, A., 1980, Amino acid inhibition of the autophagic/lysosomal pathway of protein degradation in isolated rat hepatocytes, *Biochim. Biophys. Acta* **630:**103–118.

Shiman, R., Mortimore, G. E., Schworer, C. M., and Gray, D. W., 1982, Regulation of phenylalanine hydroxylase activity by phenylalanine *in vivo, in vitro,* and in perfused rat liver, *J. Biol. Chem.* **257:**11213–11216.

Solheim, A. E., and Seglen, P. O., 1980, Subcellular distribution of proteolytically generated valine in isolated rat hepatocytes, *Eur. J. Biochem.* **107:**587–596.

Sommercorn, J. M., and Swick, R. W., 1981, Protein degradation in primary monolayer cultures of adult rat hepatocytes, *J. Biol. Chem.* **256:**4816–4821.

Surmacz, C. A., Wert Jr., J. J., and Mortimore, G. E., 1983a, Role of particle interaction on distribution of liver lysosomes in colloidal silica, *Am. J. Physiol.* **245:**C52–C60.

Surmacz, C. A., Wert Jr., J. J., and Mortimore, G. E., 1983b, Metabolic alterations and distribution of rat liver lysosomes in colloidal silica, *Am. J. Physiol.* **245:**C61–C67.

Surmacz, C. A., Pösö, A. R., and Mortimore, G. E., 1987, Regulation of lysosomal fusion during deprivation-induced autophagy in perfused rat liver, *Biochem. J.* **242:**453–458.

Takeshige, K., Baba, M., Tsuboi, S., Noda, T., and Ohsumi, Y., 1992, Autophagy in yeast demonstrated with protease-deficient mutants and conditions for its induction, *J. Cell Biol.* **19:**301–311.

Tam, J. P., 1989, High-density multiple antigenic-peptide system for preparation of antipeptide antibodies, in *Methods in Enzymology,* Vol. 168 (P. M. Conn, ed.), pp. 7–15, Academic Press, New York/London.

Tischler, M. E., Desautels, M., and Goldberg, A. L., 1982, Does leucine, leucyl-tRNA, or some metabolite of leucine regulate protein synthesis and degradation in rat skeletal and cardiac muscle? *J. Biol. Chem.* **257:**1613–1621.

Tooze, J., Hollinshead, M., Ludwig, T., Howell, K., Hoflack, B., and Kern, H., 1990, In exocrine pancreas, the basolateral endocytic pathway converges with the autophagic pathway immediately after the early endosome, *J. Cell Biol.* **111:**329–345.

Tsao, T. C., Shi, J. D., Mortimore, G. E., Cragoe Jr., E. J., and Rabkin, R., 1990, Modulation of kidney cell protein degradation by insulin, *J. Lab. Clin. Med.* **116:**369–376.

Vandenburgh, H., and Kaufman, S., 1980, Protein degradation in embryonic skeletal muscle. Effects of medium, cell type, inhibitors, and passive stretch, *J. Biol. Chem.* **255:**5826–5833.

Venerando, R., Miotto, G., Kadowaki, M., Siliprandi, N., and Mortimore, G. E., 1994, Multiphasic control of proteolysis by leucine and alanine in the isolated rat hepatocyte, *Am. J. Physiol.* **266:**C455–C461.

Ward, W. F., Cox, J. R., and Mortimore, G. E., 1977, Lysosomal sequestration of intracellular protein as a regulatory step in hepatic proteolysis, *J. Biol. Chem.* **252:**6955–6961.

Wert, Jr., J. J., Miotto, G., Kadowaki, M., and Mortimore, G. E., 1992, 4-Amino-6-methylhept-2-enoic acid: A leucine analogue and potential probe for localizing sites of proteolytic control in the hepatocyte, *Biochem. Biophys. Res. Commun.* **186:**1327–1332.

Wheatley, D. N., 1984, Intracellular protein degradation: Basis of a self-regulating mechanism for the proteolysis of endogenous proteins, *J. Theor. Biol.* **107:**127–149.

Woodside, K. H., and Mortimore, G. E., 1972, Suppression of protein turnover by amino acids in the perfused rat liver, *J. Biol. Chem.* **247:**6474–6481.

Woodside, K. H., Ward, W. F., and Mortimore, G. E., 1974, Effects of glucagon on general protein degradation and synthesis in perfused rat liver, *J. Biol. Chem.* **249:**5458–5463.

Yokota, S., 1993, Formation of autophagosomes during degradation of excess peroxisomes induced by administration of dioctyl phthalate, *Eur. J. Cell Biol.* **61:**67–80.

Young, V. R., and Munro, H. N., 1978, N^τ-methylhistidine (3-methylhistidine) and muscle protein turnover: An overview, *Fed. Proc. Fed. Am. Soc. Exp. Biol.* **37:**2291–2300.

Chapter 5

Selective Proteolysis: 70-kDa Heat-Shock Protein and Ubiquitin-Dependent Mechanisms?

R. John Mayer and Fergus J. Doherty

1. SELECTIVE LYSOSOMAL PROTEOLYSIS

1.1. Enhanced Proteolysis on Serum Withdrawal

The proteolysis of the vast majority of cytoplasmic proteins, once inside the lysosome compartment, would be expected to occur rapidly; therefore, the rate-limiting step in the lysosomal degradation of cytoplasmic proteins must be their initial sequestration. Indeed, the initial sequestration serves to remove cytoplasmic proteins from their functional site in the cell. Bulk sequestration of cytoplasm (macro-autophagy), described elsewhere in this volume (Chapter 4), is considered essentially nonselective. Microautophagy has been described by a number of investigators over the years (see Chapter 4) and may provide a mechanism for the selective sequestration of cytoplasmic proteins into the lysosome system. However, controversy surrounds the whole topic of selective lysosomal proteolysis, with some investigators probably of the view that the lysosome system is entirely nonselective. However, recent work from the laboratory of Dice (Dice, 1990) describes a selective mechanism of sequestration of cy-

R. John Mayer and **Fergus J. Doherty**   Department of Biochemistry, Queens Medical Center, Nottingham NE7 2UH, United Kingdom.

Subcellular Biochemistry, Volume 27. Biology of the Lysosome, edited by Lloyd and Mason. Plenum Press, New York, 1996.

tosolic proteins into lysosomes for degradation in a process regulated by serum and involving a member of the 70-kDa heat-shock protein family.

It has been known for many years that when animal cells in culture are deprived of serum there is an increase in protein degradation on the order of 50–300%, depending on the cell type and the culture conditions (Amenta *et al.,* 1976; Hendil, 1977; Slot *et al.,* 1986). Enhanced proteolysis caused by serum deprivation is prevented by lysosomotropic agents such as ammonium chloride (Amenta *et al.,* 1978), implicating lysosomes in this process. Vinblastine, an inhibitor of autophagic vacuole-lysosome fusion (Amenta *et al.,* 1976) also inhibits this enhanced proteolysis, suggesting the involvement of autophagic processes. The effect of serum deprivation is probably due to the loss of growth factors and hormones present in serum (e.g., insulin) that are known to suppress lysosomal protein degradation (Ballard *et al.,* 1980). The effect of serum deprivation is also transient, lasting about 24 hr, after which degradation rates return to previous values (Slot *et al.,* 1986) in the continued absence of serum.

1.1.1. Degradation of a Subset of Cytosolic Proteins

While the involvement of lysosomes in serum deprivation-enhanced proteolysis suggests a nonselective increase in intracellular protein degradation, it appears that only cytosolic proteins are degraded more rapidly during serum withdrawal and possibly only a subset of 25–30% of these (Slot *et al.,* 1986). The transient nature of this enhanced proteolysis could therefore be due to depletion of available substrates. The degradation of hemoglobin microinjected into IMR90 fibroblasts is increased on serum withdrawal and the increased degradation of this exogenous protein is sustained over 50 hr, perhaps due to the quantity introduced into the recipient cell, as compared to 24 hr for endogenous proteins. This would support the notion that endogenous substrates for this pathway are exhausted in cells after about a day without serum (Slot *et al.,* 1986). The degradation of radiolabeled exogenous proteins introduced into culture cells by red cell-mediated microinjection also reveals that not all proteins are subject to enhanced lysosomal proteolysis on prolonged serum withdrawal. One microinjected exogenous protein which is degraded more rapidly in the absence of serum is pancreatic ribonuclease A (RNAse) (Dice *et al.,* 1985). Although this protein is not normally resident in the cystosol, studies employing microinjected RNAse were instrumental in allowing Dice and co-workers to dissect the mechanisms and specificity of this serum deprivation-enhanced pathway.

1.1.2. KFERQ and KFERQ-like Motifs Target Proteins
for Lysosomal Sequestration

Microinjected RNAse is subject to serum withdrawal-enhanced lysosomal proteolysis (Dice *et al.,* 1985). Tagging RNAse with the membrane impermeant label [³H]raffinose revealed that all the labeled breakdown products ([³H]raffinose-lysine) accumulate in the lysosome compartment, both in the absence and presence of serum, and that the rate of RNAse sequestration by lysosomes increases in response to serum withdrawal (Dice et al., 1985). Treatment of RNAse with subtilisin results in a single cleavage of a 20-amino acid fragment from the amino-terminus (S-peptide, residues 1–20) and the degradation of this fragment, but not the caboxyl-terminal S-protein, is enhanced on serum withdrawal (Backer et al., 1983). Furthermore, chemical conjugation of the S-peptide to unrelated proteins conferred sensitivity to serum withdrawal when the modified proteins were microinjected into cells, as compared to their unmodified counterparts (Backer and Dice, 1986). Further experiments revealed that residues 7–11 of the S-peptide (KFERQ) confer proteolytic sensitivity to serum withdrawal (Dice et al., 1986). This pentapeptide sequence is restricted to the RNAse family of proteins, however antibodies to KFERQ precipitate 25% of cytosolic polypeptides found in cultured human fibroblasts (Chiang and Dice, 1988). While KFERQ itself is restricted to RNAse it is suggested that the cytosolic proteins immunoprecipitated by this antibody contain sequences sufficiently similar to be recognized by anti-KFERQ polyclonal antibody. Cytosolic proteins immunoprecipitated by antibodies to unrelated sequences do not exhibit serum withdrawal-enhanced degradation (Chiang and Dice, 1988).

Total human fibroblast cytosolic polypeptides exhibit a half-life of approximately 80 hr in the presence of serum; this drops to 30 hr in the absence of serum. In contrast, while total polypeptides precipitated with anti-KFERQ antibody have a half-life of 80 hr in the presence of serum, this drops to 12 hr in the absence of serum (Chiang and Dice, 1988). These results indicate not only that there are many cytosolic polypeptides in human fibroblasts which must contain sequences similar to KFERQ, but also that such polypeptides are selectively degraded in the absence of serum. Starvation of rats also induces enhanced lysosomal degradation of anti-KFERQ immunoreactive proteins of liver, kidney, and heart, while enhanced proteolysis of the less abundant immunoreactive proteins in brain and testis is not observed (Wing et al., 1991).

Examination by Dice and co-workers of the sequences of the small number of proteins whose intracellular half-lives are known to decrease in the presence of serum revealed that KFERQ-like sequences are present in these proteins. However, the consensus sequence can be reduced to a glutamine flanked by a tetrapeptide, on either side, consisting only of very basic, acidic, or hydrophobic residues. Many other proteins appear to express KFERQ-like pentapeptide motifs, as found by a search of sequence databases, but their intracellular half-lives are unknown. In addition, some of these are localized to intracellular organelles and therefore would not be expected to exhibit serum withdrawal-enhanced proteolysis (Dice and Chiang, 1989).

1.1.3. A 73-kDa Protein, prp73, Recognizes Polypeptides Carrying KFERQ-like Motifs

Further investigations by Dice and co-workers sought to establish the mechanism for the selective sequestration of cytosolic polypeptides on serum withdrawal and to identify any proteins required for uptake. A 73-kDa KFERQ-binding protein, prp73, was isolated from cell cytosol. The purified protein stimulates the import of RNAse A into isolated lysosomes. Hydrolysis of ATP is also required (Chiang *et al.*, 1989). As might be expected, cellular levels of prp73 increase on serum withdrawal (Chiang *et al.*, 1989).

1.1.4. Prp73 Is Related to 70-kDa Heat-Shock Proteins

The requirement for ATP and a protein of around 70-kDa is reminiscent of the import of nuclear encoded polypeptides into mitochondria (Stuart *et al.*, 1994), which requires ATP and a member of the 70-kDa heat-shock family of proteins or chaperones. This prompted Dice and co-workers to look for similarities between prp73 and the 70-kDa heat-shock proteins. Monoclonal antibodies to members of this family cross-react with prp73, and several internal peptide sequences obtained from prp73 are similar to sequences in the 70-kDa heat-shock proteins (Chiang *et al.*, 1989). Heat-shock proteins were first discovered as a set of proteins that were upregulated after exposing cells to elevated temperatures (39–42 °C for mammalian cells). Among these proteins are polypeptides with molecular weights of 68–73 kDa that comprise a family of related proteins with ATPase and polypeptide binding activities (Ellis and Vandervies, 1991). Although commonly referred to as heat-shock proteins, some members of the 70-kDa family are expressed under normal conditions, are only poorly inducible by heat, and are now known to be involved in a wide variety of processes in the cell under normal conditions. The term *heat shock* or *stress* protein is therefore somewhat misleading and these proteins are now often called chaperones because of their polypeptide binding activity (Ellis and Vandervies, 1991). All of the known roles of the heat-shock proteins involve interaction with other cellular proteins. Heat-shock proteins have been described as "polypeptide binding proteins" (Rothman, 1989). The basis of recognition of cellular polypeptides by members of the 70-kDa family is still not entirely clear. While different members of the 70-kDa family may share the ability to bind polypeptides, they differ in their specificity for polypeptides and presumably recognize different peptide motifs. Constitutively expressed members of the 70-kDa group include proteins that act as chaperones for import of polypeptides into mitochondria and the endoplasmic reticulum (Becker and Craig, 1994). It appears that the N-terminal signal sequence of nuclear-encoded mitochondrial polypeptides is recognized by a 70-kDa chaperone which binds to the precursor and prevents premature folding prior to mitochondrial import. This is achieved by cycles of binding and release of the precursor, accompanied by ATP hydrolysis. Once the precursor signal sequence has "docked" with a mitochondrial membrane receptor, the precursor is translocated across the membrane, driven by the electrochemical gradient, but aided by the activity of a mitochondrial 70-kDa chaperone which prevents folding until the polypeptide has been imported. Other chaperones are also involved in this complex process. Following heat shock, heat-inducible members of the 70-kDa family appear to act to prevent aggregation of thermally denatured proteins and aid their refolding into native conformations. These observations suggest that prp73 may serve to unfold cytosolic proteins containing the KFERQ recognition motif and present them in an import-competent conformation to the lysosome membrane.

A 70-kDa chaperone also acts as a clathrin-uncoating ATPase, responsible for the removal of clathrin cages from coated vesicles prior to vesicular membrane fusion with the endosome/lysosome system (Deluca-Flaherty et al., 1990). Dice has proposed that the KFERQ-like sequences in clathrin (QVDRL, VDRLQ) suggest that the clathrin-uncoating ATPase/70-kDa cell stress protein may be responsible for recognizing KFERQ-like sequences for lysosomal sequestration (Dice, 1990), however there is no data to support this. The bulk of this 70-kDa heat-shock protein may be sequestered by clathrin under normal conditions and unavailable for promoting import of polypeptides into the lysosome. Following serum withdrawal, increased levels of this heat-shock protein would presumably be available for binding to cytosolic polypeptides expressing KFERQ-like sequences and promoting lysosomal import. The corollary of this is that the bulk of the 70-kDa heat-shock protein may be sequestered by cytosolic polypeptides destined for import into lysosomes during serum withdrawal, thus compromising clathrin uncoating. Unfortunately we are not aware of any studies in the literature examining the effect of serum deprivation on clathrin uncoating.

1.1.5. Heat-Shock Proteins Are Found in Various Vesicular Cell Compartments

Interesting, anti-70-kDa heat-shock protein immunoreactivity in rat kidney indicates a redistribution of the protein following ischemia with some heat-shock protein becoming coincident with lysosomes (Vanwhy et al., 1992). It is not known, however, if this is intralysosomal or associated with the cytoplasmic face of the lysosome membrane, but it does suggest a role for this protein in lysosomal action in stress situations other than serum deprivation and starvation. The endosome/lysosome system is implicated in the fragmentation of exogenous and endogenous polypeptides for peptide antigen presentation with major histocompatibility complex (MHC) class II (Humbert et al., 1993), while fragmentation of polypeptides in the cytosol by the proteasome (Goldberg and Rock, 1992) is followed by import of the peptide into the endoplasmic reticulum and presentation at the cell surface with MHC class I. However, cells with a defective ABC cassette peptide transporter that have been transfected with a truncated cytoplasmic SV40 large T antigen present epitopes in association with MHC class I molecules in a chloroquine- and ammonium chloride-inhibitable fashion; presentation of mutant T-antigens correlates with association with a 70-kDa heat-shock protein (Schirmbeck and Reimann, 1994). This suggests that the heat-shock protein may participate in delivering T-antigen to the endosome for processing. Therefore, a 70-kDa heat-shock protein may be involved in cytosolic protein import into prelysosomal (endosomal) and lysosomal compartments. A member of the 70-kDa heat-shock protein family is also found in the *trans* cisternae of pancreatic acinar cell Golgi apparatus and colocalizes with acid phosphatase in that compartment (Velezgranell *et al.*, 1994), suggesting that the heat-shock protein could be present in acid phosphatase containing vesicles related to lysosomes into which KFERQ-containing polypeptides are imported. Additionally, a 70-kDa heat-shock protein is found in the endosome–lysosome compartment, together with ubiquitin–protein conjugates, in Epstein-Barr virus transformed lymphoblastoid cells (László *et al.*, 1991b) and in scrapie-infected mouse brain (László *et al.*, 1992). The endosome–lysosome is enriched in the viral latent membrane protein (LMP) protein and the abnormal prion protein, respectively. There is an increase in the numbers and size of endosomes and lysosomes under these conditions, which may explain why heat-shock protein 70 is observed in these organelles.

1.2. Possible Mechanisms of Lysosomal Sequestration

RNAse, RNAse S-peptide, and glyceraldehyde-3-phosphate dehydrogenase (GAPDH) bind to and are imported into isolated rat liver lysosomes (Cuervo *et al.*, 1994). Uptake of RNAse appears to involve a stalled intermediate with 2 kDa of the C-terminus still remaining outside the lysosome. This suggests that import, at least in the case of RNAse, starts with the amino terminus. This is also the case with the import of nuclear encoded polypeptides into the mitochondria, which is mediated by chaperone proteins including members of the 70-kDa heat-shock family. In these recent experiments it is notable that appreciable import occurs in the absence of prp73, particularly with GAPDH, and the effect of added prp73 and ATP is most marked for the import of added [³H]leucine–labeled total cytosolic proteins (Cuervo *et al.*, 1994). This may indicate the presence of a prp73-independent import process. Nonhydrolyzable analogs of ATP do not stimulate import in these systems (Cuervo *et al.*, 1994); therefore, ATP hydrolysis is required, as for polypeptide import into mitochondria. Pretreatment of isolated lysosomes with trypsin blocks import presumably by compromising a receptor or pore in the lysosome membrane.

All the evidence points to a translocation mechanism for uptake of selected cytosolic polypeptides into the lysosome similar to that for mitochondrial import, with recognition and binding of cytosolic polypeptides by prp73, which acts as a chaperone presenting the polypeptide to lysosomes (Fig. 1). The cytosolic chaperone, which aids import of mitochondrial precursor polypeptides, probably binds

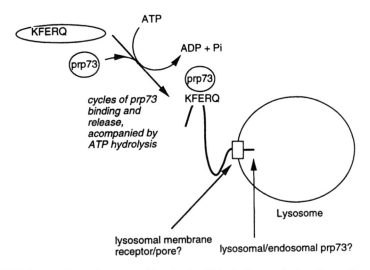

FIGURE 1. Scheme illustrating the possible role of prp73 in the import of selected cytosolic proteins into the lysosome following serum withdrawal. Proteins bearing KFERQ-like motifs are recognized by prp73. Following the analogy with import of polypeptides into the mitochondrion the selected protein must be unfolded in some way. This may be accomplished by prpr73 acting as an "unfoldase" with the hydrolysis of ATP. Prp73 would presumably reassociate with the unfolded protein, with continuing ATP hydrolysis until the N-terminus of the selected protein docks with "receptor" in the lysosomal membrane and the protein is translocated into the lumen of the lysosome. In this way prpr73 facilitates import and prevents refolding which could block translocation across the lysosomal membrane. If the organelle involved is an endosome it is possible that, as with import into the mitochondrion mediated by chaperones, there could be an endosomal prp73 to maintain the protein in its unfolded, import-competent state as it is translocated into the organelle.

polypeptides prior to their folding. If prp73 acts in an analogous fashion it must either be capable of unfolding selected mature cytosolic polypeptides, or another so far unknown factor may act as an "unfoldase" prior to prp73 binding. In the case of mitochondrial polypeptide import, a mitochondrial 70-kDa chaperone is also required to complete import of polypeptides (Stuart *et al.,* 1994). If the analogy is pursued then the question arises, is there a requirement for an intraorganellar prp73 for import of cytosolic polypeptides into the lysosome as is found for import into mitochondria (Leustek *et al.,* 1989)? In addition, is ATP hydrolysis inside the organelle required to complete import (Stuart *et al.,* 1994)? These criteria may be unlikely given the inhospitable environment of the lysosomal milieu to cytosolic proteins. Perhaps rapid degradation of the incoming polypeptide or the low pH in the lysosome prevents blocking of the putative import pore, obviating the need for a chaperone in the lysosome to prevent premature folding, as is required in the mito chondrion. However, the pH in some early compartments of the endosome–lysosome system is not likely to be so inhospitable (see Chapter 10). The intravacuolar pH of early components of the endosomal pathway in *Dictyostelium discoideum* is 6.2 (Aubry *et al.,* 1993), while early endosomes in neurons appear to be neutral (Overly *et al.,* 1995). Therefore, it is not impossible that intraluminal prp73 homologues and ATP may be present to facilitate protein import/unfolding for controlled protein fragmentation (e.g., for antigen presentation) and degradation.

Why is this selective mechanism of lysosomal sequestration activated on serum withdrawal and during starvation in the liver? Macroautophagy would seem to be rapidly increased on serum withdrawal and starvation, and this is presumably nonselective. Dice has suggested that the initial rapid increase in nonselective macroautophagy seen on serum withdrawal and inhibitable by vinblastine (Amenta *et al.,* 1976), which takes place in a matter of minutes, cannot be allowed to persist or the cell would eliminate too many essential proteins. Prolonged serum withdrawal or starvation (a few hours and more) may lead to reduced macroautophagy and activation of the selective mechanism, which may then degrade nonessential proteins, releasing their amino acids for the synthesis of essential proteins. This hypothesis may reconcile the evidence supporting a role for macroautophagy and selective microautophagic lysosomal proteolysis in enhanced lysosomal proteolysis in starvation and serum depletion.

2. UBIQUITIN AND THE LYSOSOMAL SYSTEM

2.1. Stress-Enhanced Lysosomal Proteolysis

2.1.1. Ubiquitin Activation and Lysosomal Proteolysis

The rapid increase in autophagy in response to serum withdrawal is also seen in cell starvation and heat stress (Gropper *et al.,* 1991) and surprisingly is dependent on an active ubiquitination system.

Several cell lines have been established that harbor a temperature-sensitive mutant of E1, the ubiquitin-activating enzyme responsible for the first step in the ubiquitin pathway. The E1 enzyme catalyzes the formation of ubiquitin adenylate from ubiquitin monomer and ATP, with the ubiquitin moiety subsequently transferred to an E1 cysteine residue to form a thiol ester between the C-terminus of ubiquitin and the enzyme thiol. The ubiquitin is subsequently transferred to one of a growing family of E2 enzymes, forming a thiol ester with an E2 cysteine. Finally the ubiquitin is ligated to a target protein by formation of an isopeptide bond between the C-terminus of ubiquitin and an ε amino group of a lysine residue in the target protein; this process is frequently mediated by E3 enzymes, ubiquitin-protein ligases (Fig. 2). Further copies of ubiquitin can be ligated, with isopeptide bond formation between the carboxyl terminus of one ubiquitin with the ε amino group of lysine 48 on the preceding ubiquitin to form a multi-ubiquitin chain on the protein target. Multi-ubiquitin chains are recognized by a protease of large

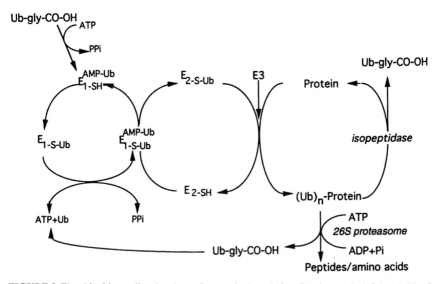

FIGURE 2. The ubiquitin-mediated pathway for protein degradation. Starting on the left-hand side of the figure ubiquitin forms a complex with AMP and ubiquitin activating enzyme, E1, following ATP hydrolysis. The ubiquitin is then transferred to form a thiol ester with a cysteine of E1; at the same time another ubiquitin forms a complex with AMP and E1. In the middle portion of the scheme, ubiquitin is transferred from the thiol ester position of E1 to E2. Then, E2 enzymes transfer ubiquitin to the target protein; this may be facilitated by a recognition factor, E3, which binds E2-ubiquitin and the target protein. When multi-ubiquitin chains have been formed on the target protein, the target protein is degraded by the 26S proteasome and ubiquitin is recycled. Ubiquitin may also be removed intact from target proteins by the action of a ubiquitin C-terminal hydrolase, without degradation of the target protein, providing a possible "proofreading" mechanism.

molecular weight, the 26S proteasome, which degrades target proteins to small peptides and amino acids in an ATP-dependent fashion, while the ubiquitin is recycled (for recent reviews, see Finley and Chau, 1991; Hershko, 1991; Hershko and Ciechanover, 1992; Jentsch, S 1992; Ciechanover, 1994).

Cells expressing temperature-sensitive E1 enzymes (ts85, ts20) degrade short-lived proteins much more slowly than their wild-type counterparts (FM3A, E36) at the nonpermissive temperature (39–43 °C) (Finley *et al.*, 1984; Gropper *et al.*, 1991). Residual degradation may be due to incomplete inactivation of the E1. Degradation of short-lived proteins is not markedly enhanced at the higher temperature in wild-type cells, even though these temperatures induce a heat-shock response and would be expected to lead to some heat-induced damage to intracellular proteins. In contrast, the degradation of long-lived proteins is enhanced some twofold at the elevated temperature (39.5 °C) in the wild-type cells, and this enhanced proteolysis, not basal proteolysis at the lower temperature, is inhibited by lysosomotropic agents. The enhanced proteolysis in wild-type cells is also inhibited by 3-methyladenine, which inhibits autophagic sequestration of cytoplasmic proteins by an unknown mechanism, implicating autophagic processes. The mutant cell lines fail to exhibit enhanced lysosomal proteolysis of long-lived proteins at the nonpermissive elevated temperature (Gropper *et al.*, 1991). These observations led Ciechanover and co-workers to suggest that ubiquitin activation, and presumably ubiquitin ligation to some unknown protein target(s), is required for heat stress-induced autophagic-lysosomal proteolysis. Serum and amino acid deprivation further elevates the lysosomal degradation of long-lived proteins in wild-type cells at the elevated temperature, and once again this is not observed in the mutant cells due to the inactive E1 under these conditions (Gropper *et al.*, 1991). Thus it would seem that multiple stresses rapidly induce an autophagic-lysosomal proteolytic pathway that is dependent on an active ubiquitination system.

Heat stress results in a shift in the distribution of β-hexosaminidase in Percoll density gradients, with the appearance of less dense vesicles containing β-hexosaminidase activity. This has been attributed to the rapid formation of autophagic vacuoles (Schwartz *et al.*, 1992). Formation of these less dense acid hydrolase-containing vesicles occurs, even in the mutant cells (ts20) expressing the temperature-sensitive E1, at the nonpermissive temperature. If these more buoyant vesicles are autophagic vacuoles then it would appear that ubiquitin activation is not required for their formation and the ubiquitin system is required for some other step in enhanced autophagy. Further joint studies by this group and Lenk and Dunn examined the volume fraction of various component organelles of the autophagic-lysosomal system in wild-type and mutant cells at the nonpermissive temperature. Electron microscopy revealed a large increase in the number of autolysosomes in mutant cells compared to the wild-type cells (Lenk *et al.*, 1992). Autolysosomes were characterized as acidic, containing hydrolases, but unlike autophagic vacuoles, also containing partially degraded cytoplasmic material. The accumulation of autolysosomes and not of autophagic vacuoles, as shown by electron microscopy, suggests that the block in lysosomal proteolysis, which occurs when the ubiquitin system is inactivated, is at the level of the maturation of autolysosomes into residual bodies, that is, sequestration into autophagic vacuoles is unaffected but complete proteolysis of sequestered cytoplasmic proteins is compromised. While autolysosomes in wild-type cells were found to contain ubiquitin or ubiquitinated proteins by immunogold electron microscopy, these were absent from the autolysosomes of mutant cells at the nonpermissive temperature (Lenk *et al.*, 1992). The authors postulate that either ubiquitination of target proteins is required for their degradation in autolysosomes or that ubiquitination of some regulatory protein(s) is required for lysosome function. The origin of the ubiquitinated proteins in the autolysosome compartment remains obscure. It is difficult to imagine that ubiquitination of proteins could occur in the harsh acidic and proteolytic environment of the lysosome system, yet these ubiquitinated proteins do not appear to be taken up by autophagy as they are absent from autophagic vacuoles (Lenk *et al.*, 1992). Alternative routes might include microautophagy into autolysosomes or delivery from primary lysosomes on fusion with autophagic vacuoles to form autolysosomes. Nevertheless, these studies would seem to indicate a previously unexpected connection between protein ubiquitination and lysosomal proteolysis.

2.1.2. Ubiquitin in the Lysosomal System

The suggested link between protein ubiquitination and the lysosomal apparatus is strengthened by the observations from a number of laboratories, including our own, that ubiquitinated proteins are concentrated in the lysosome apparatus of several different cell types as compared to the surrounding cytoplasm. Free ubiquitin is detectable in the lysosomal apparatus of hepatoma cells using an antibody that is specific for the unconjugated monomer by immunogold electron microscopy (Schwartz *et al.*, 1988). Ubiquitin also appears to be secreted from some cell types, suggesting that it is present in a vesicular cell compartment which communicates with the cell surface (Apfel *et al.*, 1992). Levels of recombinant human elastase secretion by yeast vary with the levels of polyubiquitin gene expression: In yeast cells overexpressing ubiquitin, ubiquitin accumulates in the periplasmic space (Chen *et al.*, 1994), supporting a role for ubiquitin in protein secretion and the presence of ubiquitin in the secretory apparatus. Overexpression of ubiquitin also rescues clathrin-deficient mutant yeast, although in this case this may be due to an increased demand for ubiquitin for the degradation of proteins miscompartmentalized by a compromised secretory apparatus (Nelson and Lemmon, 1993).

A complex pattern of ubiquitinated proteins is detectable in the lysosomal apparatus of fibroblasts treated with lysosomal protease inhibitors (Doherty *et al.*, 1989a). The presence of numerous ubiquitinated proteins in the lysosomal apparatus under these circumstances would perhaps be unexpected if ubiquitination of a single or a small number of lysosomal proteins was important in lysosome function. Fibroblasts also accumulate ubiquitinated proteins in the lysosomal apparatus in the absence of lysosomal inhibitors. Immunogold electron microscopy suggests that some of these ubiquitinated proteins may gain entry from the cytoplasm via microautophagy (László *et al.*, 1990). What appear to be primary lysosomes are also immunoreactive for ubiquitinated proteins, as shown by immunogold electron microscopy (László *et al.*, 1990; Löw *et al.*, 1993), and ubiquitinated proteins appear to be concentrated in the primary granules of polymorphonuclear neutrophils (László *et al.*, 1991a). Ubiquitinated proteins are detectable in the lysosome-related vacuoles of nonmammalian cells such as mutant yeast lacking active vacuolar proteases (Simeon *et al.*, 1992) and in the vacuole of the plant *A. thaliana* (Beers *et al.*, 1992). In both yeast and fibroblasts, ubiquitinated lysosomal proteins are insoluble and may be highly aggregated when lysosomal proteases are compromised (Doherty *et al.*, 1989a; Simeon *et al.*, 1992). This aggregation may be due to disulfide cross-linking (Doherty *et al.*, 1989b). However, immunogold electron microscopy of insect Sf9 cells suggests that at least some of the ubiquitinated proteins may be associated with lysosomal membrane complexes (Löw et al., 1993). The mechanism by which ubiquitinated proteins associated with lysosomal membranes is not clear, but viral associated ubiquitin suggests some possibilities. Studies on *Bacculovirus*, which infects Sf9 cells, have shown that virions contain free ubiquitin and a "modified ubiquitin." The latter molecular species turns out to be phosphatidyl ubiquitin (Guarino et al., 1995). The virions of African swine fever virus (ASFV) contain an incompletely characterized similar form of "modified ubiquitin," and the genome of ASFV also codes for a ubiquitin conjugating enzyme (UBCv1) (Hingamp et al., 1992). A membrane-anchored ubiquitin may serve as the source of ubiquitin for UBCv1 to conjugate to proteins early in viral infection (Hingamp et al., 1995). The existence of ubiquitin anchored to membranes by moieties like phosphatidic acid means that some of the ubiquitin adducts seen in endosome–lysosomes immunohistochemically may be lipid-anchored ubiquitin.

2.1.3. Are the Lysosomal and Cytosolic Ubiquitin-Related Proteolytic Systems Connected?

What then are we to make of these combined observations? The studies with cell lines defective in ubiquitin activation at elevated temperatures indicate that there is a functional link between protein ubiquitination and lysosomal proteolysis, at least stress-induced proteolysis. The presence of ubiquitinated proteins in organelles of the lysosomal apparatus could be due to a requirement for ubiquitination prior to sequestration and degradation of some cytoplasmic proteins. Alternatively, ubiquitination of some regulatory proteins, or proteins otherwise necessary for lysosome function, could be required for stress-induced autophagy, and the resulting ubiquitinated proteins accumulate in the lysosomal apparatus. If the latter is true, then it must be that numerous ubiquitinated species are required given the complexity of ubiquitinated proteins or protein fragments seen on immunoblots of proteins from lysosomal fractions separated by SDS-PAGE. It is possible that different ubiquitinated proteins may enter the endosome–lysosome system via different routes. Ubiquitinated receptors may enter the lysosome following internalization (see section 2.2), while the primary lysosome may contain a different repertoire of ubiquitinated cytosolic proteins. Both would contribute to the complex pattern seen in crude lysosomal subcellular fractions. Indeed, we have recently observed in a multivesicular, probably endosome-related, fraction derived from insect Sf9 cells a few discrete ubiquitinated polypeptides, while a lysosome-enriched fraction contains a complex pattern of many ubiquitinated species (Löw et al., 1995). This suggests a different role for ubiquitination late in the lysosomal pathway (Lenk et al., 1992). Is ubiquitination only required for stress-induced lysosomal proteolysis? This is possible, however ubiquitinated proteins appear to be present in the lysosome apparatus of apparently nonstressed cells. The necessity of heat-stressing the available mutant E1 cell line to inactivate ubiquitin activation, and the lack of measurable lysosomal proteolysis under nonstressed conditions precludes their use in investigating this point.

2.2. Internalization and Degradation of Cell Surface Proteins

2.2.1. Ligand Binding Triggers Ubiquitination of Cell Surface Receptors

Most studies of ubiquitin-mediated proteolysis have employed soluble proteins as test substrates or have investigated the ubiquitin-mediated proteolysis of endogenous soluble proteins. However, it is becoming clear that ubiquitination of membrane-bound proteins occurs, and in particular, cell surface receptors can be targets for ubiquitination. The first ubiquitinated cell surface receptor to be identified was the lymphocyte homing receptor (Siegelman et al., 1986). Lymphocytes enter the lymphoid organs via the high endothelial venules (HEV). Recognition of the HEV of peripheral nodes by lymphocytes is blocked by the monoclonal antibody MEL-14, which recognizes the C-terminal 13 residues of ubiquitin (St. John et al., 1986), and this antibody precipitates an 85–95-kDa cell surface glycoprotein (Siegelman et al., 1986). Peptide sequencing of the intact, highly purified lymphocyte homing receptor generated two amino-terminal sequences, one of which is identical to ubiquitin (Siegelman et al., 1986). This work did not determine whether the ubiquitin is conjugated to the cytoplasmic or extracellular domains of the receptor, but given that the ubiquitinating system is located in the cytosol, it would seem most likely that the putative cytoplasmic domain (residues 356–372) of this transmembrane-spanning protein (Dowbenko et al., 1991) is ubiquitinated. This domain contains four lysine residues that could serve as acceptors of ubiquitin. However, MEL-14 blocks lymphocyte homing and binds to intact cells, suggesting that the ubiquitin epitope is accessible to the extracellular space and is therefore attached to the extracellular receptor domain. This conflict awaits clarification.

Sequencing of highly purified growth hormone receptor indicates that at least some receptor molecules are ubiquitinated, while the soluble serum growth hormone binding protein is not (Spencer et al., 1988). Electrophoretic and immunological studies of the platelet-derived growth factor receptor, the T-cell antigen receptor, and the high-affinity IgE receptor indicates that these receptors are ubiquitinated following ligand binding (Cenciarelli et al., 1992; Mori et al., 1992; Paolini and Kinet, 1993). Multi-ubiquitinated forms of the T-cell receptor ζ subunit appear minutes after receptor occupancy, and both phosphorylated and nonphosphorylated versions of the receptor are ubiquitinated (Cenciarelli et al., 1992). Ubiquitination of the IgE receptor occurs seconds after ligand binding and is again independent of receptor phosphorylation (Paolini and Kinet, 1993). This ubiquitination is rapidly reversed by ligand displacement from the receptor. In contrast, ubiquitination of the platelet-derived growth factor receptor is dependent both on receptor activation and on tyrosine autophosphorylation (Mori *et al.*, 1993). Mutant receptor lacking the critical tyrosine residues for phosphorylation are much less efficiently ubiquitinated and also degraded more slowly (Mori *et al.*, 1993). However, the degradation of the ligand-bound mutant receptors is not completely abolished, and degradation of both the wild-type and mutant receptor is partially inhibited by the lysosomotropic agent chloroquine (Mori *et al.*, 1993). These results implicate ubiquitination in the degradation of occupied receptors as well as a lysosomal component to receptor degradation.

2.2.2. Ubiquitination and Internalization of Other Cell Surface Proteins

The ABC transporter Ste6 of yeast is a short-lived protein involved in the secretion of pheromone α-factor. It is associated with the plasma membrane, although most Ste6 molecules are found in the Golgi apparatus and the proteins appear to be transported to the cell surface prior to degradation (Kolling and Hollenberg, 1994). The half-life of Ste6 increases from 13 to 41 min in yeast lacking UBC4 and UBC5. The UBC4 and UBC5 enzymes are implicated in the degradation of normally short-lived proteins in yeast (Seufert and Jentsch, 1990). In addition, ubiquitinated forms of Ste6 are detectable at the cell surface in yeast defective in endocytosis, which suggests that ubiquitination may occur when the protein is located in the plasma membrane. However, Ste6 is stabilized to a greater extent ($t\frac{1}{2}$ = 2 hr) in yeast lacking Pep4, a vacuolar protease, than in $ubc4^-ubc5^-$ mutants. Degradation of Ste6 is also compromised in cells with a defective secretory pathway (Kolling Hollenberg, 1994). Thus, the greater portion of Ste6 would appear to be degraded in the vacuolar apparatus. Kolling and Hollenberg suggest that a portion of Ste6 protein may be degraded by the ubiquitin-mediated pathway, independent of the vacuolar pathway responsible for the bulk of Ste6 degradation. However, different-sized intermediate breakdown products of Ste6 would be expected to accumulate in UBC4/UBC5 mutants and Pep4 mutants if this occurred and these have not been reported so far (Kolling and Hollenberg, 1994). Alternatively, they propose that ubiquitination is required for Ste6 vacuolar degradation but that UBC4 and UBC5 are not the only ubiquitin conjugating enzymes that can ubiquitinate Ste6. The topology of Ste6 is currently unknown and possible sites for ubiquitination have not been identified.

Uracil permease is stable under normal conditions but is rapidly degraded ($t\frac{1}{2}$ = 100 min) during nitrogen starvation. This enhanced proteolysis is dependent on endocytosis, and uracil permease accumulates in Pep4 mutant cells, suggesting that degradation occurs in the yeast vacuole (Volland *et al.*, 1994). Uracil permease contains a sequence, RIALGSLTD, similar to the "destruction box," RXALGXIXN, identified in the cyclins. Cyclins are rapidly degraded by the ubiquitin-mediated pathway and targeted for ubiquitination by the presence of the destruction box, although ubiquitin is conjugated to a lysine elsewhere on the protein. Substitution of alanine for the essential arginine of the putative destruction box in uracil permease stabilizes the protein ($t\frac{1}{2} = 225$ min) during nitrogen star-vation (Galan *et al.,* 1994). The putative destruction box of uracil permease con-sists of residues 294–302, located between two predicted membrane-spanning domains (268–288, 310–330) (Jund *et al.,* 1988) and therefore could be available to a cytosolic recognition system. The cytoplasmic domain does not include a ly-sine residue. Thus, although recognition may be mediated by this sequence, ubiq-uitination must occur elsewhere. Ubiquitination of uracil permease has not been directly demonstrated so far. Although endocytosis appears to be an absolute re-quirement for permease degradation, the destruction box mutant is only partially stabilized. Once again it is not possible to pinpoint the connection, if any, between endocytosis and vacuolar degradation and ubiquitination. It is possible that endo-cytosis can occur in the absence of permease ubiquitination (hence the partial block of permease degradation in the destruction-box mutant) but that endocyto-sis is required for degradation of ubiquitinated permease, hence the block in per-mease degradation in endocytosis mutants.

The choline reuptake transporter in rat brain appears to be ubiquitinated (Meyer *et al.,* 1987), which may extend these findings to mammalian cells. Degra-dation of the yeast inositol permease in the presence of inositol, while dependent on endocytosis and vacuolar proteolytic activity, is not affected in ubiquitination mutants (Lai *et al.,* 1995). This suggests that ubiquitination may not be universally involved in the degradation of receptors and cell surface transporters. Perhaps a significant change in the topology of the cytosolic domain acts as a signal for ubiq-uitination of a receptor or transporter, and this may explain why ubiquitination of some receptor tails does not occur.

2.2.3. Cooperation between the Ubiquitin-Mediated and Lysosomal Proteolytic Systems?

It is now clear that a number of cell surface proteins that appear to be de-graded by the lysosomal or yeast vacuolar system are also ubiquitinated. Unfortu-nately, the accumulated data is not complete enough for us to propose a definitive reason for this. There are a number of possibilities. First, cell surface proteins may be ubiquitinated on their cytoplasmic tails, and while the membrane-spanning and extracellular domains of these proteins are degraded in the lysosome, the cyto-plasmic tail is attacked from the cytosolic side (possibly by the 26S proteasome; Ciechanover, 1994) to effect the complete degradation of the protein by collabo-ration between nonlysosomal and lysosomal systems (Fig. 3, scheme B). If this is the case it should be possible to identify degradative fragments when one of the collaborating pathways has been blocked. Second, molecules of cell surface pro-teins may be subject to alternative degradative fates, with perhaps a smaller por-tion subject to ubiquitin-mediated degradation as opposed to lysosomal degradation (Fig. 3, scheme A). The problem with this scenario is that while the

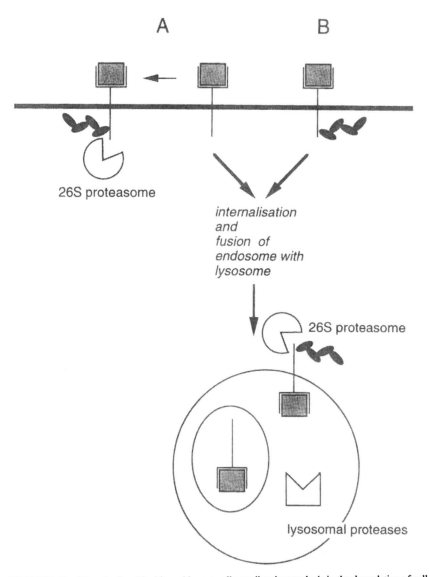

FIGURE 3. Possible roles for ubiquitin and lysosomally mediated proteolysis in the degradation of cell surface receptors. This scheme represents the possible pathways for degradation of cell surface receptors following ligand binding. In scheme A the receptor has alternate fates. Ubiquitination of receptor molecules on the cytoplasmic tail results in the complete degradation of the receptor by the 26S proteasome. This scheme poses the problem of how the cytosolic proteasome may gain access to the membrane spanning and extracellular domains of the receptor. The alternate pathway involves internalization and degradation of the receptor in the endosome/lysosome compartment in the absence of ubiquitination. To effect complete degradation by lysosomal proteases the receptor may have to be released into the lumen of the lysosome, or lysosomal/endosomal membrane regions carrying cell surface proteins could microinvaginate to form intralysosomal vesicles that, following breakdown of the membrane, would present the whole cell surface protein to lysosomal proteases (as shown). In scheme B the ubiquitin system and the lysosomal system cooperate to degrade the receptor domains to which they have access. In this case all receptors would presumably be ubiquitinated following ligand binding.

large and complex multisubunit proteasome could gain access to the cytosolic tails of ubiquitinated receptors, it is difficult to rationalize how it could attack the membrane-spanning and extracellular domains, unless the proteasome can act as a "ratchet" with successive proteolysis in the accessible intracellular domain of a cell surface protein resulting in stepwise movement of the surface protein through the membrane. Similarly, if cell surface proteins are delivered to lysosomes by fusion of endosome membranes with the lysosomal membrane, lysosomal enzymes would have ready access to the cytoplasmic portion of the protein but be unable to degrade the cytoplasmic tails, unless selected proteins could be removed from the membrane for degradation in the lumen of the lysosome. Alternatively, microinvagination of lysosomal membrane regions containing receptors and other cell surface proteins, followed by degradation of the vesicle membrane, would present all of the proteins to lysosomal proteases (Fig. 3). Third, ubiquitination may be involved in targeting a cell surface protein for internalization, endocytosis, and finally degradation. Interestingly, ubiquitinated Ste6 does accumulate at the cell surface in endocytosis mutants (Kolling and Hollenberg, 1994), providing some support for this model.

2.3. Neurodegeneration

The interplay between molecular pathology and cell biology has played a fundamental role in making new biochemical findings, for example, the discovery of microsomes in muscles in control experiments using centrifugation to study the Rous sarcoma virus. Similarly, immunohistochemical studies on ubiquitin in human and animal chronic neurodegenerative diseases led to the discovery of the relationship between ubiquitin and endosome–lysosomes.

Ubiquitin immunohistochemistry indicated two independent avenues in research into neurodegenerative diseases. Ubiquitinated proteins are part of intraneuronal filamentous inclusions characteristic of a number of neurodegenerative diseases, such as neurofibrillary tangles in Alzheimer's disease (Mori *et al.*, 1987), Lewy bodies in Lewy body dementia, and filamentous inclusions in amyotrophic lateral sclerosis (Lowe *et al.*, 1988a,b). Additionally, ubiquitin immunocytochemistry detected ubiquitin in areas of granulovacuolar degeneration in hippocampal pyramidal neurons in Alzheimer's disease (Lowe *et al.*, 1988a). Although still not formally proven in Alzheimer's disease, the observation of ubiquitin conjugates in granules in intraneuronal vacuoles led to the notion that the ubiquitinated materials were in endosome–lysosomes. This led to the direct biochemical and electron microscopic detection of ubiquitin–protein conjugates in lysosome-related organelles in fibroblasts and other cell types (Doherty *et al.*, 1989a,b; László *et al.*, 1990, 1991a).

There is evidence of ubiquitination involvement in another class of important neurodegenerative diseases, the transmissible prion encephalopathies. The most widely known of these diseases is scrapie. This is a disease of sheep where the infectious agent (prion) is though to be a proteinaceous particle, with no nucleic acid involvement (Prusiner, 1989; Weissmann *et al.*, 1993). Inoculation of prions into the brains of mice that subsequently develop scrapie has led to the development of a small animal model of scrapie, which is highly advantageous given the long incubation time of the disease, especially in larger animals. Immunogold electron microscopic studies are much easier to perform on brains of mice infected with scrapie during disease progression than on brain from human neurodegenerative diseases. This work has shown that ubiquitin conjugates are present in endosome–lysosomes in axosomatic processes abutting onto nerve cell bodies (László *et al.*, 1992). Recently, immunogold electron microscopy with antibodies to the cation-independent mannose 6-phosphate receptor has shown that ubiquitin conjugates are in late endosomes together with the lysosomal enzyme β-glucuronidase and the abnormal infective form of the prion protein that transmits disease (Arnold *et al.*, 1995). The number of late endosomes increases in an exponential manner during the terminal stages of disease progression in the transmissible encephalopathies (Kenward *et al.*, 1992).

The results of ubiquitin immunocytochemistry together with the experimental molecular biological studies described above underpins the "bioreactor hypothesis, which proposes that endosome–lysosomes serve as chambers for the conversion of the normal cellular prion protein (PrP^c) into the abnormal infective form (PrP^{sc}) (Arnold *et al.*, 1995). Endosome–lysosomes, identified in part by ubiquitin immunocytochemistry, also appear to be involved in proteolytic processing of the Alzheimer precursor protein (Mayer *et al.*, 1994) and in the response of lymphoblastoid cells to the expression of the latent membrane protein during infection with Epstein-Barr virus (László *et al.*, 1991b). The internalization and proteolytic processing of different types of surface proteins in endosome–lysosomes containing ubiquitin–protein conjugates and possibly other forms of ubiquitinated moieties seem to be central to chronic neurodegenerative and viral diseases. These discoveries were all dependent on the finding of ubiquitin conjugates in lysosome-related organelles.

3. REFERENCES

Amenta, J. S., Baccino, F. M. and Sargus, M. J., 1976, Cell protein degradation in cultured rat embryo fibroblasts, *Biochim. Biophys. Acta* **451**:511–516.
Amenta, J. S., Hlivko, T. J., MacBee, A. G., Shinozuka, H., and Brocher, S., 1978, Specific inhibition by NH₄Cl of autophagy-associated proteolysis in cultured fibroblasts, *Exp. Cell Res.* **115**:357–366.
Apfel, R., Lottspeich, F., Hoppe, J., Behl, C., Durr, G., and Bogdahn, U., 1992, Purification and analysis of growth-regulating proteins secreted by a human-melanoma cell-line, *Melanoma Res.* **2**:327–336.
Arnold, J. E., Tipler, C., László, L., Hope, J., Landon, M., and Mayer, R. J., 1995, The abnormal form of the prion protein accumulates in endosome/lysosome-like organelles in scrapie-infected mouse brain, *J. Pathol.* **177**:403–411.
Aubry, L., Klein, G., Martiel, J. L., and Satre, M., 1993, Kinetics of endosomal pH evolution in *Dictyostelium discoideum* amoebas—study by fluorescence spectroscopy, *J. Cell Sci.* **105**:861–866.

Backer, J., Bourret, L., and Dice, J., 1983, Regulation of degradation of microinjected ribonuclease A requires the amino-terminal 20 amino acids, *Proc. Natl. Acad. Sci. USA* **80**:2166–2170.

Backer, J. M. and Dice, J. F., 1986, Covalent linkage of ribonuclease S-peptide to microinjected proteins causes their intracellular degradation to be enhanced during serum withdrawal, *Proc. Natl. Acad. Sci. USA* **83**:5830–5834.

Ballard, F. J., Knowles, S. E., Wong, S. S. C., Bodner, J. B., Wood, C. M., and Gunn, J. M., 1980, Inhibition of protein breakdown in cultured cells is a consistent response to growth factors, *FEBS Lett.* **114**:209–212.

Becker, J., and Craig, E. A., 1994, Heat shock proteins as molecular chaperones, *Eur. J. Biochem.* **219**:11–23.

Beers, E. P., Moreno, T. N., and Callis, J., 1992, Subcellular localization of ubiquitin and ubiquitinated proteins in *Arabidopsis thaliana, J. Biol. Chem.* **267**:15432–15439.

Cenciarelli, C., Hou, D., Hsu, K. C., Rellahan, B. L., Wiest, D. L., Smith, H. T., Fried, V. A., and Weissman, A. M., 1992, Activation-induced ubiquitination of the T-cell antigen receptor, *Science* **257**:795–797.

Chen, Y. P., Pioli, D., and Piper, P. W., 1994, Overexpression of the gene for polyubiquitin in yeast confers increased secretion of a human-leukocyte protease inhibitor, *Biotechnology* **12**:819–823.

Chiang, H.-L., and Dice, J. F., 1988, Peptide sequences that target proteins for enhanced degradation during serum withdrawal, *J. Biol. Chem.* **263**:6797–6805.

Chiang, H.-L., Terlecky, S. T., Plant, C. P., and Dice, J. F., 1989, A role for a 70-kilodalton heat shock protein in lysosomal degradation of intracellular proteins, *Science* **246**:382–384.

Ciechanover, A., 1994, The ubiquitin-mediated proteolytic pathway: Mechanisms of action and cellular physiology, *Biol. Chem. Hoppe-Seyler* **375**:565–581.

Cuervo, A. M., Terlecky, S. R., Dice, J. F., and Knecht, E., 1994, Selective binding and uptake of ribonuclease-A and glyceraldehyde-3-phosphate dehydrogenase by isolated rat-liver lysosomes, *J. Biol. Chem.* **269**:26374–26380.

Deluca-Flaherty, C., McKay, D. B., Parham, P., and Hill, B. L., 1990, Uncoating protein (Hsc70) binds a conformationally labile domain of clathrin light chain LCa to stimulate ATP hydrolysis, *Cell* **62**:875–887.

Dice, J. F., 1990, Peptide sequences that target cytosolic proteins for lysosomal proteolysis, *Trends Biochem. Sci.* **15**:305–309.

Dice, J. F., and Chiang, H.-L., 1989, Lysosomal degradation of microinjected proteins, in *Current Trends in the Study of Protein Degradation I* (E. Knecht and S. Grisolia, eds.), pp. 13–33, Springer International, Bilbao.

Dice, J. F., Chiang, H.-L., Spencer, E. P., and Backer, J. M., 1985, Lysosomal degradation of ribonuclease-A and ribonuclease-S-protein microinjected into the cytosol of human fibroblasts, *J. Biol. Chem.* **260**:1986–1993.

Dice, J. F., Chiang, H.-L., and Spencer, E., 1986, Regulation of catabolism of microinjected ribonuclease A: Identification of residues 7–11 as the essential pentapeptide *J. Biol. Chem.* **261**:6853–6859.

Doherty, F. J., Osborn, N. U., Wassell, J. A., Heggie, P. E., László, L., and Mayer, R. J., 1989a, Ubiquitin-protein conjugates accumulate in the lysosomal system of fibroblasts treated with cysteine proteinase inhibitors, *Biochem. J.* **263**:47–55.

Doherty, F. J., Osborn, N. U., Wassell, J. A., László, L., and Mayer, R. J., 1989b, Insoluble disulfide cross-linked polypeptides accumulate in the functionally compromised lysosomes of fibroblasts treated with the cysteine protease inhibitor E-64, *Exp. Cell Res.* **185**:506–518.

Dowbenko, D. J., Diep, A., Taylor, B. A., Lusis, A. J., and Lasky, L. A., 1991, Characterization of the murine homing receptor gene reveals correspondence between protein domains and coding exons, *Genomics* **9**:270–277.

Ellis, R. J., and Vandervies, S. M., 1991, Molecular chaperones, *Annu. Rev. Biochem.* **60**:321–347.

Finley, D., and Chau, V., 1991, Ubiquitination, *Annu. Rev. Cell Biol.* **7**:25–69.

Finley, D., Ciechanover, A., and Varshavsky, A., 1984, Thermolability of ubiquitin-activating enzyme from the mammalian cell cycle mutant ts85, *Cell* **37**:43–55.

Galan, J. M., Volland, C., Urbangrimal, D., and Haguenauertsapis, R., 1994, The yeast plasma-membrane uracil permease is stabilized against stress-induced degradation by a point mutation in a cyclin-like destruction box, *Biochem. Biophys. Res. Commun.* **201**:769–775.

Goldberg, A. L., and Rock, K. L., 1992, Proteolysis, proteasomes and antigen presentation, *Nature* **357**:375–379.

Gropper, R., Brandt, R. A., Elias, S., Bearer, C. F., Mayer, A., Schwartz, A. L., and Ciechanover, A., 1991, The ubiquitin-activating enzyme, E1, is required for stress-induced lysosomal degradation of cellular proteins, *J. Biol. Chem.* **266**:3602–3610.

Guarino, L. A., Smith, G., and Dong, W., 1995, Ubiquitin is attached to membranes of baculovirus particles by a novel type of phospholipid anchor, *Cell* **80**:301–309.

Hendil, K., 1977, Intracellular protein degradation in growing, in density-inhibited and in serum restricted fibroblast cultures, *J. Cell. Physiol.* **92**:353–364.

Hershko, A., 1991, The ubiquitin pathway of protein degradation and proteolysis of ubiquitin-protein conjugates, *Biochem. Soc. Trans.* **19**:726–729.

Hershko, A., and Ciechanover, A., 1992, The ubiquitin system for protein degradation, *Annu. Rev. Biochem.* **61**:761–807.

Hingamp, P., Arnold, J. E., Mayer, R. J., and Dixon, L. K., 1992, A ubiquitin conjugating enzyme encoded by African swine fever virus, *EMBO J.* **11**:361–366.

Hingamp, P. M., Leyland, M. L., Webb, J., Twigger, S., Mayer, R. J., and Dixon, L. K., 1995, Characterisation of a ubiquitinated protein which is externally located in African swine fever virus, *J. Virol.* **69**:1785–1793.

Humbert, M., Bertolino, P., Forquet, F., Rabourdincombe, C., Gerlier, D., Davoust, J., and Salamero, J., 1993, Major histocompatibility complex class II-restricted presentation of secreted and endoplasmic reticulum resident antigens requires the invariant chains and is sensitive to lysosomotropic agents, *Eur. J. Immunol.* **23**:3167–3172.

Jentsch, S., 1992, The ubiquitin-conjugation system, *Annu. Rev. Genet.* **26**:179–207.

Jund, R., Weber, E., and Chevallier, M. R., 1988, Primary structure of the uracil transport protein of *Saccharomyces-cerevisiae, Eur. J. Biochem.* **171**:417–424.

Kenward, N., Fergusson, J., Landon, M., Hope, J., McDermott, H., McQuire, D., Lowe, J., and Mayer, R. J., 1992, Early detection of ubiquitin-protein conjugate immunoreactivity in scrapie, *J. Cell Biochem.* **16E**:212.

Kolling, R., and Hollenberg, C. P., 1994, The ABC-transporter Ste6 accumulates in the plasma-membrane in a ubiquitinated form in endocytosis mutants, *EMBO J.* **13**:3261–3271.

Lai, K., Bolognese, C. P., Swift, S., and McGraw, P., 1995, Regulation of inositol transport in *saccharomyces-cerevisiae* involves inositol-induced changes in permease stability and endocytic degradation in the vacuole, *J., Biol. Chem.* **270**:2525–2534.

László, L., Doherty, F. J., Osborn, N. U., and Mayer, R. J., 1990, Ubiquitinated protein conjugates are specifically enriched in the lysosomal system of fibroblasts, *FEBS Lett.* **261**:365–368.

László, L., Doherty, F. J., Watson, A., Self, T., Landon, M., Lowe, J., and Mayer, R. J., 1991a, Immunogold localization of ubiquitin-protein conjugates in primary (azurophilic) granules of polymorphonuclear neutrophils, *FEBS Lett.* **279**:175–178.

László, L., Tuckwell, J., Self, T., Lowe, J., Landon, M., Smith, S., Hawthorne, J. N., and Mayer, R. J., 1991b, The latent membrane protein-1 in Epstein-Barr virus-transformed lymphoblastoid cells is found with ubiquitin-protein conjugates and heat-shock protein-70 in lysosomes oriented around the microtubule organizing centre, *J. Pathol.* **164**:203–214.

László, L., Lowe, J., Self, T., Kenward, N., Landon, M., McBride, T., Farquhar, C., McConnell, I., Brown, J., Hope, J., and Mayer, R. J., 1992, Lysosomes are key organelles in the pathogenesis of prion encephalopathies, *J. Pathol.* **166**:333–341.

Lenk, S. E., Dunn, W. A., Trausch, J. S., Ciechanover, A., and Schwartz, A. L., 1992, Ubiquitin-activating enzyme, E1, is associated with maturation of autophagic vacuoles, *J. Cell Biol.* **118**:301–308.

Leustek, T., Dalie, B., Amirshapira, D., Brot, N., and Weissbach, H., 1989, A member of the hsp70 family is localized in mitochondria and resembles *Escherichia-coli* DNAK, *Proc. Natl. Acad. Sci. USA* **86**:7805–7808.

Löw, P., Doherty, F. J., Sass, M., Kovacs, J., Mayer, R. J., and László, L., 1993, Immunogold localisation of ubiquitin protein conjugates in Sf9 insect cells—implications for the biogenesis of lysosome-related organelles, *FEBS Lett.* **316**:152–156.

Löw, P., Doherty, F. J., Fellinger, E., Sass, M., Mayer, R. J., and László, L., 1995, Related organelles of the endosome-lysosome system contain a different repertoire of ubiquitinated proteins in Sf9 insect cells, *FEBS Lett.* **368**:125–131.

Lowe, J., Blanchard, A., Morrell, K., Lennox, G., Reynolds, L., Billett, M., Landon, M., and Mayer, R. J., 1988a, Ubiquitin is a common factor in intermediate filament inclusion bodies of diverse type in man, including those of Parkinson's disease, Pick's disease, and Alzheimer's disease, as well as Rosenthal fibres in cerebellar astrocytomas, cytoplasmic bodies in muscle, and Mallory bodies in alcoholic liver disease, *J. Pathol.* **155**:9–15.

Lowe, J., Lennox, G., Jefferson, D., Morrell, K., McQuire, D., Gray, T., Landon, M., Doherty, F. J., and Mayer, R. J., 1988b, A filamentous inclusion body within anterior horn neurones in motor neurone disease defined by immunocytochemical localisation of ubiquitin, *Neurosci. Lett.* **94**:203–210.

Mayer, R. J., Tipler, C., László, L., Arnold, J., Lowe, J., and Landon, M., 1994, Endosome-lysosomes and neurodegeneration, *Biomed. Pharmacother.* **48**:282–286.

Meyer, E. M., West, C. M., Stevens, B. R., Chau, V., Nguyen, M. T., and Judkins, J. H., 1987, Ubiquitin-directed antibodies inhibit neuronal transporters in rat brain synaptosomes, *J. Neurochem.* **49**:1815–1819.

Mori, H., Kondo, J., and Ihara, Y., 1987, Ubiquitin is a component of paired helical filaments in Alzheimer's disease, *Science* **235**:1641–1644.

Mori, S., Heldin, C. H., and Claesson-Welsh, L., 1992, Ligand-induced polyubiquitination of the platelet-derived growth factor beta-receptor, *J. Biol. Chem.* **267**:6429–6434.

Mori, S., Heldin, C. H., and Claessonwelsh, L., 1993, Ligand-induced ubiquitination of the platelet-derived growth factor beta-receptor plays a negative regulatory role in its mitogenic signaling, *J. Biol. Chem.* **268**:577–583.

Nelson, K. K., and Lemmon, S. K., 1993, Suppressors of clathrin deficiency—overexpression of ubiquitin rescues lethal strains of clathrin-deficient *Saccharomyces-cerevisiae*, *Mol. Cell. Biol.* **13**:521–532.

Overly, C. C., Lee, K. D., Berthiaume, E., and Hollenbeck, P. J., 1995, Quantitative measurement of intraorganelle pH in the endosomal lysosomal pathway in neurons by using ratiometric imaging with pyranine, *Proc. Natl. Acad. Sci. USA* **92**:3156–3160.

Paolini, R., and Kinet, J.-P., 1993, Cell surface control of the multi-ubiquitination and de-ubiquitination of high-affinity immunoglobulin E receptors, *EMBO J.* **12**:779–786.

Prusiner, S. B., 1989, Scrapie prions, *Ann. Rev. Microbiol.* **43**:345–374.

Rothman, J. E., 1989, Polypeptide chain binding proteins—catalysts of protein folding and related processes in cells, *Cell* **59**:591–601.

Schirmbeck, R., and Reimann, J., 1994, Peptide transporter-independent, stress protein-mediated endosomal processing of endogenous protein antigens for major histocompatibility complex class-I presentation, *Eur. J. Immunol.* **24**:1478–1486.

Schwartz, A. L., Ciechanover, A., Brandt, R. A., and Geuze, H. J., 1988, Immunoelectron microscopic localization of ubiquitin in hepatoma cells, *EMBO J.* **7**:2961–2966.

Schwartz, A. L., Brandt, R. A., Geuze, H., and Ciechanover, A., 1992, Stress-induced alterations in autophagic pathway—relationship to ubiquitin system, *Am. J. Physiol.* **262**:C1031–C1038.

Seufert, W., and Jentsch, S., 1990, Ubiquitin-conjugating enzymes Ubc4 and Ubc5 mediate selective degradation of short-lived and abnormal proteins, *EMBO J.* **9:**543–550.

Siegelman, M., Bond, M. W., Gallatin, W. M., John, T. S., Smith, H. T., Fried, V. A., and Weissman, I. L., 1986, Cell surface molecule associated with lymphocyte homing is a ubiquitinated branched-chain glycoprotein, *Science* **231:**823–829.

Simeon, A., Vanderklei, I. J., Veenhuis, M., and Wolf, D. H., 1992, Ubiquitin, a central component of selective cytoplasmic proteolysis, is linked to proteins residing at the locus of non-selective proteolysis, the vacuole, *FEBS Lett.* **301:**231–235.

Slot, L. A., Lauridsen, A. M. B., and Hendil, K. B., 1986, Intracellular protein-degradation in serum-deprived human-fibroblasts, *Biochem. J.* **237:**491–498.

Spencer, S. A., Hammonds, R. G., Henzel, W. J., Rodriguez, H., Waters, M. J., and Wood, W. I., 1988, Rabbit liver growth hormone receptor and serum binding protein. Purification, characterization, and sequence, *J. Biol. Chem.* **263:**7862–7867.

St. John, T., Gallatin, W. M., Siegelman, M., Smith, H. T., Fried, V. A., and Weissman, I. L., 1986, Expression cloning of a lymphocyte homing receptor cDNA—ubiquitin is the reactive species, *Science* **231:**845–850.

Stuart, R. A., Cyr, D. M., and Neupert, W., 1994, Hsp70 in mitochondrial biogenesis: From chaperoning nascent polypeptide chains to facilitation of protein degradation, *Experientia* **50:**1002–1011.

Vanwhy, S. K., Hildebrandt, F., Ardito, T., Mann, A. S., Siegel, N. J., and Kashgarian, M., 1992, Induction and intracellular-localization of hsp-72 after renal ischemia, *Am. J. Physiol.* **263:**F769–F775.

Velezgranell, C. S., Arias, A. E., Torresruiz, J. A., and Bendayan, M., 1994, Molecular chaperones in pancreatic tissue—the presence of cpn10, cpn60 and hsp70 in distinct compartments along the secretory pathway of the acinar cells, *J. Cell Sci.* **107:**539–549.

Volland, C., Urbangrimal, D., Geraud, G., and Haguenauertsapis, R., 1994, Endocytosis and degradation of the yeast uracil permease under adverse conditions, *J. Biol. Chem.* **269:**9833–9841.

Weissmann, C., Bueler, H., Sailer, A., Fischer, M., Aguet, M., and Aguzzi, A., 1993, Role of PrP in prion diseases, *Br. Med. Bull.* **49:**995–1011.

Wing, S. S., Chiang, H. L., Goldberg, A. L., and Dice, J. F., 1991, Proteins containing peptide sequences related to lys-phe-glu-arg-gln are selectively depleted in liver and heart, but not skeletal muscle, of fasted rats, *Biochem. J.* **275:**165–169.

Chapter 6

Lysosomal Metabolism of Proteins

Robert W. Mason

1. INTRODUCTION

The lysosome is a major site of mammalian protein degradation and is calculated to be capable of degrading as much protein as the digestive tract (Barrett and Kirschke, 1981). Detailed characteristics are known for several of the enzymes involved in lysosomal proteolysis, although recent data suggest that the current catalog of enzymes involved in protein degradation is by no means complete. Although the enzymes are the necessary tools for hydrolysis, they are only effective in the right environment. In this chapter the relationship between the different enzymes and their environment is discussed, including mechanisms by which lysosomal proteolysis is controlled. Delivery of substrates to the lysosome, the pH in the lysosome, and removal of products from the lysosome are discussed in depth elsewhere in this volume (Chapters 3, 4, 5, 10, and 11), so discussion of these topics will be restricted to how they relate specifically to protein turnover. This chapter focuses on how the proteases function to hydrolyze proteins in the lysosome (for information on the pathological roles and the chemistry of the lysosomal proteases, see Lee and Marzella, 1994, and Mason and Wilcox, 1993).

Robert W. Mason   Division of Developmental Biology, Nemours Research Programs, Wilmington, Delaware 19899, and Department of Pediatrics, Jefferson Medical College, Philadelphia, Pennsylvania 19107.

Subcellular Biochemistry, Volume 27: Biology of the Lysosome, edited by Lloyd and Mason. Plenum Press, New York, 1996.

2. THE ENZYMES

Enzymes that degrade proteins can be divided into endopeptidases and ex-opeptidases. The former hydrolyze internal peptide bonds while the latter hydrolyze amino acids or peptides from the N- or C-terminal ends of proteins. Collectively these can be termed peptidases, but, as this word is usually used to describe enzymes that act only on short peptides, the term protease will be used in this chapter. There are four known catalytic mechanisms for peptide bond cleavage, which provides a means of classifying the proteases. The catalytic residues involved in cleavage are serine, cysteine, aspartic acid, or a metal ion (usually Zn^{2+}). The specificity of a protease is determined by the amino acid sequence around the cleavage site. The most widely used terminology assigns amino acids within the substrate protein toward the N-terminus of the cleavage site as P_1, P_2, P_3, etc., and amino acids toward the C-terminus as P_1', P_2', P_3', etc. (Schecter and Berger, 1967). The related binding sites in the protease are similarly termed S_1, S_2, S_3, etc. (for review, see Barrett, 1994).

Protein hydrolysis in the lysosome is performed by an array of endo- and exopeptidases working together to degrade the proteins to small peptides and amino acids. The earliest known lysosomal proteases were discovered before their intracellular location was known. In 1941, Fruton and colleagues described four enzymes derived from crude tissue homogenates and named them cathepsins A, B, C, and D. These four enzymes represent the major types of proteases typically found in mammalian lysosomes. The collective name "cathepsin" was chosen to indicate that they were enzymes found primarily in the tissues rather than in extracellular fluids. Most of the cathepsins have subsequently been shown to function principally in the lysosome, with the possible exceptions of cathepsins G and E.

The lysosomal proteases are all active at acidic pH (although this is not necessarily the pH optimum). They are synthesized as prepro enzymes, with the signal peptide directing them into the endoplasmic reticulum and the propeptide controlling expression of their activity. The propeptide is cleaved during or shortly after packaging into the lysosomes.

The aspartic proteases are represented by two enzymes, cathepsins D and E. The cysteine proteases are represented by several enzymes, including cathepsins B, C, H, I, J, K, L, M, N, O, P, S, T, and X. Some of these letters may not represent unique gene products because the names were given to describe activities that may have been due to combinations of the better characterized members of this group of enzymes (see Mason and Wilcox, 1993).

The better characterized enzymes will be discussed in more detail below. Other activities have been given the names cathepsins A, F, G, and R, although of these only cathepsins A and G have been sufficiently well characterized to merit further discussion. Cathepsins A and G are both serine proteases. Cathepsin A is an exopeptidase that cleaves amino acids from the carboxyl terminus of proteins. It is commonly known as lysosomal carboxypeptidase A, a name that more accurately describes its catalytic action. Cathepsin G is an endopeptidase found in the azurophil granules of neutrophils. These granules are lysosome-like compartments within neutrophils, and cathepsin G is thought to be involved primarily in proteolysis during inflammation. The enzyme acts outside the cell, and thus the azurophil granules are not normally considered to be true lysosomes but, rather, storage granules for secreted enzymes.

2.1. Endopeptidases

Cathepsins B and D are the best known lysosomal endopeptidases. The early discovery of these enzymes was possible because both enzymes are rather stable (at least at pH 6 and below) and because there are no endogenous tightly binding inhibitors of these enzymes in mammalian tissues. Over time, as more advanced techniques for protein isolation and detection have been developed, other endopeptidases have been discovered. Perhaps the most significant feature of the lysosomal endopeptidases is that they do not exhibit the exquisite specificities seen for many of the serine proteases and seem to be able to accommodate a variety of amino acid side chains in their "specificity pockets." Thus the different endopeptidases show widely overlapping cleavage sites in model substrates (Bromme *et al.,* 1989). It is likely that binding of and cleavage by the endopeptidases occur as a result of the net effect of multiple interactions of residues in the substrate with binding sites in the enzymes. However, it is clear that in some proteins there are regions of peptides that are resistant to some proteases and not others, suggesting that multiple endopeptidases offer advantages for efficient cleavage of the vast array of proteins that are degraded in lysosomes.

2.1.1. Cathepsin D

Cathepsin D is an aspartic endopeptidase with two aspartic acid residues in its active site. It is related in catalytic mechanism to the gastric enzyme pepsin, and, like pepsin, it has an acidic pH optimum for hydrolysis of a range of proteins. It is proposed that the two carboxylate side chains of the catalytic aspartic acid residues of these endopeptidases bind a proton and a water molecule and that this binding is essential for catalysis to proceed (Dunn, 1989). The low pK of these carboxyl groups suggests that protonation, and hence enzymic activity, will only occur at low pH. While this is true for pepsin, it is clearly not the case for the AIDS viral protease and renin, both of which are active at neutral pH. It has been shown that amino acids in both the enzyme and substrate contribute to increasing the pK of the side chains of the aspartic acid residues in these enzymes (Hyland *et al.,* 1991; Green *et al.,* 1990). Almost all the data on the activity of cathepsin D indicate that it is only active below pH 6.0. The crystal structure of cathepsin D reveals a long substrate-binding cleft that is able to accommodate hydrophobic amino acids in synthetic substrates or inhibitors (Baldwin *et al.*, 1993; Scarborough *et al.*, 1993). Cathepsin D has no reported exopeptidase activity, and its preference for hydrophobic amino acids might imply that this enzyme is primarily involved in the degradation of denatured proteins in the lysosome.

2.1.2. Cathepsin B

Cathepsin B is a cysteine endopeptidase, reflecting the essential catalytic residue in the active site of the enzyme. Cathepsin B was sequenced in 1983 by Takio *et al.*, who showed that it was related by sequence to the plant enzyme papain.

The papain superfamily of cysteine proteases contains the largest group of related enzymes that have been found in the lysosome (Rawlings and Barrett, 1994). The major lysosomal cysteine endopeptidases and their distinguishing features are summarized in Table I. Their catalytic mechanism utilizes two conserved residues, a cysteine and a histidine. These two residues form an ion pair; as a result, the pK of the side-chain thiol of the cysteine is lowered to 4.0 and the pK of the imidazole group of histidine is raised to 8.0. The enzymes are thus active in the pH range of 4 to 8. The apparent acidic pH optimum for the hydrolysis of proteins by the lysosomal cysteine endopeptidases arises partly due to the unusual instability of some of these enzymes above pH 6.0, and partly due to the increased accessibility of peptide bonds in the substrate proteins caused by acidic denaturation.

Using a range of synthetic substrates and inhibitors, cathepsin B has been shown to be able to accommodate a variety of amino acids in specificity pockets P_3 to P_1, although it shows some preference for large hydrophobic amino acids in P_2 and P_3 (reviewed in Mason and Wilcox, 1993). Of the known lysosomal cysteine endopeptidases, cathepsin B has a unique capability to hydrolyze peptides with positively charged amino acids in P_2. This makes benzyloxycarbonyl-Arg-

Table I
Cysteine Endopeptidases

Enzyme	Unique properties
Cathepsin B	Hydrolyzes Z-Arg-Arg-NHMec
Cathepsin C	Hydrolyzes Gly-Phe-NHMec
Cathepsin H	Hydrolyzes Arg-NHMec
Cathepsin L	None
Cathepsin O	Unknown
Cathepsin S	Stable at pH 7
Cathepsin K	Primarily expressed in osteoclasts
Cathepsin Z	Reacts with Fmoc-Leu-Leu[^{125}I]Tyr-CHN$_2$ but not Fmoc-[^{125}I]Tyr-Ala-CHN$_2$

Arg-7-amino-4-methylcoumarin (Z-Arg-Arg-NHMec) a somewhat selective substrate for cathepsin B. The crystal structure of cathepsin B indicates that the enzyme is able to accommodate arginine in S_2 because of the presence of a negatively charged residue in the binding pocket (Musil *et al.*, 1991). A site-directed mutagenesis study with cathepsin S has shown that this related enzyme can increase its ability to hydrolyze Z-Arg-Arg-NHMec if a similarly charged residue is introduced into this binding site (Bromme *et al.*, 1994). The ability of cathepsin B to react with inhibitors containing basic amino acids in P_1 and P_2 led to the suggestion that cathepsin B might function as a pro-insulin processing enzyme (Docherty *et al.*, 1984). However, cathepsin B does not seem to prefer dibasic residues in proteins, and a more specific enzyme for prohormone processing has been discovered (Smeekens and Steiner, 1990).

Despite the early discovery of cathepsin B as an enzyme that degrades proteins, its endopeptidolytic activity has been questioned. Although it acts primarily as an exopeptidase against a variety of native proteins (see Section 2.2.4), it can cleave internal peptide bonds in proteins such as the β-chain of insulin and $α_2$-macroglobulin (McKay *et al.*, 1983; Mason, 1989). Thus cathepsin B is clearly an endopeptidase, but the extent to which it cleaves internal peptide bonds compared to its peptidyl dipeptidase hydrolytic activity remains controversial. Cathepsin B is found almost ubiquitously in mammalian cells and tissues and is therefore proposed to have a general role in protein turnover. In terms of its ability to degrade proteins, cathepsin B is considered to be a moderately active enzyme that has restricted activity against a number of native proteins and only poorly acts upon insoluble proteins such as elastin and collagen. The effect of the nature of the protein substrates on rates of hydrolysis will be discussed later (Section 3.2).

2.1.3. Cathepsin H

Cathepsin H is a much weaker endopeptidase than cathepsin B, but it can cleave the general endopeptidase substrates azocasein and $α_2$-macroglobulin (Schwartz and Barrett, 1980; Mason, 1989). The activity of cathepsin H against azocasein is 20 times less than the activity of cathepsin B against this substrate (Schwartz and Barrett, 1980). The enzyme is capable of degrading N-terminally blocked synthetic substrates such as benzoyl-Arg-N-naphthylamide (Bz-Arg-NNap) and Bz-Phe-Val-Arg-NHMec (Schwartz and Barrett, 1980; Xin *et al.*, 1992). Clearly cathepsin H is a poor general endopeptidase, even though it is able to cleave some specific internal peptide bonds. It has been argued that cathepsin H, like cathepsin B, is primarily an exopeptidase (see Section 2.2.1). It is certainly the case that cathepsin H can be most easily identified by its aminopeptidase activity against substrates such as Arg-NHMec, even though it is equally capable of hydrolyzing analogous substrates that are N-terminally blocked.

2.1.4. Cathepsin C

Cathepsin C is primarily considered to be an exopeptidase (see Section 2.2.5). However, a recent report suggests that this enzyme can cleave internal peptide bonds of lysozyme, N-acetylated tryptic peptides of lysozyme, and the synthetic substrate Z-Phe-Arg-NHMec (Kuribayashi et al., 1993). Peptides that are blocked at the N- and C-termini are not normally considered to be substrates for exopeptidases, suggesting that cathepsin C has endopeptidolytic activity. Hydrolysis of Z-Phe-Arg-NHMec has been confirmed using an alternative purification scheme whereby this substrate was used to purify an activity dubbed "cathepsin J" (Nikawa et al., 1992). This activity is actually due to cathepsin C. It was previously shown that the endopeptidolytic substrate Bz-Arg-NH₂ could inhibit hydrolysis of the dipeptidyl peptidase substrate Gly-Phe-NH₂ by cathepsin C (Metrione and MacGeorge, 1975). This inhibition was rationalized by proposing that the guanidino group of the arginine side chain could bind to the enzyme at the site that usually recognizes the N-terminal amino group of the exopeptidase substrate, and it is possible that this could also be responsible for the binding and hydrolysis of Z-Phe-Arg-NHMec (R. Metrione, personal communication). This proposal does not explain the hydrolysis of the N-acetylated tryptic peptides which have a variety of amino acids in P_2 and P_1, none of which are positively charged (Kuribayashi et al., 1993). It has previously been shown that highly purified cathepsin C can still contain low levels of other contaminating hydrolytic activities (McDonald et al., 1972), although the most recent study appears to rule out the possibility of such contamination (Kuribayashi et al., 1993). The relevance of the endopeptidase activity of cathepsin C is not clear as the exopeptidase activity is certainly the most prominent in the cleavage of proteins (McDonald et al., 1972). This work and the preceding discussion of the relative endo- and exopeptidase activities of cathepsins B and H do help to emphasize the difficulty in defining the physiologically significant functions of individual enzymes.

2.1.5. Cathepsin L

Cathepsin L was discovered many years after cathepsin B, primarily because of its tight inhibition by a sequence-related group of naturally occurring cysteine protease inhibitors called cystatins. These inhibitors are found both extracellularly and in the cytoplasm of cells. The cytoplasmic cystatins are poor inhibitors of cathepsin B but they bind very tightly to cathepsin L (Barrett, 1987). The first unequivocal identification of this enzyme was achieved by first isolating large quantities of lysosomes prior to purification of endopeptidases (Kirschke et al., 1977). Cathepsin L has proved to be one of the most powerful endopeptidases in cells. This enzyme has no significant exopeptidase activity and is able to hydrolyze a wide variety of both native and denatured proteins. It is able to hydrolyze fibrous proteins such as elastin and collagen as well as almost every other protein tested. Using peptide substrates and inhibitors, cathepsin L appears to have a very broad specificity, accommodating a variety of hydrophobic amino acids in subsites S_3 to S_1 (reviewed in Mason and Wilcox, 1993). Like cathepsin B, it is active in the pH range 4–8, but, being unstable at neutral pH, it appears to degrade most proteins optimally below pH 6.0. Its catalytic efficiency against most substrates is 10 to 100 times greater than that of cathepsin B. Although it was originally thought to be a minor member of the cysteine protease family, it now seems that levels of expression of cathepsin L are very high and at least equivalent to those of cathepsin B in several cell types (see Section 2.3).

2.1.6. Cathepsin S

Cathepsin S is another powerful lysosomal cysteine endopeptidase, comparable in activity to cathepsin L against proteins such as collagen and elastin (Kirschke et al., 1989; Xin et al., 1992). In fact the similarities between these two enzymes frustrated the purification of this enzyme for many years. It is now clear that cathepsin S is a different gene product and is expressed in a limited number of tissues, including spleen and kidney (Shi et al., 1992). The unique properties of cathepsin S include its stability and activity at neutral pH and its rather limited tissue distribution. In most cells, cathepsin S is considerably less abundant than cathepsins B or L. Like cathepsin L, cathepsin S does not seem to have any significant exopeptidase activity. The characteristics of cathepsin S make it clear that neither instability at neutral pH nor a housekeeping role can be considered as general features of lysosomal proteases.

2.1.7. Other Possible Endopeptidases

Molecular cloning has revealed the existence of several other proteins that are related to cathepsins B, C, H, L, and S by sequence. These have been named cathepsin O (Velasco et al., 1994), cathepsin K (also identified as a gene called OC-2 and a protein that was named cathepsin O; Inaoka et al., 1995; Tezuka et al., 1994; Shi et al., 1995), cathepsin L–related protein (Conliffe et al., 1995), and testin (Grima et al., 1995).

Cathepsin O was expressed in E. coli, and the purified, renatured protein exhibited limited activity against the endopeptidase substrates Z-Phe-Arg-NHMec and Z-Arg-Arg-NHMec, suggesting that this enzyme could be an endopeptidase (Velasco et al., 1994). Very little is known about the specificity of this enzyme, which was cloned from a human breast carcinoma cDNA library and appears to be expressed in most human tissues.

Cathepsin K was transfected into COS-7 cells and shown to be capable of degrading fibrinogen even more efficiently than cells transfected with cathepsin S

(Shi *et al.*, 1995). This work suggests that cathepsin K has comparable activity to the better known powerful endopeptidase cathepsin S. The specificity of cathepsin K has not yet been established. Cathepsin K is expressed in most tissues, although expression is very high in osteoclasts.

Testin is a protein encoded by a cDNA derived from rat sertoli cells (Grima *et al.*, 1995). Northern and Western blotting indicate that the protein is expressed primarily in sertoli or ovary cells. Localization studies in sertoli cells show that testin is located primarily in tight junctions between cells, and thus it is proposed to be a junctional protein and not a lysosomal enzyme. The sequence is 58% identical to that of rat cathepsin L, but there is a serine residue instead of cysteine in the proposed catalytic site. It has been reported that the protein has no proteolytic activity, although it does apparently cleave itself slowly at acidic pH. This latter observation implies some proteolytic capacity and therefore warrants closer investigation. If it proves to have activity, it will be an unusual serine protease that does not seem to have the typical catalytic triad that includes an aspartic acid residue as well as serine and histidine.

Cathepsin L-related protein was identified as a sequence from a placental cDNA library (Conliffe *et al.*, 1995). Although there are some errors in the published sequence (K. T. Shiverick, personal communication), the sequence is clearly different from that of any other known cysteine protease. The mRNA is expressed in placenta but not liver and kidney, suggesting a unique role for the protein. The protein has not been purified and thus it is not known whether the encoded protein is an endo- or exopeptidase.

Inhibitor studies have indicated that there may be other cysteine proteases that have not yet been characterized. A radiolabeled cysteine protease inhibitor, N-[N-(L-3-*trans*-ethoxycarbonyloxirane-2-carbonyl)-L-leucyl]-3-methylbut-lamine (E-64d), was shown to bind to a variety of unidentified proteins in human epidermoid carcinoma A431 cells (Shoji-Kasai *et al.*, 1988). The compounds E-64, N-[N-(L-3-*trans*-carboxyoxirane-2-carbonyl)-L-leucyl]amino-4-guanidinobutane, and related analogues have been shown to be rather specific for cysteine proteases, evidence that some of the labeled proteins represent previously unidentified enzymes. The compound Z-Leu-Leu-[^{125}I]Tyr-CHN$_2$ has been shown to react with a number of proteins in cells, and several of these have been shown not to be any of the known cysteine endopeptidases (Anagli *et al.*, 1991). One of these proteins has been purified and termed cathepsin Z (DeCourcy, 1995). It is a major protein expressed by a number of human cell lines. Although sequence data for the protein is not yet available, it has been purified as an inactive protein, and all of its characteristics are consistent with it being a novel lysosomal cysteine protease.

2.2. Exopeptidases

The endopeptidases are undoubtedly important for the initiation of degradation of proteins, but as protein degradation proceeds, new C- and N-terminal

Table II
Lysosomal Exopeptidases

Enzyme	Type and units cleaved
Cathepsin A (lysosomal carboxypeptidase A)	Carboxypeptidase, one
Lysosomal carboxypeptidase B	Carboxypeptidase, one
Prolylcarboxypeptidase	Carboxypeptidase, one
Cathepsin B[a]	Carboxypeptidase, two
Peptidyl dipeptidase B	Carboxypeptidase, two
Cathepsin H[a]	Aminopeptidase, one
Dipeptidyl peptidase II	Aminopeptidase, two
Cathepsin C (dipeptidyl peptidase I)[a]	Aminopeptidase, two
Tripeptidyl peptidase I	Aminopeptidase, three

[a]Enzyme is also an endopeptidase.

groups are generated to increase the number of potential substrates for the exopeptidases. In addition, when peptides are less than eight amino acids long, they make fewer contacts with the extended binding sites of some of the endopeptidases and consequently become poorer substrates. The exopeptidases include some of the endopeptidases discussed above. Exopeptidases in lysosomes have been reported that can cleave one, two, or even three amino acids from the C- or N-terminal ends of proteins (Table II). Many of the lysosomal exopeptidases await complete characterization (see McDonald and Barrett, 1986). It would seem likely that several enzymes await discovery, although the array of enzymes currently known show that lysosomes are capable of hydrolyzing the peptides produced by endopeptidase action down to smaller units.

2.2.1. Cathepsin H

Cathepsin H is an aminopeptidase, a fact that has provided unique substrates for characterizing the enzyme, in particular Arg-NHMec. When porcine spleen cathepsin H was incubated with a range of peptide substrates, only aminopeptidase activity could be detected, and the reduced and carboxymethylated proteins soybean trypsin inhibitor and aldolase were not degraded at all (Takahashi *et al.*, 1988). Although it is clear that cathepsin H does exhibit some endopeptidase activity (see Section 2.1.3), the relative contributions of the two different activities of the enzyme are not known. No other lysosomal aminopeptidases have been fully characterized, and cathepsin H may be the only lysosomal enzyme with this activity (Bohley and Seglen, 1992).

2.2.2. Cathepsin A

Cathepsin A is a carboxypeptidase more commonly known as lysosomal carboxypeptidase A. As mentioned earlier, it is a serine protease with an acidic pH optimum. In addition to hydrolyzing C-terminal peptide bonds, cathepsin A can remove amide groups from the C-termini of a range of peptidyl amides (Jackman *et al.,* 1990). It is an intriguing enzyme, because it is the only lysosomal protease implicated in a lysosomal storage disease (Okamura-Oho *et al.,* 1994). In the disease galactosialoidosis, there is a mutation in a gene that encodes a serine carboxypeptidase that had previously been termed "protective protein" (Galjart *et al.,* 1988). It appears that this protein is most likely cathepsin A (Jackman *et al.,* 1990; Galjart *et al.,* 1991). However, the major metabolic defect noted for this mutation is a lysosomal accumulation of sialyloligosaccharides that is apparently not related to the catalytic activity of cathepsin A. Protective protein, or cathepsin A, is associated with two glycosidases, β-galactosidase and neuraminidase, protecting them from proteolysis. In the genetic disease, it appears that the glycosidases are made normally but are rapidly degraded in the lysosome. There may be other effects directly due to a defect in cathepsin A, but as yet there is no clear evidence that this is the case.

2.2.3. Lysosomal Carboxypeptidase B

Lysosomal carboxypeptidase B is a cysteine protease and only acts on proteins by exopeptidase action. It does however have some esterase and amidase activity against low-molecular-weight substrates (Lipperheide and Otto, 1986). The sequence of the enzyme has not yet been determined, but since it is inhibited by E-64, there is a strong possibility that it is related to the papain superfamily of proteases.

2.2.4. Cathepsin B

Cathepsin B has long been known to have peptidyl dipeptidase activity, that is, it sequentially removes dipeptides from the C-terminus of proteins. The crystal structure of the purified enzyme has helped to explain this activity (Musil *et al.,* 1991). The substrate binding site is occluded at the S_3' binding site by two histidine residues. It is proposed that these form an electrostatic interaction with the carboxyl end of peptide substrates, allowing two amino acids to lie across the S_1' and S_2' binding sites. While this can explain exopeptidase activity of this enzyme, it does pose a problem for explaining the endopeptidase activity of the enzyme. Using a range of peptide substrates and the proteins soybean trypsin inhibitor and aldolase (reduced and carboxymethylated) Takahashi *et al.* (1986) could only demonstrate peptidyl dipeptidase activity of porcine spleen cathepsin B. The assay pH value of 5.0 would have favored protonation of the above-mentioned histidine residues that bind free carboxyl groups, so it is possible that low pH makes the exopeptidolytic activity of cathepsin B most pronounced. It is certainly true that cathepsin B is a less powerful endopeptidase than the other related endopeptidases, and it may be that the endopeptidase activity depends on the conformation of the substrate proteins as well as vesicular pH.

2.2.5. Cathepsin C

Cathepsin C, lysosomal dipeptidyl peptidase I, is primarily considered to be an exopeptidase, cleaving dipeptides from the N-terminus of proteins. In addition to its probable role in lysosomal protein turnover, cathepsin C may have a more physiologically important role in removing dipeptide propeptides from the N-termini of a range of serine endopeptidases of inflammatory cells (McGuire *et al.*, 1993; Dikov *et al.*, 1994).

2.2.6. Tripeptidylpeptidase I

Although the majority of the exopeptidases remove one or two amino acids from the ends of proteins, the lysosome does contain an enzyme that removes three amino acids from the N-terminus of proteins. This enzyme is tripeptidyl peptidase I (McDonald *et al.*, 1985). An enzyme with similar activity has been purified from a human osteoclastoma, and it is proposed that this enzyme could be responsible for the resorption of bone protein by these tumors (Page *et al.*, 1993).

2.2.7. Prolylcarboxypeptidase

A major obstacle to protein hydrolysis is a cleavage of peptides containing proline, presumably because the restricted flexibility in the N–C bond of this amino acid hinders productive binding of adjacent amino acids to proteases. In the lysosome the degradation of such peptides is enabled by the action of enzymes such as prolylcarboxypeptidase, which cleaves amino acids from the C-terminal side of proline residues. Prolylcarboxypeptidase is a serine protease and has limited sequence identity to cathepsin A (Tan *et al.*, 1993). The lysosome is reported to have other activities that are capable of cleaving peptides containing proline residues, although these enzymes have not yet been fully characterized.

2.2.8. Dipeptidases and Tripeptidases

There are a few reports of lysosomal enzymes that appear to hydrolyze only di- and tripeptides. Dipeptidyl aminopeptidase II only acts upon tripeptides and not larger peptides (McDonald *et al.*, 1968). The enzyme was purified over 1000-fold from bovine anterior pituitary gland. Enzymic activity capable of yielding free amino acids from dipeptides has been demonstrated in lysosomal extracts (Bouma *et al.*, 1976). McDonald *et al.* (1972) also reported a dipeptidase that cleaves ser-met that had an acidic pH optimum. This protein was a contaminant of a cathepsin C preparation and was inhibited by EDTA. No further characterization was performed, so its subcellular origin is not clear.

Uncharacterized dipeptides produced by the hydrolysis of globin were not degraded by lysosomal extracts, but were by cytoplasmic peptidases (Coffey and deDuve, 1968), suggesting that many dipeptides are not hydrolyzed by lysosomal enzymes. Since these early studies, many new proteases have been discovered, and better techniques have been devised for purifying lysosomes. However, a specific lysosomal dipeptidase has not been purified and it would seem that this area of research warrants reinvestigation.

2.3. Enzyme Concentrations

The concentrations of cathepsins B and D in lysosomes have been estimated to be approximately 1 mM, based upon the yield of enzymes purified from liver (Dean and Barrett, 1976). This would be equivalent to about 10% of the soluble protein in lysosomes for each enzyme. Clearly all of the 50–60 lysosomal enzymes cannot exist in the same high concentrations. Similar calculations using data from different purifications of cathepsin L give wildly different results. The problem with trying to do such calculations with cathepsin L is that a true measure of enzymic activity in original homogenates is not possible because of inactivation of the enzyme by cystatins. An approach to estimate intracellular concentrations of cathepsin L has been made using a membrane-permeant, active site–directed irreversible inhibitor. It was found that the intracellular labeling of cathepsin L in KNIH 3T3 fibroblasts by Z-[^{125}I]Tyr-Ala-CHN$_2$ was equivalent to that of cathepsin B (Mason *et al.,* 1989). A study using titrated inhibitor has enabled more precise calculation of the amounts of active cathepsins B and L in cells. In a human macrophage-like cell line, cathepsins B and L have been both calculated to exist in the range of 1 mM in lysosomes (Mason and Sander, unpublished results). By contrast, cathepsin S concentrations were much lower than cathepsin B in most cells studied. The concentrations of most of the exopeptidases are at least 10-fold lower than cathepsin B, D, and L. The reason for the high endopeptidase concentrations are not immediately obvious, but as the substrates for endopeptidases are considerably more complex, it is likely that initial cleavages will be less efficient. More enzyme would help overcome this inefficiency.

3.   SUBSTRATE DEGRADATION

It is clear from the previous section that lysosomes contain a wide range of proteases that are capable of degrading proteins down to smaller units. The ability of the enzymes to degrade any protein will depend upon the structure of the substrate. In the lysosome this will be occurring in a unique environment where the reaction conditions will be very different from those normally used in laboratory studies.

3.1. Substrate Concentration

Substrate concentration in any individual lysosome is hard to determine, partly due to heterogeneity of lysosomes and partly due to each protein having large numbers of peptide bonds that can potentially be cleaved by each of the enzymes. The concentration of peptide bonds due just to enzymes within a lysosome is in excess of 1 M. Even though most of these bonds have presumably evolved to be resistant to proteolysis, it is possible to envisage that some amino acid side chains have affinity for the active sites of the enzymes, resulting in enzyme inhibition. Substrate proteins taken up into lysosomes must compete with these "inhibitory" interactions in order to be hydrolyzed.

What then are the relative concentrations of enzymes, inhibitors, and substrates? Phagocytosis of bacteria or large particles delivers large volumes of substrates to the endosomal-lysosomal system (see Chapter 3). Pinocytosis delivers much smaller volumes of substrates to mature lysosomes. These two mechanisms of delivery of substrates will have very different effects on concentrations of enzymes and substrates. Thus, phagolysosomes will contain a high concentration of substrate and consequently low concentration of enzyme (and inhibitor). In mature lysosomes the enzyme (and inhibitor) concentration will be very high (see Section 2.3 above). *In vitro* experiments typically use concentrations of enzymes that are several orders of magnitude lower than found in lysosomes and substrate concentrations usually exceed enzyme concentrations. Phagolysosomes are therefore more similar to *in vitro* conditions than are mature lysosomes. It would seem that hydrolysis of proteins taken up into mature lysosomes could actually be catalytically inefficient, with a large proportion of the proteases being nonproductively bound to other lysosomal components and not actually being involved in proteolysis. There is some experimental evidence to suggest that this is the case. Using radiolabeled, active site-directed inhibitors to label cathepsins B and L in isolated lysosomes, it was found that the rate of labeling the enzymes was considerably slower than that seen *in vitro* (Wilcox and Mason, 1992). This can only be partly explained by the non-ideal condition of using an inhibitor concentration lower than the lysosomal enzyme concentration. Binding of the proteases to non-hydrolyzable proteins in the lysosome would be expected to reduce the rate of interaction of the proteases with the inhibitor. These considerations imply that more dilute proteases in phagolysosomes will be more catalytically efficient.

In vitro experiments with millimolar concentrations of the lysosomal proteases have not been performed, owing to the difficulty in obtaining sufficient enzyme. However, mechanisms are available to dilute lysosomal enzyme concentrations in living cells. Sucrose is accumulated in lysosomes, and this causes the organelles to become enlarged, leading to dilution of lysosomal protein. Sucrose-induced lysosomal enlargement has been demonstrated to reduce pinocytosis by macrophages (Swanson *et al.*, 1986; Montgomery *et al.*, 1991). A limited study of the degradation of radiolabeled albumin (complexed with IgG) taken up by Fc receptor-mediated endocytosis in these macrophages concluded that proteolysis is not affected by sucrose loading of lysosomes (Montgomery *et al.*, 1991). However, it was also concluded that the proteolysis was not occurring in the sucrose-loaded vacuoles because the protein was not delivered to these organelles. An earlier study had shown that the activities of some lysosomal enzymes in fibroblasts are induced by sucrose loading, although this was not related to intracellular activity and induction did not occur in direct response to lysosome enlargement (Kato *et al.*, 1984). The long time period (up to 13 days) required for this induction could have been due either to increased synthesis or decreased degradation of the individual enzymes. The effect of sucrose loading on the intracellular activities of individual proteases has not yet been studied, but it is well known that weak ionic association of lysosomal components does occur, and this has been proposed as a possible mechanism by which the enzyme activities might be controlled (Koenig, 1962).

It is thus clear that the rate of hydrolysis of any peptide bond will be regulated by the concentration of competing peptide bonds. These peptide bonds can be derived from the lysosomal enzymes themselves or from other substrates. If the affinity of a protease for a binding site in a substrate is not significantly greater than the affinity for binding to other lysosomal proteins, then clearly hydrolysis of the substrate will be slow.

3.2. Molecular Forms of the Substrates

It is clear from the preceding section that proteases are unusual enzymes in that they are themselves potential substrates. Most proteases have evolved to be resistant to proteolysis by becoming tightly packed, stable structures. The interaction of a protein with a protease requires that amino acids of the substrate bind to the enzyme such that the scissile peptide bond fits productively across the active site of the enzyme. In a native protein, the conformation of the polypeptide is usually restricted to a single conformation. This can result in very specific cleavage of a protein that precisely fits the active site of a protease, such as occurs between the proteases and their respective substrates in the blood coagulation cascade. At the other extreme, small peptides can usually adopt several conformations and thus bind to any active site that can accommodate the amino acid side chains. However, the decrease in conformational flexibility when the peptide binds is energetically unfavorable, and thus peptides frequently bind only weakly to proteases. For this reason, the K_M of factor X_a for a range of synthetic substrates is on the order of 0.1 mM, while the K_M for its natural substrate, prothrombin, is as low as 0.4 μM (Lottenberg *et al.*, 1981). It is thus quite clear that the conformation of a protein will have a profound influence on its susceptibility to proteolysis.

The proteins delivered to the lysosome for degradation are many and varied, and it is chiefly the environment in the lysosome that influences their rate of hydrolysis. Proteins that are degraded primarily in lysosomes (as compared to degradation extracellularly or in the cytoplasm) include those that are taken up by receptor-mediated endocytosis (see Chapter 3) or selective autophagy (see Chapter 5). There is little evidence to suggest that such proteins are any more or less susceptible to hydrolysis by lysosomal enzymes than other proteins. Proteins that would be more susceptible to hydrolysis are misfolded proteins, because these are more likely to have less well structured exposed loops. Berg *et al.* (1984) showed that defective collagen is degraded rapidly in the lysosome. The extent to which the lysosome is involved in the degradation of newly synthesized proteins is unclear, as it seems that many misfolded proteins are degraded in the endoplasmic reticulum and not delivered to lysosomes (Stafford and Bonifacino, 1991). It is proposed that the enzymes involved in this degradation could be cysteine proteases, and two candidates are proteins called ER60 and ERp72 (Urade *et al.*, 1992, 1993). The purified proteins were shown to degrade a variety of proteins, albeit only slowly and to a limited extent. Both ER60 and ERp72 are related in sequence and activity to protein disulfide isomerase (Rupp *et al.*, 1994), and thus the significance of the proteolytic activity of preparations of the proteins is unclear. Proteins that are newly inserted into the endoplasmic reticulum will initially be unfolded and thus be good substrates for proteases. Therefore, some sort of selective recognition or packaging of proteins to prevent all of them being hydrolyzed would seem to be necessary. It may be that the enzymes responsible for the degradation seen in the endoplasmic reticulum are in fact lysosomal enzymes that are activated in a subcompartment of the endoplasmic reticulum.

It is obviously important that some proteins in the lysosome are not degraded. These include all of the lysosomal enzymes. In endosomes, which also seem to contain active lysosomal enzymes (see Section 4.1), it is also important that recycling receptors and major histocompatibility complex (MHC) molecules are not degraded. Still other molecules require partial proteolysis especially antigens for processing. The simplest ways of achieving such selectivity would seem to be by selective packaging and altering the environment of the substrates such that they become unstable. These are the topics of the next few sections.

3.3. Influence of pH on Degradation

The effect of pH on the hydrolysis of intact proteins is profound. Much of this hydrolysis is related to the denaturation of the proteins. Evolution has favored stabilization of proteins at their normal physiological pH. As the pH is lowered, the net charge of a protein is altered such that intermolecular charge–charge stabilization is decreased, promoting unfolding. At very low pH (2–3), many proteins are completely denatured; at the pH of the lysosome, estimated to be on the order of pH 5.0, only partial denaturation would be expected. Substrates that are delivered to lysosomes will typically go through a pH gradient, with the pH dropping gradually from 7.2 outside the cells down to 6.5 to 6.0 in endosomes and then eventually to 5.0 or less in the lysosome (see Chapter 10). Even a small weakening of the stability of a protein will enhance the chance that at least a portion of substrate protein will unfold sufficiently to be recognized by a lysosomal protease. Once a newly exposed peptide bond is cleaved, the native structure of a polypeptide is less likely to reform after unfolding and thus more of the polypeptide is likely to unfold and be degraded further. For the lysosomal enzymes, evolution has favored stabilization of the enzymes at low pH, such that these do not unfold at pH 5.0, and even though many of them are clipped by proteases, this does not seem to affect their overall stability. Human cathepsin L has a half-life of 80 sec at neutral pH (Turk et al., 1993). This slow inactivation of an enzyme that retains catalytic potential at neutral pH would not be an effective mechanism for the control of the activity of the enzyme in vivo should it escape from the lysosomal environment. A more likely control mechanism would be inhibition by natural inhibitors such as cystatins and α_2-macroglobulin (Barrett, 1987). It is likely that many of the lysosomal enzymes are unstable at neutral pH because there has been no evolutionary pressure for them to be stable outside an acidic environment. The lysosomal enzymes are initially folded as proenzymes in the endoplasmic reticulum, where the pH is closer to neutral. Work with procathepsin L has shown that this form of the enzyme is stable at neutral pH and that the propeptide of the enzyme is probably important for the correct folding and stability of this enzyme in the endoplasmic reticulum (Mason et al., 1987; Smith and Gottesman, 1989). In contrast, it has been reported that cathepsin S can be expressed as a fusion protein without its propeptide and that the protein folds to give active enzyme (Petanceska and Devi, 1992). This might argue against a direct role of the propeptide in folding and pH stabilization, although it should be remembered that unlike cathepsin L, the mature form of cathepsin S is stable at neutral pH.

Low pH denaturation of substrates and high pH denaturation of enzymes led to the erroneous belief that lysosomal endopeptidases all have acidic pH optima. With the advent of small synthetic substrates, short incubation times were possible, which showed that the pH activity profiles of the lysosomal cysteine endopeptidases are similar to those of papain, with activity over a broad range of pH 4–8 (Mason et al., 1985). Collagen is a highly ordered protein and is resistant to the activity of many endopeptidases. The pH optima for the degradation of collagen fibers by cathepsins B and L are below 4.0, even though it is quite clear that these enzymes are catalytically competent at neutral pH (Kirschke et al., 1977). The low pH optimum is attributed to the denaturation of the collagen molecules at reduced pH. By contrast, the insoluble protein elastin can be degraded by cathepsin S at neutral pH, and the pH dependence of this degradation is more closely related to the pH dependence of degradation of the synthetic substrate Z-Phe-Arg-NHMec (Xin et al., 1992).

As mentioned above, cathepsin D has little activity above pH 6.0 and is optimally active against most proteins at pH 3–4 (Barrett, 1970; Cunningham and Tang, 1976). The low pH optimum is related to the pK of the side chains of the active site aspartic acid residues. Limited activity of cathepsin D above pH 6.0 has been reported in endosomes, but this is controversial (see Section 4.1). Although limited degradation of proteins by cathepsin D may occur above pH 6.0, it is certain that the enzyme will be much more efficient at lysosomal pH values.

Activation of proforms of the lysosomal proteases occurs at reduced pH. There are essentially two different types of mechanisms of activation. One involves processing by preexisting active enzymes, while the other involves auto-activation. Intracellular activation of cathepsins B and D was suggested to be caused by preexisting active enzymes, using protease inhibitors (Hara *et al.,* 1988; Nishimura *et al.,* 1989; Samarel *et al.,* 1989). A mechanism of autoactivation was suggested when the proform of cathepsin L was purified (Gal and Gottesman, 1986). Procathepsin L was shown to autoactivate at pH 3.0 to yield an active mature form of the enzyme. It was later discovered that this autoactivation could be catalyzed by interaction of the proenzyme with negatively charged surfaces (Mason and Massey, 1992). This activation could occur at a more physiological pH of 6.0, and was clearly second-order. Procathepsin B has also been shown to be capable of autoactivation below pH 6.0, and this was proposed to be a first-order activation process (Mach *et al.,* 1994). Activation *in vitro* above pH 6.0 either does not occur or occurs very slowly, indicating that the proenzymes may not be activated if they are delivered to compartments that do not reduce the pH to 6.0 or lower. Cathepsin D can undergo autoactivation at pH 3.5, although the cleavage that occurs *in vitro* may not occur *in vivo* (Richo and Conner, 1994).

The idea that lysosomal proteases are only active in acidic degradative compartments clearly has to be modified to include the possibility of limited hydrolysis of proteins pH values closer to neutral pH, especially in endosomes. It is however very clear that acidic pH will strongly favor proteolysis.

3.4. Thiol Oxidation/Reduction

In the cytoplasm of cells, a high concentration of reduced glutathione (5 mM) ensures that a reductive environment is maintained, resulting in few proteins with disulfide bonds. Proteins that are targeted for secretion or intracellular packaging often have multiple disulfide bonds, and thus the environment in the endoplasmic reticulum where these proteins are initially folded must be oxidizing (Young *et al.,* 1993). In the lysosome, most of the enzymes have disulfide bonds, implying that the environment is not so reducing as to break these bonds. Also, in the disease cystinosis, the lack of a functional transporter for cystine results in accumulation of this disulfide in lysosomes, suggesting that the environment in the lysosome is oxidizing (Gahl and Tietze, 1987). The lysosomal cysteine proteases all need reducing agent in order to express activity *in vitro* (e.g., Barrett, 1980). How then do the cysteine proteases function in the lysosome? There is a transporter for cysteine that is capable of delivering this amino acid into lysosomes (Pisoni *et al.,* 1990). This amino acid will be able to provide reducing power to enable disulfide bonds within substrate proteins to be broken and should also help maintain the active sites of cysteine endopeptidases in a reduced state. Lack of the transporter of cystine cannot be alleviated by uptake of cysteine, because this will not change the concentration of cystine (Lloyd, 1986). In isolated lysosomes, added cysteine does not significantly increase the reactivity of cathepsins B and L (Wilcox and Mason, 1992). These results imply that the environment in the lysosome is sufficiently reducing to prevent oxidation of the active site of the lysosomal cysteine proteases. Furthermore, there is no evidence that the activities of cysteine proteases are impaired in cystinosis, a genetic disease that results in lysosomal accumulation of cystine. Data are available to suggest that the environment in endosomes is not reducing. Radio-iodinated tyrosine that was cross-linked to the endosome-restricted protein transferrin by a reducible disulfide bond was not released from the conjugate, while a similarly ligated tyrosine conjugated to the lysosomally targeted protein α_2-macroglobulin was released (Collins et al., 1991). The degree of reduction of the disulfide bonds in these experiments may be more related to the amount of time that the conjugates are resident in the different organelles; a major difference in the uptake processes for the two proteins is that while α_2-macroglobulin will accumulate in cells over time, transferrin will be rapidly released. These results do suggest that the lysosomal system has the ability to reduce disulfide bonds. An earlier study using a nonhydrolyzable compound linked by a disulfide bond to a low-molecular-weight ligand also concluded that the environment in endosomes is not reducing (Feener et al., 1990). However, this study also failed to demonstrate that the lysosomal environment is reducing, and the authors concluded that reduction probably occurs in the Golgi apparatus. A major difference between the two studies was that Feener et al. (1990) measured total conjugated ligand, whereas Collins et al. (1991) only measured ligand that was immunoprecipitated with antibodies to the carrier protein. The former study would not distinguish between ligand attached to the original carrier and ligand exchanged to another compound. The latter study only showed that ligand was removed from the carrier protein, but does not necessarily mean that the ligand is released as a reduced compound. Both results are compatible with disulfide exchange occurring, which does not necessarily require a significantly reducing environment.

From the above discussion, it would seem that the state of oxidation of any given sulfhydryl group of a protein in the lysosome will be dependent on the relative concentrations of cysteine and cystine, as well as the pK of the sulfhydryl group. The degree of oxidation/reduction is likely to vary between the different compartments of the endosomal/lysosomal system, but it is unlikely that all sulfhydryls will be either fully oxidized or fully reduced.

4. INTRACELLULAR SITES OF DEGRADATION

4.1. Endosomes

The protease content of endosomes is a controversial issue. This may be due in part to difficulties in precisely defining the identity of the endosome. Transferrin is a molecule that is responsible for the delivery of Fe^{2+} into cells, without being degraded itself. It must therefore either deliver the iron to a protease-deficient environment or itself be resistant to proteolysis. Other polypeptides, such as antigens that require processing and presentation to antibody-producing cells, must be degraded, but only to a limited extent.

Isolated endosomes from mouse peritoneal macrophages have been shown to contain a cathepsin D-like protease that can degrade radiolabeled albumin at acidic pH (Diment and Stahl, 1985). It is not clear that albumin can be degraded at normal endosomal pH in these macrophages. Runquist and Havel (1991) showed that although cathepsin D could be found in rat liver endosomes, it was not capable of degrading apolipoprotein B-100 *in vivo*. It may be that cathepsin D exhibits limited activity above pH 6.0 and that this activity will depend on the nature of the substrate. Casciola-Rosen and Hubbard (1991) showed that radiolabeled albumin was not degraded in the endosomes of rat liver cells and that these organelles did not contain any cathepsin B–like activity. These results suggest that not all endosomes have strong proteolytic capacities and that although proteases may be present in endosomes, they may not be proteolytically competent. Ricin A-chain can be hydrolyzed rapidly by a mouse macrophage cell line, and this hydrolysis is inhibited by both cysteine and aspartic protease inhibitors (Blum *et al.*, 1991). When endosomes were isolated from these cells, they were shown to be able to degrade ricin A-chain at neutral pH, and this activity could be inhibited by a range of protease inhibitors, including pepstatin and leupeptin. These results suggest that endosomes from these cells contain aspartic and cysteine proteases that are active at neutral pH. Epidermal growth factor appears to be partially processed in the endosomes of mouse fibroblasts and rat liver, but the processing is limited to carboxypeptidase activity, and complete degradation only occurs in lysosomes (Schaudies *et al.*, 1987; Renfrew and Hubbard, 1991).

The pH in the endosome is generally considered to be 6.0 or higher (see Chapter 10). This is expected to restrict the activity of cathepsin D. Also, the endocytosed proteins are less likely to be denatured above pH 6.0 and thus would be less susceptible to proteolysis. The concentrations of lysosomal enzymes in these compartments are not known but are probably much lower than in mature lysosomes, so again the hydrolytic capacity of endosomes is presumably impaired. The major question seems to be whether the quantities of enzyme present are significant and whether their activity against substrates is significant in endosomes.

4.2. Specialized Organelles

In the past few years it has become apparent that several mammalian cells contain a discrete set of organelles that are involved in packaging peptides for antigen presentation. Ultrastructural analysis has demonstrated that cathepsins B and D, antigen, and the proteins of the MHC class II complex can be found in a single organelle after only 2 min of antigen uptake (Guagliardi *et al.*, 1990). A set of organelles have been isolated in a number of laboratories based upon differences in charge and density. These organelles contain MHC class II proteins and processed peptides. The MHC molecules appeared in these organelles 30 min after synthesis and finally moved to the plasma membrane during the next 1–3 hr (Amigorena *et al.*, 1994). These organelles do not contain the MHC invariant chain, which is processed by cysteine and aspartic proteases (Maric *et al.*, 1994), but they appear to contain proteases that continue to process antigen (West *et al.*, 1994; Geuze, 1994). Thus, although antigens are rapidly packaged with the proteins necessary for processing and presentation, the processing and assembly of the presentation complex probably occurs in an organelle that can be distinguished from bulk endosomes and lysosomes. A variety of endosomal and lysosomal "markers" have been detected in these organelles and it remains to be established whether these organelles contain a unique complement of lysosomal proteases. The length of time that substrates reside in these compartments could allow an ideal organelle to carry out more extensive digestion of proteins than when proteins are transiently in endosomes, but not as extensive as possible in the very acidic lysosomes.

4.3. Mature Lysosomes

The mature lysosome is characterized as a dense organelle that contains mature lysosomal enzymes and substrates, but not the mannose 6-phosphate transporter proteins. These organelles receive substrates from the cytoplasm and the extracellular milieu by endocytosis and autophagy (see Chapters 3, 4, and 5). This continuous delivery of materials by membrane fusion is presumably accompanied by membrane budding to return the membrane components to endosomes and autophagic vacuoles. Soluble proteins can enter the lysosome in the vesicles that fuse and presumably exit the lysosome in the budding vesicles. It is not yet clear how substrates and enzymes are retained in the lysosome.

Aggregation and tight association with other proteins seem to be common with the lysosomal enzymes. Cathepsin D has been shown to be tightly associated with saponin C, an activator of β-glucosidase (Zhu and Conner, 1994), and cathepsin C self-associates to give a multimeric protein (Nikawa *et al.*, 1992). As discussed above, cathepsin A is tightly associated with two other lysosomal enzymes. It was also shown that β-hexosaminidase, β-galactosidase, α-fucosidase, and cathepsin C activities were pelletted by high-speed centrifugation after lysis of lysosomes at lysosomal pH (Buckmaster *et al.*, 1988). It is likely that the high protein concentration within lysosomes will lead to other protein-protein interactions, resulting in a "lysosomal matrix" (Koenig, 1962). Such a matrix has been proposed to explain the ability of lysosomes to retain their enzyme contents in repeated cycles of fusion and budding of vesicles delivering proteins between endosomes and lysosomes. Clearly, association of substrates with such a matrix would also result in their retention in lysosomes.

The mature lysosomes is often considered to be a terminal degradative compartment where proteins are degraded down to amino acids and peptides. The lysosomes appear to be interconnected, and are able to exchange contents (Deng *et al.*, 1991). It has also been shown that proteins that enter lysosomes are able to transfer back to endosomes, so the term "terminal compartment" would appear to be an oversimplified description of this compartment (Wilson *et al.*, 1993). Jahraus *et al.* (1994), showed that sucrose accumulated in lysosomes is capable of retrograde traffic to the endosomes and thus may be able to slowly exit the cell by a reversal of the endocytic process. Under normal culture conditions, lysosomal endopeptidases rarely escape from the cell, so these must be retained in mature lysosomes by some mechanism. Aggregation seems to be a plausible explanation for retention, and any molecules that do not aggregate, such as sucrose, might be expected to escape from the lysosome.

5. THE FINAL PRODUCTS

5.1. Production of Peptides and Amino Acids

The final products of protein hydrolysis by lysosomal proteases are amino acids and peptides. In 1968, Coffey and deDuve demonstrated that lysosomal extracts could digest acid-denatured globins and that the major products were amino acids and small peptides. The conditions used for hydrolysis were pH 4.4 without reducing agents, and added reducing agent at this pH was reported not to enhance the percentage of degradation. Reducing agents have been shown to be essential for *in vitro* activity of cathepsins B, H, L, and S, and thus the endopeptidase activity measured by Coffey and deDuve would primarily be due to cathepsin D. By measuring release of free N-termini, it was concluded that 30% of the globin peptide bonds were resistant to hydrolysis and that most of the resistant products were dipeptides. Bestatin, an inhibitor of a range of cytosolic aminopeptidases, has been shown to cause the accumulation of peptides derived from cellular protein turnover in the cytosol, even though it is presumed that the proteins are hydrolyzed by the autophagic process (Botbol and Scornik, 1989). This work suggests that most of the aminopeptidase activity of cells is cytosolic. It could be concluded that dipeptidases are not necessary in lysosomes, as dipeptides can be transported out of lysosomes and hydrolyzed by the cytosolic aminopeptidases. However, it seems that not all dipeptides can be transported out of the lysosome (see Chapter 11), and thus those that cannot may be degraded by the lysosomal dipeptidases.

The relative importance of exopeptidases and endopeptidases in proteolysis is difficult to ascertain. Units that are resistant to endopeptidases can be determined by examining the products of protein hydrolysis. The most widely used peptide substrate for determining specificity of proteases is the β-chain of insulin. This 30-amino acid peptide can be hydrolyzed by proteases and the peptides produced identified by HPLC and amino acid analysis. The cleavage products imply that cathepsin S requires at least two amino acids toward the N- and C-terminus of the cleavage site (Bromme *et al.*, 1989). The same appears to be true for cathepsin B (McKay *et al.*, 1983), and cathepsin L may require these three amino acids on each side of the cleavage site (Kirschke *et al.*, 1977). Thus the smallest unit that these endopeptidases act upon would be a tetrapeptide, and they will not yield free amino acids. The cleavage of tetrapeptides by endopeptidases is probably rare, and in the β-chain of insulin there is a segment of eight amino acids that is completely resistant to the action of cathepsins S and L. It is thus clear that considerably less than 50% of the peptide bonds within proteins can be cleaved by the endopeptidases, and the products of endopeptidase action alone will yield peptides containing four to ten amino acids. These peptides most probably cannot exit the lysosome (see Chapter 11), and so must be hydrolyzed further by exopeptidases to prevent lysosomal storage problems.

The relative importance of endo- and exopeptidases in protein hydrolysis can be demonstrated by considering that an intact protein of 200 amino acids has 197 potential endopeptidolytic cleavage sites, whereas it has only one aminopeptidase site and one carboxypeptidase site. The number of potential exopeptidase sites is increased when the dipeptidyl and tripeptidyl peptidase and peptidyl dipeptidase cleavage sites are considered, but these are considerably less than the potential endopeptidase sites. However, every endopeptidase cleavage doubles the number of exopeptidase sites, and as demonstrated in Fig. 1, as the peptides get smaller, the exopeptidases become increasingly important in peptide bond hydrolysis.

Another important consideration in protein turnover is that in the intact protein a large proportion of the peptide bonds are buried in the structure and not accessible to endopeptidase. Thus although there are a large number of potential cleavage sites in a protein, only a few of these will be cleaved efficiently. A particularly extreme example of this is the cleavage of α_1-proteinase inhibitor (a protein that has over 500 peptide bonds) by the powerful lysosomal endopeptidase cathepsin L. In 1 hr, only two peptide bonds are cleaved (Johnson *et al.*, 1986). The endopeptidases are likely to be the rate-limiting enzymes in protein turnover, and most of the peptide bonds in proteins are probably hydrolyzed by exopeptidases.

The relative importance of the individual enzymes in degrading proteins down to amino acids or dipeptides will depend on the rates of hydrolysis of the individual enzymes, as well as the concentrations of the enzymes. Of the endopeptidases, cathepsins L and S seem to be the most catalytically efficient. Cathepsin D is very efficient at low pH, whereas cathepsins B and H are much poorer endopeptidases. From what is known about the concentrations of individual enzymes in lysosomes, endopeptidases seem to be more abundant than exopeptidases (see Section 2.1).

For the bifunctional enzymes, cathepsins B and H, it is difficult to determine relative exo- and endopeptidase activities. For instance, for synthetic substrates, k_{cat}/K_m values for the exopeptidase activity of cathepsin H against Arg-NNap is six times greater than against Bz-Arg-NNap (Schwartz and Barrett, 1980). However, a similar comparison with the exopeptidase substrate Arg-NHMec compared to

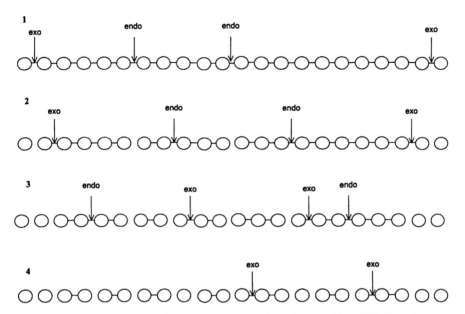

FIGURE 1. Scheme for the degradation of proteins by endo- and exopeptidases. This figure demonstrates the relative contributions of endo- and exopeptidases in the degradation of a protein. Assuming that an endopeptidase can only cleave bonds that have at least two amino acids on each side of the bond cleaved and that an exopeptidase can only cleave terminal residues from peptides at least three amino acids long, we can calculate the number of cleavages performed in the digestion of a protein. Initially, a polypeptide of 22 amino acids has two potential exopeptidase sites and 19 potential endopeptidase sites (line 1). After two cleavages by endo- and exopeptidases, two free amino acids are generated, along with 11 potential endopeptidase sites and six potential exopeptidase sites (line 2). After two more cleavages by each type of protease, the remaining substrates have four potential endopeptidase sites and eight potential exopeptidase sites (line 3). After two more cleavages only four potential exopeptidase sites remain, of which only two can be cleaved (line 4). In this model, exopeptidases have hydrolyzed eight peptide bonds, and endopeptidases have hydrolyzed six peptide bonds. The final products will be eight amino acids and seven dipeptides. Although this model is a great simplification of the likely *in vivo* situation, it clearly demonstrates that as polypeptides are digested exopeptidases become increasingly important in peptide bond hydrolysis and that the final products will not all be free amino acids.

the endopeptidase substrate Bz-Phe-Val-Arg-NHMec suggested that the exopeptidase and endopeptidase activities are similar (Xin *et al.*, 1992). Clearly, the relative rates of cleavage of the different substrates will depend on the amino acid side chains that fit into the specificity pockets of the enzyme. A similar study for cathepsin B has documented a peptidyl dipeptidase activity that is 250 times lower than endopeptidase-like activity, although the amino acids in the specificity pockets were not the same (Polgar and Csoma, 1987). The exopeptidase activity of cathepsin B was optimal below pH 5, which is consistent with a requirement of protonation of the occluding histidine residues as discussed earlier (Musil *et al.*, 1991). By contrast, the endopeptidase activity was optimal above pH 6. These results suggest that cathepsin B might primarily be an endopeptidase in less acidic environments (such as endosomes) and an exopeptidase in lysosomes.

There are many lysosomal storage diseases reported that are due to defects in carbohydrate degradation, but none that result from inappropriate protein degradation. It is unlikely that the proteases are less susceptible to mutation than other enzymes. One possible reason for the lack of build-up of insoluble protein due to a putative defect in a protease could be that the overlapping specificities of the enzymes enable amino acids and dipeptides to be produced even if one protease is missing. While this might be true for the endopeptidases, it may not be the case for the exopeptidases, particularly the few that act upon prolyl peptides. It may be that defects in some of the proteases do occur, but the incompletely processed peptides are not retained in the lysosome. If the peptides were not substrates for any of the lysosomal enzymes, they would certainly bind less tightly to the lysosomal proteins and thus may escape from the lysosome to the endosome much the same way as sucrose can (as described in Section 4.3). Mechanisms for the release of peptides (or conversely retention of proteins) are poorly understood, thus it is not known whether defects in lysosomal hydrolysis occur. The use of low-molecular-weight inhibitors has shown that protein turnover can be reduced when lysosomal proteases are inhibited, but this inhibition does not seem to result in excessive intracellular protein accumulation. For instance, E-64 and Z-Phe-Ala-CHN$_2$ result in an intracellular build-up of proteolytic intermediates of the β-amyloid precursor (Siman et al., 1993). However, the loss of these intermediates is only delayed, not prevented, indicating that either the inhibitors are not 100% effective, or the degradation/loss of the peptides is achieved by an alternative pathway. This apparent ability of cells to use alternative proteolytic pathways could help to explain why there are not yet any documented cases of genetic deficiencies of the endopeptidases that result in any recognizable disease. A mouse deficient for cathepsin D has been shown to develop normally up to 2 weeks postnatal, and the phenotypic changes seen later are not apparently related to bulk protein turnover (Saftig et al., 1995). It would seem that the lethal effect of deletion of cathepsin D is due to a more specific function that regulates differentiation of epithelial cells. It is thus possible that although deficiencies of any one lysosomal protease can be compensated for by the other lysosomal proteases to prevent the appearance of lysosomal storage problems, unique cleavages of individual proteins by any one of the proteases could be physiologically crucial.

5.2. Insoluble and Indigestible Products

It is very clear that several proteins resist proteolysis in the lysosome, usually due to insolubility and aggregation of the proteins. A number of striking examples of indigestible products include the protein deposited in Batten disease (ATPase subunit c), the β-amyloid peptide in Alzheimer's disease, and the deposition of prion proteins in diseases such as scrapie.

In Batten disease there is evidence that the ATPase subunit c accumulates in lysosomes and it may be that this protein is resistant to proteolysis. This accumulation may be caused by inappropriate methylation of the lysine residues in subunit c, leading to hydrophobic aggregation of the protein, rendering it resistant to lysosomal proteolysis (Katz et al., 1994). The genetic defect in Batten disease has not yet been elucidated, but aberrant fatty acid metabolism is implicated, suggesting that the accumulation of subunit c is secondary to the disease process.

In scrapie, the major infectious agent appears to be a protein that accumulates in lysosomes as an insoluble aggregate (Prusiner, 1991; McKinley et al., 1991; Laszlo et al., 1992). The defective conformation of the infectious agent may associate with naturally produced protein to change its conformation to that of the infectious agent. The normal host protein is a lipid-anchored protein that is turned over rapidly, but the infectious particles are resistant to proteolysis. It would appear therefore that the primary defect is not due to the proteolytic enzymes but rather to the ability of the protein to aggregate and become insoluble. These peptides have been shown to accumulate in lysosome-like vacuoles in cells, and it is proposed that the disease can progress through cycles of aggregation, followed by excretion (or cell death), to yield more infectious particles that can be taken up by other cells to repeat cycles of damage. This could result in increased cell death and deposition of extracellular aggregates of protein.

Alzheimer's disease results in the deposition of aggregate of protein that contain a fragment (called $A\beta$) of another membrane protein that has been termed the β-amyloid precursor protein (APP; Marx, 1993). The fragment is 40–43 amino acids long and spans a portion of the APP that is not normally produced in vivo. The precise peptide has not been found in lysosomes, but intermediates that could eventually be processed to the extracellular form have been detected (Dreyer et al., 1994). Synthetic $A\beta$ is taken up by cultured cells, and at high concentrations it forms protease-resistant deposits in cells, presumed to be in endosomes or lysosomes (Knauer et al., 1992). There is no evidence to suggest that the $A\beta$ peptides are infectious, so production of $A\beta$ may well prove to be due to a combination of aberrant processing as well as aggregation. Lysosomal proteases are implicated both in the production and degradation of $A\beta$, so it is unlikely that complete blocking of lysosomal proteolysis will generate excess $A\beta$ (Siman et al., 1993; Ladror et al., 1994). Cathepsin D is the prime candidate enzyme for $A\beta$ production as it is capable of cleaving at both the N- and C-terminal sites that produce $A\beta$ (Ladror et al., 1994). However other enzymes such as cathepsin S are also implicated (see Lemere et al., 1995, and references therein). Both cysteine and aspartic proteases can further degrade soluble forms of $A\beta$ to smaller peptides (Siman et al., 1993; Ladror et al., 1994). If lysosomal proteases are involved in the generation of the $A\beta$ fragments, then it is likely due to an imbalance of the lysosomal proteolysis, rather than a general defect in lysosomal hydrolysis. Early onset forms of Alzheimer's disease have been linked to genetic defects in the APP. These mutated amino acids are near the sites where $A\beta$ peptide is processed and are degraded more rapidly by cathepsin D, supporting the hypothesis that this protein and proteases that process it are the primary causative agents in this disease (Dreyer *et al.*, 1994). Given that Alzheimer's disease affects such a large percentage of the elderly population, it may be that there are many alternative pathways to generate the Aβ peptide.

These three diseases suggest that aberrant lysosomal proteolysis can have dire consequences for mammals. Interestingly, all three diseases result primarily in the deposition of a single protein. The deposition is presumably caused by elevation of the concentrations of the peptides to a critical concentration where aggregation occurs. This increase could be achieved by an imbalance between the biosynthetic and degradative pathways for the peptides. Therapies that redress this imbalance will depend on the identification of the enzymes involved both in generation and degradation of the pathological forms of the amyloid deposits.

Residual bodies, which are composed of undigested material, are often found during the ultrastructural examination of cells. It is not clear to what extent these represent particles in the process of digestion or particles that cannot be digested. Although it is generally accepted that lysosomes will reduce proteins down to amino acids and peptides, the identity of the largest sized fragments that remain indigestible is less clear. It is certainly reasonable to suggest that some tri- and tetrapeptides would be resistant to proteolysis and thus it is not clear whether these would accumulate in the lysosome.

6. FUTURE DIRECTIONS

The complexity of proteases in lysosomes seems to be enormous, but the flurry of papers describing newly discovered enzymes suggests that the list of enzymes is by no means complete. It is likely that the newly discovered enzymes will have more restricted tissue distributions and have more unique functions than those already fully characterized. These will add to the complexity of the lysosomal system, and they demonstrate that it is inappropriate to consider all lysosomes as equal. Clearly, the complement of enzymes in most lysosomes is able to degrade most proteins down to amino acids and dipeptides for delivery to the cytoplasm for further processing and use in protein synthesis, but the fate of incompletely processed peptides is less clear. It has been many years since the products of lysosomal proteolysis were studied in any detail, and given that several diseases have now been identified that involve inappropriate processing of normal proteins, it may be worthwhile to reinvestigate the products of lysosomal proteolysis. For the major diseases that seem to involve abnormal lysosomal proteolysis, scrapie and Alzheimer's disease, it is likely that there will be considerable activity in attempts to identify the mechanisms and enzymes involved. This is likely to stimulate the study of the relationship between the different lysosomal proteases as well as pharmacological means for controlling the activity of individual enzymes. Antigen processing is another area which will be receiving detailed study in the next few years. In particular, it will be of interest to discover why antigens that are presented by cells of the immune system escape delivery to mature lysosomes and whether these are processed by a unique subset of proteases.

ACKNOWLEDGMENTS. The work in the author's laboratory was supported in part by grant #MCB9304109 from the National Science Foundation and by Nemours Research Programs. I thank Herbert Jackson for bringing the work on cathepsin A to my attention, and Robert Metrione for valuable discussions about cathepsin C.

7. REFERENCES

Amigorena, S., Drake, J. R., Webster, P., and Mellman, I., 1994, Transient accumulation of new class II MHC molecules in a novel endocytic compartment in B lymphocytes, *Nature* **369:**113–120.

Anagli, J., Hagmann, J., and Shaw, E., 1991, Investigation of the role of calpain as a stimulus-response mediator in human platelets using new synthetic inhibitors, *Biochem. J.* **274:**497–502.

Baldwin, E. T., Bhat, T. N., Gulnik, S., Hosur, M. V., Sowder II, R. C., Cachau, R. E., Collins, J., Silva, A. M., and Erickson, J. W., 1993, Crystal structures of native and inhibited forms of human cathepsin D: Implications for lysosomal targeting and drug design, *Proc. Natl. Acad. Sci. USA* **90:** 6796–6800.

Barrett, A. J., 1970, Cathepsin D: Purifcation of isoenzymes from human and chicken liver, *Biochem. J.* **117:**601–607.

Barrett, A. J., 1980, Fluorimetric assays for cathepsin B and cathepsin H with methylcoumarylamide substrates, *Biochem. J.* **187:**909–912.

Barrett, A. J., 1987, The cystatins: A new class of peptidase inhibitors, *Trends Biochem. Sci.* **12:**193–196.

Barrett, A. J., 1994, Classification of peptidases, *Methods Enzymol.* **244:**1–18.

Berg, R. A., Schwartz, M. L., Rome, L. H., and Crystal, R. G., 1984, Lysosomal function on the degradation of defective collagen in cultured lung fibroblasts, *Biochemistry* **23:**2134–2138.

Blum, J. S., Fiani, M. L., and Stahl, P. D., 1991, Proteolytic cleavage of ricin A chain in endosomal vesicles. Evidence for the action of endosomal proteases at both neutral and acidic pH, *J. Biol. Chem.* **266:**22091–22095.

Bohley, P., and Seglen, P. O., 1992, Proteases and proteolysiss in the lysosome, *Experimentia* **48:**151–157.

Botbol, V., and Scornik, O. A., 1989, Role of bestastin-sensitive exopeptidases in the intracellular degradation of hepatic proteins, *J. Biol. Chem.* **264:**13504–13509.

Bouma, J. M. W., Scheper, A., Duursma, A., and Gruber, M., 1976, Localization and some properties of lysosomal dipeptidases in rat liver, *Biochim. Biophys. Acta* **444:**853–862.

Bromme, D., Steinert, A., Friebe, S., Fittkau, S., Wiederanders, B., and Kirschke, H., 1989, The specificity of bovine spleen cathepsin-S—A comparison with rat liver cathepsin-L and cathepsin-B, *Biochem. J.* **264:**475–481.

Bromme, D., Bonneau, P. R., Lachance, P., and Storer, A. C., 1994, Engineering the S2 subsite specificity of human cathepsin S to a cathepsin L- and cathepsin B-like specificity, *J. Biol. Chem.* **269:** 30238–30242.

Buckmaster, M. J., Ferris, A. L., and Storrie, B., 1988, Effects of pH, detergent and salt on aggregation of chinese-hamster-ovary-cell lysosomal enzymes, *Biochem. J.* **249:**921–923.

Casciola-Rosen, L. A., and Hubbard, A. L., 1991, Hydrolases in intracellular compartments of rat liver cells. Evidence for selective activation and/or delivery, *J. Biol. Chem.* **266:**4341–4347.

Coffey, J. W., and deDuve, C., 1968, Digestive activity of lysosomes: The digestion of proteins by extracts of rat liver lysosomes, *J. Biol. Chem.* **243:**3255–3263.

Collins, D. S., Unanue, E. R., and Harding, C. V., 1991, Reduction of disulfide bonds within lysosomes is a key step in antigen processing, *J. Immunol.* **147:**4054–4059.

Conliffe, P. R., Ogilvie, S., Simmen, R. C. M., Michel, F. J., Saunders, P., and Shiverick, K. T., 1995, Cloning and expression of a rat placental cDNA encoding a novel cathepsin L-related protein, *Mol. Reprod. Dev.* **40:**146–156.

Cunningham, M., and Tang, J., 1976, Purification and properties of cathepsin D from porcine spleen, *J. Biol. Chem.* **251:**4528–4536.

Dean, R. T., and Barrett, A. J., 1976, Lysosomes, *Essays Biochem.* **12:**1–40.

DeCourcy, K. R., 1995, Purification and characterization of a novel cysteine protease, Ph.D. Dissertation, Virginia Polytechnic Institute and State University, Blacksburg, Virginia.

Deng, Y. P., Griffiths, G., and Storrie, B., 1991, Comparative behavior of lysosomes and the prelysosome compartment (PLC) in *in vivo* cell fusion experiments, *J. Cell Sci.* **99:**571–582.

Dikov, M. M., Springman, E. B., Yeola, S., and Serafin, W. E., 1994, Processing of procarboxypeptidase A and other zymogens in murine mast cells, *J. Biol. Chem.* **269:**25897–25904.

Diment, S., and Stahl, P. D., 1985, Macrophage endosomes contain proteases which degrade endocytosed protein ligands, *J. Biol. Chem.* **260:**15311–15317.

Docherty, K., Hutton, J. C., and Steiner, D. F., 1984, Cathepsin B-related proteases in the insulin secretory granule, *J. Biol. Chem.* **259:**6041–6044.

Dreyer, R. N., Bausch, K. M., Fracasso, P., Hammond, L. J., Wunderlich, D., Wirak, D. O., Davis, G., Brini, C. M., Buckholz, T. M., Koenig, G., Kaarck, M. E., and Tamburini, P. P., 1994, Processing of the pre-beta-amyloid protein by cathepsin D is enhanced by a familial Alzheimer's disease mutation, *Eur. J. Biochem.* **224:**265–271.

Dunn, B. M., 1989, Determination of protease mechanism, in *Proteolytic Enzymes, A Practical Approach* (R. J. Beynon and J. S. Bond, eds.), pp. 57–81, IRL Press, Oxford.

Feener, E. P., Shen, W. C., and Ryser, H. J. P., 1990, Cleavage of disulfide bonds in endocytosed macromolecules—A processing not associated with lysosomes or endosomes, *J. Biol. Chem.* **265:**18780–18785.

Fruton, J. S., Irving, G. W., and Bergmann, M., 1941, On the protelytic enzymes from animal tissues: III. The proteolytic enzymes of beef kidney, beef kidney and swine kidney. Classification of the cathepsins, *J. Biol. Chem.* **141:**763–774.

Gahl, W. A., and Tietze, F., 1987, Lysosomal cystine transport in cystinosis variants and their parents, *Pediatr. Res.* **21:**193–196.

Gal, S., and Gottesman, M. M., 1986, The major excreted protein of transformed fibroblasts is an activable acid-protease, *J. Biol. Chem.* **261:**1760–1765.

Galjart, N. J., Gillemans, N., Harris, A., van der Horst, G. T. J., Verheijen, F. W., Galjaard, H., and d'Azzo, A., 1988, Expression of cDNA encoding the human "protective protein" associated with lysosomal B-galactosidase and neuramidase: Homology to yeast proteases, *Cell* **54:**755–764.

Galjart, N. J., Morreau, H., Willemsen, R., Gillemans, N., Bonten, E. J., and d'Azzo, A., 1991, Human lysosomal protective protein has cathepsin A-like activity distinct from its protective function, *J. Biol. Chem.* **266:**14754–14762.

Geuze, H., 1994, EJCB-Lecture. A novel lysosomal compartment engaged in antigen presentation, *Eur. J. Cell. Biol.* **64:**3–6.

Green, D. W., Aykent, S., Gierse, J. K., and Zupec, M. E., 1990, Substrate specificity of recombinant human renal renin: Effect of histidine in the P2 subsite on pH dependence, *Biochemistry* **29:**3126–3133.

Grima, J., Zhu, L. J., Zong, S. D., Catterall, J. F., Bardin, C. W., and Cheng, C. Y., 1995, Rat testin is a newly identified component of the junctional complexes in various tissues whose mRNA is predominantly expressed in the testis and ovary, *Biol. Reprod.* **52:**340–355.

Guagliardi, L. E., Koppelman, B., Blum, J. S., Marks, M. S., Cresswell, P., and Brodsky, F. M., 1990, Co-localization of molecules involved in antigen processing and presentation in an early endocytic compartment, *Nature* **343:**133–139.

Hara, K., Kominami, E., and Katunuma, N., 1988, Effect of proteinase inhibitors on intracellular processing of cathepsin B, H and L in rat macrophages, *FEBS Lett.* **231**:229–231.

Hyland, L. J., Tomaszek, T. A., Roberts, G. D., Carr, S. A., Magaard, V. W., Bryan, H. L., Fakhoury, S. A., Moore, M. L., Minnich, M. D., and Culp, J. S., 1991, Human immunodeficiency virus-1 protease. 1. Initial velocity studies and kinetic characterization of reaction intermediates by 180 isotope exchange, *Biochemistry* **30**:8441–8453.

Inaoka, T., Bilbe, G., Ishibashi, O., Tezuka, K., Kumegawa, M., and Kokubo, T., 1995, Molecular cloning of human cDNA for cathepsin K: Novel cysteine proteinase predominantly expressed in bone, *Biochem. Biophys. Res. Commun.* **206**:89–96.

Jackman, H. L., Tan, F. L., Tamei, H., Beurling-Harbury, C., Li, X. Y., Skidgel, R. A., and Erdos, E. G., 1990, A peptidase in human platelets that deamidates tachykinins. Probable identity with the lysosomal "protective protein," *J. Biol. Chem.* **265**:11265–11272.

Jahraus, A., Storrie, B., Griffiths, G., and Desjardins, M., 1994, Evidence for retrograde traffic between terminal lysosomes and the prelysosomal/late endosome compartment, *J. Cell Sci.* **107**:145–157.

Johnson, D. A., Barrett, A. J., and Mason, R. W., 1986, Cathepsin L inactivates alpha-1-proteinase inhibitor by cleavage in the reactive site region, *J. Biol. Chem.* **261**:14748–14751.

Kato, T., Okada, S., Yutaka, T., and Yabuuchi, H., 1984, The effects of sucrose loading on lysosomal hydrolases, *Mol. Cell. Biochem.* **60**:83–98.

Katz, M. L., Christianson, J. S., Norbury, N. E., Gao, C. L., Siakotos, A. N., and Koppang, N., 1994, Lysine methylation of mitochondrial ATP synthase subunit c stored in tissues of dogs with hereditary ceroid lipofuscinosis, *J. Biol. Chem.* **269**:9906–9911.

Kirschke, H., Langner, J., Wiederanders, B., Ansorge, S., and Bohley, P., 1977, Cathepsin L: A new proteinase from rat-liver lysosomes, *Eur. J. Biochem.* **74**:293–301.

Kirschke, H., Wiederanders, B., Bromme, D., and Rinne, A., 1989, Cathepsin-S from bovine spleen—Purification, distribution, intracellular localization and action on proteins, *Biochem. J.* **264**:467–473.

Knauer, M. F., Soreghan, B., Burdick, D., Kosmoski, J., and Glabe, C. G., 1992, Intracellular accumulation and resistance to degradation of the Alzheimer amyloid A4/beta-protein, *Proc. Natl. Acad. Sci. USA* **89**:7437–7441.

Koenig, H., 1962, Histological distribution of brain gangliosides: Lysosomes as glycoliprotein granules, *Nature* **195**:782–784.

Kuribayashi, M., Yamada, H., Ohmori, T., Yanai, M., and Imoto, T., 1993, Endopeptidase activity of cathepsin-C, dipeptidyl aminopeptidase-I, from bovine spleen, *J. Biochem.* **113**:441–449.

Ladror, U. S., Snyder, S. W., Wang, G. T., Holzman, T. F., and Krafft, G. A., 1994, Cleavage at the amino and carboxyl termini of Alzheimer's amyloid-beta by cathepsin D, *J. Biol. Chem.* **269**:18422–18428.

Laszlo, L., Lowe, J., Self, T., Kenward, N., Landon, M., McBride, T., Farquhar, C., McConnell, I., Brown, J., Hope, J., and Mayer, R. J., 1992, Lysosomes as key organelles in the pathogenesis of prion encephalopathies, *J. Pathol.* **166**:333–341.

Lee, H.-K., and Marzella, L., 1994, Regulation of intracellular protein degradation with special reference to lysosomes: Role in cell physiology and pathology, *Int. Rev. Exp. Pathol.* **35**:39–147.

Lemere, C. A., Munger, J. S., Shi, G. P., Natkin, L., Haass, C., Chapman, H. A., and Selkoe, D. J., 1995, The lysosomal cysteine protease, cathepsin S, is increased in Alzheimer's disease and Down syndrome brain. An immunocytochemical study, *Am. J. Pathol.* **146**:848–860.

Lipperheide, C., and Otto, K., 1986, Improved purification and some properties of bovine lysosomal carboxypeptidase B, *Biochim. Biophys. Acta* **880**:171–178.

Lloyd, J. B., 1986, Disulfide bond reduction in lysosomes: The role of cysteine, *Biochem. J.* **237**:271.

Lottenberg, R., Christensen, U., Jackson, C. M., and Coleman, P. L., 1981, Assay of coagulation proteases using peptide chromogenic and fluorogenic substrates, *Methods Enzymol.* **80**:341–361.

Mach, L., Mort, J. S., and Glossl, J., 1994, Maturation of human procathepsin B. Proenzyme activation and proteolytic processing of the precursor to the mature proteinase, in vitro, are primarily unimolecular processes, *J. Biol. Chem.* **269**:13030–13035.

Maric, M. A., Taylor, M. D., and Blum, J. S., 1994, Endosomal aspartic proteinase are required for invariant-chain processing, *Proc. Natl. Acad. Sci. USA* **91**:2171–2175.

Marx, J., 1993, Alzheimer's pathology begins to yield its secrets, *Science* **259**:457–458.

Mason, R. W., 1989, Interaction of lysosomal cysteine proteinases with alpha-2-macroglobulin: Conclusive evidence for the endopeptidase activities of cathepsin B and cathepsin H, *Arch. Biochem. Biophys.* **273**:367–374.

Mason, R. W., and Massey, S. D., 1992, Surface activation of pro-cathepsin L, *Biochem. Biophys. Res. Commun.* **189**:1659–1666.

Mason, R. W., and Wilcox, D., 1993, Chemistry of lysosomal proteinases, in *Advances in Cell and Molecular Biology of Membranes,* Vol. 1 (B. Storrie and R. Murphy, eds), pp. 81–116, JAI Press, New York.

Mason, R. W., Green, G. D. J., and Barrett, A. J., 1985, Human liver cathepsin L., *Biochem. J.* **226:** 233–241.

Mason, R. W., Johnson, D. A., Barrett, A. J., and Chapman, H. A., 1986, Elastinolytic activity of human cathepsin L, *Biochem. J.* **233:**925–927.

Mason, R. W., Gal, S., and Gottesman, M. M., 1987, The identification of the major excreted protein (MEP) from a transformed mouse fibroblast cell line as a catalytically active precursor form of cathepsin L, *Biochem. J.* **248**:449–454.

Mason, R. W., Wilcox, D., Wikstrom, P., and Shaw, E. N., 1989, The identification of active forms of cysteine proteinases in Kirsten-virus-transformed mouse fibroblasts by use of a specific radiolabeled inhibitor, *Biochem. J.* **257**:125–129.

McDonald, J. K., Leibach, F. H., Grindeland, R. E., and Ellis, S., 1968, Purification of dipeptidyl aminopeptidase II (dipeptidyl arylamidase II) of the anterior pituitary gland: Peptidase and dipeptide esterase activities, *J. Biol. Chem.* **243**:4143–4150.

McDonald, J. K., Zeitman, B. B., and Ellis S., 1972, Detection of a lysosomal carboxypeptidase and a lysosomal dipeptidase in highly purified dipeptidyl aminopeptidase I (cathepsin C) and the elimination of their activities from preparations used to sequence peptides, *Biochem. Biophys. Res. Commun.* **46**:62–70.

McDonald, J. K., Hoisington, A. R., and Eisenhauer, D. A., 1985, Partial purification and characterization of an ovarian tripeptidyl peptidase: A lysosomal exopeptidase that sequentially releases collagen-related (Gly-Pro-X) triplets, *Biochem. Biophys. Res. Commun.* **126**:63–71.

McDonald, J. K., and Barrett, A. J., 1986, Mammalian Proteases: A Glossary and Bibliography, Vol. 2, *Exopeptidases,* Academic Press, London.

McGuire, M. J., Lipsky, P. E., and Thiele, D. L., 1993, Generation of active myeloid and lymphoid granule serine proteases requires processing by the granule thiol protease dipeptidyl peptidase-I, *J. Biol. Chem.* **268**:2458–2467.

McKay, M. J., Offermann, M. K., Barrett, A. J., and Bond, J. S., 1983, Action of human liver cathepsin B on the oxidized insulin B chain, *Biochem. J.* **213**:467–471.

McKinley, M. P., Taraboulos, A., Kenaga, L., Serban, D., Stieber, A., DeArmond, S. J., Prusiner, S. B., and Gonatas, N., 1991, Ultrastructural localization of scrapie prion proteins in cytoplasmic vesicles of infected cultured cels, *Lab. Invest.* **65**:622–630.

Metrione, R. M., and MacGeorge, N. L., 1975, The mechanism of action of dipeptidyl aminopeptidase. Inhibition by amino acid derivatives and amines; activation by aromatic compounds, *Biochemistry* **14**:5249–5252.

Montgomery, R. R., Webster, P., and Mellman, I., 1991, Accumulation of indigestible substances reduces fusion competence of macrophage lysosomes, *J. Immunol.* **147**:3087–3095.

Musil, D., Zucic, D., Turk, D., Engh, R. A., Mayr, I., Huber, R., Popovic, T., Turk, V., Towatari, T., Katunuma, N., and Bode, W., 1991, The refined 2.15-A X-ray crystal structure of human liver cathepsin-B—the structural basis for its specificity, *EMBO J.* **10**:2321–2330.

Nikawa, T., Towatari, T., and Katunuma, N., 1992, Purification and characterization of cathepsin-J from rat liver, *Eur. J. Biochem.* **204**:381–393.

Nishimura, Y., Kawabata, T., Furuno, K., and Kato, K., 1989, Evidence that aspartic proteinase is involved in the proteolytic event of procathepsin L in lysosomes, *Arch. Biochem. Biophys.* **271**:400–406.

Okamura-Oho, Y., Zhang, S., and Callahan, J. W., 1994, The biochemistry and clinical features of galactosialidosis, *Biochim. Biophys. Acta* **1225**:244–254.

Page, A. E., Fuller, K., Chambers, T. J., and Warburton, M. J., 1993, Purification and characterization of a tripeptidyl peptidase I from human osteoclastomas: Evidence for its role in bone resorption, *Arch. Biochem. Biophys.* **306**:354–359.

Petanceska, S., and Devi, L., 1992, Sequence analysis, tissue distribution, and expression of rat cathepsin-S, *J. Biol. Chem.* **267**:26038–26043.

Pisoni, R. L., Acker, T. L., Lisowski, K. M., Lemons, R. M., and Thoene, J. G., 1990, A cysteine-specific lysosomal transport system provides a major route for the delivery of thiol to human fibroblast lysosomes: Possible role in supporting lysosomal proteolysis, *J. Cell Biol.* **110**:327–335.

Polgar, L., and Csoma, C., 1987, Dissociation of ionizing groups in the binding cleft inversely controls the endo- and exopeptidase activities of cathepsin B, *J. Biol. Chem.* **262**:14448–14453.

Prusiner, S. B., 1991, Molecular biology of prion diseases, *Science* **252**:1515–1522.

Rawlings, N. D., and Barrett, A. J., 1994, Families of cysteine peptidases, *Methods Enzymol.* **244**:461–486.

Renfrew, C. A., and Hubbard, A. L., 1991, Sequential processing of epidermal growth factor in early and late endosomes of rat liver, *J. Biol. Chem.* **266**:4348–4356.

Richo, G. R., and Conner, G. E., 1994, Structural requirements of procathepsin D activation and maturation, *J. Biol. Chem.* **269**:14806–14812.

Runquist, E. A., and Havel, R. J., 1991, Acid hydrolases in early and late endosome fractions from rat liver, *J. Biol. Chem.* **266**:22557–22563.

Rupp, K., Birnbach, U., Lundstrom, J., Van, P. N., and Soling, H. D., 1994, Effects of CaBP2, the rat analog of ERp72, and of CaBP1 on the refolding of denatured reduced proteins. Comparison with protein disulfide isomerase, *J. Biol. Chem.* **269**:2501–2507.

Saftig, P., Hetman, M., Schmahl, W., Weber, K., Heine, L., Mossmann, H., Koster, A., Hess, B., Evers, M., von Figura, K., and Peters, C., 1995, Mice deficient for the lysosomal proteinase cathepsin D exhibit progressive atrophy of the intestinal mucosa and profound destruction of lymphoid cells, *EMBO J.* **14**:3599–3608.

Samarel, A. M., Ferguson, A. G., Decker, R. S., and Lesch, M., 1989, Effects of cysteine protease inhibitors on rabbit cathepsin-D maturation, *Am. J. Physiol.* **257**:C1069–C1079.

Scarborough, P. E., Guruprasad, K., Topham, C., Richo, G. R., Conner, G. E., Blundell, T. L., and Dunn, B. M., 1993, Exploration of subsite binding specificity of human cathepsin D through kinetics and rule-based molecular modeling, *Protein Science.* **2**:264–276.

Schaudies, R. P., Gorman, R. M., Savage, C. R., and Poretz, R. D., 1987, Proteolytic processing of epidermal growth factor within endosomes, *Biochem. Biophys. Res. Commun.* **143**:710–715.

Schecter, I., and Berger, A., 1967, On the active site of proteases. I. Papain, *Biochem. Biophys. Res. Commun.* **27**:157–161.

Schwartz, W. N., and Barrett, A. J., 1980, Human cathepsin H, *Biochem. J.* **191**:487–497.

Shi, G. P., Munger, J. S., Meara, J. P., Rich, D. H., and Chapman, H. A., 1992, Molecular cloning and expression of human alveolar macrophage cathepsin S, an elastinolytic cysteine protease, *J. Biol. Chem.* **267**:7258–7262.

Shi, G. P., Chapman, H. A., Bhairi, S. M., Deleeuw, C., Reddy, V. Y., and Weiss, S. J., 1995, Molecular cloning of human cathepsin O, a novel endoproteinase and homologue of rabbit OC2, *FEBS Lett.* **357**:129–134.

Shoji-Kasai, Y., Senshu, M., Iwashita, S., and Imahori, K., 1988, Thio protease-specific inhibitor E-64 arrests human epidermoid carcinoma A431 cells at mitotic metaphase, *Proc. Natl., Acad. Sci. USA* **85**:146–150.

Siman, R., Mistretta, S., Durkin, J. T., Savage, M. J., Loh, T., Trusko, S., and Scott, R. W., 1993, Processing of the beta-amyloid precursor. Multiple proteases generate and degrade potentially amyloidogenic fragments, *J. Biol. Chem.* **268**:16602–16609.

Smeekens, S. P., and Steiner, D. F., 1990, Identification of a human insulinoma cDNA encoding a novel mammalian protein structurally related to the yeast dibasic processing protease Kex2, *J. Biol. Chem.* **265**:2997–3000.

Smith, S. M.,. and Gottesman, M. M., 1989, Activity and deletion analysis of recombinant human cathepsin L expressed in Escherichia-coli, *J. Biol. Chem.* **264**:20487–20495.

Stafford, F. J. and Bonifacino, J. S., 1991, A permeabilized cell system identifies the endoplasmic reticulum as a site of protein degradation, *J. Cell Biol.* **115**:1225–1236.

Swanson, J., Yirinec, B., Burke, E., Bushnell, A., and Silverstein, S. C., 1986, Effect of alterations in the size of the vacuolar compartment on pinocytosis in J774.2 macrophages, *J. Cell. Physiol.* **128**:195–201.

Takahashi, T., Dehdarani, A. H., Yonezawa, S., and Tang, J., 1986, Porcine spleen cathepsin B is an exopeptidase, *J. Biol. Chem.* **261**:9375–9381.

Takahashi, T., Dehdarani, A. H., and Tang, J., 1988, Porcine spleen cathepsin H hydrolyses oligopeptides solely by aminopeptidase activity, *J. Biol. Chem.* **263**:10952–10957.

Takio K, Towatari, T., Katunuma, N., Teller, D. C., and Titani, K., 1983, Homology of amino acid sequences of rat liver cathepsins B and H with that of papain, *Proc. Natl. Acad. Sci. USA* **80**:3666–3670.

Tan, F., Morris, P. W., Skidgel, R. A., and Erdos, E. G., 1993, Sequencing and cloning of human prolylcarboxypeptidase (angiotensinase C). Similarity to both serine carboxypeptidase and prolylendopeptidase families, *J. Biol. Chem.* **268**:16631–16638.

Tezuka, K., Tezuka, Y., Maejima, A., Sato, T., Nemoto, K., Kamioka, H., Hakeda, Y., and Kumegawa, M., 1994, Molecular cloning of a possible cysteine proteinase predominantly expressed in osteoclasts, *J. Biol. Chem.* **269**:1106–1109.

Turk, B., Dolenc, I., Turk, V., and Bieth, J. G., 1993, Kinetics of the pH-induced inactivation of human cathepsin L, Biochemistry **32**:375–380.

Urade, R., Nasu, M., Moriyama, T., Wada, K., and Kito, M., 1992, Protein degradation by the phosphoinositide-specific phospholipase C-Â family from rat liver endoplasmic reticulum, *J. Biol. Chem.* **267**:15152–15159.

Urade, R., Takenaka, Y., and Kito, M., 1993, Protein degradation by ERp72 from rat and mouse liver endoplasmic reticulum, *J. Biol. Chem.* **268**:22004–22009.

Velasco, G., Ferrando, A. A., Puente, X. S., Sanchez, L. M., and Lopez-Otin, C., 1994. Human cathepsin O. Molecular cloning from a breast carcinoma, production of the active enzyme in Escherichia coli, and expression analysis in human tissues, *J. Biol. Chem.* **269**:27136–27142.

West, M. A., Lucocq, J. M., and Watts, C., 1994, Antigen processing and class II MHC peptide-loading compartments in human B-lymphoblastoid cells, *Nature* **369**:147–151.

Wilcox, D., and Mason, R. W., 1992, Inhibition of cysteiene proteinases in lysosomes and whole cells, *Biochem. J.* **285**:495–502.

Wislon, R. B., Mastick, C. C., and Murphy, R. F., 1993, A Chinese hamster ovary cell line with a temperature-conditional defect in receptor recycling is pleiotropically defective in lysosome biogenesis, *J. Biol. Chem.* **268**:25357–25363.

Xin, X. Q., Gunesekera, B., and Mason, R. W., 1992, The specificity and elastinolytic activities of bovine cathepsin S and cathepsin H, *Arch. Biochem. Biophys.* **299**:334–339.

Young, J., Kane, L. P., Exley, M., and Wileman, T., 1993, Regulation of selective protein degradation in the endoplasmic reticulum by redox potential, *J. Biol. Chem.* **268**:19810–19818.

Zhu, Y., and Conner, G. E., 1994, Intermolecular association of lysosomal protein precursors during biosynthesis, *J. Biol. Chem.* **269**:3846–3851.

Chapter 7

Lysosomal Metabolism of Glycoconjugates

Bryan G. Winchester

1. INTRODUCTION

A variety of intracellular and extracellular glycoconjugates are degraded in the lysosomes of animal cells. The main groups are glycosaminoglycans, glycolipids, and glycoproteins. A portion of the cellular glycogen is also turned over in the lysosomes as a result of autophagy of cytoplasmic glycogen. The importance of the lysosomal catabolism of glycoconjugates for the normal functioning of cells and tissues is dramatically illustrated by the severe and often fatal lysosomal storage diseases that result from a genetic defect in the lysosomal catabolism of glycoconjugates. Under certain circumstances exogenous carbohydrate-containing material may be delivered to the lysosomes. This material is catabolized if the glycosidic linkages are susceptible to hydrolysis by the endogenous glycosidases.

The capacity to degrade glycoconjugates in lysosomes is universal in animal cells, and the enzymic pathways by which this is achieved are essentially the same in all species. There are, however, some minor but important differences in the pathways between species. The catabolic pathways for the different glycoconjugates have been established by a combination of (1) substrate specificity studies

Bryan G. Winchester   Division of Biochemistry and Genetics, Institute of Child Health, London WC1N 1EH, United Kingdom.
Subcellular Biochemistry, Volume 27. Biology of the Lysosome, edited by Lloyd and Mason. Plenum Press, New York, 1996.

using natural substrates and purified enzymes or preparations of lysosomes *in vitro*, (2) metabolic labeling of intermediates in cells in culture, and (3) characterization of the storage products that accumulate in lysosomal storage diseases. The carbohydrate moiety of glycolipids is completely degraded while still attached to the noncarbohydrate component, ceramide, which is subsequently degraded by specific enzymes (see Chapter 8). In contrast the protein moieties of glycoproteins and glycosaminoglycans are completely or partially degraded before the catabolism of the glycans. The lysosomal enzymes involved in the breakdown of the protein and lipid components of glycoconjugates are discussed in detail in Chapters 6 and 8.

The lysosomal pathways for the breakdown of glycans are highly ordered. Monosaccharides are released sequentially by the action of exoglycosidases, predominantly from the nonreducing end(s). Other types of enzymes are also involved. Specific enzymes cleave the different protein–oligosaccharide linkages, and endoglycosidases participate in the release of the carbohydrate moiety. Sulfatases and an N-acetyltransferase act at specific steps in the pathways to generate the unsubstituted glycosyl residues that are recognized by the glycosidases. The resultant monosaccharides leave the lysosomes by diffusion or specific transport processes, as do the catabolic products of the noncarbohydrate moieties (see Chapter 11). Although some parallel steps have been observed in the breakdown of high-mannose N-linked glycans *in vitro* (Al Daher *et al.*, 1991), alternative catabolic routes do not appear to be very active *in vivo*. There is some evidence that the accumulating substrates in lysosomal storage disease can be degraded or modified by alternative enzymes to a limited extent. This is probably due to the high concentration of a substrate forcing low-specificity reactions or transglycosylation by glycosidases by the law of mass action. The relevance of these alternative reactions to normal metabolism is not known.

Over 20 different enzymes are involved in the catabolism of the glycans of glycoconjugates in lysosomes. Several of the enzymes are involved in more than one pathway, contributing to the heterogeneity of the storage products in lysosomal storage diseases. It has not been established definitely whether the enzymes in a particular pathway are physically associated or whether their activities are regulated coordinately, but many observations suggest that there is some organization of the enzymes in the pathways. For example, β-galactosidase (EC 3.2.1.23), which participates in the catabolism of three types of glycoconjugates, forms a ternary complex with two other lysosomal enzymes, α-neuraminidase and a protective protein (cathepsin A), which protects it against degradation and promotes its aggregation (Hoogeven *et al.*, 1983) and requires the presence of an activator protein (saposin C) when acting on the lipid substrate G_{M1}-ganglioside (Li *et al.*, 1973). This suggests that association with other proteins can modify the specificity of β-galactosidase or sequester a proportion of the enzyme for different catabolic functions. Other lysosomal hydrolases associate with nonenzymic proteins, and the formation of noncovalent, multimeric complexes may be a way of organizing and coordinating the action of soluble lysosomal hydro-lases in a pathway.

The genes for lysosomal glycosidases and sulfatases are scattered over the genome. It is not known whether expression of the genes for the enzymes in a lyso-somal pathway is coordinated or regulated. Some genes have upstream regions found in promoters of housekeeping genes, while others do not (Neufeld, 1991). The frequently observed phenomenon of hyperactivity of other lysosomal en-zymes when there is a deficiency of a lysosomal hydrolase suggests that some feedback might affect the expression of more than one lysosomal enzyme gene. An alternative explanation of this phenomenon of secondary hyperactivity is the enlargement of the lysosomal system in lysosomal storage disorders, perhaps en-hanced by stabilization of related enzymes by the accumulated storage products. Although it is possible to overload the lysosomal system in genetic defects and ar-tificially in cells in culture, under normal conditions cells seem to be able to con-trol the activity of the lysosomal pathways to meet physiological demands.

1.1. Lysosomal Glycosidases

The lysosomal exoglycosidases, generally just called glycosidases, are highly specific but they share some common properties and structural features. They are typical lysosomal hydrolases with acidic pH optima. They are all glycoproteins and are transported to lysosomes via the mannose 6-phosphate pathway (see Chap-ter 2), except for β-glucocerebrosidase (β-glucosidase, EC 3.2.1.45), which is as-sociated with the lysosomal membrane (Willemsen et al., 1987; Aerts et al., 1988). In some cells acid maltase or α-glucosidase (EC 3.2.1.20) may be transported to the lysosomes via the plasma membrane (Tsuji et al., 1988).

Lysosomal glycosidases show absolute specificity for the anomeric linkage and, with one exception, for the monosaccharide released. The exception is β-N-acetylhexosaminidase A (EC 3.2.1.52), which can release N-acetylglucosamine from N-linked glycans of glycoproteins and N-acetylgalactosamine from glyco-lipids. The dual specificity of this enzyme can be explained by the fact that it is a heterodimer of two genetically distinct subunits, α and β, each with an active site of different specificity (Neufeld, 1989). The active site in the α subunit can catalyze the hydrolysis of β-hexosaminidic linkages in neutral and negatively charged substrates such as gangliosides whereas the β subunit active site can only act on neutral substrates, such as those found in the catabolism of N-linked gly-cans. Hexosaminidase B, which consists of two β subunits can only act on neu-tral substrates. Neither subunit is active on its own, demonstrating further the importance of protein–protein interactions in these pathways. The glycosidases may have more relaxed specificities toward synthetic substrates, which although useful for diagnostic purposes, can be misleading. α-N-Acetylgalactosaminidase (EC 3.2.1.49), which was formerly called α-galactosidase B, can hydrolyze the α-galactosidic linkages in synthetic substrates but not in natural substrates. The accumulation of glycolipids terminating in α-linked galactose in Fabry disease, in which there is a deficiency of α-galactosidase but not of α-galactosaminidase, is a poignant illustration of the specificity of these enzymes.

Specificity toward the aglycon is not so strict because a sugar may be linked by the same anomeric linkage to different molecules or to different hydroxyl groups of the same sugar. The major lysosomal α-D-mannosidase (EC 3.2.1.24) can catalyze the hydrolysis of $\alpha 1 \rightarrow 2$, $\alpha 1 \rightarrow 3$ and $\alpha 1 \rightarrow 6$ α-mannosidic linkages in the N-linked glycans of glycoproteins. As already mentioned lysosomal β-galactosidase is multifunctional and can act on Gal $\beta 1 \rightarrow 4$ GlcNAc linkages in N-linked glycans, keratan sulfate, and glycolipids; and Gal $\beta 1 \rightarrow 3$ GalNAc and Gal $\beta 1 \rightarrow 4$Glc linkages in glycosphingolipids. In contrast, the Gal $\beta 1 \rightarrow$ ceramide linkage in glycosphingolipids is hydrolyzed *in vivo* by a genetically distinct β-galactosidase, galactocerebrosidase (EC 3.2.1.46), although the broad specificity β-galactosidase can hydrolyze this linkage *in vitro*.

Most mammalian lysosomal glycosidases have been purified to homogeneity, cloned, and their amino acid sequences determined directly or deduced. There do not appear to be any characteristic cDNA or amino acid sequences associated with lysosomal enzymes, except for a similar signal peptide sequence. The homology between the amino acid sequences of two human lysosomal glycosidases is relatively low, 15–20% identity and 40–45% similarity, even when they are in the same catabolic pathway. The homologies between human α-fucosidase and other enzymes in the N-glycan degradative pathway are α-mannosidase, 20%; α-galactosidase, 17%; α-N-acetylgalactosaminidase, 19%; β-N-acetylglucosaminidase α and β subunits, both 17%. However, the homology between human α-fucosidase and the rat and slime mold lysosomal α-fucosidases is 79% and 44% identity, respectively (Fisher and Aronson, 1989). Furthermore, some of the exons of canine α-fucosidase have identical sequences to the corresponding human ones (Skelly *et al.,* 1996). The homology between species is true for other glycosidases (Neufeld, 1991). When two lysosomal enzymes catalyze very similar reactions the homology increases. Both α-galactosidase and α-N-acetylgalactosaminidase show 60% homology at the gene level and 47% for the deduced amino acid sequences (Wang *et al.,* 1990). There is 60% homology between the α and β subunits of N-acetyl-β-hexosaminidase (Neufeld, 1989). These observations suggest that a particular glycosidase activity must have evolved from a common ancestral gene. The structural features determining the precise specificity must be very subtle and involve multiple interactions in the binding site. Analysis of the mutations in patients with lysosomal storage diseases has shown that defects affecting all aspects of the structure, biosynthesis, or mechanism of action of a lysosomal glycosidase can lead to an intralysosomal deficiency of that enzyme. Seemingly innocuous amino acid changes, especially in conserved sequences, can have very profound and quite different effects on the activity of an enzyme.

All the glycosidases (except β-glucocerebrosidase) must have a conformational motif recognized by the GlcNAc-1-phosphotransferase to form the mannose

6-phosphate lysosomal recognition marker (Dustin *et al.,* 1995; Cuozzo *et al.,* 1995). A highly redundant consensus sequence, YXX(Y,W, or F) has been suggested as a marker for phosphorylation of mannose in lysosomal glycosidases from an examination of amino acid sequences (Barnes and Wynn, 1988). Details of this marker and of other important structural features must await x-ray analysis of crystallized lysosomal glycosidases. The tertiary structure and mechanism of action of over 20 nonmammalian glycosidases have been established (McCarter and Withers, 1994). This information should be highly relevant to understanding the corresponding mammalian enzymes in view of the interspecies homology. Human α-galactosidase has been crystallized and preliminary x-ray analysis carried out (Murali *et al.,* 1994). Similar studies on other mammalian lysosomal glycosidases cocrystallized with substrate analogues will hopefully reveal whether there is a common carbohydrate-recognizing domain and catalytic mechanism and what determines substrate specificity at a molecular level.

1.2. Lysosomal Sulfatases

The lysosomal sulfatases are also typical acidic hydrolases and share common features with the glycosidases. They show very precise specificities toward sulfated glycosyl residues in the catabolism of glycosaminoglycans and glycolipids. Several mammalian lysosomal sulfatases with distinct specificities and functions have been cloned (Kreysing *et al.,* 1990; Peters *et al.,* 1990; Wilson *et al.,* 1990; Tomatsu *et al.,* 1991; Robertson *et al.,* 1992). Their gene structures and predicted amino acid sequences have been examined to see if there is a common sulfate-recognizing motif. The arrangements of exons and introns are different and they do not have identical signal peptides, but there are several highly conserved amino acid sequences. The same features are present in microsomal sterol sulfatase (EC 3.1.6.2) (Yen *et al.,* 1987) and sea urchin embryo sulfatase (Sasaki *et al.,* 1988), suggesting that they have also evolved from a common ancestral gene. A mutation in one of the conserved regions of a lysosomal sulfatase generally leads to a severe form of the corresponding lysosomal storage disease.

In the genetic disorder multiple sulfatidosis, there are deficiencies of at least seven sulfatases and the patients have symptoms of both a mucopolysaccharidosis and metachromatic leukodystrophy, which is due to a deficiency of cerebroside sulfatase (EC 3.1.6.8) (Basner *et al.,* 1979). It has been shown very recently that a conserved cysteine in normal recombinant arylsulfatase A and B is posttranslationally modified to 2-amino-3-oxopropionic acid but that this modification does not occur in the same sulfatases in cells from patients with multiple sulfatidosis. It is suggested that this modification is necessary for the generation of catalytic activity and that it is defective in multiple sulfatidosis (Schmidt *et al.,* 1995). Overexpression of a sulfatase in mammalian cells leads to a deficiency of other lysosomal sulfatases, suggesting exhaustion of an essential factor by the overexpressed enzyme.

2. GLYCOSAMINOGLYCANS

2.1. Structure, Occurrence, and Function of Glycosaminoglycans

Proteoglycans consist of a core polypeptide chain to which are attached from one to over 100 acidic polysaccharide chains called glycosaminoglycans and N- and/or O-linked oligosaccharides (Roden, 1980; Poole, 1986; Kjellen and Lindahl, 1991). The number and composition of the attached sugar chains and the size of the core protein determine the properties and function of the different proteogly- cans. Glycosaminoglycans (GAGs), which were formerly called mucopolysac- charides, are large straight-chain acidic polysaccharides composed of a repeating disaccharide unit of a hexuronic acid and a hexosamine (Fig. 1). The hexuronic acid is D-glucuronic acid in hyaluronic acid (HA) and chondroitin sulfate (CS), and a mixture of D-glucuronic acid and L-iduronic acid in heparan sulfate/heparin (HS) and dermatan sulfate (DS). The proportion and distribution of the two hex- uronic acids vary along the chains of DS and HS. D-Galactose replaces the hex- uronic acid in keratan sulfate (KS). The hexosamine is N-acetylglucosamine in HA, HS, and KS and N-acetylgalactosamine in CS and DS. All of the gly- cosaminoglycans, except hyaluronic acid, are sulfated, giving them a high nega- tive charge at physiological pH. Sulfation occurs at the C2, C3, C4, and C6

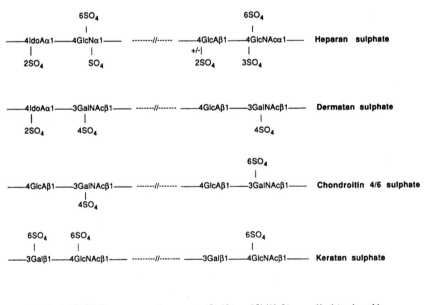

FIGURE 1. Repeating disaccharide units of glycosaminoglycans. The distribution of glucuronic and iduronic acids in DS and HS and the position of sulfation in GAGs are irregular but not random.

hydroxyl group of the sugars and replaces acetylation at some of the amino groups of N-acetylglucosamine in heparan sulfate/heparin. The degree and positions of sulfation vary within and between each class of glycosaminoglycan. Chondroitin sulfate, dermatan sulfate, and heparin/heparan sulfate are linked to the core protein by an O-glycosidic linkage between serine on the protein and xylosyl residue in a GlcAβ1 → 3 Galβ1 → 3Galβ1 → 4Xyl sequence at the reducing end of the chains. The linkage region in KS is different. Skeletal KS is linked by an α-O-glycosidic linkage between serine and an N-acetylgalactosamine at the reducing terminal, as in mucins, whereas corneal KS is linked to the protein by a β-N-glycosidic link between asparagine and N-acetylglucosamine at the reducing terminal as in glycoproteins. Asparagine-linked chondroitin sulfates and heparin/heparan sulfates have been found on cell surface glycoproteins (Sundblad *et al.,* 1988). Hyaluronic acid also differs from the other GAGs in not appearing to be linked to a core protein in its functional form in the extracellular matrix.

Virtually all animal cells have the capacity to synthesize proteoglycans and they occur in the extracellular matrix, on the cell surface, and intracellularly in secretory granules, for example, heparin in mast cells. Free HS (Fedarko and Conrad, 1986) and DS (Hiscock *et al.,* 1994) have been found in the nucleus. Glycosaminoglycans range widely in size and composition, reflecting their different locations and functions (Poole, 1986; Kjellen and Lindahl, 1991; Heinegard and Oldberg, 1993). Aggrecan, a large proteoglycan found in cartilage, contains about 100 chains of CS of 40–50 repeating disaccharide units, about 30 chains of KS of 5–10 repeating disaccharide units, and O- and N-linked oligosaccharides linked to a core protein of over 200 kDa. It has the ability to form aggregates of M_r greater than 3×10^9 by binding noncovalently to hyaluronate in the presence of a linker protein. The anionic proteoglycans can absorb a large volume of water and their presence within the collagenous network enables it to resist compression.

Smaller proteoglycans, such as biglycan (PG-S1), decorin (PG-S2), and fibromodulin are also present in the extracellular matrix but do not form aggregates. Their roles include modulation of collagen fibril formation and growth factor binding. The core protein of biglycan has an M_r of about 40,000 and is substituted by only two GAG chains of chondroitin sulfate or dermatan sulfate, depending on the tissue of origin. Decorin has only one GAG chain, again chondroitin or dermatan sulfate depending on the tissue. In contrast, fibromodulin does not contain chondroitin or dermatan sulfate chains but has N-linked keratan sulfate chains as in corneal proteoglycans. Heparan sulfate side chains are found on proteoglycans at cell surfaces (Hook *et al.,* 1986) and in basement membranes (Paulsson, 1987). The cell surface-associated heparan sulfate proteoglycans are believed to play an important role in cell growth, migration, and morphogenesis through their interaction with growth factors and extracellular matrix proteins (Turnbull and Gallagher, 1993; Spillmann and Lindahl, 1994). Perlecan, a basement membrane proteoglycan containing heparan sulfate, has core protein domains homologous to several proteins of well-defined functions (Iozzo *et al.,* 1994). The negatively charged membrane-spanning proteoglycans may also act as a barrier to plasma proteins. The putative functions of proteoglycans are largely attributed to the binding of proteins to specific structural motifs of the GAGs or core protein (Jackson *et al.*, 1991; Wight *et al.*, 1992), but the precise molecular basis of most interactions is only just beginning to be understood (Spillmann and Lindahl, 1994). The anticoagulant properties of heparin are due to a specific pentasaccharide sequence that binds to lysyl residues on antithrombin, enhancing its neutralization of serine proteases of the coagulation cascade (Lindahl, 1989).

2.2. Turnover of Proteoglycans

Evidence for the continuous turnover of proteoglycans comes from metabolic labeling of cells and explants and from the massive excretion in the urine and accumulation in tissues of partially catabolized GAGs in patients with a defect in the lysosomal catabolism of GAGs. There is normally a balance in the extracellular matrix between the synthesis and secretion of newly synthesized material by the resident cells and the degradation of previously secreted material (Murphy and Reynolds, 1993). Very little is known about the regulation of this homeostasis, but inflammation or the presence of tumor cells can disrupt the balance, leading to changes in the concentration or composition of the extracellular matrix. More than half of newly synthesized proteoglycan is not secreted, but is diverted to the lysosomes for catabolism *in vitro* and *in vivo* (Burkhart and Wiesmann, 1987). The signals controlling the division of the flux of newly synthesized proteoglycans to degradation or secretion are not known, but the lysosomal system is clearly an integral part of the homeostasis of the extracellular matrix. Proteoglycan synthesis is initiated by the formation of the xylosyl–serine linkage in the endoplasmic reticulum or Golgi apparatus. This is followed in the Golgi by the sequential addition of monosaccharides to form the GAG chains. Epimerization of some glucuronic acid to iduronic acid and sulfation occur before secretion. Aggregation by binding to hyaluronic acid occurs extracellularly, although the hyaluronic acid may still be bound to the cell surface, where it is synthesized by hyaluronate synthase (Prehm, 1983).

The degradation of matrix proteoglycans is initiated extracellularly by proteolytic cleavage of the core protein by proteases secreted by connective tissue cells (Nguyen *et al.*, 1989). Matrix proteinases act first at a site near the region responsible for binding to hyaluronate to release almost intact proteoglycans. Subsequent proteolytic cleavage gives rise to short peptides with up to 12 GAG chains attached. These highly hydrophilic molecules are no longer entrapped in the collagen network and can diffuse to surrounding fluids and tissues. Several cell types, including the local chondrocytes and liver endothelial cells, can bind and endocytose proteoglycans and deliver them to lysosomes for catabolism. Binding can be via the core protein fragment (Schmidt *et al.*, 1990) or the GAG chain (Smedsrod *et al.*, 1984). GAGs, with the exception of some HS chains, are not degraded extracellularly. Although this process has largely been studied for aggrecan and chondroitin sulfates, it is believed to be similar for other matrix proteoglycans as well. Hyaluronic acid is turned over at the same rate as aggrecan and is broken down intracellularly after endocytosis (Truppe *et al.,* 1977), although some prior free radical-mediated depolymerization may occur extracellularly (Ng *et al.,* 1992).

Cell surface-associated HS proteoglycans are probably internalized intact, with the rate of degradation very sensitive to extracellular Ca^{2+} (Takeuchi *et al.,* 1990). Some heparan sulfates are partially catabolized extracellularly by heparatinase, an endoglucuronidase produced by platelets (Oosta *et al.,* 1982), T lymphocytes, and some tumor cells. This process may be important in cell signaling because HS is involved in the binding of cells to other cells and extracellular proteins.

2.3. Intracellular Catabolism of Glycosaminoglycans

The subsequent intracellular digestion of the endocytosed GAGs may be initiated in endosomes, but the majority of it takes place in the acidic lumen of the lysosomes. The pathways for the catabolism of each class of GAG have been elucidated (Hopwood and Morris, 1990; Neufeld and Muenzer, 1995). Some enzymes function in more than one pathway because their substrate is part of the structure of more than one GAG, while other enzymes are unique to a pathway for a particular GAG. This is reflected in the composition of the GAGs that are secreted in the urine and accumulate in the tissues of patients with defects in the lysosomal catabolism of GAGs, the mucopolysaccharidoses (MPS). A brief summary of these disorders is given in Table I, (see Hopwood and Morris, 1990; Leroy and Wiesmann, 1993; and Neufeld and Muenzer, 1995 for more detailed accounts).

2.3.1. Heparan Sulfate

Endocytosed HS proteoglycan is proteolytically digested in early endosomes to produce single HS chains, which are broken down by a specific endoglucuronidase to oligosaccharides of approximately 5 kDa as they pass through the endosomal system to the lysosomes (Klein and von Figura, 1979; Hopwood, 1989). Within the lysosomes the oligosaccharides are degraded to their constituent monosaccharides from the nonreducing end by the concerted action of exoglycosidases, sulfatases, and a transferase. The distribution of the two hexuronic acids along the HS chain and the pattern of sulfation of the sugars are irregular. Therefore, the precise sequence of enzymic steps will be determined by the structure of the disaccharide unit at the nonreducing terminal. However, the hexuronic acids have to be desulfated before the hexuronidic linkage is hydrolyzed. Likewise, the 2-amino group of glucosamine has to be desulfated and reacetylated and the O-esterified sulfates removed, before the hexosaminidic linkage is cleaved. Figure 2 shows all the possible enzymic steps, commencing arbitrarily with a sulfated

Table I
The Mucopolysaccharidoses[a]

McKusick's classification	Eponym	Enzyme deficiency	Storage products	Gene location	Organs and functions mainly affected
IH	Hurler	α-L-iduronidase (EC 3.2.1.76)	DS, HS	4p16.3	CNS, skeleton, viscera
IS	Scheie	α-L-iduronidase	DS, HS		Skeleton (mild relative to Hurler), viscera (mild relative to Hurler)
IH/S	Hurler/Scheie	α-L-iduronidase	DS, HS		Phenotype intermediate between Hurler and Scheie
II	Hunter	Iduronate 2-sulfatase (EC 3.1.6.13)	DS, HS	Xq27-Xq28	CNS, skeleton, viscera (recessive, mild, and severe forms)
IIIA	Sanfilippo A	Heparan N-sulfamidase (EC 3.10 1.1)	HS	17q25.3	CNS, hyperactivity
IIIB	Sanfilippo B	α-D-N-acetylglucosaminidase (EC 3.2.1.50)	HS	17q21	Phenotype similar to IIIA
IIIC	Sanfilippo C	α-glucosamine-N-acetyltransferase (EC 2.3.1.3)	HS	—	Phenotype similar to IIIA
IIID	Sanfilippo D	α-N-acetylglucosamine-6-sulfatase (EC 3.1.6.14)	HS, GlcNAc 6S	12q14	Phenotype similar to IIIA
IVA	Morquio A	Galactose-6-sulfatase (EC 3.1.6.4)	KS, CS, GalNAc 6S	16q24	Skeleton
IVB	Morquio B	β-Galactosidase (EC 3.2.1.23)	KS	3p21cen	Skeleton
VI	Maroteaux-Lamy	N-acetylgalactosamine 4-sulfatase [aryl sulfatase B (EC 3.1.6.1)]	DS, CS, GalNAc 4S GalNAc 4,6 diS	5q13-q14	Skeleton, normal intelligence
VII	Sly	β-Glucuronidase (EC 3.2.1.31)	DS, HS, CS	7q21.1-q22	CNS, skeleton, viscera

[a]GAGS, glycosaminoglycans; CNS, central nervous system; DS, dermatan sulfate; HS, heparan sulfate; KS, keratan sulfate; CS, chondroitin sulfate.

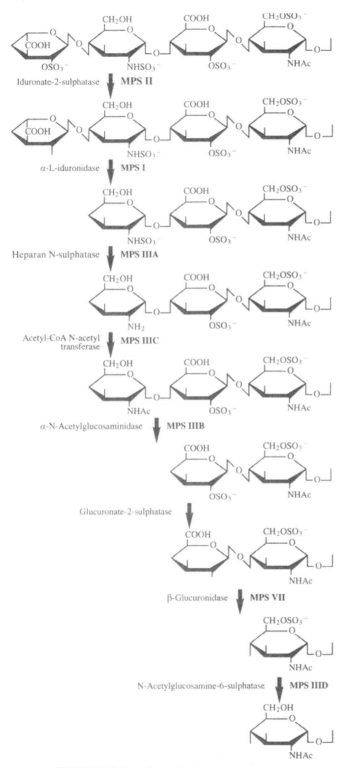

FIGURE 2. Pathway for catabolism of heparan sulfate.

iduronic acid at the nonreducing end. The desulfation of the terminal iduronic acid is catalyzed specifically by iduronate-2-sulfate sulfatase to generate the substrate for the next enzyme, α-L-iduronidase, which brings about the removal of the iduronic acid. Both of these enzymes also occur in the catabolic pathway of dermatan sulfate. When the hexuronic acid is D-glucuronic acid, it is released by the sequential action of glucuronate-2-sulfate sulfatase and β-D-glucuronidase. Glucuronate-2-sulfatase (Shaklee et al., 1985) is unique to the catabolic pathway for HS, but patients lacking this enzyme have not been reported. The β-D-glucuronidase is involved in the catabolism of several GAGs, and a deficiency of this enzyme leads to the characteristic severe clinical symptoms and storage products of MPS VII (Sly disease). If the new terminal N-acetylglucosamine is sulfated on the C6 position, the sulfate is removed by another sulfatase, N-acetylglucosamine-6-sulfatase. This enzyme can act on α-linked GlcNS.6S or GlcNAc.6S in HS and the β-linked GlcNAc.6S in KS. Heparan sulfate also contains a small amount of glucosamine O-sulfated at the C3 position (Kusche et al., 1988). It is not known which sulfatase is responsible for the hydrolysis of this sulfate ester. The O-sulfates have to be removed before the desulfation of the amino group of glucosamine, catalyzed by sulfamidase. The amino group then has to be reacetylated by acetyl-CoA:α-glucosaminide N-acetyltransferase to create a terminal N-acetylglucosamine, which is released by α-N-acetylglucosaminidase. The transferase is an integral lysosomal membrane protein, which is consistent with its function of transferring an acetyl group from acetyl-CoA in the cytosol to the desulfated 2-amino group of glucosamine in the lumen of the lysosome. A deficiency of any one of the four enzymes involved in the breakdown and release of the sulfated glucosamine leads to the accumulation and excretion of HS and the characteristic symptoms of MPS III (Sanfilippo disease).

2.3.2. Dermatan Sulfate

Details of the initial stages of the catabolism of DS are not well documented, but it is assumed that single chains of dermatan sulfate conjugated to a short peptide are produced by proteolysis extracellularly and then endocytosed. The polysaccharide chains are presumed to be broken down by endoglycosidase activity to oligosaccharides that are susceptible to hydrolysis by the sequential action of exoglycosidases and sulfatases in the lysosomes. It has been reported that the endohexosaminidase hyaluronidase (EC 3.2.1.35) can cleave dermatan sulfate, but its role in the normal catabolism of DS is not known (Roden, 1980). It may provide an alternative route when the usual pathway is defective.

The lysosomal degradation of the DS oligosaccharides follows a sequential pattern similar to that of HS oligosaccharides. (Fig. 3). The removal of sulfate from iduronate and the subsequent release of iduronic acid are catalyzed by iduronate-2-sulfate sulfatase and α-L-iduronidase, respectively, as in the degradation of HS. This explains the concomitant accumulation of DS and HS in MPS types I and II,

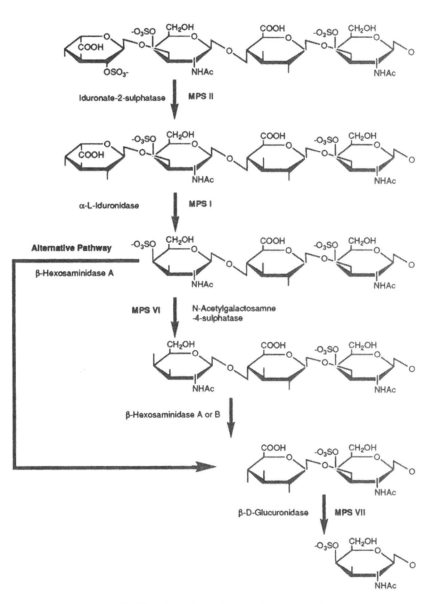

FIGURE 3. Pathway for catabolism of dermatan sulfate.

which are due to deficiencies of α-L-iduronidase and iduronate-2-sulfate sulfatase, respectively. The next substrate in the pathway, N-acetylgalactosamine-4-sulfate, is desulfated by a specific sulfatase, N-acetylgalactosamine-4-sulfate sulfatase (arylsulfatase B). The genetic deficiency of this enzyme leads to MPS VI or

Maroteaux-Lamy disease in which DS is the only GAG to accumulate. N-Acetyl-galactosamine is then believed to be released by the action of β-hexosaminidase A or B, but there is little accumulation of DS in Sandhoff disease, in which both isoenzymes of β-hexosaminidase are deficient due to a defect in the β subunit of the enzyme. The final step in the pathway is the release of glucuronic acid by β-D-glucuronidase, another enzyme in common with the HS degradative pathway. There is tissue accumulation and urinary excretion of DS and HS, as well as chondroitin sulfates, in patients with a deficiency of β-D-glucuronidase (MPS VII). An alternative route for the desulfation and release of N-acetylgalactosamine may occur by the action of hexosaminidase A, which can act on charged substrates, followed by the desulfation of the released monosaccharide by the N-acetylgalac-tosamine-4-sulfatase. This could explain the accumulation of galactosamine-4-sulfate and N-acetylgalactosamine-4,6-disulfate (GalNAc4,6diS) in MPS VI patients (Hopwood and Elliott, 1985).

Genetic disorders of DS catabolism fall into two classes: MPS I and II, in which the enzyme defect affects the catabolism of other glycoconjugates, and MPS VI, in which the defective enzyme is unique to the DS catabolic pathway. The spectrum of clinical symptoms in MPS VI reflects the distribution of DS and the cells involved in its turnover. There is little involvement of the central nervous system but there are skeletal abnormalities, organ enlargement, and problems with vision and the heart.

2.3.3. Keratan Sulfate

Chains of KS are released from cartilage and corneas by proteolysis extra-cellularly and broken down in the lysosomes by the sequential action of exogly-cosidases and sulfatases (Fig. 4). Mammalian cells do not contain endoglycosidase activity toward KS. The enzyme that removes the sulfate from C6 of galactose, galactose-6-sulfatase, can also act on N-acetylgalactosamine-6-sulfate found in chondroitin-6-sulfate. A deficiency of this enzyme therefore leads to the accu-mulation of KS and chondroitin-6-sulfate in MPS IVA (Morquio A disease). Hydrolysis of the β-D-galactosidic linkage is brought about by lysosomal β-D-galactosidase, which can also hydrolyze the same linkage in glycolipids and glycoproteins. A complete deficiency of β-D-galactosidase leads to G_{M1}-gan-gliosidosis, but an allelic variant in which the activity toward the β-galactosidic linkage in KS is selectively destroyed gives rise to MPS IVB, (Paschke and Kresse, 1982). N-Acetylglucosamine is desulfated and released by the sequential action of N-acetylglucosamine-6-sulfatase and β-hexosaminidase A and B. N-Acetylglucosamine-6-sulfatase catalyzes the same reaction in the catabolism of HS, but a deficiency of this enzyme in MPS IIID only leads to the accumulation of HS and not KS, because β-linked N-acetylglucosamine-6-sulfate (GlcNAc6S) in KS can be released by β-hexosaminidase A, whereas the α-linked GlcNAc6S

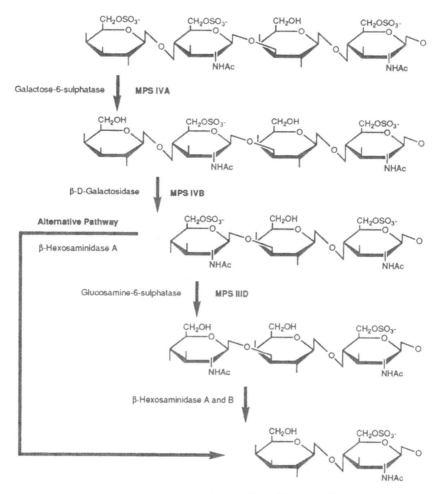

FIGURE 4. Pathway for catabolism of keratan sulfate.

in HS cannot. Patients with MPS IIID excrete large amounts of GlcNAc6S, but the intralysosomal accumulation of GlcNAc6S does not exacerbate the clinical condition because patients with MPS IIID have clinical problems similar to patients with MPS IIIA-C, who only accumulate HS. Under normal circumstances the release of β-linked N-acetylglucosamine is catalyzed by either β-hexosaminidase A or B. A defect in the β subunit of β-hexosaminidase, as in Sandhoff disease, leads to a deficiency of both hexosaminidase A and B and the accumulation of G_{M2}-ganglioside and fragments of N-linked glycans but not of KS. It is possible that the αα–homodimer of β–hexosaminidase (β-hexosaminidase S) can release GlcNAc6S directly *in vivo* as well as *in vitro* (Fuchs *et al.*, 1985).

2.3.4. Chondroitin Sulfates

The lysosomal catabolism of chondroitin sulfates involves the desulfation of N-acetylgalactosamine-4-sulfate and N-acetylgalactosamine-6-sulfates by N-acetyl-galactosamine-4-sulfatase and N-acetylgalactosamine-6-sulfatase, respectively, followed by the action of β-hexosaminidase A or B and β-glucuronidase. All of these enzymes participate in the catabolism of other GAGs. However, there is not massive excretion and accumulation of chondroitin-4-sulfate in MPS VI, as might be expected for a deficiency of GalNAc-4-sulfatase. Similarly, there is some excreted chon-droitin-6-sulfate in MPS IVA, which is due to a deficiency of GalNAc-6-sulfatase, but it is not excessive. The increase in excreted chondroitin-4-sulfate and chon-droitin-6-sulfate is also very low in Sandhoff disease and MPS VII, which are the dis-eases due to deficiencies of the other two enzymes in the pathway, β-hexosaminidase A and B and β-glucuronidase, respectively. The probable explanation of these ob-servations is the digestion of chondroitin sulfates by the lysosomal hyaluronidase, which is present in some cells and serum (Aronson and Davidson, 1967; Sampson *et al.*, 1992). Some chondroitin sulfates are a normal component of human urine.

2.3.5. Hyaluronic Acid

As well as providing alternative routes for the catabolism of dermatan sulfate and the chondroitin sulfates, hyaluronidase is involved in the catabolism of hyaluronic acid (or hyaluronan), the most abundant GAG in the body (Evered and Whelan, 1989). Hyaluronic acid can be endocytosed by resident cells of the ex-tracellular matrix and degraded within lysosomes (Ng *et al.*, 1992). There are also specific receptors for HA on liver endothelial cells and this may be the main route for clearance of circulating HA (Smedsrod *et al.*, 1984). The catabolism of HA by lysosomal hyaluronidase in lung fibroblasts is regulated by cytokines, possibly through altering the binding of HA to the fibroblasts (Sampson *et al.*, 1992). The main products of the digestion of HA by hyaluronidase are a tetrasaccharide and larger oligosaccharides (Aronson and Davidson, 1967). A deficiency of hyaluronidase has been reported in the serum of children with some characteristic features of an MPS but without mental retardation or organ enlargement (Fiszer-Szafarz *et al.*, 1991). Hyaluronidase activity is widely distributed in animal tissues and it has been purified and characterized from rat liver (Aronson and Davidson, 1967). The acidic lysosomal enzyme is kinetically and physically distinct from the abundant testicular enzyme, but is the same as the serum activity, which may ac-count for the digestion of some accumulated GAGs in the mucopolysaccharidoses. Pig liver hyaluronidase has been cloned and found to have a predicted amino acid sequence identical to that of hemopexin, a serum heme-binding protein (Zhu *et al.*, 1994). Hemopexin has a domain that is highly conserved in stromelysins, colla-genases, and other enzymes involved in the turnover of the extracellular matrix, suggesting that it also might play a part in this process.

2.3.6. Hydrolysis of the Linkage Region

The presence of galactose and xylose at the reducing terminals of human urinary GAGs (Endo *et al.*, 1980; Matsue and Endo, 1987) suggests that endo-β-galactosidase and endo-β-xylosidase exist in human tissues. There is evidence for the presence of both of these activities in rabbit liver lysosomes (Takagaki *et al.*, 1988a). As both xylose and galactose were found terminally, it is possible that the two enzymes act independently or act on different linkage structures. The C2 of xylose can be phosphorylated in heparan sulfate (Fransson *et al.*, 1985) and chondroitin sulfate (Oegama *et al.*, 1984). An endo-β-glucuronidase capable of hydrolyzing the glucuronyl–galactosyl linkage of chondroitin sulfate chains that are linked to very small peptides but not to intact proteoglycans has also been purified from rabbit liver (Takagaki *et al.*, 1988b). The precise roles of these endoglycosidases in the lysosomal catabolism of GAGs remain to be established. The O-glycosidic and N-glycosidic protein linkages in skeletal and corneal KS, respectively, are believed to be cleaved by the same enzymes that act on these bonds in glycoproteins.

3. GLYCOPROTEINS

3.1. Structure, Occurrence, and Function of Glycoproteins

The majority of proteins in eukaryotic cells contain covalently linked oligosaccharide chains or single monosaccharides. The glycans of glycoproteins fall into two classes. N-Glycans are linked by an N-glycosidic linkage between N-acetylglucosamine at the reducing end of the glycan and the amide nitrogen of asparagine. O-Glycans are linked by an O-glycosidic bond, usually between N-acetylgalactosamine at the reducing end of the glycan and the side-chain hydroxyl group of serine or threonine.

All N-linked glycans share a common core structure, Man $\alpha1 \to 6$ (Man $\alpha1 \to 3$) Man $\beta1 \to 4$ GlcNAc $\beta1 \to 4$ GlcNAc1 $\to$ Asn because they are all derived from a common precursor oligosaccharide, which is transferred *en bloc* from a lipid carrier, dolicholpyrophosphate, to asparagine residues in a sequon of Asp.X.Thr/Ser, where X is any amino acid except proline (Kornfeld and Kornfeld, 1985). Subsequent trimming and elongation of the oligosaccharide, called processing, by glycosidases and glycosyltransferases, respectively in the endoplasmic reticulum and Golgi, generates the enormous structural diversity found in N-linked glycans. Representative structures of the three main classes of N-linked glycans, high-mannose, complex, and hybrid, are shown in Fig. 5. Straight or branched poly-N-acetyllactosamine chains made up of the repeating unit Gal $\beta1 \to$ 4 GlcNAc $\beta1 \to 3$ can also be attached to the core. Fucose linked $\alpha1 \to 6$ to the N-acetylglucosamine attached to the asparagine and a bisecting N-acetylglucosamine

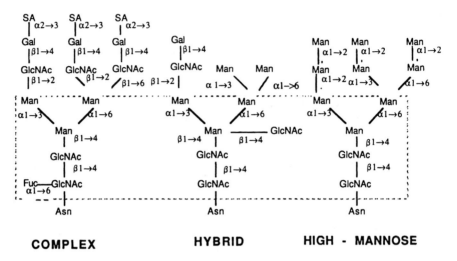

FIGURE 5. Structures of representative high-mannose, hybrid, and complex N-linked glycans.

linked β1 → 4 to the β-linked mannose can occur in all N-linked glycans except high-mannose ones. Up to five branches or antennae are found in complex glycans. The complement of glycosidases and glycosyltransferases expressed in a cell determines its glycosylating and processing capacity and is called the glycotype (Rademacher *et al.,* 1988). The factors regulating the processing of an individual glycan at a particular position in a protein are not well understood, but the conformation of the glycan-peptide presented to the processing enzyme as a substrate is crucial. Individual proteins even when synthesized in a single cell type show microheterogeneity in their N-linked glycans. Glycoproteins with the same polypeptide chain but different glycans are called glycoforms. This heterogeneity suggests that in many cases it is the general type of glycosylation rather than a specific structure of an N-linked glycan that is important for function. N-Linked glycans occur on soluble lysosomal hydrolases, integral membrane proteins, secreted proteins, and protein components of the extracellular matrix (Kobata, 1992). Endogenous glycoproteins and enzymes for the biosynthesis and processing of N-linked glycans have been reported to be present in mitochondria (Levral *et al.,* 1990; Gasnier *et al.,* 1992).

O-Linked glycans do not have a common core structure but occur as a single monosaccharide, N-acetylgalactosamine, disaccharides, or larger glycans with a distinct core, backbone, and nonreducing terminus (Kobata, 1992). Some representative structures with the more common core structures are shown in Fig. 6. The nonreducing terminal residues can be D-galactose, N-acetyl-D-galactosamine, L-fucose, sialic acid, or N-acetylglucosamine-3/4-sulphate. O-Glycans are synthesized in the Golgi by the sequential addition of monosaccharides from nucleoside phosphate donors to the N-acetylgalactosamine linked to serine/threonine. The pathways are highly ordered, with the addition of a monosaccharide in a par-

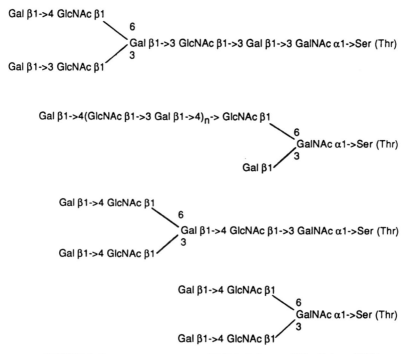

FIGURE 6. Common core structures of O-linked glycans. (After Kobata, 1992.)

ticular linkage determining the subsequent steps in the assembly of the glycan (Schachter, 1991; Schachter and Brockhausen, 1992). O-Linked glycans were originally characterized from mucins and are often called mucin-type, but they are present in many glycoproteins with very diverse functions. They include mucins, the "antifreeze" glycoprotein of antarctic fish, fetuin, blood group substances, epiglycanin, the oligosaccharide ligands for L- and P-selectins, the core proteins of connective tissue proteoglycans, and other glycosylated proteins of the extracellular matrix (Carraway and Hull, 1991).

Galactose and the disaccharide $Glc\alpha 1 \rightarrow 2Gal$ are O-glycosidically linked to the hydroxylysine residues within the triple-helical domains of vertebrate collagens (Butler, 1978). The extent of glycosylation varies among the different collagen types, within the same collagen type in different tissues, and between species. There is an inverse relationship between the diameter of collagen fibrils and the carbohydrate content, suggesting that glycosylation may, in part, regulate fibril diameter (Kielty *et al.,* 1993). Very little is known about the degradation of these glycans, but it is assumed that they are degraded inside lysosomes along with the internalized fragments of collagen.

Recently, several novel forms of protein glycosylation have been described (Hayes and Hart, 1994). The addition of a single N-acetylglucosamine to the hydroxyl group of serine and threonine is particularly prevalent in a diverse range of cytoplasmic and nuclear proteins (Haltiwanger *et al.,* 1992). This form of glycosylation appears to be transient, and it is possible that the regulatory function of most proteins modified in this way is mediated in a similar fashion to phosphorylation. A cytosolic β-hexosaminidase with a neutral pH optimum can catalyze the removal of the O-linked N-acetylglucosamine, which may also be a signal for degradation of the protein. Therefore, deglycosylation of these proteins in lysosomes is probably not important. Another cytosolic glycosylation in yeast, the addition of glucose 1-phosphate to O-linked mannose(s) residues, appears to be restricted to a single protein, phosphoglucomutase (Marchase *et al.,* 1993). O-Linked fucose, which can be processed to a tetrasaccharide, has been found on epidermal growth factor homologous regions. The trisaccharide (Xyl $\alpha 1 \rightarrow 3)_{1-2}$ Glc β1 → 0-Ser has also been identified in these regions. It is not known whether these unusual O-linked glycans are catabolized in lysosomes. β-Glucosylasparagine has been found on rat and kidney laminin, but little is known about its metabolism (Schreiner *et al.,* 1994). The most common "unusual" glycosylation is the presence at the nonreducing end of N-linked glycans in glycoprotein hormones of N-acetylgalactosamine residues, which are often sulfated or linked to sialic acid (Baenziger, 1994). A hepatocyte receptor specific for sulfated N-acetylgalactosamine mediates the clearance of these proteins from the circulation, presumably for lysosomal degradation, thereby regulating their half-lives.

3.2. Turnover of Glycoproteins

Glycoproteins, like other glycoconjuates, are being turned over continuously. There is much experimental evidence that this takes place predominantly in the lysosomes. Many glycoproteins reach the lysosomes by the endoctyic pathway. This may be by receptor-mediated endocytosis, fluid-phase pinocytosis, or as components of the endosomal membrane system. The half-lives of proteins in the circulation are determined by the degree of sialylation, as loss of terminal sialic acids leads to their removal by receptor-mediated endocytosis by the asialoglycoprotein receptor particularly abundant on hepatocytes. Other receptors specific for different terminal monosaccharides or glycans mediate the removal of specifically modified N-glycosylated proteins (Neufeld and Ashwell, 1980). Excess secretory proteins are diverted to lysosomes by crinophagy. Cytosolic proteins may be delivered to lysosomes by microautophagy, which forms multivesicular bodies or by macroautophagy or engulfment of organelles such as the Golgi or endoplasmic reticulum, which contain many glycoproteins (Dunn, 1994). The latter process is enhanced during starvation or deprivation of serum in cell cultures (Mayer and Doherty, 1986). The proportion of intracellular protein degraded in the lysosomes depends on the cell type and its physiological state. Certain long-lived proteins may be preferentially degraded in lysosomes (Ahlberg *et al.,* 1985) after delivery by a receptor or binding protein that recognizes the sequence KFERQ or a closely related one (Dice, 1990). The uptake into lysosomes and subsequent degradation of these proteins is mediated by the heat-shock protein hsc73 in a manner similar to the delivery of proteins by chaperones to other cytoplasmic organelles (Cuervo *et al.*, 1994). Chapters 4 and 5 of this volume provide detailed information on these autophagic processes.

3.3. Lysosomal Catabolism of Glycoprotein Glycans

The complete catabolism of glycoproteins in lysosomes has been demonstrated in perfused organs, cells in culture, and in isolated lysosomes (Rome and Hill, 1986). Glycoproteinoses (defects in the lysosomal catabolism of glycoproteins) are due to failures to degrade the carbohydrate moiety and not to failure of proteolysis (Table II) (for reviews of glycoproteinoses, see Durand and O'Brien, 1982; Hancock and Dawson, 1989; Cantz and Ullrich-Bott, 1990; Thomas and Beaudet, 1995). Although there have been several reports of defects in lysosomal proteases or cathepsins, no lysosomal storage disease has been attributed to a defect in a specific lysosomal protease. It is assumed that the deficiency of one protease can be compensated for by the overlapping specificity of another cathepsin or that failure of proteolysis is fatal *in utero*. The lysosomal catabolism of proteins has been reviewed extensively by Glaumann and Ballard (1987) and is

Table II
The Glycoproteinoses

Disorder	Defective gene product and enzyme defect	Gene location
Sialidosis (Mucolipidosis I)	N-Acetyl-α-neuraminidase (sialidase) (EC 3.2.1.18)	10pter-q23
α-Mannosidosis	α-D-Mannosidase (EC 3.2.1.24)	19p13.2-q12
β-Mannosidosis	β-D-Mannosidase (EC 3.2.1.26)	4q25-4q22
Fucosidosis	α-L-Fucosidase (EC 3.2.1.51)	1q34
G_{MI}-gangliosidosis	β-D-Galactosidase (EC 3.2.1.23)	3p21-3pter
Aspartylglucosaminuria	1-Aspartamido β-N-GlcNAc-amidohydrolase (aspartyl-glucosaminidase) (EC 3.5.1.26)	4q32-q33
Mucolipidosis II (I-cell disease)	Glc-NAc-1-phosphotransferase, recognition marker defect	4q21-q23
Mucolipidosis III (pseudo-Hurler polydystrophy)	of lysosomal enzymes (EC 2.7.8.17)	
α-N-acetylgalactosaminidase deficiency (Schindler disease)	α-N-Acetylgalactosaminidase (α-galactosidase B) (EC 3.2.1.49)	22q13.1-13.2
Galactosialidosis	32-kDa protective protein, combined decrease of β-galactosidase and N-acetyl-α-neuraminidase	20q13.1

discussed in detail in Chapter 6. In the present chapter the catabolism of the conjugated glycans is described and compared with the catabolism of the glycans in other glycoconjugates.

3.3.1. O-Linked Glycans

No systematic studies on the catabolism of O-linked glycans have been reported. Since the structures of O-linked glycans are similar to the glycans in glycosphingolipids, it is assumed that the same exoglycosidases are responsible for their sequential breakdown (see Section 4). Oligosaccharides and glycopeptides corresponding to O-linked glycans are found among the storage products in the diseases resulting from a deficiency of these glycosidases, supporting this assumption. It is not known whether proteolysis precedes the breakdown of the glycans or whether endoglycosidases release the glycans at any stage. Evidence that α-N-acetylgalactosaminidase (EC 3.2.1.49) catalyzes the cleavage of the core GalNAcα1 → 0-Ser/Thr linkage is provided by the storage products in patients with a deficiency of α-N-acetylgalactosaminidase. (Van Diggelen *et al.,* 1987; Kanzaki *et al.,* 1989). In addition to GalNAcα1 → 0-Ser/Thr, several glycopeptides with the core linkage intact and terminating in neuraminic acid at the nonreducing end have been isolated from the urine of patients with this disorder (Linden *et al.,* 1989; Hirabayashi *et al.,* 1990). Very similar or identical glycopeptides are excreted by patients with an isolated deficiency of α-neuraminidase (sialidosis or mucolipidosis I) (Lecat *et al.,* 1984) or with the combined deficiency of α-neuraminidase and β-galactosidase (galactosialidosis) that is due to a defect in the protective protein (Takahashi *et al.,* 1991). This suggests that these O-linked sialo-oligosaccharides are degraded intralysosomally by a multienzyme complex and that a genetic defect in any one of its constituents can disrupt its activity. α-N-Acetylgalactosaminidase activity is found in association with the α-neuraminidase/β-galactosidase/protective protein complex (Warner *et al.,* 1990). Skeletal keratan sulfate contains the core GalNAcα1 → 0-Ser/Thr linkage, but there is no evidence of disturbance of GAG metabolism in the patients studied so far. N-Acetylgalactosamine also occurs glycosidically linked to sugars in the glycans of glycosphingolipids, for example, in the blood group A substance and the Forsmann antigen, but again no glycolipid accumulation or consequent clinical symptoms of a sphingolipidosis have been noted in the patients (Klima *et al.,* 1992). However, exogenous glycolipid with a terminal α-linked N-acetylgalactosamine is not metabolized by cultured fibroblasts from a patient with the disease, suggesting that the enzyme does normally act on these compounds. Patients with blood group A excrete the blood group A trisaccharide, GalNAcα1 → 3[Fucα1 → 2]Gal in their urine (Linden *et al.,* 1989). It is possible that patients with different mutations will display a wider or different spectrum of storage products and throw more light on the role of this enzyme in the catabolism of O-linked and other conjugated glycans.

3.3.2. N-Linked Glycans

Most studies *in vitro* on the digestion of intact glycoproteins by lysosomal glycosidases have suggested that extensive degradation of the polypeptide chain takes place before breakdown of the N-linked glycans by the action of exoglycosidases and endoglycosidases. This does not preclude the removal of exposed glycosyl residues from the nonreducing terminal during or before proteolysis. Similarly, the complete catabolism of the glycan and cleavage of the GlcNAc–asparagine linkage is not necessary for the degradation of the polypeptide. The administration to perfused rat liver of 5-diazo-4-oxo-L-norvaline, a specific inhibitor of aspartylglucosaminidase, the enzyme that cleaves the aspartylamido linkage, leads to the accumulation of GlcNAcβ1 → 4GlcNAcβ1 → Asn (Kuranda and Aronson, 1987). In addition to showing that degradation of the polypeptide did not require removal of the carbohydrate, this experiment demonstrated that partial but not complete breakdown of the glycan can occur while it is still attached to the protein. The storage products in human fucosidosis, which results from a deficiency of the exoglycosidase α-fucosidase, contain glycans still linked to asparagine (Tsay *et al.*, 1976). In contrast, the storage products in all the other human glycoproteinoses due to defects in exoglycosidases have a single N-acetylglucosamine at the reducing terminal, indicating that cleavage of the protein–carbohydrate linkage and removal of some of the glycosyl residues at the reducing terminal can occur even though digestion by exoglycosidases at the nonreducing end is blocked (Durand and O'Brien, 1982; Thomas and Beaudet, 1995). These observations led to the proposal that the catabolism of N-linked glycans is bidirectional and highly ordered (Brassart *et al.*, 1987; Kuranda and Aronson, 1987). Subsequent studies on the digestion of natural substrates using purified lysosomal enzymes have confirmed this hypothesis for human (Al Daher *et al.*, 1991), rat (Haeuw *et al.*, 1991a), cat, and cattle (DeGasperi *et al.*, 1991); they also revealed important differences between species in the catabolism of the core region (Abraham *et al.*, 1983).

The first step in the catabolism of the N-glycans of glycoproteins is proteolysis of the polypeptide to produce glycoasparagines. The next step for complex glycans (Fig. 7) is the removal of any fucose α1 → 6 linked to the N-acetylglucosamine bonded to the asparagine, catalyzed by α-fucosidase. Fucose residues linked α1 → 2,3 to sugars at the nonreducing end are probably also removed at this stage (Barker *et al.*, 1988). The removal of the core fucose and lack of substitution of the α-amino and carboxy groups of asparagine are prerequisites for the action of lysosomal aspartyl-N-acetyl-β-D-glucosaminidase, which releases the glycan from asparagine (Kaartinen *et al.*, 1992). A deficiency of this enzyme leads to the lysosomal storage disorder aspartylglycosaminuria. In humans and rats the released glycan is then degraded by an endo-N-acetyl-β-D-glucosaminidase to produce oligosaccharides containing only one N-acetylglucosamine at the reducing end. This enzyme, which is also called chitobiase, has been purified from

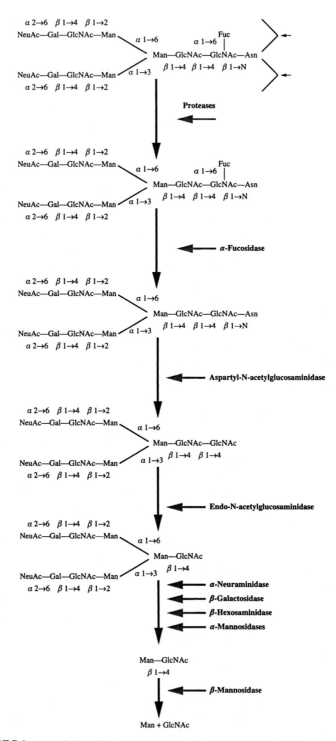

FIGURE 7. Lysosomal degradation of a typical complex N-linked glycan of glycoproteins.

human (Stirling, 1974) and rat (Aronson *et al.*, 1989) liver and human kidney (DeGasperi *et al.*, 1989). It has been cloned from human placenta and rat liver (Fisher and Aronson, 1992). Chitobiase is expressed in primates and rodents but not in species such as cat, dog, goat, pig, and cattle (Song, *et al.*, 1987). Consequently, the storage products from glycoproteinoses in these species have chitobiose as opposed to a single N-acetylglucosamine at the reducing end (Abraham *et al.*, 1983). The exoglycosidase β-N-acetylglucosaminidase probably hydrolyzes the chitobiosidic linkage in these species as the last step in the catabolism of complex glycans. It is likely that it can carry out a similar function in humans because patients with aspartylglycosaminuria excrete GlcNAc1 → Asn rather than Glc-Nacβ1 → 4 GlcNAc1 → Asn, which might have been expected with a deficiency of aspartylglucosaminidase. The released complex oligosaccharide, with one or two N-acetylglucosamine residues at the reducing end, is then broken down to its constituent monosaccharides by the sequential release of glycosyl groups from the nonreducing end, catalyzed by exoglycosidases (Fig. 7).

The lysosomal catabolism of the high-mannose N-linked glycans is also highly ordered and bidirectional (Fig. 8), and falls into two stages. Analysis of the storage products from cases of α-mannosidosis in different species (Strecker *et al.*, 1976; Abraham *et al.*, 1983) suggested that the high-mannose glycans were released from the protein and partially catabolized at the reducing end in the same way as the complex glycans. The absence of chitobiase in some species was again indicated by the presence of chitobiose at the reducing end of stored high-mannose glycans. Direct confirmation of the first steps was obtained by *in vitro* studies in the rat (Baussant *et al.*, 1986; Haeuw *et al.*, 1991a). An endo-β-N-acetyl-glucosaminidase with an acidic pH optimum and a preference for the cleavage of the chitobiosidic linkage in high-mannose oligosaccharides has been isolated from human kidney (DeGasperi *et al.*, 1989). This suggests that the hydrolysis of the chitobiosidic linkage is not only species-specific but also depends on the structure of the oligosaccharide. The structural and genetic relationship of this activity to other chitobiosidases is not known.

In the second stage, the released high-mannose oligosaccharides are broken down in a highly ordered manner by the major lysosomal α-mannosidase, which can hydrolyze α1 → 2, α1 → 3, and α1 → 6 mannosidic linkages (Fig. 8). The pathways are essentially the same in man (Al Daher *et al.*, 1991), rat (Michalski *et al.*, 1990), and bovine and feline (DeGasperi *et al.*, 1991) liver. The precise route of catabolism is determined by the structure of the first substrate but they all converge on a key intermediate, Manα1 → 6 Manβ1 → 4 GlcNAc. This is the common intermediate for the catabolism of complex, hybrid, and high-mannose glycans (Al Daher *et al.*, 1991). The major lysosomal α-mannosidase has low activity toward the remaining core α1 → 6 mannosidic linkage, which is probably hydrolyzed by another specific lysosomal α1 → 6 mannosidase in man (Cenci di Bello *et al.*, 1983; Daniel *et al.*, 1992; DeGasperi *et al.*, 1992a) and rat (Haeuw *et al.*, 1994). Interestingly the α1 → 6 mannosidase cannot act on oligosaccharides

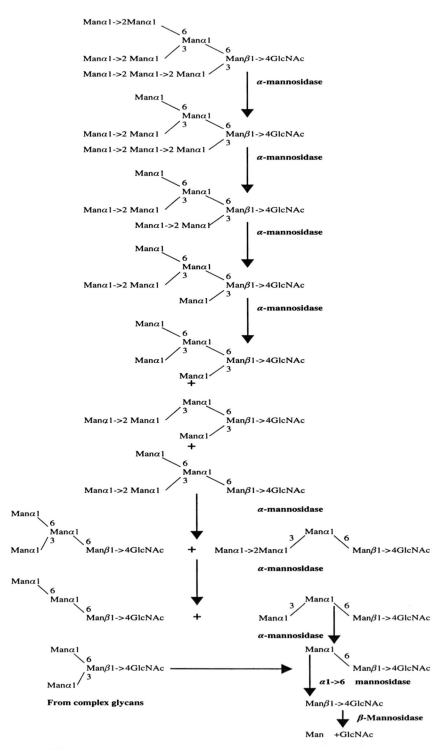

FIGURE 8. Lysosomal breakdown of a high-mannose N-linked glycan of glycoproteins.

with chitobiose at the reducing end and requires the prior action of chitobiase (Haeuw *et al.*, 1994). There is probably a functional relationship between these two activities, as they are only expressed in the same tissues and species. There is evidence that the early-acting α-fucosidase and aspartylglucosaminidase are structurally and functionally linked in dogs (Barker *et al.*, 1988). Thus it is possible that the breakdown of the core region of N-linked glycans in lysosomes is mediated by functional multienzyme complexes, which have evolved differently in different species.

3.3.3. Relationship between Lysosomal and Nonlysosomal Breakdown of Glycoproteins

Glycosidase activity is also present in other compartments of most mammalian cells. The glycosidases in the brush border membrane of the epithelial cells of the intestine are involved in the breakdown of ingested starch and glycogen. Glycosidases in the ER and Golgi are involved in the processing of the N-linked glycans of glycoproteins (Kornfeld and Kornfeld, 1985). The cytosol of many cells contains β-hexosaminidase (Braidman *et al.*, 1974), α-mannosidase (Phillips *et al.*, 1974; Shoup and Touster, 1976), a broad-specificity β-galactosidase/β-glucosidase (Glew *et al.*, 1976), and an endo-β-N-acetylhexosaminidase (Pierce *et al.*, 1979) with neutral pH optima. The accumulation of oligosaccharides and glycolipids in genetic and chemically induced blocks in the lysosomal catabolism of glycoproteins demonstrates that these neutral cytosolic glycosidases cannot substitute for the lysosomal system. A role for the cytosolic neutral β-hexosaminidase in the rapid removal of transient O-linked N-acetylglucosamine residues has recently been established (Dong and Hart, 1994). The broad-specificity β-galactosidase can hydrolyze the β-glucosidic linkage between glucose and natural endogenous lipid alcohols and may provide an alternative mechanism for the hydrolysis of β-glucocerebroside in Gaucher disease (Vanderjagt *et al.*, 1994).

A role for the cytosolic neutral α-mannosidase and endo-β-hexosaminidase in the proofreading or editing of incompletely assembled or incorrectly folded glycoproteins has been proposed (Haeuw *et al.*, 1991a,b; Daniel *et al.*, 1994). This is based on the careful analysis of the storage products in genetic and swainsonine-induced α-mannosidosis in several species (Cenci di Bello *et al.*, 1983; Daniel *et al.*, 1989) and the specificity of neutral cytosolic α-mannosidase toward natural substrates (Haeuw *et al.*, 1991b; Al Daher *et al.*, 1992; DeGasperi *et al.*, 1992b). Human and rat liver cytosolic α-mannosidase digests high-mannose glycans in a highly ordered pathway to a limit digestion product, Man α1 → 2Man α1 → 2 Man α1 → 3[Man α1 → 6] Man β1 → 4GlcNAc, which resembles an intermediate in the assembly of the dolichol oligosaccharide precursor rather than a structure in the N-glycan processing or catabolic pathways. This compound is found among the storage products in α-mannosidosis in species lacking the lysosomal endo-β-glucosaminidase and in the cytosol of cells growing in the presence of swainsonine (Tulsiani and Touster, 1992; Kang *et al.*, 1993). Its formation has also been demonstrated in the cytosol of normal cells by metabolic labeling (Moore and Spiro, 1994). Oligomannose glycans with a single N-acetylglucosamine at the reducing end are produced in the ER or a closely related compartment as a result of the degradation of newly synthesized glycoproteins (Villers *et al.*, 1994). It is feasible that these oligosaccharides are formed by the action of the cytosolic β-endohexosaminidase and digested to the limit product by the neutral α-mannosidase prior to delivery to the lysosomes by autophagocytosis or specific translation across the lysosomal membrane. Dolichol intermediates not required for protein glycosylation could be degraded by the same route (Cacan *et al.*, 1992). This route is only revealed when there is a genetic or chemically induced deficiency of the lysosomal α-mannosidase, but it is postulated to be the normal route for the catabolism of N-glycans from excess dolichol oligosaccharides and newly synthesized glycoproteins (Daniel *et al.*, 1994).

4. GLYCOSPHINGOLIPIDS

4.1. Structure, Occurrence, and Function of Glycosphingolipids

Glycosphingolipids are important and ubiquitous components of eukaryotic cell membranes (Hakomori, 1981; Rademacher *et al.*, 1988; Hakomori, 1993). They are amphipathic molecules containing a hydrophilic oligosaccharide chain linked glycosidically to ceramide, which is made up of the long-chain amino alcohol sphingosine, substituted with an acyl fatty acid (Fig. 9). Over 150 glycosphingolipids with different glycans have been isolated from mammalian tissues. They have been classified according to the structure of the core oligosaccharide (Fig. 10). Glycosphingolipids with glycan chains containing sialic acid are called gangliosides, reflecting their abundance in neural tissue (Ando, 1983).

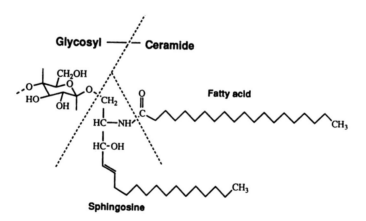

FIGURE 9. Structure of a glycosphingolipid.

Ganglio-series   Galβ1->3GalNAcβ1->4Galβ1->4Glcβ1->Cer
                                          |2->3
                                        NeuAc

Globo-series   GalNAcβ1->3Galα1->4Galβ1->4Glcβ1->Cer

Isoglobo-series   GalNAcβ1->3Galα1->3Galβ1->4Glcβ1->Cer

Lacto-series   Galβ1->3GlcNAcβ1->3Galβ1->4Glcβ1->Cer
                                       | 1->3
                                       Fuc

Neolacto-series   Galβ1->4GlcNAcβ1->3Galβ1->4Glcβ1->Cer
                                       | 1->3
                                       Fuc

FIGURE 10. Glycosphingolipid core oligosaccharide structures. The substitutions shown on the tetrasaccharide cores are representative.

There is a much higher concentration of glycosphingolipids in the plasma membrane than in the intracellular membranes of most cells, but some cells do have high intracellular concentrations of glycosphingolipids (Symington *et al.,* 1987). It has been suggested that some of the intracellular glycosphingolipids may be associated with cytoskeletal proteins (Gillard *et al.,* 1993). Glycosphingolipids are located in the outer leaflet of the bilayer of plasma membranes, with the ceramide portion acting as an anchor and the oligosaccharide protruding from the cell surface. They are not distributed uniformly over the surface but aggregate in clusters (Rock *et al.,* 1990). The conformations of the glycans are probably determined by these interactions, expanding the repertoire of potential functions for a particular structure.

The glycosphingolipid composition of membranes varies markedly from one cell type to another, reflecting the diverse functions of glycosphingolipids in cell growth, differentiation, transformation, adhesion, antigenicity, and interaction with hormones and tissues (Schnaar, 1991; Hakomori, 1993). In some instances the function of the sphingolipid may be indirect by modulating the effect of a protein receptor.

4.2.   Turnover of Glycosphingolipids

The composition of the membrane glycosphingolipids also changes with the physiological state of the cell (Ando, 1983; Hakomori, 1990). This is brought about by changes in the expression of glycosyltransferases and *de novo* synthesis or remodeling of existing glycosphingolipids and the catabolism of glycosphingolipids in lysosomes. The plasma membrane is internalized continuously. The components destined for catabolism are delivered to the lysosomes via the endosome system

(Griffiths *et al.,* 1988), in which sorting takes place, with other components being returned to the plasma membrane or directed to the Golgi (Wessling-Resnick and Braell, 1990; Kok and Hoekstra, 1994). The precise route by which plasma membrane glycosphingolipids reach the lumen of the lysosomes is not fully understood (see Chapter 3). It has been suggested that during sorting they are sequestered into intraendosomal vesicles, which are delivered to the lumen of the lysosome by membrane fission and fusion (Sandhoff *et al.,* 1992; Furst and Sandhoff, 1992; Sandhoff and Klein, 1994). In this way glycosphingolipids on the outer leaflet of the plasma membrane would be on the outer surface of intralysosomal vesicles and exposed to the catabolic hydrolases. Experimental evidence for this hypothesis is provided by the observation of multivesicular bodies in normal, early, and late endosomes (Hopkins *et al.,* 1990) and especially of multivesicular storage bodies in Kupffer cells of patients with a deficiency of the sphingolipid activator precursor (Harzer *et al.,* 1989; Schnabel *et al.,* 1992). Although the intracellular membrane glycosphingolipid content is low, some glycosphingolipid is delivered to lysosomes by autophagy. A major source of the glycosphingolipids that are degraded in lysosomes is the plasma membrane of senescent or damaged cells, which are phagocytosed by macrophages and neutrophils. Awareness of the relative importance of the catabolism of endogenous and exogenous glycosphingolipids in different cells is extremely important for understanding the nature of the storage products, and the pathogensis and progress of the sphingolipidoses, the lysosomal storage diseases resulting from defects in the catabolism of glycosphingolipids.

It is difficult to study the turnover of individual glycosphingolipids because of the recycling of the released labeled products. Differences in turnover would be expected to correlate with function, and it has been shown that G_{M1}, the main ganglioside of myelin, is turned over more slowly than the total gangliosides in the cerebrum (Suzuki, 1970).

4.3.   Lysosomal Catabolism of the Glycans of Glycosphingolipids

The characterization of the lipids that accumulate in the sphingolipidoses was a great stimulus to the elucidation of the pathways for the catabolism of the glycan moieties of glycosphingolipids (Fig. 11). As with the glycans in other glycoconjugates, the oligosaccharide chains of glycosphingolipids are degraded by the sequential action of exoglycosidases from the nonreducing end. For glycosphingolipids this takes place while the oligosaccharides are still conjugated to the lipid moiety, ceramide. Thus, the pathways for the different glycosphingolipids converge on ceramide. The catabolism of ceramide and sphingomyelin is discussed in detail in Chapter 8. Figure 11 only shows the pathways for the breakdown of common representatives of the major groups of glycosphingolipids in animal cells. The presence of less abundant glycosphingolipids among the storage products for a particular enzyme defect suggests that the same complement of enzymes can break down all known glycosphingolipids. This does not preclude the discovery of novel

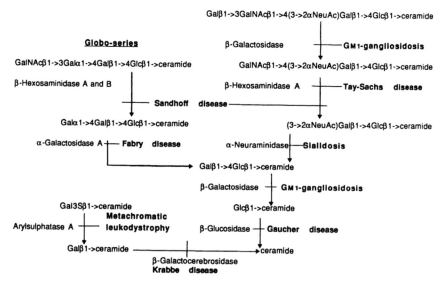

FIGURE 11. Pathway for the catabolism of representative glycosphingolipids.

enzymes specific for the breakdown of the rare and often tissue-specific sphin-golipid structures continually being discovered. The glycosidases can hydrolyze the same linkage to several different sugar aglycons in glycosphingolipids and of-ten in other glycoconjugates. For example, acid β-galactosidase hydrolyzes Galβ1 → 3GalNAc linkages in glycosphingolipids and O-glycans in glycopro-teins, Galβ1 → 3GlcNac and Galβ1 → 4Glc linkages in glycosphingolipids, and Galβ1 → 4GlcNAc linkages in glycosphingolipids, glycosaminoglycans, and N-glycans in glycoproteins.

A different β-galactosidase, β-galactosylceramidase (EC 3.2.1.46), hy-drolyzes Galβ1 → ceramide or galactocerebroside, which is found predomi-nantly in myelin and oligodendroglia and is the precursor of cerebroside sulfate. This enzyme can also hydrolyze the β-galactosidic linkage in lactosylceramide, monogalactosyl diglyceride, and, significantly, the cytotoxic galactosylsphingo-sine (psychosine), but not in G_{M1}. There is a deficiency of β-galactosylcerami-dase in Krabbe disease (globoid cell leukodystrophy) but galactocerebroside does not accumulate. It has been suggested that the accumulation of galactosyl-sphingosine is the causative agent in the pathology of Krabbe disease (Miyatake and Suzuki, 1972). Galactosylceramide can also be hydrolyzed by the acidic G_{M1}-β-galactosidase under certain assay conditions *in vitro,* and it has been pos-tulated that this might happen *in vivo,* explaining the absence of accumulation of galactocerebroside in Krabbe disease (Kobayashi *et al.,* 1985). Galactosylce-ramidase has proved very difficult to purify because of its highly hydrophobic nature and low abundance, but recently it has been purified and cloned from two human tissues (Chen *et al.,* 1993; Sakai *et al.,* 1994).

The enzyme that cleaves the β-glucosyl-ceramide core linkage found in most extraneural glycosphingolipids, β-glucocerebrosidase (EC 3.2.1.45), is different from the other glycosidases in that it is associated with the lysosomal membrane and is not transported to the lysosomes by the mannose 6-phosphate pathway (Aerts *et al.,* 1988). N-Glycosylation of β-glucocerebrosidase is, however, essential for attainment of the active conformation (Berg-Fussman *et al.,* 1993). The association of the enzyme with the membrane may be related to the hydrophobic nature of the product ceramide.

A genetic deficiency of any of the enzymes in these pathways leads to a characteristic pattern of storage of glycosphingolipids in different cells and a specific sphingolipidosis (Fig. 11). (For an up-to-date account of the biochemistry, molecular genetic basis, and clinical symptoms of the sphingolipidoses, the relevant chapter of Scriver *et al.,* 1995, should be consulted.) Most of the human enzymes have now been cloned and mutation analysis carried out for patients. All of the disorders are heterogeneous genetically, although there are some common mutations in ethnic groups. The different mutations in a particular enzyme have provided important information about the relationship between the structure and specificity, stability, intracellular transport, and interaction with other proteins of the enzymes.

4.4. Sphingolipid Activator Proteins (Saposins)

The breakdown of the glycans of glycosphingolipids while they are still linked to ceramide poses the problem of accessibility of the soluble lysosomal glycosidases to these amphipathic molecules. Studies on the hydrolysis of its natural substrate, cerebroside sulfate, by purified arylsulfatase A showed that it required an added detergent. It was discovered that the detergent could be replaced by a nonenzymic protein, which was called sulfatide activator (Mehl and Jatzkewitz, 1964). It is now known that several lysosomal hydrolases require the assistance of small nonenzymic glycoproteins for the hydrolysis of glycosphingolipids (Table III). To date, two genes are known to encode these sphingolipid activator proteins or saposins (Furst and Sandhoff, 1992; Kishimoto *et al.,* 1992). One encodes the G_{M2} activator protein, which facilitates the action of hexosaminidase A on ganglioside G_{M2} (Conzelmann and Sandhoff, 1978) and causes the AB variant of G_{M2}-gangliosidosis when it is genetically defective. The other gene encodes prosaposin (or sap-precursor) (O'Brien *et al.,* 1988), which is proteolytically processed sequentially from the N-terminal end in the lysosomes to four homologous saposins A–D with specificities for different sphingolipids (Table III). A deficiency of prosaposin leads to the accumulation of a range of glycosphingolipids (Harzer *et al.,* 1989). Prosaposin is transported to the lysosomes by the mannose 6-phosphate pathway, possibly in a complex with other lysosomal enzyme precursors (Zhu and Conner, 1994). The G_{M2}-activator protein and sap B, which has a wide specificity

Table III
Sphingolipid Activator Proteins (Saposins)

Protein	Size (amino acids)	Function		Deficiency/disease
		Enzyme	Substrate(s)	
G_{M2}-activator (sap 3)	162	β-Hexosaminidase A	G_{M2} (GA2)	AB variant of G_{M2} gangliosidosis
Prosaposin (sap precursor)	524	Unknown—lipid carrier?		Prosaposin deficiency
Saposin A	84	β-Glucocerebrosidase β-Galactocerebrosidase	Glucocerebroside Galactocerebroside	
Saposin B (sap 1)	80	Arylsulfatase A α-Galactosidase Sphingomyelinase β-Galactosidase	Sulfatide Globotriaosylceramide Sphingomyelin G_{M1}	MLD-like storage disorders
Saposin C (sap 2)	80	β-Glucocerebrosidase β-Galactocerebrosidase	Glucocerebroside Galactocerebroside	Variant of Gaucher disease
Saposin D	78	Acid sphingomyelinase	Sphingomyelin	Not known

(Wenger and Inui, 1984; Vogel *et al.*, 1991), act by binding to glycosphingolipids to bring them into aqueous solution and make them accessible to the soluble lyso-somal glycosidases. In contrast, sap C interacts with the membrane-associated β-glucocerebrosidase directly to form a more active complex by inducing a con-formational change in the enzyme (Berent and Radin, 1981). Sap C also activates the hydrolysis of galactosylceramide and sphingomyelin. A genetic defect in sap C causes a variant juvenile form of Gaucher disease (Schnabel *et al.*, 1991) but does not affect sphingomyelin catabolism. Acidic lipids can also bind to β-gluco-cerebrosidase, causing aggregation and stimulation of activity (Glew *et al.*, 1988). The binding sites for acidic lipids and sap C are distinct and the stimulatory effects synergistic (Morimoto *et al.*, 1990a).

Saposins have been found in all tissues examined and have been localized to the lysosomes. They also accumulate in several lysosomal and other storage dis-orders but are not tightly associated with the storage products, suggesting perhaps compensatory overproduction (Morimoto *et al.*, 1990b; Tayama *et al.*, 1992). Prosaposin seems to exist in two pools, one as the precursor for the lysosomal saposins and the other for secretion or retention as an integral membrane protein independent of the mannose 6-phosphate pathway. Prosaposin has been found in various human and rat extracellular fluids (Hineno *et al.*, 1991; O'Brien *et al.*, 1988; Sylvester *et al.*, 1989). It is also widely distributed as a neuronal cell sur-face membrane component in human adult and fetal brain (Fu *et al.*, 1994) and rat brain (Kondoh *et al.*, 1993), suggesting it has a role in neuronal function and de-velopment. Prosaposin binds gangliosides with the same or greater avidity than the saposins, and it is probable that activator proteins play an important role in intra-cellular transport of glycosphingolipids (Kishimoto *et al.*, 1992; Hiraiwa *et al.*, 1993; Sandhoff and Klein, 1994; Kuwana *et al.*, 1995).

5. GLYCOGEN

5.1. Structure, Occurrence, and Turnover of Glycogen

Glycogen is the storage polysaccharide of animals. It is a highly branched polymer of $\alpha 1 \rightarrow 4$-linked glucose residues, with the branches formed by $\alpha 1 \rightarrow 6$ linkages every eight to ten residues of glucose. Glycogen is synthesized in the cytosol conjugated to the protein glycogenin (Rodriguez and Whelan, 1985; Smythe and Cohen, 1991). The first step is the formation of a glucosidic linkage between glucose and the hydroxyl group of the phenolic side chain of tyrosine 194 of glycogenin, catalyzed by a specific glucosyl transferase (Campbell and Cohen, 1989), possibly glycogenin itself (Cao *et al.*, 1993). The addition of six or seven α-linked glucose residues to the first glucose is then autocatalyzed by glyco-genin. This process provides the primer for the synthesis of the $\alpha 1 \rightarrow 4$-linked chains catalyzed by glycogen synthase by the successive addition of glucose from

UDP-Glc. The $\alpha 1 \to 6$ branch points are formed by transfer of part of the $\alpha 1 \to 4$ chain to the 6 position of a glucose, catalyzed by the branching enzyme amylo $(\alpha 1 \to 4) \to (\alpha 1 \to 6)$ transglucosidase. Recent studies have shown that glycogen may contain an appreciable number of glucose residues phosphorylated at the 6 position, which are formed by the transfer of glucose 1-phosphate to form a phosphodiester followed by removal of the phosphoester-linked glucose (Marchase *et al.*, 1993; Lomako *et al.*, 1993). Liver glycogen commonly contains glucosamine (Kirkman *et al.*, 1989). Glycogen is stored in the cytosol of muscle cells in the form of β-particles, which consist of molecules containing up to 60,000 D-glucose residues with a core of glycogenin. The α-particles or rosettes seen in the cytosol of liver cells are much larger and have a higher glucose-to-glycogenin ratio. Cytosolic glycogen is broken down (glycogenolysis) by the concerted action of glycogen phosphorylase and the debranching enzyme, which has both transferase and amylo-1,6-glucosidase activity.

Although the metabolism of glycogen takes place predominantly in the cytosol, glycogen can be observed normally in lysosomes in cells. Over 60 years ago, Pompe observed that there was a dramatic increase of glycogen in vacuoles in all tissues of a child who had died suddenly at 7 months of heart failure with a grossly enlarged heart (Pompe, 1932). It was not until 1963 that the biochemical and cellular basis of this disorder was discovered when Hers and colleagues showed that there was a deficiency of a lysosomal acid maltase, α-glucosidase (EC 3.2.1.20), in the cells of patients with Pompe disease or glycogen storage disease type II (Hers, 1963). This was the first demonstration of a deficiency of a lysosomal hydrolase leading to accumulation of indigestible substrate in the lysosomes and was the origin of the concept of lysosomal storage diseases. It is presumed that the glycogen enters the lysosomes by the process of autophagocytosis of the cytoplasm and that the lysosomal acid maltase normally breaks it down to glucose. The proportion of cellular glycogen turned over by lysosomes is probably very small and not quantitatively significant for the energy metabolism of the cell. It certainly cannot compensate for genetic defects in the cytoplasmic pathways of glycogen metabolism, which cause the other forms of glycogen storage diseases in which glycogen accumulates in the cytoplasm. However, it has been suggested that there might be some connection between the lysosomal and cytosolic metabolism of glycogen in both liver and muscle, perhaps mediated by a feedback mechanism related to intralysosomal accumulation of glycogen (Geddes and Taylor, 1985; Geddes and Chow, 1994). As the accumulation of glycogen and the loss of recycled glucose is not significant and no toxic side products are formed, Pompe disease must result directly from the hypertrophy of the lysosomal system. This is an important deduction for our understanding of the pathogenesis of lysosomal storage diseases.

Patients with the classic phenotype of GSD II, as described by Pompe, have no or very little residual α-glucosidase activity. Glycogen accumulates intralysosomally in all their tissues, including white blood cells (Fig. 12), but is particularly marked in heart and skeletal muscle and the liver. In the milder forms of the

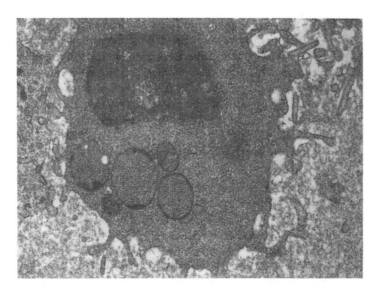

FIGURE 12. White blood cell from patient with Pompe disease (glycogen storage disease type II) showing intralysosomal accumulation of glycogen.

disease with juvenile or adult onset, there is residual α-glucosidase activity, and glycogen accumulation is essentially restricted to skeletal muscle. The ensuing clinical symptoms are progressive muscle weakness and respiratory problems, which can lead to early death (for a review of GSD II, see Reuser *et al.,* 1993; Hirschhorn, 1995). Several patients have been described with a cardiomyopathy and lysosomal accumulation of glycogen but without a deficiency of acidic α-glucosidase (Danon *et al.,* 1981). The biochemical basis of these cases is not known but there may be more than one cause because the mode of inheritance appears to vary among the affected families. One possibility is that there is a deficiency of the protein that has been shown to activate hydrolysis of glycogen by acid α-glucosidase (Radin *et al.,* 1989).

5.2. Lysosomal α-Glucosidase (Acid Maltase)

5.2.1. Substrate Specificity

The glycogen that accumulates in the lysosomes of patients with Pompe disease is structurally normal, occurring largely as β particles. This suggests that α-glucosidase can degrade glycogen completely to glucose by hydrolyzing both the $\alpha 1 \rightarrow 4$ and $\alpha 1 \rightarrow 6$ linkages in its natural, branched polysaccharide substrate. Lysosomal α-glucosidase has been purified from various sources and been shown to hydrolyze $\alpha 1 \rightarrow 4$ linkages in glycogen and starch and in natural and synthetic low-molecular-weight substrates at an acidic pH. It can also hydrolyze the $\alpha 1 \rightarrow 6$

linkage in isomaltose, albeit at a slower rate. This contrasts strongly with the depolymerization and degradation of cytoplasmic glycogen, which requires the action of several enzymes. It is presumed that any C6 phosphate groups are removed by acid phosphatase and that the core glucosyl–tyrosine linkage can be cleaved by an acid lysosomal enzyme. In the cytosol, the glucosyl–tyrosine linkage is probably hydrolyzed by the amylo-1,6-glucosidase activity of the debranching enzyme as is the case for the nonmammalian isoamylase (Lomako *et al.*, 1992).

5.2.2. Biosynthesis

Acid maltase is synthesized as a high-molecular-weight precursor which undergoes posttranslational modification en route to the lysosomes (Wisselaar *et al.*, 1993). Interestingly, the signal peptide is not removed cotranslationally, as with other lysosomal hydrolases, and this may play a role in the unusual distribution and intracellular transport of this acidic hydrolase. Although the enzyme is delivered to lysosomes predominantly by the mannose 6-phosphate pathway, in some cells it is transported to the lysosomes via the plasma membrane by a mechanism independent of mannose 6-phosphate receptors (Tsuji *et al.*, 1988). Furthermore, in polar epithelial cells α-glucosidase can be detected at and is secreted from the apical surface, like the homologous sucrase:isomaltase (Fransen *et al.*, 1988). This process is also independent of mannose 6-phosphate receptors. The purpose of these alternative routes and destinations is not understood, but it does suggest that some cells can regulate the distribution of their capacity to hydrolyze α-glucosidic linkages at an acidic pH. This is especially interesting as most cells possess at least two other genetically and structurally distinct α-glucosidases with different pH optima, substrate specificities, and subcellular locations (see Hirschhorn, 1995).

5.2.3. Molecular Genetics of GSD II

The human gene has been cloned (Hoefsloot *et al.*, 1990; Martiniuk *et al.*, 1991) and localized to chromosome 17q23-25 (Solomon and Barker, 1989), consistent with the autosomal inheritance of a deficiency of α-glucosidase. There is considerable homology between the human enzyme and α-glucosidases and α-amylases of yeast and fungi and the intestinal sucrase:isomaltase from humans and other animals, including the putative active site region (Hoefsloot *et al.*, 1988). Mutation analysis of human α-glucosidase has revealed several single-base polymorphisms in the coding region, one of which has been shown to be responsible for an electrophoretic variant with diminished affinity for high-molecular-weight substrates (Martiniuk *et al.*, 1990). Missense, nonsense, and splice site mutations and partial deletions have been found in patients with a deficiency of α-glucosidase, confirming the heterogeneity of the disease. Three mutations have been found in more than one Caucasian patient, and there is a strong correlation between the genotype and phenotype in patients with these three mutations (Kroos *et al.*, 1995).

6. CONCLUSION

The structures and functions of glycoconjugates are quite diverse, but the lysosomal catabolism of their constituent glycans shows many common features. The glycans are broken down in highly ordered pathways by the sequential release of monosaccharides catalyzed by exoglycosidases. Any substituents on the monosaccharides are removed prior to the action of the glycosidases. As might be expected, specific enzyme mechanisms break down the linkage between the carbohydrate and noncarbohydrate parts for each type of glycoconjugate. Where the same glycosidic linkage occurs in glycans derived from different glycoconjugates, it is hydrolyzed by the same enzyme. The involvement of some glycosidases in more than one pathway invites questions about the physical organization of enzymes in a pathway. There is increasing evidence that ancillary proteins can modify the activity of an enzyme or sequester a subpopulation of the enzyme for a particular function. It is possible that the lumen of the lysosome is as highly organized as the cytosol. The genes encoding lysosomal glycosidases and sulfatases are widely distributed over the chromosomes, but it would be surprising if there was not a regulatory mechanism for the expression of the activities in a pathway of the lysosomal system. Interaction between the metabolism of glycans inside and outside the lysosome is a fascinating aspect of the regulation of the lysosomal system. Although the biochemical features of the lysosomal catabolism of glycoconjugates are largely understood, much remains to be discovered at the molecular cellular level.

ACKNOWLEDGMENTS. I acknowledge the contribution by discussion and collaboration over many years of my colleagues at the Institute of Child Health and at Queen Elizabeth College, London, to the content of this article. I am particularly grateful to Arnold Reuser (Rotterdam) and Elisabeth Young (London), who read the sections on glycogen and glycosaminoglycans, respectively and Brian Lake (London) for Figure 12. Finally, I thank Paulette Lewis and Kate Lewis for their help in producing the manuscript.

7. REFERENCES

Abraham, D., Blakemore, W. F., Jolly, R. D., Sidebotham, R., and Winchester B., 1983, The catabolism of mammalian glycoproteins: Comparison of the storage products in bovine, feline and human mannosidosis, *Biochem. J.* **215**:573–579.

Aerts, J. M. F. G., Schram, A. W., Strijland, A., van Weely, S., Johnson, L. M. V., Tager, J. M., Sorrell, S. H., Ginns, E. I., Barranger, J. A., and Murray, G. J., 1988, Glucocerebrosidase, a lysosomal enzyme that does not undergo oligosaccharide phosphorylation, *Biochim. Biophys. Acta* **964**: 303–308.

Ahlberg, J., Berkenstam, A., Henell, F., and Glaumann, H., 1985, Degradation of short and long-lived proteins in isolated rat liver lysosomes; effects of pH, temperature and proteolytic inhibitors, *J. Biol. Chem.* **260**:5847–5854.

Al Daher, S., De Gasperi, R., Daniel, P., Hall, N., Warren, C. D., and Winchester, B., 1991, The substrate-specificity of human lysosomal α-D-mannosidase in relation to genetic α-mannosidosis, *Biochem. J.* **277**:743–751.

Al Daher, S., De Gasperi, R., Daniel, P., Hirani, S., Warren, C., and Winchester, B., 1992, Substrate specificity of human liver neutral α-mannosidase, *Biochem. J.* **286**:47–53.

Ando, S. 1983, Gangliosides in the nervous system, *Neurochem, Int.* **5**:507–537.

Aronson, N. N., and Davidson, E. A., 1967, Lysosomal hyaluronidase from rat liver, *J. Biol. Chem.* **242**:437–440.

Aronson, N. N., Backes, M., and Kuranda, M. J., 1989, Purification of rat liver chitobiosidase, *Arch. Biochem. Biophys.* **272**:290–300.

Baenziger, J., 1994, Protein-specific glycosyltransferases: How and why they do it! *FASEB J.* **8**: 1019–1025.

Barker, C., Dell, A., Rogers, M., Alhadeff, J. A., and Winchester, B., 1988, Canine α-L-fucosidase in relation to the enzymic defect and storage products in canine fucosidosis, *Biochem. J.* **254**:861–868.

Barnes, A. K., and Wynn, C. H., 1988, Homology of lysosomal enzymes and related proteins: Prediction of posttranslational modification of sites including phosphorylation of mannose and potential epitopic and substrate binding sites in the α- and β-subunits of hexosaminidases, α-glucosidase and rabbit and human isomaltase, *Proteins* **4**:182–189.

Basner, R., Von Figura, K., Glossl, J., Klein, U., Kresse, H., and Mlekusch, W., 1979, Multiple deficiency of mucopolysaccharide sulfatases in mucosulfatidosis, *Pediatr. Res.* **13**:1316–1318.

Baussant, T., Strecker, G., Wieruszeski, J-M., Montreuil, J., Michalski, J-M., 1986, Catabolism of glycoprotein glycans: Characterization of a lysosomal endo-N-acetyl-β-D-glucosaminidase specific for glycans with a terminal chitobiose residue, *Eur. J. Biochem.* **159**:381–385.

Berent, S. L., and Radin, N. S., 1981, Mechanism of activation of glucocerebrosidase by Co-B (glucosidase activator protein), *Biochim. Biophys. Acta* **664**:572–582.

Berg-Fussman, A., Grace, M. E., Ioannou, Y., and Grabowski, G. A., 1993, Human acid β-glucosidase: N-glycosylation site occupancy and the effect of glycosylation on enzymatic activity, *J. Biol. Chem.* **268**:14861–14868.

Braidman, I., Carroll, M., Dance, N., Robinson, D., Poenaru, L., Weber, A., Dreyfus, J. C., Overdijk, B., Hooghwinkel, G. J. M., 1974, Characterisation of human N-acetyl-β-hexosaminidase C, *FEBS Lett.* **41**:181–184.

Brassart, D., Baussant, T., Weiruszeski, J-M., Strecker, G., Montreuil, J., and Michalski, J-C., 1987, Catabolism of N-glycosylprotein glycans: Evidence for a degradation pathway of sialylglycoasparagines resulting from the combined action of the lysosomal aspartylglucosaminidase and endo-N-acetyl-β-D-glucosaminidase. A 400-MHz ¹H-NMR study, *Eur. J. Biochem.* **169**:131–136.

Burkhart, T., and Wiesmann, U., 1987, Sulphated glycosaminoglycan (GAG) in the developing mouse brain. Quantitative aspects on the metabolism of total and individual sulphated GAGS in vivo, *Dev. Biol.* **120**:447–456.

Butler, W. T., 1978, Carbohydrate moieties of the collagens and collagen-like proteins in health and disease, in *Glycoproteins and Glycolipids in Disease Process* (E. F. Walboorg, ed.), pp. 213–226, American Chemical Society, Washington, D.C.

Cacan, R., Villers, C., Belard, M., Kaiden, A., Krag, S. S., and Verbert, A., 1992, Different fates of the oligosaccharide moieties of lipid intermediates, *Glycobiology* **2**:127–136.

Campbell, D. G., and Cohen, P., 1989, The amino acid sequence of rabbit skeletal muscle glycogenin, *Eur. J. Biochem.* **185**:119–125.

Cantz, M., and Ulrich-Bott, B., 1990, Disorders of glycoprotein degradation, *J. Inher. Metab. Dis.* **13**: 523–537.

Cao, Y., Mahrenholz, A. M., De Paloi-Roach, A. A., and Roach, P. J., 1993, Characterization of rabbit skeletal muscle glycogenin, *J. Biol. Chem.* **268**:14687–14693.

Carraway, K. L., and Hull, S. R., 1991, Cell surface mucin-type glycoproteins and mucin-like domains, *Glycobiology* **1**:131–138.

Cenci di Bello, I., Dorling, P., and Winchester, B., 1983, The storage products in genetic and swainsonine-induced human mannosidosis, *Biochem. J.* **215:** 693–696.

Chen, Y. Q., Rafi, M. A., de Gala, G., and Wenger, D. A., 1993, Cloning and expression of cDNA
encoding human galactocerebrosidase, the enzyme deficient in globoid cell leukodystrophy, *Hum.
Mol. Genet.* **2:**1841–1845.

Conzelman, E., and Sandhoff, K., 1978, Deficiency of a factor necessary for stimulation of hexosaminidase A-catalyzed degradation of ganglioside GM_2 and glycolipid A_2, *Proc. Natl. Acad.
Sci. USA* **75:**3979–3983.

Cuervo, A. M., Terlecky, S. R., Dice, J. K., and Knecht, E., 1994, Selective binding and uptake of
ribonuclease A and glyceraldehyde-3-phosphate dehydrogenase by isolated rat liver lysosomes,
J. Biol. Chem. **269:**26374–26380.

Cuozzo, J. W., Tao, K., Wu, Q. L., Young, W., and Sahagian, G. G., 1995, Lysine-based structure in the
proregion of cathepsin L is the recognition site for mannose phosphorylation, *J. Biol. Chem.*
270:15611–15619.

Daniel, P. F., Warren, C. D., James, L. F., and Jolly, R. D., 1989, A comparison of swainsonine-induced
and genetic α-mannosidosis in Aberdeen Angus cattle, in *Swainsonine and Related Glycosidase
Inhibitors* (L. F. James, A. D. Elbein, R. J. Molyneux, and C. D. Warren, eds.), pp. 331–343, University of Iowa Press, Ames, Iowa.

Daniel, P. F., Evans, J. E., De Gasperi, R., Winchester, B., and Warren, C. D., 1992, A human lysosomal α(1 → 6) mannosidase active on the branched trimannosyl core of complex glycans, *Glycobiology* **2:**327–336.

Daniel, P. F., Winchester, B., and Warren C. D., 1994, Mammalian α-mannosidases—multiple forms
but a common purpose? *Glycobiology* **4:**551–566.

Danon, M. J., Oh, S. J., DiMauro, S., Manaligod, J. R. R., Eastwood, A., Naidu, S., and Schlisefeld,
L. H., 1981, Lysosomal glycogen storage disease with normal acid maltase, *Neurology* **31:**51–57.

De Gasperi, R., Li, Y-T., Li, S-C., 1989, Presence of two endo-β-N-acetylglucosaminidases in human
kidney, *J. Biol. Chem.* **264:**9329–9334.

De Gasperi, R., Al Daher, S., Daniel, P. F., Winchester, B. G., Jeanloz, R. W., and Warren, C. D., 1991,
The substrate specificity of bovine and feline lysosomal α-D-mannosidases in relation to α-
mannosidosis, *J. Biol. Chem.* **266:**16556–16563.

De Gasperi, R., Daniel, P. F., and Warren, C. D., 1992a, A human lysosomal α-mannosidase specific
for the core of complex glycans, *J. Biol. Chem.* **267:**9706–9712.

De Gasperi, R., Al Daher, S., Winchester, B. G., and Warren, C. D., 1992b, Substrate specificity of the
bovine and feline neutral α-mannosidases, *Biochem. J.* **286:**55–63.

Dice, J. F., 1990, Peptide sequences that target cytosolic proteins for lysosomal proteolysis, *Trends
Biochem. Sci.* **15:**305–309.

Dong, L-Y. D., and Hart, G. W., 1994, Purification and characterization of an O-GlcNAc selective
N-acetyl-β-D-glucosaminidase from rat spleen cytosol, *J. Biol. Chem.* **269:**19321–19330.

Dunn, W. A., 1994, Autophagy and related mechanisms of lysosome-mediated protein degradation,
Trends Cell Biol. **4:**139–143.

Durand, P., and O'Brien, J. S., 1982, *Genetic Errors of Glycoprotein Metabolism,* Springer-
Verlag, Berlin.

Dustin, M. L., Baranski, T. J., Sampath, D., and Kornfeld, S., 1995, A novel mutagenesis strategy
identified distantly spaced amino acid sequences that are required for the phosphorylation of both
the oligosaccharides of procathepsin D by N-acetylglucosamine 1-phosphotransferase, *J. Biol.
Chem.* **270:**170–179.

Endo, M., Yamamoto, M., Munakata, H., Yamamoto, R., Namiki, O., and Yosizawa, A., 1980, *Tohoku
J. Exp. Med.* **133:**355–361.

Evered, D., and Whelan, J., 1989, *The Biology of Hyaluronan,* John Wiley, Chichester.

Fedarko, N. S., and Conrad, H. E., 1986, A unique heparan-sulfate in the nuclei of hepatocytes–
structural changes with the growth-state of the cells *J. Cell Biol.* **102:**587–599.

Fisher, K. J., and Aronson, N. N., 1989, Isolation and sequence analysis of a cDNA encoding rat liver α-L-fucosidase, *Biochem. J.* **265**:695–701.

Fisher K. J., and Aronson, N. N., 1992, Cloning and expression of the cDNA sequence encoding the lysosomal glycosidase di-N-acetylchitobiase, *J. Biol. Chem.* **267**:19607–19616.

Fiszer-Szafarz, B., Czartoryska, B., Tylki-Szymanska, A., 1991, Evidence for the existence of a human serum hyaluronidase deficiency, *Proceedings 8th Workshop of the European Study Group on Lysosomal Diseases, Annecy, France.*

Fransen, J. A. M., Ginsel, L. A., Cambier, P. H., Klumperman, J., Oude Elferink, R. P. J., and Tager, J. M., 1988, Immunocytochemical demonstration of the lysosomal enzyme α-glucosidase in the brush border of human intestinal epithelial cells, *Eur. J. Cell Biol.* **47**:72–80.

Fransson, L-A., Silverberg, I., and Carlsledt, I., 1985, Structure of the heparan sulphate–protein linkage region, *J. Biol. Chem.* **260**:14722–14726.

Fu, Q., Carson, G. S., Hiraina, M., Grate, M., Kishimoto, Y., and O'Brien, J. S., 1994, Occurrence of prosaposin as a neuronal surface membrane component, *J. Mol. Neurosci.* **5**:59–67.

Fuchs, W., Beck, M., and Kresse, H., 1985, Intralysosomal formation and metabolic fate of N-acetyl-glucosamine-6-sulfate from keratan sulfate, *Eur. J. Biochem.* **151**:551.

Furst, W., and Sandhoff, K., 1992, Activator proteins and topology of lysosomal sphingolipid catabolism, *Biochim. Biophys. Acta* **1126**:1–16.

Gasnier, F., Rousson, R., Lerme, F., Vagnanay, E., Louisot, F., and Gateau-Rosch, O., 1992, Mitochondrial dolichyl-phosphate mannose synthase, *Eur. J. Biochem.* **206**:853–858.

Geddes, R., and Chow, J. C., 1994, Differing patterns of carbohydrate metabolism in liver and muscle, *Carbohydr. Res.* **256**: 139–147.

Geddes, R., and Taylor A., 1985, Lysosomal glycogen storage induced by acarbose, a 1,4-α-glucosidase inhibitor, *Biochem. J.* **228**:319–324.

Gillard, B. K., Thurmon, L. T., and Marcus, D. M., 1993, Variable subcellular localization of glycosphingolipids, *Glycobiology* **3**:57–67.

Glaumann, H., and Ballard, F. J., 1987, *Lysosomes: Their Role in Protein Breakdown,* Academic Press, London.

Glew, R. H., Peters, P. S., and Christopher, A. R., 1976, Isolation and characterization of β-glucosidase from the cytosol of rat kidney cortex, *Biochim. Biophys. Acta* **422**:179–199.

Glew, R. H., Basu, A., LaMarco, K. L., and Prence, E. M., 1988, Mammalian glucocerebrosidase: Implications for Gaucher's disease, *Lab. Invest.* **58**:5–25.

Griffiths, G., Hoflack, B., Simons, K., Mellman, I., and Kornfeld, S., 1988, The mannose 6-phosphate receptor and the biogenesis of lysosomes, *Cell* **52**:329–341.

Haeuw, J-F., Michalski, J-C., Strecker, G., Spik, G., and Montreuil, J., 1991a, Cytosolic glycosidases: Do they exist?, *Glycobiology* **1**:487–492.

Haeuw, J. F., Strecker, G., Wieruszeski, J. M., Montreuil, J., and Michalski, J. C., 1991b, Substrate specificity of rat liver cytosolic α-D-mannosidase–novel degradation pathway for oligomannosidic type glycans, *Eur. J. Biochem.* **202**:1257–1268.

Haeuw, J-F., Grard, T., Alonso, C., Strecker, G., and Michalski, J.-C., 1994, The core-specific lysosomal α(1-6)-mannosidase activity depends on aspartamidohydrolase activity, *Biochem. J.* **297**: 463–466.

Hakomori, S., 1981, Glycosphingolipids in cellular interaction, differentiation and oncogenesis, *Annu. Rev. Biochem.* **50**:733–764.

Hakomori, S., 1990, Biofunctional role of glycosphingolipids, modulators for transmembrane signaling and mediators for cellular interactions, *J. Biol. Chem.* **265**:18713–18716.

Hakomori, S., 1993, Structure and function of sphingoglycolipids in transmembrane signalling and cell-cell interactions, *Biochem. Soc. Trans.* **21**:583–595.

Haltiwanger, R. S., Kelly, W. G., Roquemore, E. P., Blomberg, M. A., Dong, L. Y.-D., Kreppel, L., Chou, T.-Y., and Hart, G. W., 1992, Glycosylation of nuclear cytoplasmic proteins is ubiquitous and abundant, *Biochem. Soc. Trans.* **20**:264–269.

Hancock, L. W., and Dawson, G., 1989, Lysosomal degradation of glycoproteins and glycosamino-glycans, in *Neurobiology of Glycoconjugates* (R. V. Margolis and R. K. Margolis, eds.), pp 187–218, Plenum Press, New York.

Harzer, K., Paton, B. C., Poulos, A., Kustermann-Kuhn, B., Roggendoff, W., Grisar, T., and Popp, M., 1989, Sphingolipid activator protein deficiency in a 16-week-old atypical Gaucher disease patient and his fetal sibling: Biochemical signs of combined sphingolipidoses, *Eur. J. Pediatr.* **149**:31–39.

Hayes, B. K., and Hart, G. W., 1994, Novel forms of protein glycosylation, *Curr. Opin. Struct. Biol.* **4**:692–696.

Heinegard, D., and Oldberg, A., 1993, Glycosylated matrix proteins, in *Connective Tissue and its Heritable Disorders* (P. M. Royce and B. Steinmann, eds.), pp. 189–209, Wiley-Liss, New York.

Hers, H. G., 1963, Alpha-glucosidase deficiency in generalized glycogen-storage disease (Pompe's Disease), *Biochem. J.* **86**:11–16.

Hineno, T., Sano, A., Kodoh, K., Ueno, S.-I., Kakomoto, Y., Yoshida, K.-I., 1991, Secretion of sphingolipid hydrolase activator precursor, prosaposin, *Biochem. Biophys. Res. Commun.* **176**:668–674.

Hirabayashi, Y., Marumoto, Y., Matsumoto, M., Toida, T., Iida, N., Matsubara, T., Kanzaki, T., Yokota, M., and Ishizuka, I., 1990, Isolation and characterization of major urinary amino acid O-glycosides and a dipeptide O-glycoside from a new lysosomal storage disorder (Kanzaki disease), *J. Biol. Chem.* **265**:1693–1701.

Hiraiwa, M., O'Brien, J. S., Kishimoto, Y., Goldzicka, M., Fluharty, A. L., Ginns, E. I., and Martin, B. M., 1993, Isolation, characterization and proteolysis of human prosaposin, the precursor of saposins (sphingolipid activator proteins), *Arch. Biochem. Biophys.* **304**:110–116.

Hirschhorn, R., 1995, Glycogen storage disease type II: acid α-glucosidase (acid maltase) deficiency, in *The Metabolic and Molecular Bases of Inherited Disease* (C. R. Scriver, A. L. Beaudet, W. S. Sly, and D. Valle, eds.), pp. 2443–2464, McGraw-Hill, New York.

Hiscock, D. R. R., Yanagishita, M., and Hasall, V. G., 1994, Nuclear localization of glycosaminoglycans in rat ovarian granulosa cells, *J. Biol. Chem.* **269**:4539–4564.

Hoefsloot, L. H., Hoogeven-Westerveld, M., Kroos, M. A., Van Beeumer, J., Reuser, A. J. J., and Oostra, B., 1988, Primary structure and processing of lysosomal α-glucosidase: Homology with the intestinal sucrase:isomaltase complex, *EMBO J.* **7**:1697–1704.

Hoefsloot, L. H., Hoogeven-Westeveld, M., Reuser, A. J. J., and Oostra, B. A., 1990, Characterization of the human lysosomal α-glucosidase gene, *Biochem. J.* **272**: 493–497.

Hoogeven, A. J., Verheijen, F. W., and Galjard, H., 1983, The relation between human lysosomal β-galactosidase and its protective protein, *J. Biol. Chem.* **258**:12143–12146.

Hook, M., Woods, A., Johansson, S., Kjellen, L., and Couchman, J. R., 1986, Functions of proteoglycans at the cell surface, *Ciba Found. Symp.* **124**:143–157.

Hopkins, C. R., Gibson, A., Shipman, M., and Miller, K., 1990, Movement of internalized ligand receptor complexes along a continuous endosomal reticulum, *Nature* **346**:335–339.

Hopwood, J. J., 1989, Enzymes that degrade heparin and heparan sulphate, in *Heparin: Chemical and Biological Properties, Clinical Applications* (D. W. Lane and U. Lindahl, eds.), pp. 191–229, Edward Arnold, London.

Hopwood, J., and Elliott, H., 1985, Urinary excretion of sulphated N-acetylhexosamines in patients with various mucopolysaccharidoses, *Biochem. J.* **229**:579–586.

Hopwood, J. J., and Morris, C. P., 1990, The mucopolysaccharidoses: Diagnosis, molecular genetics and treatment, *Mol. Biol. Med.* **7**:381–404.

Iozzo, R. V., Cohen, I. R., Grasell, S., and Murdoch, A. D., 1994, The biology of perlecan: The multifaceted heparan sulphate proteoglycan of basement membranes and pericellular matrices, *Biochem. J.* **302**:625–639.

Jackson, R. L., Busch, S. J., and Cardin, A. D., 1991, Glycosaminoglycans: Molecular properties, protein interactions and role in physiological processes, *Physiol. Rev.* **71**:481–539.

Kaartinen, V., Mononen, T., Laatikainen, R., and Mononen, I., 1992, Substrate specificity and reaction mechanism of human glycoasparaginase, *J. Biol. Chem.* **267**:6855–6858.

Kang, M. S., Bowlin, T. L., Vijay, I. K., and Sunkara, S. P., 1993, Accumulation of pentamannose oligosaccharides in human mononuclear leukocytes by action of swainsonine, an inhibitor of glycoprotein processing, *Carbohydr. Res.* **248**:327–337.

Kanzaki, T., Yokota, M., Mizuno, N., Matsumoto, Y., and Hirabayashi, Y., 1989, Novel lysosomal glycoaminoacid storage disease with angiokeratoma corporis diffusum, *Lancet* 875–876.

Kielty, C. M., Hopkinson, I., and Grant, M. E., 1993, The collagen family: Structure, assembly and organization in the extracellular matrix, in *Connective Tissue and Its Heritable Disorders* (P. M. Royce and B. Steinmann, eds.), pp. 103–147, Wiley-Liss, New York.

Kirkman, B. R., Whelan, W. J., and Bailey, J. M. 1989, The distribution of glucosamine in mammalian glycogen from different species, organs and tissues, *Biofactors* **2**:123.

Kishimoto, Y., Hiraiwa, M. and O'Brien, J. S., 1992, Saposins: Structure, function, distribution, and molecular genetics, *J. Lipid Res.* **33**:1255–1267.

Kjellen, L., and Lindahl, U., 1991, Proteoglycans: Structures and interactions, *Annu. Rev. Biochem.* **60**:443–475.

Klein, U., and von Figura, K., 1979, Substrate specificity of a heparan-degrading endoglucuronidase from human placenta, *Hoppe-Seyler's Z. Physiol. Chem.* **360**:1465.

Klima, B., Pohlenz, G., Schindler, D., and Egge, H., 1992, An investigation into the glycolipid metabolism of α-N-acetylgalactosaminidase-deficient fibroblasts using native and artificial glycolipids, *Biol. Chem. Hoppe-Seyler* **373**:989–999.

Kobata, A., 1992, Structures and functions of the sugar chains of glycoproteins, *Eur. J. Biochem.* **209**: 483–500.

Kobayashi, T., Shinnon, N., Goto, I., and Kuroiwa, Y., 1985, Hydrolysis of galactosylceramide is catalyzed by two genetically distinct acid β-galactosidases, *J. Biol. Chem.* **260**:14982–14987.

Kok, J. W., and Hoekstra, D., 1994, Glycosphingolipid trafficking in the endocytic pathway *Curr. Top. Membr.* **40**:503–557.

Kondoh, K., Sano, A., Kakimoto, Y., Matsuda, S., and Sakanata, M., 1993, Distribution of prosaposin-like immunoreactivity in rat brain, *J. Comp. Neurol.* **334**:590–602.

Kornfeld, R., and Kornfeld, S., 1985, Assembly of asparagine-linked oligosaccharides, *Annu. Rev. Biochem.* **54**:631–664.

Kreysing, J., von Figura, K., and Gieselmann, V., 1990, Structure of the arylsulfatase A gene, *Eur. J. Biochem.* **191**:627–631.

Kroos, M. A., Van der Kraan, M., Van den Boogaard, M. J., Ausens, M. G. E. M., Ploos van Amstel, H. K., Poenaru, L., Nicolino, M., Wevers, R., Van Diggelen, O., Kleijer, W., and Reuser, A. J. J., 1995, Glycogen storage disease type II: The frequency of 3 common mutant alleles and their associated clinical phenotypes studied in 121 patients, *J. Med. Genet.* **32**:836–837.

Kuranda, M. J., and Aronson, N. N., 1987, A di-N-acetylchitobiase activity is involved in the lysosomal catabolism of asparagine-linked glycoproteins in rat liver, *J. Biol. Chem.* **261**: 5803–5809.

Kusche, M., Backstrom, G., Riesfeld, J., Petitou, M., Chosy, J., and Lindahl, U., 1988, Biosynthesis of heparin: O-sulfation of the anti-thrombin binding region, *J. Biol. Chem.* **263**:15474–15484.

Kuwana, T., Mullock, B. M., and Luzio, J. P., 1995, Identification of a lysosomal protein causing lipid transfer, using a fluorescence assay designed to monitor membrane fusion between rat liver endosomes and lysosomes, *Biochem. J.* **308**:937–946.

Lecat, D., Lemmonier, M., Derappe, C., Lhermitte, M., Halbeek, H., Dorland, I., and Vliegenthart, J. F. G., 1984, The structure of sialoglycopeptides of the O-glycosidic types isolated from sialidosis (muco-lipidosis I) urine, *Eur. J. Biochem.* **140**:415–420.

Leroy, J. G., and Wiesmann, U., 1993, Disorders of lysosomal enzymes, in *Connective Tissue and its Heritable Disorders* (P. M. Royce and B. Steinmann, eds.), pp. 613–639, Wiley-Liss, New York.

Levral, C., Adrial, D., and Louisot, P., 1990, Comparative study of the N-glycosylation synthesis through dolichol intermediates in mitochondria, Golgi apparatus-rich and endoplasmic-rich fraction, *Int. J. Biochem.* **22**:287–293.

Li, Y.-T., Mazzotta, Y., War, C. C., Orth, R., and Li, S.-C., 1973, Hydrolysis of Tay-Sachs ganglioside by β-hexosaminidase A of human liver and urine, *J. Biol. Chem.* **248:**7511–7515.

Lindahl, U., 1989, Biosynthesis of heparin and related structures, in *Heparin, Chemical and Biological Properties, Clinical Applications* (D. A. Lane and U. Lindahl, eds.), pp. 159–189, Edward Arnold, London.

Linden, H.-U., Klein, R. A., Egge, H., Peter-Katalinic, J., Dabrowski, J., and Schindler, D., 1989, Isolation and structural characterization of sialic acid–containing glycopeptides of the O-glycosidic type from the urine of two patients with a hereditary deficiency in α-N-acetylgalactosaminidase activity, *Biol. Chem. Hoppe-Seyler* **370:**661–672.

Lomako, J., Lomako, W. M., and Whelan, W. J., 1992, The substrate specificity of isoamylase and the preparation of apo-glycogenin, *Carbohydr. Res.* **227:**331–338.

Lomako, J., Lomako, W. M., Whelan, W. J., and Marchase, R. B., 1993, Glycogen contains phosphodiester groups that can be introduced by UDPGlucose-glycogen glucose 1-phosphotransferase *FEBS Lett.* **329:**263–267.

Marchase, R. B., Bounelis, P., Brumley, L. M., Dey, N., Browne, B., Auger, D., Fritz, T. A., Kulesza, P., and Bedwell, D. M., 1993, Phosphoglucomutase in *saccharomyces cerevisiae* is a cytoplasmic glycoprotein and the acceptor for a Glc-phosphotransferase, *J. Biol. Chem.* **268:**8341–8349.

Martiniuk, F., Bodkin, M., Taall, S., and Hirschhorn, R., 1990, Identification of base-pair substitution responsible for a human acid alpha-glucosidase allele with lower affinity for glycogen (GAA 11) and transient gene expression in deficient cells, *Am. J. Hum. Genet.* **47:**440–445.

Martiniuk, F., Bodkin, M., Tzall, S., and Hirschhorn, R., 1991, Isolation and partial characterization of the structural gene for human acid alpha glucosidase (GAA), *DNA Cell Biol.* **10:**283–292.

Matsue, H., and Endo, M., 1987, Heterogeneity of reducing terminals of urinary chondroitin sulfates, *Biochim. Biophys. Acta.* **923:**470–477.

Mayer, R., and Doherty, F., 1986, Intracellular protein catabolism: State of art, *FEBS Lett.* **198:**181–193.

McCarter, J. D., and Withers, S. G., 1994, Mechanisms of enzymatic glycoside hydrolysis, *Curr. Opin. Struct. Biol.* **4:**885–892.

Mehl, E., and Jatzkewitz, H., 1964, Eine cerebrosid-sulfatase aus schweiniere, *Hoppe-Seyler's Z. Physiol. Chem.* **339:**260–276.

Michalski, J.-C., Haeuw, J.-F., Wieruszeski, J.-M., Montreuil, J., and Strecker, G., 1990, In vitro hydrolysis of oligomannosyl oligosaccharides by the lysosomal α-D-mannosidases, *Eur. J. Biochem.* **189:**369–379.

Miyatake, T., and Suzuki, K., 1972, Globid cell leukodystrophy: Additional deficiency of psychosine galactosidase, *Biochem. Biophys. Res. Commun.* **48:**538.

Moore, S. E. H., and Spiro, R. G., 1994, Intracellular compartmentalization and degradation of free polymannose oligosaccharides released during glycoprotein biosynthesis, *J. Biol. Chem.* **269:**12715–12721.

Morimoto, S., Kishimoto, Y., Tomich, J., Weiler, S., Ohashi, T., Barranger, J. A., Kretz, K. A., and O'Brien, J., 1990a, Interaction of saposins, acidic lipids and glycosyl-ceramidase, *J. Biol. Chem.* **265:**1933–1937.

Morimoto, S., Yamamoto, Y., O'Brien, J. S., and Kishimoto, Y., 1990b, Distribution of saposin proteins (sphingolipid activator proteins) in lysosomal storage and other diseases, *Proc. Natl. Acad. Sci. USA* **87:**3493–3497.

Murali, R., Ioannou, Y. A., Desnick, R. J., and Bunett, R. M., 1994, Crystallization and preliminary X-ray analysis of human α-galactosidase A complex, *J. Mol. Biol.* **239:**578–580.

Murphy, G., and Reynolds, J. J., 1993, Extracellular matrix degradation, in *Connective Tissue and Its Heritable Disorders* (P. M. Royce and B. Steinmann, eds.), pp. 287–316, Wiley-Liss, New York.

Neufeld, E. F., 1989, Natural history and inherited disorders of a lysosomal enzyme, β-hexosaminidase, *J. Biol. Chem.* **264:**10927–10930.

Neufeld, E. F., 1991, Lysosomal storage diseases, *Annu. Rev. Biochem.* **60:**257–280.

Neufeld, E. F., and Ashwell, G., 1980, Carbohydrate recognition systems for receptor-mediated pinocytosis, in *The Biochemistry of Glycoproteins and Proteoglycans* (W. J. Lennarz, ed.), pp. 241–266, Plenum Press, New York.

Neufeld, E. F., and Muenzer, J., 1995, The mucopolysaccharidoses, in *The Metabolic and Molecular Bases of Inherited Disease* (C. R. Scriver, A. L. Beaudet, W. S. Sly, and D. Valle, eds.), pp. 2465–2494, McGraw-Hill, Inc., New York.

Ng, C. K., Handley, C. J., Preston, B. N., and Robinson, H. C., 1992, The extracellular processing and catabolism of hyaluronan in cultured adult articular cartilage explants, *Arch. Biochem. Biophys.* **298:**70–79.

Nguyen, Q., Murphy, G., Roughley, P. J., and Mort, J. S., 1989, Degradation of proteoglycan aggregates by a cartilage metalloproteinase. Evidence for the involvement of stromelysin in the generation of link protein heterogeneity *in situ*, *Biochem. J.* **259:**61–67.

O'Brien, J. S., Kretz, K. A., Dewji, N. N., Wenger, D. A., Esch, F., and Fluharty, A. L., 1988, Coding of two sphingolipid activator proteins (SAP-1 and SAP-2) by same genetic locus, *Science* **241:** 1098–1101.

Oegama, T. R., Kraft, E. L., Jourdian, G. W., and Van Valen, T. R., 1984, Phosphorylation of chondroitin sulphate in proteoglycans from the swarm rat chondrosarcoma, *J. Biol. Chem.* **259:**1720–1726.

Oosta, G. M., Faureau, L. V., Beeler, D. L., and Rosenberg, R. D., 1982, Purification and properties of human platelet heparatinase, *J. Biol. Chem.* **257:**11249–11255.

Paschke, E., and Kresse, H., 1982, Morquio disease type B: Activation of GM$_1$-β-galactosidase by GM$_1$ activator, *Biochem. Biophys. Res. Commun.* **109:**568–578.

Paulsson, M., 1987, Noncollagenous proteins of basement membranes, *Collagen Relat. Res.* **7:**443–461.

Peters C., Schmidt, B., Rommerskirch, W., Rupp, K., Zuhlsdorft, M., Vingron, M., Meyer, H. E., Pohlmann, R., and von Figura, K., 1990, Phylogenetic conservation of arylsulfatases cDNA cloning and expression of human arylsulfatase B, *J. Biol. Chem.* **265:**3374–3381.

Phillips, N. C., Robinson, D., and Winchester, B. G., 1974, Human liver α-mannosidase, *Clin. Chim. Acta* **65:**11–19.

Pierce, R. J., Spik, G., and Montreuil, J., 1979, Cytosolic location of endo-N-acetyl-β-D-glucosaminidase activity in rat liver and kidney, *Biochem. J.* **180:**673–676.

Pompe, J.-C., 1932, Over idiopatische hypertrophie van het hart, *Ned. Tijdschr. Geneeskd.* **76:**304–311.

Poole, A. R., 1986, Proteoglycans in health and disease: Structures and functions, *Biochem. J.* **236:**1–14.

Prehm, P., 1983, Synthesis of hyaluronate in differentiated teratocarcinoma cells. Characterization of the synthase, *Biochem. J.* **211:**181–198.

Rademacher, T. W., Parekh, R. B., and Dwek, R. A., 1988, Glycobiology, *Annu. Rev. Biochem.* **57:** 785–838.

Radin, N. S., Shukla, A., Shukla, G. S., and Sano, A., 1989, Heat-stable protein that stimulates acid alpha-glucosidase, *Biochem. J.* **264:**845–849.

Reuser, A. J. J., Kroos, M. A., Hermans, M. P. P., Bijcoet, A. G. A., Verbet, M. A., Van Diggelen, O., Kleijer, W. J., and Van der Brughe, A., 1993, Glycogenosis type II (acid maltase deficiency), *Muscle Nerve* **18:**561–569.

Robertson, D. A., Freeman, C., Morris, C. P., and Hopwood, J. J., 1992, A cDNA clone for human glucosamine-6-sulfatase reveals differences between arylsulfatase and non-arylsulfatase, *Biochem. J.* **288:**539–544.

Rock, P., Allietta, M., Young, W. W., Thompson, T. E., and Tillack, T. W., 1990, Organization of glycosphingolipids in phosphatidyl choline bilayers: Use of antibody molecules and Fab fragments as morpholigic markers, *Biochemistry* **29:**8488–8490.

Roden, L., 1980, Structure and metabolism of connective tissue proteoglycans, in *The Biochemistry of Glycoproteins and Proteoglycans* (W. Lennarz, ed.), pp. 267–371, Plenum Press, New York.

Rodriguez, I. R., and Whelan, W. J., 1985, A novel glycosyl-amino acid linkage: Rabbit muscle glycogen is covalently linked to a protein via tyrosine, *Biochem. Biophys. Res. Commun.* **132:** 829–836.

Rome, L. H., and Hill, D. F., 1986, Lysosomal degradation of glycoproteins and glycosaminoglycans: Efflux and recycling of sulphate and N-acetylhexosamines, *Biochem. J.* **235**:707–713.

Sakai, N., Inui, K., Fujii, N., Fukushima, H., Nishimoto, J., Yanagihara, I., Isegawa, Y., Iwamatsu, A., and Okada, S., 1994, Krabbe disease: Isolation and characterization of a full-length cDNA for human galactocerebrosidase, *Biochem. Biophys. Res. Commun.* **198**:485–491.

Sampson, P. M., Rochester, C. L., Freundlich, B., and Elias, J. A., 1992, Cytokine regulation of human lung fibroblast hyaluronan (hyaluronic acid) production: Evidence for cytokine-regulated hyaluronan (hyaluronic acid) degradation and human lung fibroblast-derived hyaluronidase, *J. Clin. Invest.* **90**:1492–1503.

Sandhoff, K., and Klein, A., 1994, Intracellular trafficking of glycosphingolipids: Role of sphingolipid activator proteins in the topology of endocytosis and lysosomal digestion, *FEBS Lett.* **346**:103–107.

Sandhoff, K., van Echten, G., Schroder, M., Schnabel, D., and Suzuki, K., 1992, Activators and inhibitors of glycosidases and glycosyltransferases, *Biochem. Soc. Trans.* **20**:695–699.

Sasaki, H., Yamada, K., Akasaka, K., Kawasaki, H., Suzuki, K., Saito, A., Sato, M., and Shimada, H., 1988, cDNA cloning, nucleotide sequence and expression of the gene for arylsulfatase in the sea urchin (Hemicentrotus pulcherrimus) embryo, *Eur. J. Biochem.* **177**:9–13.

Schachter, H., 1991, Enzymes associated with glycosylation, *Curr. Opin. Struct. Biol.* **1**:755–765.

Schachter, H., and Brockhausen, I., 1992, The biosynthesis of serine (threonine)-N-acetylgalactosamine-linked carbohydrate moieties, in *Glycoconjugates, Composition, Structure and Function* (H. J. Allen and E. C. Kisalius, eds.), pp. 263–332, Marcel Dekker, Inc., New York.

Schmidt, B., Selmer, T., Ingendoh, A., and von Figura, K. 1995, A novel amino acid modification in sulfatases that is defective in multiple sulfatase deficiency, *Cell* **82**:271–278.

Schmidt, G., Hausser, H., and Kresse, H., 1990, Extracellular accumulation of small dermatan sulphate proteoglycan II by interference with the secretion-recapture pathway, *Biochem. J.* **266**:591–595.

Schnaar, R. L., 1991, Glycolipids cell surface recognition, *Glycobiology* **1**:477–485.

Schnabel, D., Schroder, M., and Sandhoff, K., 1991, Mutation in the sphingolipid activator protein 2 in a patient with a variant of Gaucher disease, *FEBS Lett.* **284**:57–59.

Schnabel, D., Schroder, M., Furst, W., Klein, A., Hurwitz, R., Zeur, T., Weber, J., Harzer, K., Paton, B. C., Poulos, A., Suzuki, K., and Sandhoff, K., 1992, Simultaneous deficiency of sphingolipid activator proteins 1 and 2 is caused by a mutation in the initiation code of their common gene, *J. Biol. Chem.* **267**:3312–3315.

Schreiner, R., Schnabel, E., and Wieland, F., 1994, Novel N-glycosylation in eukaryotes: Laminin contains the linkage unit β-glucosylasparagine, *J. Cell Biol.* **124**:1071–1081.

Scriver, C. R., Beaudet, A. L., Sly, W. S., and Valle, D., 1995, *The Metabolic and Molecular Bases of Inherited Disease,* 7th Ed., McGraw-Hill, New York.

Shaklee, P. N., Glaser, J. H., and Conrad, H. E., 1985, A sulfatase specific for glucuronic acid 2-sulfate residues in glycosaminoglycans, *J. Biol. Chem.* **260**:9146–9149.

Shoup, V. A., and Touster, O., 1976, Purification and characterization of the α-D-mannosidase of rat liver cytosol, *J. Biol. Chem.* **251**:3845–3852.

Skelly, B., Sargan, D., Herrtage, M., and Winchester, B., 1996, The molecular defect underlying canine fucosidosis, *J. Med. Genet.* **33**:284–288.

Smedsrod, B., Pertoff, H., Ericsson, S., Fraser, J. R. E., and Laurent, T. C., 1984, Studies in vitro on the uptake and degradation of sodium hyaluronate in rat liver endothelial cells, *Biochem. J.* **223**: 617–626.

Smythe, C., and Cohen, P., 1991, The discovery of glycogenin and the priming mechanism for glycogen biogenesis, *Eur. J. Biochem.* **200**:625–631.

Solomon, E., and Barker, D. F., 1989, Report of the committee on the genetic constitution of chromosone 17, *Cytogenet. Cell. Genet.* **51**:319–337.

Song, Z., Li, S.-C., Li, Y.-Y., 1987, Absence of endo-β-N-acetylglucosaminidase activity in the kidneys of sheep, cattle and pig, *Biochem. J.* **284**:145–149.

Spillmann, D., and Lindahl, U., 1994, Glycosaminoglycan–protein interactions: A question of specificity, *Curr. Opin. Struct. Biol.* **4**:667–682.

Stirling, J. L., 1974, Human N-acetyl-β-hexosaminidases: Hydrolysis of *N,N*-diacetylchitobiose by a low molecular weight enzyme, *FEBS Lett.* **39**:171–175.

Strecker, G., Fournet, B., Bouquelet, S., Montreuil, J., Dhondt, J. L., and Farriaux, J.-P., 1976, Etude chimique de mannosides urinaires excretes au cours de la mannosidose, *Biochimie* **58**:579–586.

Sundblad, G., Holojda, S., Roux, L., Varki, A., and Freeze, H. H., 1988, Sulfated N-linked oligosaccharides in mammalian cells, identification of glycosaminoglycan-like chains attached to complex-type glycans, *J. Biol. Chem.* **263**:8890–8896.

Suzuki, K., 1970, Formation and turnover of myelin gangliosides, *J. Neurochem.* **17**:209–213.

Sylvester, S. R., Morales, C., Oko, R., and Griswold, M. D., 1989, Sulfated glycoprotein-1 (saposin precursor) in the reproductive tract of the male rat, *Biol. Reprod.* **41**:941–948.

Symington, F. W., Murray, W. A., Bearman, S. L., and Hakomori, S., 1987, Intracellular localisation of lactosylceramide, the major human neutrophil glycosphingolipid, *J. Biol. Chem.* **262**: 11356–11363.

Takagaki, K., Nakamura, T., and Endo, M., 1988a, Demonstration of an endo-β-galactosidase and an endo-β-xylosidase that degrade the proteoglycan linkage region, *Biochim. Biophys. Acta* **966**:94–98.

Takagaki, K., Nakamura, T., Majima, M., and Endo, M., 1988b, Isolation and characterization of chondroitin sulphate-degrading endo-β-glucuronidase from rabbit liver, *J. Biol. Chem.* **163**:7000–7006.

Takahashi, Y., Nakamura, Y., Tamaguchi, S., and Orii, T., 1991, Urinary oligosaccharide excretion and severity of galactosialidosis and sialidosis, *Clin. Chim. Acta* **203**:199–210.

Takeuchi, Y., Sakaguchi, K., Yanagishita, M., and Hasall, V. C., 1990, Heparan sulphate proteoglycans on rat parathyroid cells recycled in low Ca^{2+} medium, *Biochem. Soc. Trans.* **18**:816–818.

Tayama, M., O'Brien, J. S., and Kishimoto, Y., 1992, Distribution of saposins (sphingolipid activator proteins) in tissues of lysosomal storage disease patients, *J. Mol. Neurosci.* **3**:171–175.

Thomas, G. H., and Beaudet, A. L., 1995, Disorders of glycoprotein degradation and structure, in *The Metabolic and Molecular Bases of Inherited Disease* (C. R. Scriver, A. L. Beaudet, W. S. Sly, and D. Valle, eds.), pp. 2529–2561, McGraw-Hill, New York.

Tomatsu, S., Fukuda, S., Masue, M., Sukegawa, K., Fukao, T., Yamagisshi, A., Hori, T., Iwata, H., Ogawa, T., Nakashima, Y., Hanyu, Y., Hashimoto, T., Titani, K., Oyama, R., Suzuki, M., Yagi, K., Hayashi, Y., and Orii, T., 1991, Morquio disease: Isolation characterization and expression of full-length cDNA for human N-acetylgalactosamine-6-sulfate sulfatase, *Biochem. Biophys. Res. Commun.* **181**:677–683.

Truppe, W., Basner, R., von Figura, K., and Kresse, H., 1977, Uptake of hyaluronate by cultured cells, *Biochem. Biophys. Res. Commun.* **78**:713–719.

Tsay, G. C., Dawson, G., and Sung, S.-S. J., 1976, Structure of the accumulating oligosaccharide in fucosidosis, *J. Biol. Chem.* **251**:5852–5859.

Tsuji, A., Omura, K., and Suzuki, Y., 1988, Intracellular transport of acid α-glucosidase in human fibroblasts: Evidence for involvement of phosphomannosyl receptor-independent system, *J. Biochem. (Tokyo)* **104**:276–278.

Tulsiani, D. R. P., and Touster, O., 1992, Evidence that swainsonine pretreatment of rats leads to the formation of autophagic vacuoles and endosomes with decreased capacity to mature to, or fuse with, active lysosomes, *Arch. Biochem. Biophys.* **296**:556–561.

Turnbull, J. E., and Gallagher, J. J., 1993, Heparan sulphate: Functional role as a modulator of fibroblast growth factor activity, *Biochem. Soc. Trans.* **21**:477–482.

Vanderjagt, D. J., Fry, D. E., and Glew, R. H., 1994, Human glucocerebrosidase catalyses transglucosylation between glucocerebroside and retinol, *Biochem. J.* **300**:309–315.

Van Diggelen, O. P., Schindler, D., Kleijer, W. J., Huijmans, J. M. G., Galjaard, H., Linden, H. U., Peter-Kalanic, J., Egge, H., Dabrowski, U., and Cantz, M., 1987, Lysosomal α-N-acetylgalactosaminidase deficiency: A new inherited metabolic disease, *Lancet* **2**:804.

Villers, C., Cacan, R., Mir, A.-M., Labau, O., and Verbert, A., 1994, Release of oligosaccharide-type glycans as a marker of the degradation of newly synthesized glycoproteins, *Biochem. J.* **298:** 135–142.

Vogel, A., Schwarzmann, G., and Sandhoff, K., 1991, Glycosphingolipid specificity of the human sulfatide activator protein, *Eur. J. Biochem.* **200:**591–597.

Wang, A. M., Bishop, D. F., and Desnick, R. J., 1990, Human α-N-acetylgalactosaminidase-molecular cloning, nucleotide sequence, and expression of a full-length cDNA, *J. Biol. Chem.* **265:**21859–21866.

Warner, T. G., Louie, A., and Potier, M., 1990, Photolabeling of the α-neuraminidase/β-galactosidase complex from human placenta with a photoreactive neuraminidase inhibitor, *Biochem. Biophys. Res. Commun.* **173:**13–19.

Wenger, D. A., and Inui, K., 1984, Studies on the sphingolipid activator protein for the enzymatic hydrolysis of GM1 ganglioside and sulfatide, in *Molecular Basis of Lysosomal Storage Diseases* (R. O. Brady and J. Barranger, eds.), pp, 1–18, Academic Press, New York.

Wessling-Resnick, M., and Braell, W. A., 1990, The sorting and segregation mechanism of the endocytic pathway is functional in a cell-free system, *J. Biol. Chem.* **265:**690–699.

Wight, T. N., Kinselk, M. G., and Qwarnstrom, E. E., 1992, The role of proteoglycans in cell adhesion, migration and proliferation, *Curr. Opin. Cell Biol.* **4:**793–801.

Willemsen, R., van Dongen, J. M., Ginns, E. I., Schram, A. W., Tager, J. M., Barranger, J. A., and Reuser, A. A. J., 1987, Ultrastructural localization of glucocerebrosidase in cultured Gaucher's disease fibroblasts by immunocytochemistry, *J. Neurol.* **234:**44–51.

Wilson, P. J., Morris, C. P., Anson, D. S., Occhiodoro, T., Bielicki, J., Clements, P. R., and Hopwood, J. J., 1990, Hunter syndrome, isolation of an iduronate-2-sulfatase cDNA clone and analysis of patient DNA, *Proc. Natl. Acad. Sci. USA* **87:**8531–8535.

Wisselaar, H. A., Kross, M. A., Hermans, M. M. P., van Beeumen, J., and Reuser, A. J. J., 1993, Structural and functional changes of lysosomal acidic α-glucosidase during intracellular transport and maturation, *J. Biol. Chem.* **268:**2223–2231.

Yen, P. H., Allen, E., Marsh, B., Mohandas, T., Wang, N., Taggart, R. T., and Shapiro, L. J., 1987, Cloning and expression of steriod sulfatase cDNA and the frequent occurrence of deletions in STS deficiency, implications for X-Y interchange, *Cell* **49:**443–454.

Zhu, Y., and Conner, G. E., 1994, Intermolecular association of lysosomal protein precursors during biosynthesis, *J. Biol. Chem.* **269:**3846–3851.

Zhu, L., Hopes, T. J., Hall, J., Davies, A., Stern, M., Muller-Eberhard, U., Stern, R., and Parslow, T. G., 1994, Molecular cloning of a mammalian hyaluronidase reveals identity with hemopexin, a serum heme-binding protein, *J. Biol. Chem.* **269:**32092–32097.

Chapter 8

Lysosomal Metabolism of Lipids

William J. Johnson, Gregory J. Warner, Patricia G. Yancey, and George H. Rothblat

1. INTRODUCTION

Mammalian cells are very active in lipid metabolism, constantly acquiring new lipid by synthesis and by endocytosis of exogenous materials, and constantly degrading lipid by a combination of lipolytic and oxidative processes. The lysosome is a critical component in the degradative arm of this system and is primarily responsible for the lipolysis of both exogenous lipids (e.g., the lipid acquired by endocytosis of lipoproteins) and endogenous membrane lipids that are delivered to this organelle as a result of endocytosis and lysosome biogenesis. The lipolysis of complex lipids (e.g., glycerophospholipids, cholesteryl esters, etc.) in lysosomes yields relatively simple lipid and nonlipid products, which subsequently diffuse or undergo transport through the lysosomal limiting membrane to become available for oxidation, efflux, or biosynthetic reutilization elsewhere in the cell. The degradation of lipids within lysosomes is catalyzed by several specific lipases, which have acidic pH optima and which are concentrated in this organelle. There are several lysosomal storage diseases attributable to the inherited absence of individual acid lipases. In addition, lysosomes play a central role in the turnover of plasma lipoproteins and thereby participate in processes

William J. Johnson, Gregory J. Warner, Patricia G. Yancey, and **George H. Rothblat**   Department of Biochemistry, MCP Hahnemann School of Medicine, Allegheny University of the Health Sciences, Philadelphia, Pennsylvania 19129.

Subcellular Biochemistry, Volume 27: Biology of the Lysosome, edited by Lloyd and Mason. Plenum Press, New York, 1996.

that paradoxically tend to prevent and promote atherosclerosis, a major health problem in the developed Western nations.

In this chapter our intent is to present a comprehensive summary of lysosomal lipid degradation in humans and other mammals. The topics include the lipid composition of lysosomes, the biochemical and biological features of lysosomal degradation of the different classes of complex lipids, the human metabolic diseases attributed to defective lysosomal lipolysis, and the role of lysosomes in lipoprotein metabolism and atherosclerosis.

2. LIPID COMPOSITION OF LYSOSOMES

Table I summarizes various reports of the lipid composition of purified rat liver lysosomes. In some reports the lipid composition was compared at the same time to fractions representing a more general cellular lipid composition, and these comparisons are also provided. Several different procedures have been employed to obtain lysosomes in pure form, including pretreatment of rats with Triton WR1339 (which induces accumulation of lipids and thereby reduces lysosome density), free-flow electrophoresis, density-gradient centrifugation, lectin affinity chromatography, and two-phase aqueous partitioning. The main lipids reported in all cases are phospholipids, and thus reflect the composition of the lysosome membrane. These data suggest some characteristic features of the lipid composition: (1) Phosphatidylcholine is the major phospholipid, as in other mammalian membranes; (2) Lysosomes contain a relatively high content of sphingomyelin (8–18% of total phospholipid); (3) Pure lysosomes contain little or no cardiolipin (diphosphatidylglycerol), a lipid that is characteristic of mitochondria, but which sometimes contaminates lysosomes due to the tendency of lysosomes and mitochondria to cosediment from cell homogenates; (4) Lysosomes contain significant amounts of bis(monoacylglycero)phosphate(4–23% of phospholipid).

The lysosomes resulting from treatment of rats with Triton WR1339 are called tritosomes. The treatment with detergent induces enlargement and a reduction in density of the lysosomes. The change in density induced by this agent permitted one of the earliest isolations of purified liver lysosomes, which otherwise tended to be contaminated with mitochondria and peroxisomes (Trouet, 1974). Tritosomes have been found to contain substantial amounts of triacylglycerol and cholesteryl ester. These are thought to be derived from lipoproteins, which are delivered to the lysosomes but poorly degraded due to the coating with the nondegradable detergent. The accumulation of these lipids appears to be the main cause of the low density of tritosomes (Hayashi *et al.,* 1981).

The presence of bis(monoacylglycero)phosphate in lysosomes is of interest because it is synthesized in lysosomes and considered a marker for this organelle. As reviewed by Hostetler *et al.* (1992), the diglycerophosphate backbone is derived from either cardiolipin or phosphatidylglycerol (Fig. 1). The synthesis is

Table 1
Major Lipids of Lysosomes[a]

Subcellular Fraction (Method)	% Distribution of phospholipid								PL/PR (mg/mg)	CH/PL (mol/mol)	Other significant lipids	Reference
	PC	PE	SM	PS	PI	CL	lyPL	BMP				
Rat liver tritosomes	40	14	17	5	9		2	13			TG	Weglicki et al., 1974
Homogenate	53	22	6	3	8		3	1				
Rat liver tritosomes	30	12	18	3	4	0	2	23	0.30	0.43	PG,CE	Matzuzawa and Hostetler, 1980
Homogenate	48	24	4	3	10	5	1	1	0.15	0.17		
Rat liver lysosomes (CFE)	42	27	9	0	9	0	3	4				Bleistein et al., 1980
Lysosomes and mitochondria	43	31	3	0	9	8	2	1				
Rat liver lysosomes (FFE)	46	19	9	3	10	3	2	4	0.24			Hostetler et al., 1985
Rat liver lysosomes (Lectin affinity)	40	12	19	5	9	3	2	5				Kamrath et al., 1984
Lysosomes and mitochondria	43	31	3		9	8	2	1				
Rat liver lysosomes (PEG/dextran)	40	23	8	9 (PS + PI)	4	3	4					Osada et al., 1990

[a]Abbreviations: CH, cholesterol; CE, chloresteryl ester; PL, phospholipid; PC, phosphatidylcholine; PE, phosphatidylethanolamine; SM, sphingomyelin; PS, phosphatidylserine; PI, phosphatidylinositol; CL, cardiolipin; lyPL, lysophospholipids; BMP, bis(monoacylglycero)phosphate; PR, protein; TG, triacylglycerol; PG, phosphatidylglycerol; CFE, carrier-free electrophoresis; FFE, free-flow electrophoresis; PEG, polyethylene glycol.

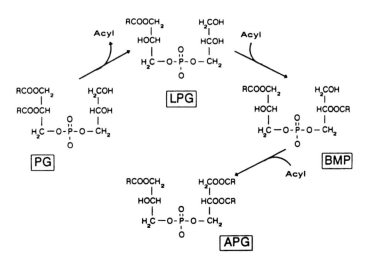

FIGURE 1. Metabolism of bis(monoacylglycero)phosphate in lysosomes. Abbreviations: PG, phosphatidylglycerol; LPG, lysophosphatidylglycerol; BMP, bis(monoacylglycero)phosphate; APG, acylphosphatidylglycerol. Assignment of positions of acyl groups in BMP and APG is provisional. (From Huterer *et al.* (1993) with permission.)

thought to proceed in the sequence: cardiolipin → phosphatidylglycerol → lysophosphatidylglycerol → bis(monoacylglycero)phosphate (Hostetler *et al.,* 1985). The acyl donor for the last reaction can be either phosphatidylinositol (Matsuzawa *et al.,* 1978) or phosphatidylcholine (Huterer and Wherrett, 1989). Properties of the transacylase catalyzing this reaction are discussed by Hostetler *et al.* (1992). The bis(monoacylglycero)phosphate product has an unusual stereoconfiguration (sn-1:sn-1'), and conversion to this form involves the intermediate formation of the sn-3:sn-1' form, followed by deacylation, rearrangement, and reacylation (Amidon *et al.,* 1995). The bis(monoacylglycero)phosphate product is subject to further acylation in lysosomes to form acylphosphatidylglycerol. In this case, phosphatidylcholine serves as the acyl donor, and the enzyme for this reaction may be lysosomal phospholipase A1 (Huterer *et al.,* 1993). The function of bis(monoacylglycero)phosphate in lysosomes is not well established. In macrophages, it may serve as a source of arachidonic acid for leukotriene synthesis (Waite *et al.,* 1990).

There is limited data on the cholesterol-to-phospholipid ratio of the lysosome membrane. This is surprising in view of the great interest in the metabolism of sterol-rich lipoproteins in this organelle and the role that this metabolism might play in atherosclerosis. The available data suggest a cholesterol-to-phospholipid ratio that is intermediate between the plasma membrane and the endoplasmic reticulum (~ 0.4 mol/mol). The presence of significant cholesterol in the lysosomal membrane might be anticipated in view of the role played by this organelle in the turnover of sterol-rich membranes and in the degradation of endocytosed sterol-rich lipoproteins. A relatively low ratio of sterol-to-phospholipid in comparison to the plasma membrane is consistent with functional studies showing rapid translocation of lipoprotein-derived cholesterol to the plasma membrane and other membranes. The mechanism of this translocation and the extent to which it may be balanced by the movement of cholesterol into the lysosome by varying routes are not established (reviewed by Liscum and Dahl, 1992; Liscum and Underwood, 1995).

The fatty acids of lysosomal phosphatidylcholine and phosphatidylethanolamine have been described by Olsson *et al.* (1991). In comparison to microsomes and mitochondria, the distribution of fatty acids in lysosomal phospholipids was not remarkable. The major fatty acids in phosphatidylcholine (in decreasing order of abundance) were 16:0, 20:4, 18:0, and 18:2. The ratio of saturated to monounsaturated to polyunsaturated acids was 48:7:46.

Lysosomes also are enriched relative to other organelles in dolichol, a class of long-chain polyisoprenoid alcohols that participate in N-linked protein glycosylation in the endoplasmic reticulum. Wong *et al.* (1982) prepared highly purified rat liver lysosomes by Metrizamide density gradient centrifugation and found levels of total dolichol as high as 15 µg/mg protein, which was 72 times the concentration of this lipid in the starting liver homogenate. Using similar methods, Olsson *et al.* (1991) also found a substantial enrichment of dolichol in liver lysosomes and that approximately 40% of the dolichol was esterified to fatty acid. Lysosomal dolichol appears to be derived from both the uptake of extracellular lipoproteins (Rip *et al.,* 1994) and turnover of endogenous membranes (Wong and Lennarz, 1982). The significance of dolichol enrichment in lysosomes is not clear. It may reflect the inability of lysosomes to rid themselves of this extremely hydrophobic substance.

The above summary indicates that apart from the accumulation of bis (monoacylglycero) phosphate and dolichol, the lipid composition of the lysosome membrane is not particularly remarkable. There is little hint in this composition of the potential for lipid degradation in lysosomes. Whether this reflects an inherent resistance of this membrane to lipolysis or efficient removal of breakdown products is not established.

3. LIPOLYTIC PROCESSES

Lysosomes participate in the hydrolytic catabolism of three major classes of complex lipids: neutral ester lipids, phosphoglycerides, and sphingolipids. For each of these classes, a distinct set of acid hydrolases is utilized. In most cases both endogenous and exogenous lipids can serve as substrates. Through biochemical and cell biological studies performed over the last 30–40 years, a great deal is known about the functional characteristics of these enzyme systems in mammals.

Detailed structural information about some of the mammalian enzymes is becoming available through the recent application of molecular cloning technology. Several distinct lysosomal lipid storage diseases are known. Nearly all of these have been attributable to a deficiency in a specific acid lipase or glycosidase, or hydrolase cofactor.

In this section, the lysosomal enzymes responsible for degradation of neutral lipids, phosphoglycerides, and the lipid portion of sphingolipids are considered. Section 4 briefly reviews the lipid storage diseases resulting from deficiencies of specific acid lipases. It should be noted that the acid glycosidases and sulfatases that participate in the catabolism of glycosphingolipids as well as the related storage diseases are reviewed in Chapter 7.

3.1. Neutral Lipids

Of the different lipid hydrolyzing enzymes located in lysosomes, the acid cholesteryl ester hydrolase (aCEH) is one of the most extensively investigated, due to its role in a variety of metabolic diseases such as Wolman disease, cholesteryl ester storage disease, and atherosclerosis. The enzyme is sometimes referred to as the acid lipase or the acid neutral lipid hydrolase. The early literature describing studies on this enzyme have been presented in the excellent review by Fowler and Brown (1984). The reader is also referred to more recent reviews by Assmann and Seedorf (1995), which focus on acid lipase deficiencies, and by Glick (1990), which address cholesteryl ester hydrolysis and the cholesteryl ester cycle.

Acid cholesteryl ester hydrolase isolated from various sources has reported molecular weights ranging from 40 kDa (rabbit liver; Imanaka *et al.*, 1984) to 50 kDa (rat liver; Klements and Lundberg, 1984). Radiation inactivation studies have placed the functional molecular weight of rat hepatic acid lipase at 48 ± 2 kDa (Erickson *et al.*, 1994). The differences in molecular weight between different preparations may, in part, be attributed to the extent of glycosylation, since cDNA sequencing has indicated that the enzyme consists of 378 amino acids, with six possible N-linked glycosylation sites (Anderson and Sando, 1991). The enzyme appears to be structurally related to gastric and lingual lipases (58% and 57% homology, respectively), but shares no homology with neutral cholesteryl ester hydrolase (Anderson and Sando, 1991).

Although there are a number of neutral lipid lipases in cells, the lysosomal enzyme is characterized as having a pH optimum of approximately 4.0, as opposed to other lipases with optima ranging from pH 6 to 8 (Fowler and Brown, 1984; Negre *et al.*, 1987; Van Berkel *et al.*, 1980; Glick, 1990). The aCEH appears to be confined exclusively to lysosomes, whereas other neutral lipid lipases are recovered from microsomal and cytoplasmic subcellular fractions (Glick, 1990; Fowler and Brown, 1984). The lysosomal aCEH, as with many cellular lipases, exhibits a broad substrate specificity including triglycerides, cholesteryl ester with different acyl groups, retinyl esters, and water-soluble substrates. A comparison of the literature on aCEH is complicated by a number of methodological difficulties. There is no standardized assay for aCEH, and a variety of substrates have been used to quantitate activity (Assmann and Seedorf, 1995). For studies on cell homogenates and enzyme preparations of varying purity both lipid and water-soluble substrates have been employed. Radiolabeled lipid substrates such as cholesteryl esters or triglycerides have been used and incorporated into reaction mixtures in a number of ways including direct acetone or ethanol addition (Deykin and Goodman, 1962), incorporation into phospholipid liposomes (Burrier and Brecher, 1983), as lipid emulsions (Lundberg *et al.*, 1990), or solubilization with detergents (Fowler and deDuve, 1969; Burton *et al.*, 1980). Artificial substrates such as esters of 4-methylumbelliferone (Coates *et al.*, 1979, 1986) have been used because their water solubility simplifies substrate presentation and quantitation. Although there have been numerous attempts to establish the acyl specificity of cholesteryl ester hydrolysis by aCEH, the order of acyl specificity remains unresolved because of potential artifacts related to the relative solubilities of different cholesteryl esters. It is apparent that the enzyme is most active against cholesteryl esters with chain lengths greater than eight carbons (Negre *et al.*, 1987). The lack of a standardized procedure for quantitating aCEH activity has made quantitative comparisons between various studies difficult. A complicating factor when comparing activity between different preparations, particularly with crude homogenates, is the presence of endogenous substrates that compete with the added, radiolabeled substrate. Thus, crude homogenates or lysosome preparations that have substantial concentrations of endogenous cholesteryl esters or triglycerides may yield artifactually low estimates of enzyme activity because of such competition.

Assays for the aCEH activity in whole-cell preparations generally involve the incorporation of radiolabeled substrates into the lysosome compartment of intact cells. The substrate delivery particle can be a lipoprotein which is incorporated by receptor mediated mechanisms (Brown and Goldstein, 1986) or lipid droplets that are taken up by phagocytosis (Wolfbauer *et al.*, 1986; Mahlberg *et al.*, 1990). The most extensively used lipoproteins are LDL, incorporated by the apo B/E receptor (Brown and Goldstein, 1986) and chemically modified LDL, such as acetyl LDL, incorporated by scavenger receptors on macrophages (Brown *et al.*, 1980a). The method of Krieger *et al.* (1978, 1979) has been used extensively for the incorporation of radiolabeled or fluorescent substrates into the core of the lipoprotein, since this procedure results in efficient labeling of the particle while still retaining its receptor recognition properties. If cell monolayers are exposed to the LDL containing the labeled substrate at temperatures below 15°C, the lipoprotein continues to bind to the receptor and undergoes incorporation into the endosome compartment. However, endosome–lysosome fusion does not occur until the cells are warmed to 37°C, at which time substrate hydrolysis is initiated (Anderson *et al.*, 1977). This temperature shift technique has been used effectively to study the kinetics of cholesteryl ester hydrolysis within the lysosomes and the subsequent transport of the generated free cholesterol out of the lysosome (Johnson *et al.*, 1990). Lipid droplets containing radiolabeled cholesteryl ester, triglyceride, or mixtures of esterified cholesterol and triglyceride can be formed by sonication of the lipid substrates together with phospholipid (Minor *et al.*, 1989). Because these lipid emulsions will float on the surface of the tissue culture medium, above the cell monolayer, phagocytic incorporation of the particles is achieved using an inverted culture technique (Wolfbauer *et al.*, 1986). The need to use inverted culture conditions limits the amount of cell material that can be studied; however, the ability to assemble lipid droplets with a wide range of substrate mixtures and physical states allows one to quantitate the rates of lysosome hydrolysis of various substrates supplied within the same particle (Minor *et al.*, 1989; Lundberg *et al.*, 1990).

In whole-cell assays of lysosomal lipase activity using radiolabeled substrates, such as [^{3}H]cholesteryl ester or [^{3}H]triglyceride, the labeled substrate generated by hydrolysis should not be reutilized in some other subcellular compartment after exiting the lysosome. For example, some of the labeled free cholesterol generated in the lysosome by cholesteryl ester hydrolysis can undergo re-esterification by acyl-CoA:cholesterol acyltransferase (ACAT) in the ER membranes. Such re-esterification will result in an underestimate of the actual extent of lysosomal hydrolysis. This complication can be particularly evident in long-term incubations lasting many hours and under conditions of net accumulation of cell cholesterol, which increases ACAT activity. Under such conditions it may be advisable to conduct the incubations in the presence of pharmacological inhibitors of the ACAT reaction (Ross *et al.*, 1984; Warner *et al.*, 1995). Similar difficulties may be encountered if the labeled substrate is triglyceride. The end products of lysosomal lipase hydrolysis of triglyceride are fatty acids and glycerol, with the following reaction sequence: triglyceride $\rightarrow$ 1,2-diglyceride $\rightarrow$ 2-monoglyceride $\rightarrow$ glycerol (Assmann and Seedorf, 1995). There is no practical inhibitor that can be used to block the resynthesis of triglycerides. However, if the substrate is labeled in the fatty acid moiety, incubations in the presence of fatty acid-free albumin can reduce the potential for recycling (Minor *et al.*, 1989).

Validation that the neutral lipid substrate being supplied to cells is undergoing hydrolysis in lysosomes is generally achieved by establishing that the hydrolysis can be inhibited by lysosomotropic agents such as chloroquine, ammonium chloride, or methylamine (Goldstein *et al.*, 1975; Mahlberg *et al.*, 1990). Continued cellular hydrolysis of cholesteryl ester in the presence of these compounds suggests that the ester is being hydrolyzed by nonlysosomal enzymes such as the cytoplasmic neutral cholesteryl ester hydrolase (nCEH; for reviews see Glick, 1990; Harrison, 1993). However, the effectiveness of inhibitors of lysosomal hydrolysis can be influenced by a variety of experimental conditions (see below). A series of diethyl phosphates and N-alkylcarbamates have been demonstrated to be effective inhibitors of the nCEH while having no effect on the aCEH (Harrison *et al.*, 1990). Thus, these inhibitors, together with the lysosomotropic agents, can be used to establish the intracellular site of cholesteryl ester hydrolysis (DeLamatre *et al.*, 1993).

The loss of both cholesteryl ester and triglyceride hydrolytic activities in cells from individuals with Wolman disease and cholesteryl ester storage disease (Assmann and Seedorf, 1995), coupled with the observations that a single, purified enzyme is capable of hydrolyzing both lipids (Sando and Rosenbaum, 1985; Klemets and Lundberg, 1986; Imanaka *et al.*, 1984) and that transfection of the lysosomal acid lipase cDNA into cells greatly enhances hydrolysis of both cholesteryl ester and triglyceride (Anderson and Sando, 1991), provides compelling evidence that a single enzyme in lysosomes is responsible for neutral lipid hydrolysis. However, the relative rates of hydrolysis and the effect of inhibitors can differ depending on a variety of physicochemical factors. As discussed by Lundberg *et al.* (1990), the enzymatic hydrolysis of cholesteryl ester or triglyceride by aCEH is a heterogeneous reaction where the water-soluble enzyme acts on an insoluble hydrophobic substrate. The enzyme–substrate interaction takes place at a lipid–water interface, and properties of the substrate such as its physical state, surface concentration, and surface charge influence the enzymatic activity. Because of the complexity of the interaction of enzyme with lipid substrates, it is often difficult to distinguish which factors influence hydrolysis, and comparisons between kinetic data obtained from neutral lipid hydrolysis studies using cell-free systems or whole cells are difficult to make. For example, in experiments where cholesteryl ester droplets were incorporated by phagocytosis into smooth muscle cells or macrophages, it was evident that the cholesteryl ester in droplets made isotropic by the inclusion of triglycerides was hydrolyzed at a rate greater than esterified cholesterol supplied as anisotropic, liquid crystalline droplets containing only cholesteryl ester (Mahlberg *et al.*, 1990; Minor *et al.*, 1989). However, in contrast to whole-cell experiments, no effect of lipid physical state was obtained if similar lipid droplets were treated exogenously with aCEH obtained from rat liver (Lundberg *et al.*, 1990). These results suggest that the physical state effect on the kinetics of lipid hydrolysis observed in whole cells may not be attributed to physical state *per se,* but is a reflection of the compositional differences used to manipulate physical state (Lundberg *et al.*, 1990). The activity of the aCEH is sensitive to such interfacial properties of the lipid droplets as surface area available to the enzyme, net surface charge, and surface solubility of the substrate molecules (Lundberg *et al.*, 1990; Bergstrom and Brockman, 1984). Although the data discussed above were obtained using lipid droplets as the substrate for the aCEH, the conclusions are probably applicable to the hydrolysis of esterified cholesterol and triglyceride delivered into the lysosomes as the core of lipoproteins.

An example of the complexities of interpreting and comparing the kinetics of hydrolysis of different lipids are the studies on the relative rates of cholesteryl ester and triglyceride hydrolysis. Both the rates of hydrolysis and the sensitivity of the hydrolysis to inhibitors depend on the nature of the particle delivering the lipids into cells. When lipid droplets consisting of equimolar cholesteryl oleate and triolein are internalized by either macrophages or smooth muscle cells, the hydrolysis of triglycerides is four to six times faster than esterified cholesterol

(Mahlberg *et al.*, 1990; Minor *et al.*, 1989). However, if equimolar amounts of the two lipids are incorporated by receptor-mediated uptake of a reconstituted LDL, the rates of hydrolysis of the lipids are similar (Minor *et al.*, 1991). In addition to these different hydrolysis rates, the sensitivity of the hydrolase to lysosomotropic agents also differs. Thus, treatment of cells with lysosomotropic agents results in a similar reduction in hydrolysis of cholesteryl ester and triglyceride if the two lipids are supplied as the core of LDL (Minor *et al.*, 1991). In contrast, if lipid droplets are used as the delivery vehicle, the inhibition of triglyceride hydrolysis is less than the inhibition of cholesteryl ester hydrolysis (Minor *et al.*, 1991). This differential sensitivity to lysosomotropic agents may be due to differences in the pH optima of the lipase for the two substrates, coupled with the lysosomal production of large amounts of fatty acids from the lipid droplets (Minor *et al.*, 1991). Although the pH optima of lysosomal acid lipase for the hydrolysis of triglyceride and cholesteryl esters are generally considered to be similar (pH 4.5 to 5), a published pH curve for the hydrolysis of cholesteryl oleate is fairly narrow, between 3.5 and 4.4, with an optimum of about 3.8 (Haley *et al.*, 1980). However, the range for triglyceride hydrolysis appears to be broader and higher (between 4.5 and 6) with an optimum of 5.6 (Goldberg and Khoo, 1988). This apparent difference in the pH optima of the lipase for the two substrates could explain the observed decreased potency of the lysosomotropic agents against triglycerides when cells phagocytose lipid droplets. In the case of lysosomotropic agents, which act to increase the lysosomal pH, the generation of large amounts of fatty acids through the hydrolysis of incorporated triglycerides might serve to partially counteract the increase in lysosomal pH produced by the inhibitors. Another explanation of the differences noted in the hydrolysis of cholesteryl ester and triglyceride is that some of the hydrolysis may occur in prelysosomal compartments, and that the extent of this processing may be dependent on the route of delivery of the lipid-carrying particle. For example, although cholesteryl ester incorporated into cells as the core of LDL is clearly hydrolyzed in lysosomes, cholesteryl ester supplied to cells from high density lipoprotein (HDL) through a selective uptake process (which does not involve whole particle uptake) is not hydrolyzed by aCEH (DeLamatre *et al.*, 1993). Similarly, although retinyl esters can serve as a substrate for aCEH, there is an apparent lack of involvement of lysosomes in the hydrolysis of chylomicron retinyl esters (Harrison *et al.*, 1995). In addition, in rat liver, the triglyceride of endocytosed VLDL appears to undergo significant hydrolysis in endosomes due to the action of hepatic lipase and at least one other lipase (Hornick *et al.*, 1992).

3.2. Lysosomal Sphingolipid Catabolism

The lysosomal lipases mediating sphingolipid degradation are acid sphingomyelinase and acid ceramidase. These two lipases are reviewed in this section. For glycosphingolipids, lysosomal catabolism also requires the participation of

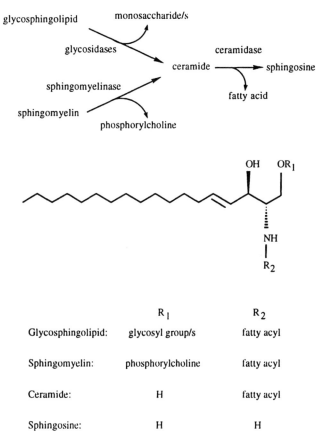

FIGURE 2. Overview of sphingolipid structures and catabolism in lysosomes.

acid glycosidases, which are reviewed in Chapter 7. An overview of lysosomal sphingolipid structures and catabolism is presented in Fig. 2.

3.2.1. Lysosomal Sphingomyelinase

Sphingomyelinases (sphingomyelin phosphodiesterases, EC 3.1.4.12) act as type C phospholipases, catalyzing hydrolysis of the phosphate ester bond at position 1 of the sphingosine backbone of sphingomyelin:

$$\text{sphingomyelin} + H_2O \rightarrow \text{ceramide} + \text{phosphorylcholine}$$

Three apparently distinct forms of sphingomyelinase have been reported in mammals: (1) an activity with a neutral pH optimum and a requirement for magnesium

ion that appears to reside in the plasma membrane, (2) the lysosomal sphingomyelinase with an acidic pH optimum, and (3) a serum activity with an acidic
pH optimum but with a requirement for zinc ion, which distinguishes it from the
lysosomal enzyme (Spence, 1993). This review addresses only the mammalian
lysosomal (acid) sphingomyelinase. Spence (1993) provided a comparative review
of the three activities. In addition, Rao and Spence (1976) presented a detailed functional comparison of the neutral and lysosomal activities in human brain.

The lysosomal sphingomyelinase is critical for sphingolipid homeostasis,
with deficiencies in this activity leading to the A and B subtypes of the Niemann-
Pick lipid storage disease (see Section 4.3.2). Over the last three decades, this enzyme has been prepared from various sources and subjected to extensive
characterization. Recently, cDNA and genomic clones of the enzyme have been
isolated, providing the primary amino acid sequence of the enzyme, its chromosomal location, and identification of mutations causing types A and B Niemann-
Pick disease.

   3.2.1a.   Enzyme Assay.   A widely used assay for acid sphingomyelinase
uses sphingomyelin substrate labeled with ^{14}C or ^{3}H in the N-methyl choline
groups (Suzuki, 1987). Labeled product and substrate are separated based on partitioning between aqueous and organic solvents, and the product in the aqueous
phase is quantified by liquid scintillation counting. Other important conditions of
the assay include pH 5.0, the inclusion of Triton X-100 for dispersal of substrate,
fairly harsh dispersal of the enzyme source (repeated freeze/thaw and sonication)
to ensure complete exposure of the enzyme, and the lack of divalent cations to ensure that any contaminating neutral or serum sphingomyelinase is inactive. Alternative substrates for activity assays include (1) derivatized sphingomyelins
carrying fluorescent and chromophoric reporter groups attached to the terminal
carbon of the N-fatty acyl group (Gatt *et al.*, 1978, 1980); and (2) fluorogenic and
chromogenic phosphodiester compounds such as bis-4-methylumbelliferyl phosphate and bis-para-nitrophenylphosphate (Jones *et al.*, 1981) and 2-hexadecanoyl-
amino-4-nitrophenylphosphorylcholine (Sakuragawa, 1982).

   An assay of sphingomyelinase that relied on the formation of an
enzyme–monoclonal antibody complex and its immobilization in a microtiter well
was developed by Freeman *et al.* (1984). This assay depends on the retention of
enzyme activity after antibody binding and allows a comparison of the immune-
relatedness of enzymes from different sources. Jobb and Callahan (1989) reported
the preparation of a polyclonal antiserum against placental sphingomyelinase that
was able to precipitate ^{35}S-labeled enzyme and was used to demonstrate the presence of normal amounts of immunoreactive protein in types A and B Niemann-
Pick fibroblasts. Reduced immunoreactive protein was found in I-cell disease
fibroblasts, indicating targeting of the enzyme by the phosphomannosyl receptor.
More recently Rousson *et al.* (1993) reported the preparation of a monoclonal antibody that could precipitate acid sphingomyelinase from detergent solution, suggesting a higher affinity for the enzyme than had been displayed by previous mon-

oclonal preparations. This antibody was used in quantitative precipitation and Western blotting assays to demonstrate once again normal levels of immunoreactive protein in fibroblasts from a type A Niemann-Pick patient.

3.2.1b. Purification. Various procedures for concentrating mammalian acid sphingomyelinase to high specific activity have been described. The sources of the enzyme for these purifications include rat liver (Heller and Shapiro, 1966; Kanfer et al., 1966; Watanabe et al., 1983), human placenta (Pentchev et al., 1977; Jones et al., 1983; Sakuragawa, 1982; Rousson et al., 1993), human brain (Yamanaka and Suzuki, 1982), and human urine (Weitz et al., 1985; Quintern et al., 1987). The solubilization of sphingomyelinase from tissue sources requires severe disruption, including detergent treatment (Kanfer et al., 1966; Pentchev et al., 1977; Jones et al., 1981; Sakuragawa, 1982; Watanabe et al., 1983), suggesting that the enzyme is membrane-associated. Most purification schemes include detergent throughout the procedures. The purification steps providing substantial increases in enzyme-specific activity included sequential anion and cation-exchange chromatographies (Pentchev et al., 1977), concanavalin A affinity and hydrophobic-interaction chromatography (Jones et al., 1981), chromatofocusing (Sakuragawa, 1982), and chromatography on blue Sepharose (Quintern et al., 1987). The molecular weight of purified enzyme as determined by denaturing polyacrylamide gel electrophoresis has been reported variously to be in the range of 28–46 kDa (Watanabe et al., 1983; Pentchev et al., 1977), 70–72 kDa (Sakuragawa, 1982; Yamanaka and Suzuki, 1982; Quintern et al., 1987), or 89–110 kDa (Jones et al., 1981; Jobb and Callahan, 1989). The recent molecular cloning data (discussed below) indicate a peptide molecular weight of 64 kDa with six potential N-glycosylation sites. Thus, in retrospect, the value of approximately 72 kDa appears to be correct. The enzyme is glycosylated, as implied by its retention on concanavalin A affinity columns (Jones et al., 1981; Yamanaka and Suzuki, 1982), its secretion from I-cell fibroblasts (Schuchman and Desnick, 1995), and the loss of approximately 11 kDa of molecular mass when treated with endoglycosidase F (Quintern et al., 1989). Gel filtration of nondenatured enzyme during various stages of purification has indicated molecular weights of 200–290 kDa, suggesting that the active enzyme may be a trimer or tetramer (Pentchev et al., 1977; Yamanaka and Suzuki, 1982; Watanabe et al., 1983; Quintern et al., 1987).

The enzyme from urine is soluble and can be purified in the absence of detergents (Weitz et al., 1985), and I-cell disease fibroblasts secrete a soluble form of the enzyme (Schuchman and Desnick, 1995). The urine form purified by Quintern et al. (1987) had a molecular weight of 70 kDa, indicating that it was not an extensively degraded fragment of the enzyme. In addition, the enzyme purified from urine was the source of tryptic peptides that led ultimately to the molecular cloning of the sphingomyelinase cDNA and gene. Thus, the soluble enzyme appears to be very similar to that isolated from tissues. The occurrence of soluble forms of sphingomyelinase, which in tissues appears to be tightly associated with the lysosomal membrane, has not been explained.

3.2.1c. Functional Characteristics. The pH optimum of acid sphingomyelinase has been reported to be 4.6–5.5, depending on the buffer used (Heller and Shapiro, 1966; Kanfer et al., 1966; Pentchev et al., 1977; Jones et al., 1981; Sakuragawa, 1982). The activity is not affected by the presence or absence of divalent cations (Rao and Spence, 1976; Pentchev et al., 1977). The V_{max} of highly purified enzyme preparations using either sphingomyelin or the efficiently hydrolyzed 2-hexadecanoyl-4-nitrophenylphosphorylcholine substrate has been in the range of 50–3200 μmols/(h × mg protein) (Pentchev et al., 1977; Jones et al., 1981; Yamanaka and Suzuki, 1982; Watanabe et al., 1983; Quintern et al., 1987), suggesting that single enzyme molecule is capable of hydrolyzing as many as 3700 substrate molecules per minute. The K_m for sphingomyelin (usually dispersed in nonionic detergent) has been in the range 0.025–0.9 mM (Heller and Shapiro, 1966; Kanfer et al., 1966; Pentchev et al., 1977; Jones et al., 1981; Yamanaka and Suzuki, 1982; Sakuragawa, 1982). As for other enzymes that act on substrates at a lipid–water interface, it is not clear if the K_m values actually reflect the affinity of the substrate for the active site of the enzyme.

With respect to substrate specificity, potential alternative substrates that undergo little or no hydrolysis include (1) synthetic sphingomyelins with the threo (rather than the naturally occurring erythro) relationship between the amide and hydroxy groups at positions 2 and 3 of the carbon backbone (Heller and Shapiro, 1966; see Merrill et al., 1993, for a clear explanation of sphingolipid stereochemistry); (2) phosphatidylcholine, phosphatidylethanolamine, and phosphatidic acid (Heller and Shapiro, 1966; Kanfer et al., 1966; Pentchev et al., 1977); (3) glycosphingolipids (Kanfer et al., 1966); and (4) sphingosylphosphorylcholine (Yamanaka and Suzuki, 1982). Activity against phosphatidylcholine, phosphatidylethanolamine, and phosphatidylglycerol could be increased to significant levels by substituting sodium taurodeoxycholate for the nonionic NonidetP-40 detergent (Quintern et al., 1987). Activity was 10–20% of that against sphingomyelin. Other phosphodiester compounds that undergo hydrolysis include (1) derivatized sphingomyelins carrying fluorescent and chromophoric reporter groups attached to the terminal carbon of the N-fatty acyl group (Gatt et al., 1978, 1980); (2) symmetric fluorogenic and chromogenic phosphodiester compounds such as bis-4-methylumbelliferyl phosphate and bis-para-nitrophenylphosphate (Jones et al., 1981); and (3) the chromogenic sphingomyelin structural analog 2-hexadecanoylamino-4-nitrophenylphosphorylcholine (Sakuragawa, 1982). Where controlled comparisons have been made between sphingomyelin and these artificial substrates, susceptibilities to hydrolysis generally rank (with some exceptions) as follows: sphingomyelin ≃ sphingomyelins with fluorescent or chromophoric reporter groups in the N-fatty acyl position ≃ 2-hexadecanoylamino-4-nitrophenylphosphorylcholine ≃ or > bis-4-methylumbelliferyl phosphate > bis-para-nitrophenylphosphate (Gatt et al., 1978, 1980; Jones et al., 1981; Sakuragawa, 1982; Yamanaka and Suzuki, 1982; Pentchev et al., 1977; Watanabe et al., 1983). In general, the studies of substrate specificity suggest the importance of the phosphodiester linkage, the correct stereochemical relationship between the amide and hydroxy groups of sphingomyelin, and the amide hydrophobic acyl group at carbon 2.

Various agents have been found to stimulate enzyme activity *in vitro*. For the most part, these agents appear to enhance substrate presentation rather than serve as specific enzyme cofactors. The activity with either crude or purified enzyme preparations is stimulated greatly by the addition of detergents. Suitable detergents include Triton-X-100 (Yedgar and Gatt, 1976), Nonidet P-40 at concentrations below 0.1% (Quintern *et al.*, 1987), taurine-conjugated bile acids (Yedgar and Gatt, 1980), and lysophosphatidylcholine (Alpert and Beaudet, 1981). The studies with bile acids suggest that the stimulation of hydrolysis begins when substrate shifts from turbid particulate form to micelles (Yedgar and Gatt, 1980). Quintern *et al.* (1987) found that considerable activity could be obtained in the absence of detergent if substrate was codispersed with certain free fatty acids, or mono, di, or triacylglycerols. The authors speculated that this effect was due to the neutral lipids functioning as spacers between substrate molecules. In work with fibroblast sphingomyelinase, the plasma apolipoproteins apoC-II and apoC-III were found to stimulate activity in the absence of detergent (Alpert and Beaudet, 1981; Ahmad *et al.*, 1986). According to Ahmad *et al.* (1986) this effect appears to be due to the ability of the apoproteins to interact with substrate aggregates and reduce their size. Thus, a detergent-like effect is proposed, which could be important in some aspects of lysosomal lipoprotein degradation; however, this hypothesis does not appear to have been subjected to direct testing using intact cells and endocytosed lipoproteins containing specific apolipoproteins. Lysosomes contain endogenous sphingolipid activator proteins (saposins) that enhance the activity of various enzymes involved in sphingolipid catabolism (see Chapter 7, Section 4.4). These are also believed to play a detergent-like role, extracting lipid molecules from organized lipid aggregates and presenting them to hydrolases (Sandhoff *et al.*, 1995). Of the saposins so far described, three of those derived from prosaposin (sap B, sap C, and sap D) have been reported to stimulate sphingomyelinase directly (Kleinschmidt *et al.*, 1988; Poulos *et al.*, 1984; Morimoto *et al.*, 1988). However, the *in vivo* relevance of these findings is doubtful, since in prosaposin deficiency (in which none of these saposins is produced), cellular acid sphingomyelinase activity and cellular levels of sphingomyelin are normal (Bradova *et al.*, 1993).

A few agents that can directly inhibit acid sphingomyelinase have been described. Octyl glucoside, high concentrations of Nonidet P-40 (another nonionic detergent with a linear octyl group), 5′-phosphate adenosine nucleotides, phosphate ion, and some phosphoinositides have been reported to produce noncompetitive inhibition (Rao and Spence, 1976; Pentchev *et al.*, 1977; Callahan *et al.*, 1983; Quintern *et al.*, 1987). Using crude fibroblast extracts as a source of enzyme, Maziere *et al.* (1981) reported significant inhibition when either cholesterol or 7-dehydrocholesterol was added at greater than 10 mol% relative to the sphingomyelin substrate. Maximal inhibition of about 80% was achieved at 50 mol%

of sterol. This effect may partially explain the ability of the sterol 7-reductase inhibitor AY-9944 to induce sphingomyelin accumulation in cells and may contribute to sphingomyelin accumulation in type C Niemann-Pick disease, which is thought to be due to a defect in sterol removal from lysosomes. The enzyme is not inhibited and possibly even stimulated by agents that derivatize sulfhydryl groups (Rao and Spence, 1976), but is inhibited when exposed to agents such as coenzyme A, dithiothreitol, and bisulfite ion, which reduce disulfide bonds (Rao and Spence, 1976; Eisen *et al.*, 1984). It has been reported that gentamicin and other aminoglycoside antibiotics induce accumulation of sphingomyelin and other phospholipids in cells and tissues (Hostetler and Hall, 1982; Ghosh and Chatterjee, 1987). This effect was attributed to direct inhibition of lysosomal phospholipases by the antibiotics (Hostetler and Hall, 1982). However, Eisen *et al.* (1984) reported that for acid sphingomyelinase, the actual inhibitor in commercial preparations of gentamicin and tobramycin appeared to be the sodium bisulfite that was added as a preservative; the pure antibiotics did not inhibit the enzyme. In cultures of human proximal tubular cells, long-term exposure to 0.3 mM gentamicin leads to decreased activity of both the acid and neutral sphingomyelinases, with the neutral activity declining to a greater extent (Ghosh and Chatterjee, 1987). Thus, the accumulation of sphingomyelin in tissues or cells exposed to these antibiotics probably is based on changes more complex than direct inhibition of lysosomal phospholipases. Another class of compounds that decreases acid sphingomyelinase activity in cells is a diverse set of cationic amphiphilic drugs that includes imipramine and other tricyclic antidepressants, the calmodulin inhibitor trifluoperazine, and the sterol reductase inhibitor AY-9944 (Yoshida *et al.*, 1985). These compounds do not appear to inhibit acid sphingomyelinase directly (Yoshida *et al.*, 1985). Hurwitz *et al.* (1994) reported that exposure of fibroblasts to desipramine, a tricyclic antidepressant, causes rapid, leupeptin-sensitive catabolism of ^{35}S-labeled acid sphingomyelinase. Thus, the effects of these compounds may be due to enhanced enzyme turnover.

    3.2.1d. Cloning and Molecular Biology. The cloning and molecular biology of mammalian acid sphingomyelinase were reviewed in detail by Schuchman and Desnick (1995). Briefly, work on this topic in the last six years has shown that the human gene encoding acid sphingomyelinase is located at chromosomal position 11p15.1 → p15.4 (Pereira *et al.*, 1991), is about 5 kb long, and is divided into six exons (Schuchman *et al.*, 1992). Selection and analysis of cDNA from human libraries demonstrated that the sphingomyelinase primary transcript is spliced into three alternative mRNA (Quintern *et al.*, 1989; Schuchman *et al.*, 1991). The alternative mRNA designated as type 1 is the most abundant form (about 90% of the primary transcripts appear to be spliced to this form), and as indicated by transient expression of cDNA in COS-1 cells, only this version codes functional acid sphingomyelinase (Schuchman *et al.*, 1991). None of the three alternative mRNA appears to code for a neutral sphingomyelinase activity (Schuchman *et al.*, 1991). The translation product of the type 1 mRNA is a protein of 629 amino acids, with an unusual hydrophobic signal peptide containing five tandem leucine–alanine pairs, and contains six potential N-glycosylation sites (Schuchman *et al.*, 1991). The length of the peptide and the number of glycosylation sites are consistent with determinations of molecular weight and sensitivity to glycosidase treatment performed with the urinary enzyme (Quintern *et al.*, 1989, 1987). Further analysis of the amino acid sequence of acid sphingomyelinase has suggested the existence of an N-terminal lipid binding region homologous to that found in saposins A–D, consisting of four amphipathic helices held in a bundle by interhelix disulfide bonds (Ponting, 1994; Munford *et al.*, 1995). The complete gene and cDNA from the mouse have also been cloned and sequenced. This gene also is divided into six exons. The sequences of the cDNA and protein showed about 80% identity with the human cDNA and protein (Newrzella and Stoffel, 1992). No alternative splicing of the primary transcript appears to occur in the mouse (Newrzella and Stoffel, 1992). The mutations in the human gene causing types A and B Niemann-Pick disease have been identified. This is discussed more fully in Section 4.3.2.

The relationship between mammalian acid sphingomyelinase and other phospholipases is not very clear. There are no reports of the cloning and sequencing of the gene for the neutral enzyme. *In situ* hybridization studies that localized the acid enzyme to chromosome 11 did not detect homologous genes at other locations. Thus, there may be little sequence similarity between the different mammalian enzymes (Pereira *et al.*, 1991). There is no reported significant homology of the acid sphingomyelinase with a prokaryotic sphingomyelinase and other phospholipases that have been cloned and sequenced (Schuchman and Desnick, 1995).

3.2.1e. Biology of Enzyme.   Subcellular fractionation studies have confirmed that acid sphingomyelinase is located in lysosomes (Ghosh and Chatterjee, 1987). As indicated in the above discussion, the enzyme is found in a variety of cells and tissues, reinforcing the view that it serves an important housekeeping function. The tissue distribution of gene expression has not been reported. In addition, there appears to be little information on whether the enzyme is regulated during development and differentiation or in response to environmental factors. The kinetics of synthesis and turnover of the enzyme are largely unknown. A limited amount of data on the decline of fibroblast activity in the presence of cycloheximide suggests a half-life of much longer than 8 hr (Yoshida *et al.*, 1985).

Sources of substrate for acid sphingomyelinase include endocytosed plasma lipoproteins (Levade *et al.*, 1991b; Dinur *et al.*, 1992; Graber *et al.*, 1994) and the plasma membrane (Koval and Pagano, 1990; Levade *et al.*, 1991b). It seems likely that lysosomes and the acid sphingomyelinase are critical for the degradation of lipoprotein-derived sphingomyelin, whereas both the lysosomal and plasma membrane enzymes participate in the turnover of endogenous membrane-derived sphingomyelin (Levade *et al.*, 1991b; Koval and Pagano, 1990). That the acid sphingomyelinase is crucial for sphingolipid homeostasis is clearly established by the Niemann-Pick A and B lipid storage diseases. Ceramide and sphingosine, products of the degradation of sphingomyelin and other sphingolipids, have been implicated as second messengers regulating cellular metabolism (reviewed by Koval and Pagano, 1991; Merrill *et al.,* 1993; Hannun, 1994). Both the acid and neutral sphingomyelinases would contribute to the production of these compounds. However, receptor control of sphingolipid degradation seems more likely at the plasma membrane. Thus, the neutral enzyme is likely to be the main player in this form of metabolic regulation. Consistent with this view, it was found that cellular responses to tumor necrosis factor and interleukin-1, which are thought to be mediated by sphingomyelin hydrolysis, were normal in fibroblasts from types A and B Niemann-Pick patients (Andrieu *et al.,* 1994).

3.2.2. Lysosomal Ceramidase

Lysosomal ceramidase is a central component of the sphingolipid degradation machinery, since ceramide is generated during the breakdown of sphingomyelin and all of the glycosphingolipids (Fig. 2). The importance of being able to dispose of ceramide in lysosomes is emphasized by the inherited storage disease Farber lipogranulomatosis, which results from the absence or severe deficiency of acid ceramidase (see Section 4.3.3).

Ceramidases (N-acylsphingosine deacylases, EC 3.5.1.23) catalyze the hydrolytic reaction:

$$ceramide + H_2O \rightarrow sphingosine + fatty\ acid$$

Mammals are reported to have three ceramidase activities, distinguishable mainly by their pH optima: (1) an acid (lysosomal) activity with a pH optimum of 4–5, (2) an alkaline activity with pH optimum of 8–9, and (3) a neutral activity with a pH optimum of about 7 (reviewed by Hassler and Bell, 1993; Moser, 1995). The acid and alkaline activities are widely distributed in the body (Spence *et al.,* 1986), whereas the neutral activity may be confined to the gastrointestinal tract, suggesting that this form of ceramidase functions mainly in nutrient absorption (Nilsson, 1969). None of these enzymes has been isolated in pure form and none has been cloned or sequenced. Thus, relatively little can be concluded with certainty about their structure and mechanism of action. As noted above, acid ceramidase deficiency appears to be the primary metabolic defect in Farber lipogranulomatosis (reviewed by Moser, 1995), suggesting that this enzyme plays a critical role in sphingolipid homeostasis.

A widely used assay for acid ceramidase is described by Suzuki (1987). The usual substrate is N-[1-^{14}C]oleoylsphingosine. The liberated [^{14}C]oleate is isolated by solvent partition and quantified by liquid scintillation counting. Other important conditions of the assay are pH 4, the inclusion of both sodium cholate and nonionic detergent in the reaction cocktail, both of which are necessary to obtain maximal activity (Gatt, 1966), and sonication of the enzyme source to eliminate latency. For the specific diagnosis of Farber disease, Momoi *et al.* (1982) demonstrated that N-lauryl(12:0)sphingosine is a more suitable substrate, since it is hydrolyzed efficiently by acid ceramidase but poorly by the alkaline enzyme.

The acid ceramidase activity has been detected in a variety of cells and tissues, including brain (Gatt, 1963), kidney (Sugita *et al.,* 1972), placenta (Chen and Moser, 1979), skin fibroblasts (Chen *et al.,* 1981; Momoi *et al.,* 1982), liver, muscle, lung (Spence *et al.,* 1986), spleen (Spence *et al.,* 1986; Al *et al.,* 1989), and virus-transformed B lymphocytes (Levade *et al.,* 1993). The activity sediments from rat brains homogenate at intermediate *g* forces, consistent with a lysosomal location (Gatt, 1966). Enrichments of 100–300-fold from tissue homogenates have been obtained (Gatt, 1963, 1966; Yavin and Gatt, 1969; Chen and Moser, 1979). Complete solubilization of the activity requires both detergent treatment and vigorous mechanical disruption, suggesting that the enzyme is an integral membrane protein (Gatt, 1966; Al *et al.,* 1989). By gel filtration, detergent-solubilized enzyme appears to have a molecular weight of about 100 kDa (Al *et al.,* 1989). It is precipitable with concanavalin A, suggesting that it is a glycoprotein (Al *et al.,* 1989). The brain activity is resistant to trypsin and chymotrypsin, providing one means of removing some protein contaminants (Yavin and Gatt, 1969). Substrate specificity studies show activity against ceramides with a variety of saturated and unsaturated N-acyl groups and against N-acylated dihydrosphingosine (Gatt, 1966; Momoi *et al.,* 1982; Spence *et al.,* 1986). Gatt (1966), however, reported no activity against N-acetyl sphingosine and N-lignoceryl(24:0)dihydrosphingosine. The relatively broad specificity of acid ceramidase contrasts with the alkaline enzyme, which shows little or no activity against ceramides with saturated N-acyl groups (Momoi *et al.,* 1982; Gatt, 1966). Semi-purified acid ceramidase does not hydrolyze the N-acyl group of sphingomyelin or cerebrosides (Gatt, 1963). Using N-oleoyl-sphingosine and assaying acid ceramidase from various sources, K_m values in the range 0.15–1.2 mM have been reported (Momoi *et al.,* 1982; Spence *et al.,* 1986; Al *et al.,* 1989).

There is strong evidence for the involvement of sphingolipid activator protein-D (sap D) in acid ceramidase activity, derived from experiments with tissues and cells of patients with sap-precursor deficiency, in which four of the five known saposins are absent. The evidence includes (1) the observation that ceramide accumulates in tissues and cells of patients with sap-precursor deficiency (Harzer *et al.,* 1989; Paton *et al.,* 1992; Klein *et al.,* 1994); (2) a partial deficiency of acid ceramidase in homogenates from these cells when assayed in the presence of detergent (Harzer *et al.,* 1989); and (3) the reversal of ceramide accumulation by addition of sap D to the culture medium of fibroblasts from affected individuals (Klein *et al.,* 1994). Klein *et al.,* (1994) also state that acid ceramidase and sap D copurify from human tissues, but they do not show the data.

An interesting feature of the acid and alkaline ceramidases is that they catalyze both the hydrolysis of ceramide and its synthesis from fatty acid and sphingosine (Gatt, 1963, 1966; Nilsson, 1969). For acid ceramidase, the synthetic reaction does not depend on addition of CoA or ATP (Gatt, 1966). Whether fatty acyl-CoA can support the synthetic reaction directly is not clear, although Yavin and Gatt (1969) present data against this possibility. The hydrolytic and synthetic reactions are perturbed similarly by heat denaturation, pH changes, protease treatment, and sulfhydryl reagents (Yavin and Gatt, 1969). Both the hydrolytic and synthetic activities at acid pH are absent in cells from Farber disease patients (Sugita *et al.*, 1972). Yavin and Gatt (1969) measured the equilibrium constant of the reaction when either ceramide or sphingosine and fatty acid were the reactants. Values substantially less than 1 were obtained in both directions; thus, the overall direction of the reaction could not be established using this approach. Stoffel *et al.* (1980) argued that the synthetic reaction was an artifact resulting from the formation of N-acylethanolamine, using ethanolamine in the assay buffer. However, others demonstrated the synthetic reaction in the absence of ethanolamine (Sugita *et al.*, 1975). Thus, the catalysis of ceramide formation by acid ceramidase appears to be real, but its physiological significance has not been established. The major route of ceramide synthesis in mammalian cells probably is by transfer of an acyl group from fatty acyl-CoA to dihydrosphingosine, followed by dehydrogenation of the resulting intermediate (Stoffel and Melzner, 1980). The accumulation of ceramide in Farber disease argues against an essential role for acid ceramidase in ceramide synthesis.

The wide tissue distribution of acid ceramidase, its broad substrate specificity, and the consequences of its absence in Farber disease indicate that this enzyme plays an important housekeeping function. Studies with fibroblasts in culture indicate that its substrate is derived from both endogenous sphingolipids (Klein *et al.*, 1994) and exogenous sphingolipids that can be supplied either in vesicles or in LDL (Chen and Decker, 1982; Kudoh and Wenger, 1982; Sutrina and Chen, 1982). The method of delivery of exogenous ceramide can have a significant influence on its site of degradation, with LDL-delivered substrate degraded largely by the acid enzyme and vesicle-delivered substrate degraded by the alkaline enzyme (Sutrina and Chen, 1982). Analogous to the different forms of sphingomyelinase, the alkaline form of ceramidase may play a role in signal transduction at the cell surface, and the acid form may serve an unregulated catabolic role in lysosomes.

3.3. Glycerophospholipids

When either exogenous or endogenous glycerophospholipids are delivered to lysosomes, the degradation process is initiated by one or more of the esterases that are part of a group of enzymes named the phospholipases. The different types of phospholipases are described here in general terms, and this is followed by a review of the specific phospholipase activities that have been localized in lysosomes.

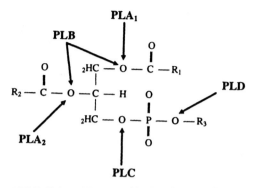

FIGURE 3. Positional specificities of phospholipases.

3.3.1. Types of Phospholipases

The phospholipase enzymes can be divided into two hydrolytic groups (Fig. 3). Phospholipases A_1, A_2, and B hydrolyze the ester bonds of the glycerophospholipids. Phospholipases A_1 and A_2 show specificity for the hydrolysis of the 1-acyl ester and the 2-acyl ester, respectively. Phospholipase B is similar to the lysophospholipase in that it is capable of hydrolyzing at both the sn-1 and sn-2 positions. The lysophospholipases would also be included in this group. Phospholipases C and D can each cleave one of the phosphodiester bonds. The phospholipases can be assigned to two broad functional groups: regulatory enzymes (Rhee *et al.*, 1989) and digestive enzymes (Verheij *et al.*, 1981). The first category refers to the role of the turnover of phospholipids in the production of lipid second messenger systems, while the second refers to the hydrolysis of exogenous phospholipid. For reviews of the range of different phospholipases and what is known about their function, the reader is referred to the reviews by Waite (1987, 1991).

3.3.2. Substrates and the Phospholipase Assay Methods

The hydrolytic activity of the phospholipases on substrates is maximal above the critical micellar concentration of the phospholipid (Roholt and Schlamowtitz, 1961). For this reason, knowledge of the physical properties of the substrate is important in assays in phospholipase activity, and as a consequence model membrane systems rather than biological membranes are often used as substrates. When biological membranes are used, however, they offer the advantage that the enzyme acts on a substrate that it actually could encounter *in vivo*. The disadvantages of using biological membranes include the heterogeneity of phospholipid species in the substrate and the possibility of competing enzymes that remain associated with the substrate (Waite, 1987). As a model membrane system, liposomes have been used as substrates for assays of lysosomal phospholipases because of the ease with which they are made and because they are a bileaflet membrane system with an aqueous center (Bangham *et al.*, 1965). Another less time-consuming choice that investigators have used as a substrate are phospholipid emulsions. Since such emulsions often use detergents, caution must be taken to determine what effect the detergent will have on both the physical state of the phospholipid complex (i.e., liposomes versus micelles) and the surface charge of the phospholipid complex (Robinson and Waite, 1983; Deems *et al.*, 1975; Bangham and Dawson, 1959). Studies have shown that the activity of most phospholipases is affected by surface charge (Waite, 1987). For a complete discussion of the various substrates that can be used to assay phospholipase activity the reader is referred to Waite (1987) and Dennis (1983).

Investigators have used a number of methods to measure phospholipase activity as described in previous reviews (Waite, 1987; Van den Bosh and Aarsman, 1979). The most common procedures use substrates consisting of phospholipids that are radiolabeled in specific functional groups of the molecule. The radiolabeled phospholipid can be prepared by chemical or biosynthetic means (Waite, 1987). The advantages of this technique are that the specific type of phospholipase activity can be determined and that it provides the greatest sensitivity when compared to other assays. One disadvantage is that the procedure is time-consuming, since quantitation of the radiolabeled products requires the use of extraction and chromatographic techniques. Also, the purchase or synthesis of the radiolabeled substrates can be expensive. An alternative approach is to measure the mass of the products of hydrolysis. This approach is not used often since it is time-consuming, but it can be useful in establishing the accuracy of isotopic techniques.

3.3.3. Phospholipase Activities Found in Lysosomes

The majority of studies have used the soluble fractions of isolated lysosomes to assess lysosomal phospholipase activity. For most studies, the soluble lysosome fraction is prepared either from lysosomes isolated by differential ultracentrifugation as described by Sawant *et al.* (1964) or from lysosomes (tritosomes) isolated from the tissues of animals injected intraperitoneally with Triton (Leighton *et al.*, 1968). Using these approaches a number of the phospholipase activities have been detected in lysosomes, including A_1, A_2, and C.

3.3.4. Phospholipase A_1

Phospholipase A_1 activity has been detected in the lysosomes of a number of mammalian tissues, including rat liver (Robinson and Waite, 1983; Hostetler *et al.*, 1982), rabbit arterial smooth muscles cells (Ishikawa *et al.*, 1988), and rabbit alveolar macrophages (Franson and Waite, 1973). Most of the lysosomal phospholipase A_1 preparations show optimal activity in the pH range 4.0–4.5 (Hostetler *et al.*, 1982; Ishikawa *et al.*, 1988; Franson and Waite, 1973). The most extensively characterized lysosomal phospholipase A_1 is of the rat liver. The enzyme is a glycoprotein consisting of D-glucosyl or D-mannopyranosyl units and is not dependent on calcium for activity (Robinson and Waite, 1983). Purification of the enzyme showed that there are up to five forms of the enzyme with molecular weights in the range of 24,000 to 90,000 daltons (Hostetler *et al.*, 1982). There do not appear to be any functional or regulatory differences (i.e., substrate specificity or surface charge requirements) between the different forms of phospholipase A_1. Rather the different forms of the enzyme may be the result of varying degrees of maturation of the carbohydrate of the enzyme (Hostetler *et al.*, 1982). The activity of the enzyme is highly dependent on the phase of the phospholipid substrate (Robinson and Waite, 1983). When phospholipid substrates are used without the addition of detergents, substrates consisting of phosphatidylethanolamine are degraded, while those containing other phospholipids, including phosphatidylcholine, phosphatidylinositol, and phosphatidylglycerol, are not efficiently hydrolyzed. Without the presence of detergent, the phosphatidylethanolamine substrate exists as hexagonal arrays, while the other phospholipid substrates exist as liposomes. Once sufficient detergent is added to the substrates consisting of phosphatidylcholine, phosphatidylinositol, or phosphatidylglycerol so that mixed micelles are formed, there is substantial hydrolysis of all these phospholipids, phosphatidylglycerol being the preferred substrate (Robinson and Waite, 1983). Thus, it appears that the lysosomal phospholipase A_1 prefers a substrate in the non-bilayer form. Through the use of divalent cations, charged amphipaths, and substrates consisting of either phosphatidylethanolamine or phosphatidylcholine, the enzyme activity was found to require a slight negative surface charge of the substrate (Robinson and Waite, 1983).

3.3.5. Phospholipase A_2

Phospholipase A_2 activity has been detected in the soluble fraction of lysosomes isolated from a number of mammalian tissues including rat liver (Franson *et al.*, 1971), rabbit alveolar macrophages (Franson and Waite, 1973), mouse peritoneal macrophages (Wightman *et al.*, 1981), rabbit smooth muscle cells (Ishikawa *et al.*, 1988), and rat heart (Franson *et al.*, 1972). The phospholipase A_2 of lysosomes exhibits optimal activity in a pH range of 4.0–4.5 and like the phospholipase A_1 does not require calcium (Franson *et al.*, 1971; Franson and Waite, 1973; Wightman *et al.*, 1981; Ishikawa *et al.*, 1988; Franson *et al.*, 1972). For the most part phosphatidylethanolamine is the preferred substrate (Wightman *et al.*, 1981), but phosphatidylcholine and other phosphatidylglycerols can be hydrolyzed when added in the appropriate substrate form. In contrast to the lysosomal phospholipase A_1, the lysosomal phospholipase A_2 activity is optimal when the substrate is a liposome rather than a mixed micelle (Wightman *et al.*, 1981). In addition, both phosphatidylethanolamine and phosphatidylcholine are hydrolyzed more efficiently when liposome substrates that contain a mixture of the two phospholipids are used (Wightman *et al.*, 1981). An extracellular phospholipase A_2 that is secreted from polymorphonuclear cells during phagocytosis was determined to originate from lysosomal granules (Traynor and Kalwant, 1981). This phospholipase A_2 has been postulated to play a role in the production of prostaglandins during the inflammation process (Traynor and Kalwant, 1981). It differs from the other lysosomal phospholipase A_2 enzymes in that it requires calcium and is active over a broad pH range (Traynor and Kalwant, 1981).

3.3.6. Phospholipase C

Phospholipase C activity has been localized in the lysosomal fractions from rat liver (Matsuzawa and Hostetler, 1980), rat kidney (Hostetler and Hall, 1982), rabbit macrophages (Eisen *et al.*, 1984), and rat intestine (Hostetler and Hall, 1980). The lysosomal phospholipase C is optimally active at pH 4.0–4.5 (Matsuzawa and Hostetler, 1980; Hostetler and Hall, 1982; Eisen *et al.*, 1984), and like the lysosomal phospholipases A_1 and A_2, it does not require calcium. The enzyme exhibits the most activity toward phosphatidylinositol, but other phospholipids including phosphatidylcholine, phosphatidylglycerol, phosphatidylserine, and phosphatidylethanolamine are also degraded by the enzyme (Matsuzawa and Hostetler, 1980; Irvine *et al.*, 1978). There are conflicting reports, however, as to how active the enzyme is toward the other phospholipids compared to activity toward phosphatidylinositol (Irvine *et al.*, 1995; Matsuzawa and Hostetler, 1980). These differences in results may be attributed to differences in the preparation of the soluble lysosome fractions (Matsuzawa and Hostetler, 1980). An alternative explanation that has been postulated is that there may be two forms of the enzyme, with one being specific for phosphatidylinositol and the other having a broad substrate specificity (Matsuzawa and Hostetler, 1980; Irvine *et al.*, 1978). Better determination of the enzyme(s) substrate specificity will require its purification. Unlike phospholipase A_1, phospholipase C activity is inhibited by the addition of either sodium deoxycholate to a phosphatidylcholine substrate (Matsuzawa and Hostetler, 1980) or Triton WR1339 to a phosphatidylinositol substrate (Irvine *et al.*, 1978).

3.3.7. Lysosomal Degradation of Different Glycerophospholipids

While, as discussed above, substantial activity of some of the different phospholipase types has been detected in the soluble fractions of lysosomes, there are only a few studies that address both the major pathways for the lysosomal degradation of the different phospholipids and the contributions the different enzymes make to this process. Early studies by Fowler and de Duve (1969) using the soluble fraction of rat liver lysosomes and model membrane substrates indicated that the degradation of the major glycerophospholipids including phosphatidylcholine, phosphatidylethanolamine, phosphatidylserine, and phosphatidylinositol proceeded via deacylation, which would be attributed to a phospholipase A. Other studies indicate that the resulting lysophospholipids are further degraded by two different lysosomal lysophospholipases (Kunze *et al.*, 1982). One enzyme, lysophospholipase A, would hydrolyze the resulting lysophospholipids to yield free fatty acids and phosphodiesters. The phosphodiesters are not hydrolyzed intralysosomally (Fowler and de Duve, 1969), and it remains unknown as to how they would exit the lysosome. Alternatively, the studies of Kunze *et al.* (1982) suggest that the lysophospholipids could also be degraded by a diacylglycerohydrolase, where the end products would be monoacylglycerol and phosphomonoesters. These investigators suggested that this lysophospholipase and phospholipase C may be the same enzyme. Whether phosphomonoesters such as phosphorylcholine are further degraded in lysosomes or transported into the cytosol is not clearly established (see Chapter 11).

In the early studies of Fowler and de Duve (1969), there was no detectable diacylglycerol as an end product of degradation of the glycerophospholipids, suggesting that there is not a major role for phospholipase C in the initial degradation of the glycerophospholipids. However, since the discovery of phospholipase C activity in lysosomes, the question of whether this enzyme plays a major role in the initial degradation of the glycerophospholipids is controversial. In contrast to the studies of Fowler and de Duve (1969), studies by Matsuzawa and Hostetler (1980) suggest that the hydrolysis of the glycerophospholipids by phospholipase C possibly could be a major route of degradation. This is particularly true for phosphatidylinositol where the activity of the enzyme was found to be 3.5- to 12-fold higher than the activity toward the other glycerophospholipids tested (Matsuzawa and Hostetler, 1980). Consistent with this are studies by Irvine *et al.* (1978) and Richards *et al.* (1979) showing that when the soluble fraction of rat liver lysosomes is incubated with labeled phosphatidylinositol, present in either a model membrane substrate or microsomal membrane, both inositol monophosphate and diacylglycerol are detected products of degradation. In addition, Matsuzawa and Hostetler (1980) showed that when [1-^{14}C]dioleoylphosphatidylcholine dispersions were incubated with the soluble fraction of rat liver lysosomes, [^{14}C]diglyceride, [^{14}C]monoglyceride, and phosphorylcholine were present in significant amounts as the end products of degradation. These investigators concluded that monoglyceride accumulated as a product because of the rapid conversion of diglyceride by a lysosomal lipase, and that the reasons for monoglycerides not being detected in previous studies was that the chromatographic techniques employed would not have separated the monoglycerides from free fatty acids (Fowler and de Duve, 1969). In contrast, Kunze *et al.* (1982) concluded that the initial deacylation of phosphatidylcholine by phospholipase A_1 would be the major route of degradation. The investigators suggested that the reason for the difference in results was that delipidation of the lysosomal fraction results in a lower recovery of the phospholipase A_1 activity (Kunze *et al.*, 1982). While the studies of Matsuzawa and Hostetler (1980) demonstrated that the specific activities of phospholipases C and A_1 plus A_2 in their delipidated lysosomal fraction were equal, these studies showed that the specific activity of phospholipase A_1 was 50-fold greater than that of phospholipase A_2 in their nondelipidated lysosomal fraction. Future experiments will be needed using purified lysosomal phospholipases to examine more accurately the major pathways of degradation for the different glycerophospholipids.

3.3.8. Lysosomal Degradation of Exogenous Phospholipids

The degradation of exogenous phospholipids by lysosomal phospholipases in whole cells can be studied using lysosomotropic agents such as chloroquine that inhibit lysosomal enzyme function in part by causing an increase in the pH of the lysosome. Studies by Hostetler *et al.* (1985) have shown that when rats were administered chloroquine for 3 days there was a 3.7-fold increase in the phospholipid content of hepatic lysosomes. Later studies showed that the increase in phospholipid content was due to competitive inhibition of phospholipase A_1 by chloroquine (Kubo and Hostetler, 1985). The results suggest that the lysosomal phospholipases may play a significant role in the catabolism of phospholipid, but it is not known from these studies whether it is the exogenous or endogenous phospholipid that is the main source of phospholipid accumulation. Few studies have been done using whole cells in culture to address the effects of chloroquine treatment on the lysosomal hydrolysis of phospholipids in either exogenous or endogenous biological substrates. A study by Ishikawa *et al.* (1989) suggested that the phospholipid in low density lipoproteins is not degraded in the lysosomes. When rabbit arterial smooth muscle cells were incubated with [^{14}C]linoleoyl phosphatidylcholine in the absence or presence of chloroquine, 25.1% and 24.8%, respectively, of the radiolabel was found as phosphatidylcholine in the lysosome fraction. In the same studies a 50% reduction in the hydrolysis of the labeled cholesteryl ester moiety of the lipoprotein was measured when the cells were incubated with chloroquine compared to when cells were not incubated with chloroquine. In earlier work, Ishikawa *et al.* (1988) had shown that there were phospholipase A_1 and A_2 activities with neutral pH optima present in the plasma membrane and thus suggested that the degradation of low density lipoprotein phospholipid may occur prelysosomally. More studies on the hydrolysis of the phospholipid portion of low density lipoproteins and other natural exogenous substrates are needed to determine how much of a role the lysosome plays in the glycerophospholipid degradation.

4. LYSOSOMAL STORAGE DISEASES RESULTING FROM DEFECTIVE LIPOLYSIS

There are three distinct groups of storage diseases resulting from defective lysosomal lipolysis: (1) Wolman and cholesteryl ester storage diseases, (2) the

Niemann-Pick diseases, and (3) Farber lipogranulomatosis. These diseases result in the accumulation of cholesteryl esters, sphingomyelin, and ceramide, respectively, and most cases can be attributed to defects in the relevant acid lipases. These groups of diseases are reviewed briefly in this section. It should be noted that all of these syndromes have been reviewed quite thoroughly by Scriver *et al.* (1995).

4.1. Wolman Disease and Cholesteryl Ester Storage Disease (CESD)

Wolman disease and cholesteryl ester storage disease (CESD) are two closely related metabolic diseases in which very large amounts of cholesteryl ester, and sometimes triglycerides, accumulate in a number of different tissues. Both of these lysosomal storage diseases have been reviewed extensively by Assmann and Seedorf (1995).

Of the two diseases Wolman disease is clearly the most clinically severe, with symptoms becoming evident within a few weeks following birth and resulting in death before the end of the first year. Massive lipid accumulation is evident in most tissues, and adrenal calcification is a hallmark of Wolman disease. In contrast, CESD is much less severe, and patients often survive beyond middle age. Tissues of patients with either type of disease show massive accumulation of cholesterol and triglycerides (Fig. 4). Although unesterified cholesterol shows some elevation, the principal lipid that accumulates is cholesteryl ester. In addition to cholesterol, a number of oxidized sterols have also been identified (Assmann *et al.*, 1975).

A particularly intriguing question has been why the lack of aCEH in both Wolman disease and CESD leads to such a dramatic difference in the severity of clinical responses. In part this may be a reflection of differences in the mutations associated with the diseases and the subsequent level of expression of aCEH in tissues. Although fibroblasts derived from both Wolman disease and CESD patients show essentially no enzyme activity, aCEH activity in the liver of CESD individuals was approximately 23% of normal values, whereas the residual activity in the specimens derived from the liver of Wolman disease patients was only 4% (Hoeg *et al.*, 1984). An explanation for this difference in clinical symptoms may be related to the nature of the mutations underlying the two diseases (Anderson *et al.*, 1994; R. Anderson, personal communication). Analysis of the DNA sequence in cells from the two types of patients indicate that in almost all CESD cases the mutation in at least one of the defective aCEH alleles is confined to a single base change at a splice junction site that leads to production of a shortened mRNA (Anderson *et al.*, 1994; Ameis *et al.*, 1995). In Wolman disease cases, a number of different mutations have been identified in protein coding sequences of the aCEH gene alleles; however, none are at the splice junction. Thus, it is possible that the liver is capable of correctly splicing a fraction of the RNA in CESD and forming mature, full-length mRNA in sufficient amounts to provide residual active enzyme, whereas there is a complete lack of catalytically active enzyme synthesis in the livers of Wolman disease patients.

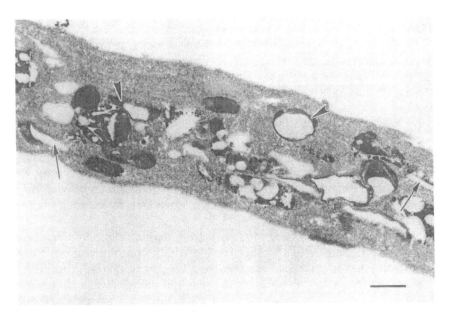

FIGURE 4. Transmission electron micrograph of a fibroblast from a Wolman-disease patient. The cell has been stained to demonstrate acid phosphatase as a marker for lysosomes. Dark acid phosphatase reaction product (arrowheads) delineates numerous lipid-engorged lysosomes. Some of the lysosomes contain cholesterol crystals (arrows). Magnification = × 19600; bar = 0.5 μm. (Provided by Dr. W. Gray Jerome, Bowman Gray School of Medicine, Winston-Salem, North Carolina.)

4.2. Niemann-Pick Diseases

The Niemann-Pick diseases constitute a related group of autosomal recessive lipid storage disorders. The first documented case was reported in 1914 by the German pediatrician Albert Niemann (Niemann, 1914). Niemann described an infant who suffered from severe hepatosplenomegaly and progressive neurodegeneration. The child died at the age of 18 months. Autopsy revealed the presence of lipid-laden "foamy cells" characteristic of other lipid storage disorders. After this initial report, the disease has been further subclassified into types A, B, and C, according to the primary lipid storage disorder. Although reports of a Niemann-Pick type D exist, this form of the disease appears to be the same as Niemann-Pick type C, but is restricted to a genetic isolate in Nova Scotia (Schuchman and Desnick, 1995).

Niemann-Pick disease types A and B (NP-A and NP-B) are the result of a biochemical defect in acid sphingomyelinase, the enzyme that catalyzes the lysosomal catabolism of sphingomyelin to ceramide and phosphorylcholine (Fig. 5). Both NP-A and NP-B patients have between 5% and 10% of normal acid sphingomyelinase, while results in the lysosomal accumulation of sphingomyelin and

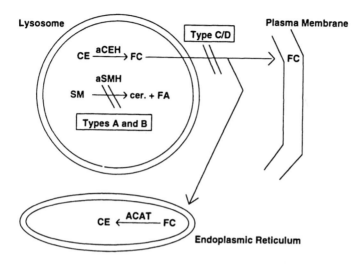

FIGURE 5. Niemann-Pick diseases: proposed biochemical defects causing different subtypes of the disease. Abbreviations: CE, cholesteryl ester; aCEH, acid cholesteryl ester hydrolase; FC, free (unesterified) cholesterol; SM, sphingomyelin; aSMH, acid sphingomyelinase; cer., ceramide, ACAT, acyl-CoA:cholesterol acyltransferase.

cholesterol, as well as other metabolically related lipids (Schuchman and Desnick, 1995). Patients with Niemann-Pick disease type C (NP-C) have normal acid sphingomyelinase activity, but still accumulate unesterified cholesterol within the lysosome. This general difference is the criterion used to distinguish the different subclasses of Niemann-Pick disease. Unlike NP-A and NP-B, the primary defect of NP-C has yet to be determined but may involve reduced ability to transport cholesterol in cells (Fig. 5).

Clinically, NP-A is characterized by massive hepatosplenomegaly, apparent within a few months after birth (Schuchman and Desnick, 1995). As the disease progresses, neurodegeneration occurs, resulting in the loss of motor skills and intellectual capacities. Patients with NP-A usually do not live more than 2–3 years. The clinical manifestations of NP-B are much more variable. Unlike NP-A, patients with NP-B do not suffer neurodegeneration, apparently because the residual sphingomyelinase activity of NP-B patients is sufficient to prevent neurodegeneration. Patients with NP-B are usually diagnosed early in childhood by hepatosplenomegaly, but survive into adulthood.

The hallmark of all Niemann-Pick disease is the foamy cell. This foamy appearance is due to numerous cytoplasmic lipid droplets that stain positive for cholesterol. The cells originate from bone marrow progenitor cells, and therefore it is unusual to find tissues without any foam cells.

Normal cells contain approximately 5–20% of total phospholipid as sphingomyelin. The absence of acid sphingomyelinase activity results in sphingomyelin concentrations comprising 70% of total phospholipids in NP-A and NP-B (Vanier, 1983; Kamoshita *et al.*, 1969). Sphingomyelin accumulates to the greatest extent in the spleen and lymph nodes, however, other tissues including the lung, brain, and kidneys also show abnormal sphingomyelin accumulation. Interestingly, while NP-A and NP-B patients show similar accumulation in the above-mentioned tissues, NP-B patients show little or no sphingomyelin accumulation in nervous tissue (Spence and Callahan, 1989). Tissues from NP-A and NP-B patients also show enhanced cholesterol (Vanier, 1983) and bis(monoacylglycero)phosphate (Rouser *et al.*, 1968) accumulation compared to normal individuals. Enhanced storage of the latter two compounds may be secondary to the primary acid sphingomyelinase deficiency in these patients.

Recently, the cDNA encoding acid sphingomyelinase has been isolated and sequenced (Quintern *et al.*, 1989). This accomplishment has led to the discovery of two mutations in the sequence coding for acid sphingomyelinase that are responsible for NP-A and NP-B (Levran *et al.*, 1991). First, a missense mutation at nucleotide 1487 leads to a predicted arginine-to-leucine substitution at position 496. This mutation appears to be more common in severe neuropathic patients (i.e., NP-A). The second mutation is a three-base deletion of codon 608, encoding an arginine residue. This mutation appears to be specific for NP-B.

Unlike NP-A and NP-B, the biochemical defect in NP-C disease has yet to be defined. As described above, NP-C patients have normal acid sphingomyelinase activity. The primary disorder appears to be a defect in the transport of unesterified cholesterol out of the lysosome, resulting in lysosomal cholesterol accumulation (Fig. 5). Mechanisms that regulate the cytosolic concentration of unesterified cholesterol are affected by lysosomal cholesterol accumulation as seen in NP-C. Normally, uptake of exogenous cholesterol via a receptor-mediated process results in downregulation of receptor-mediated LDL uptake, stimulation of intracellular cholesterol esterification, and downregulation of sterol biosynthesis (Brown and Goldstein, 1986). In NP-C, these cellular responses to LDL uptake are delayed (Pentchev *et al.*, 1987). However, the delayed response is not due to a defect in LDL degradation (Liscum and Faust, 1987). Likewise, NP-C cells have normal intracellular responses to 25-hydroxycholesterol, a compound that induces similar intracellular responses as LDL-derived cholesterol (Liscum and Faust, 1987). Therefore, it appears that NP-C cells have a defect in transport of LDL-derived cholesterol out of lysosomes, which results in abnormal regulation of intracellular cholesterol metabolism.

The accumulation of LDL-derived cholesterol within the lysosome of NP-C cells has been confirmed through both cytochemical (Blanchette-Mackie *et al.*, 1988) and cell fractionation experiments (Pentchev *et al.*, 1987; Liscum *et al.*, 1989). Liscum *et al.* (1989) showed that incubation of NP-C fibroblasts with exogenous LDL resulted in a threefold increase in lysosomal cholesterol content compared to normal fibroblasts and a delayed transport of the lysosomal cholesterol to the plasma membrane. Transport of biosynthetic cholesterol to the plasma membrane and desorption of cellular cholesterol from the plasma membrane were similar in normal and NP-C fibroblasts, a finding that supports a defect in lysosomally processed cholesterol and not an overall defect in intracellular cholesterol transport.

Interestingly, Blanchette-Mackie *et al.* (1988) showed through cytochemical measures that not only is the lysosome a site for cholesterol accumulation in NP-C, but also the Golgi complex seemed to show unusual cholesterol enrichment. Later, the cholesterol enrichment in NP-C fibroblasts was localized to the *trans*-Golgi (Coxey *et al.*, 1993), suggesting that NP-C affects not only transport of cholesterol out of the lysosome, but also transport of cholesterol through the Golgi network.

Recent work has focused on finding the gene defect responsible for NP-C. A number of compounds have been shown to inhibit exogenously derived cholesterol transport out of the lysosome, thereby mimicking the NP-C phenotype (Liscum and Faust, 1989; Rodriguez-Lafrasse *et al.*, 1990; Roff *et al.*, 1991; Butler *et al.*, 1992; Aikawa *et al.*, 1994). These compounds are now being used to induce lysosomal cholesterol accumulation in a model for NP-C in order to determine the mechanism by which cholesterol is transported out of the lysosome.

Two recent reports indicate that more than one gene may be responsible for NP-C (Steinberg *et al.*, 1994; Dahl *et al.*, 1994). Phenotypically, NP-C is very heterogeneous, leading to the hypothesis that more than one gene is involved in the disease. Somatic cell hybridization experiments of skin fibroblasts from 12 NP-C patients revealed the presence of two complementation groups: a major group, α, and a minor group, β (Steinberg *et al.*, 1994).

4.3. Farber Disease

Farber disease (Farber lipogranulomatosis) results from the absence or severe deficiency of the acid ceramidase. It is a rare disease, with only 43 reported cases, and its mode of inheritance appears to be autosomal recessive (reviewed by Moser, 1995). Seven phenotypic subtypes have been described with approximately one half of the patients exhibiting the "classic" type 1 symptoms (Moser, 1995). These symptoms begin to appear in the first few weeks of life and include swelling of joints, granulomatous thickenings at joints and other body locations, a hoarse cry, and feeding and respiratory difficulties (Farber *et al.*, 1957). Affected tissues and organs become infiltrated with macrophages and other leukocytes. Lipid-laden foam cells occur, and the leukocyte infiltrations often develop into prominent granulomas (Toppet *et al.*, 1978). At the ultrastructural level, lysosomes of cells in affected tissues contain characteristic curvilinear inclusions termed "banana bodies" (Rauch and Auböck, 1983). These exhibit two parallel dark lines of material separated by a clear space of 12–33 nm (Toppet *et al.*, 1978). These bodies are thought to contain ceramide, since similar inclusions can be produced in cultured fibroblasts by adding exogenous ceramide (Rutsaert *et al.*, 1977). At the biochemical level, affected tissues accumulate ceramide to levels that are three to ten times those in normal control tissues (Sugita *et al.*, 1974). The accumulated ceramide sometimes contains a high proportion of 2-hydroxy fatty acyl groups (Sugita *et al.*, 1974), probably reflecting the utilization of this type of ceramide in galactocerebroside synthesis (Ullman and Radin, 1972). Secondary accumulation of gangliosides has also been reported (Prensky *et al.*, 1967). Acid ceramidase activity in homogenates of skin fibroblasts or blood leukocytes from Farber patients is 0–10% of normal (Momoi *et al.*, 1982; Chen *et al.*, 1981; Antonarakis *et al.*, 1984). Activity in obligate heterozygotes is approximately 50% of normal (Momoi *et al.*, 1982; Antonarakis *et al.*, 1984). The alkaline ceramidase is normal in cells from Farber patients (Momoi *et al.*, 1982). Diagnosis of Farber disease is also possible by supplying exogenous cerebroside sulfate to cultured fibroblasts and monitoring the cellular accumulation of ceramide (Kudoh and Wenger, 1982). Prenatal diagnosis is possible by assay of acid ceramidase in amniocytes (Fensom *et al.*, 1979). In an abstract, Tiffany *et al.* (1987) described the production of a monoclonal antibody to acid ceramidase. This antibody was used to isolate a 47-kDa putative enzyme from normal human tissue and what appeared to be a poorly functional enzyme of the same molecular weight from Farber tissue. Thus, at least some forms of this disease may be due to the production of defective ceramidase. At the time of this literature review, there were no studies yet published on the molecular biology of acid ceramidase or Farber disease.

5. ROLE OF LYSOSOMES IN TURNOVER OF ENDOGENOUS LIPIDS

This section addresses the capacity of cells to degrade their own endogenous lipid within lysosomes. There are several distinct sources of lipid that may be subject to this type of degradation: (1) the lysosomal membrane, (2) organelles such as mitochondria and endoplasmic reticulum delivered to lysosomes by autophagy, (3) endocytic vesicle membrane lipids that incorporate into the lysosomal membrane in the course of endocytosis, and (4) cytoplasmic droplets of triacylglycerol and cholesteryl ester.

Regarding turnover of lysosomal membrane lipids, it is generally believed that the inner surface of the lysosomal limiting membrane must be resistant to attack by acid hydrolases. Ultrastructurally, this membrane encloses a clear "halo" of carbohydrate-rich material, which is thought to limit interaction between the membrane and the soluble hydrolases of the lysosomal lumen (Holtzman, 1989). However, it seems unlikely that such protective mechanisms are 100% effective. In addition, susceptibility to degradation may be increased by the inward blebbing of the membrane to form the multivesicular structures that are observed frequently in populations of lysosomes and late endosomes (Holtzman, 1989). For glycosphingolipids in the lysosomal membrane, saposins are believed to extract the lipid

molecules from the limiting lipid bilayer and present them to soluble glycosidases in the lysosomal lumen, thereby initiating the breakdown of this type of
endogenous lipid (Sandhoff and Klein, 1994; also see Chapter 7). Experimental
demonstrations that other lysosomal membrane lipids undergo degradation are
very limited. One type of demonstration consisted simply of incubating purified
liver lysosomes *in vitro* and then analyzing them for changes in lipid composition.
It was found that after 30 min, approximately 25% of the major phospholipids
were degraded (Weglicki *et al.*, 1975; Beckman *et al.*, 1981). Another strategy
has been to isolate lysosomes containing biosynthetically labeled lipids and then
reinject them into rats and monitor their phagocytosis and degradation by liver
Kupffer cells. Using this approach Hennell *et al.* (1983) demonstrated that
glycerol-labeled lysosomal lipids were degraded with a half-life of 2–3.5 hr. These
studies show that lysosomal lipids possess no inherent resistance to degradation.
However, it is not clear if the data can be used to establish the kinetics or selectivity of turnover of the lipids of the lysosomal membrane in intact cells. The
degradation of lysosomal membrane lipids presumably is balanced by the constant
acquisition of new membrane lipids through endocytotic membrane flow and the
generation of new primary lysosomes in the Golgi apparatus.

Lysosomal degradation of organelles following autophagocytosis is well established (Holtzman, 1989; also see Chapter 4). Most investigations of this process
have focused on ultrastructural changes or on the biochemistry of protein degradation. There is little direct biochemical information on lipid degradation.
Richards *et al.* (1979) investigated the capacity of lysosomal enzymes to degrade
endogenous membranes by isolating microsomes from livers prelabeled with
[^{32}P]phosphate or [^{3}H]inositol and incubating these with an aqueous extract of
lysosomal enzymes. Under the conditions of these studies, 50–100% of individual
phospholipid types were converted to their lyso forms within 2 hr, demonstrating
that endogenous membrane lipids can undergo efficient degradation by lysosomal
enzymes. An important model for studying lysosomal degradation of organelles
has consisted of isolating organelles in which endogenous components were prelabeled with isotopic precursors and then injecting the organelles into rats, followed
by monitoring the phagocytosis and degradation of the labeled components by
liver Kupffer cells (Marzella *et al.*, 1981). Ultrastructurally, the phagocytosed organelles appear to be degraded rapidly, often leaving a residue after 24 hr of what
appears to be lipid droplets, indicating that lipid degradation may be less efficient
than that of other organellar components (Marzella *et al.*, 1981). The biochemical
data on degradation of phagocytosed microsomal lipids suggest half-lives of a few
hours for various classes of lipids after uptake by Kupffer cells (Glaumann and
Trump, 1975; Marzella *et al.*, 1981). This appears to be similar to (Glaumann and
Trump, 1975) or slower than (Marzella *et al.*, 1981) the degradation of microsomal protein, depending on which study is consulted.

The lysosomal turnover of membrane sphingomyelin has been examined
in a few studies. In mammalian cells most of this sphingomyelin probably is

delivered to lysosomes from the plasma membrane via the endocytic pathway. Spence *et al.* (1983) addressed this issue in experiments with normal and type A Niemann-Pick (acid sphingomyelinase-deficient) fibroblasts, using exogenous [^{32}P]phosphate and [^{3}H]choline to label endogenous sphingomyelin. After pulsing cells with precursor, turnover of endogenous sphingomyelin in both types of cells occurred at the same rate, indicating that little of the normal turnover of this lipid was due to lysosomal catabolism. Koval and Pagano (1990) used exchange procedures to introduce fluorescent nitrobenzoxadiazol (NBD)-labeled sphingomyelin into the plasma membrane of normal and Niemann-Pick type A fibroblasts. The fluorescently tagged sphingomyelin analog was shown to undergo continuous endocytosis and recycling to the plasma membrane, with approximately 5% diversion of the tracer to lysosomes with each cycle. In normal cells this diverted material was degraded, whereas in Niemann-Pick type A cells it was not degraded and accumulated in lysosomes. The overall rate of lysosomal hydrolysis of NBD-sphingomyelin in normal fibroblasts was about 8% per hr. In similar studies using pyrene-tagged fluorescent sphingomyelin analogs, Levade *et al.* (1991a,b) found that when uptake of the sphingolipid depended on endocytosis via the cell-surface receptor for apolipoprotein B, degradation was entirely lysosomal, whereas when the lipid was exchanged into the plasma membrane or endocytosed by nonreceptor-dependent mechanisms, a large, if not predominant, component of the degradation became nonlysosomal. These experiments thus indicate that lysosomes participate in the turnover of plasma-membrane sphingomyelin due to the inevitable delivery of some of this lipid to lysosomes during endocytosis, but that the bulk of the turnover is due to nonlysosomal mechanisms, most likely the neutral sphingomyelinase located in the plasma membrane.

Regarding the turnover of cytoplasmic triacylglycerol and cholesteryl ester, the limited data available suggest little if any lysosomal involvement. For instance in macrophage foam cells containing large cytoplasmic stores of cholesteryl ester, chloroquine did not block the hydrolysis of this material (Brown *et al.*, 1979). Likewise, Hilaire *et al.* (1993) found that the catabolism of endogenous triacylglycerol and cholesteryl ester in Wolman disease fibroblasts (which lack the lysosomal neutral lipid lipase) was similar to that in normal fibroblasts.

6.   LIPOPROTEIN DEGRADATION AND ATHEROSCLEROSIS

Atherosclerosis is a pathological process, characterized by the development of lesions (i.e., plaques) in the vascular system, that eventually results in the occlusion of the lumen of the vessel and the formation of fibrin clots. Plaque development, particularly in the early stages, involves the infiltration of blood monocytes into the artery wall, the proliferation of arterial smooth muscle cells, the formation of extracellular matrix materials, and the deposition of lipid. One of the earliest events is the focal appearance of cells within the vessel wall that exhibit large amounts of intracellular lipid, stored as lipid droplets or inclusions (Type I lesion; Stary *et al.,* 1994). The appearance of the lipid-laden cells led to the term "foam cells," and the development of the foam cell is thought to be a hallmark of atherosclerosis. The focal initial lesion progresses into a Type II lesion, or fatty streak (Stary *et al.* 1994). These lesions are better defined and contain a variety of cell types; however, lipid-laden foam cells are present in large numbers (Stary *et al.,* 1994). Further progression results in the formation of a core consisting of extracellular lipid droplets, some cholesterol crystals, and other cellular debris. This stage has been termed the intermediate or Type III lesion (Stary *et al.,* 1994). It is at the earliest stages of the development of the atherosclerotic plaque, where the metabolism of lipid and lipoproteins in macrophages and smooth muscle cells is prominent, that the lysosome is thought to play a critical role (Fig. 6).

Through the use of monoclonal antibodies specific to either smooth muscle cells or macrophages, it has been established that both cell types can accumulate sufficient lipid to be classified as foam cells (Tsukada *et al.,* 1986; Rosenfeld and Ross, 1990). However, the sequence of appearance of smooth muscle-derived

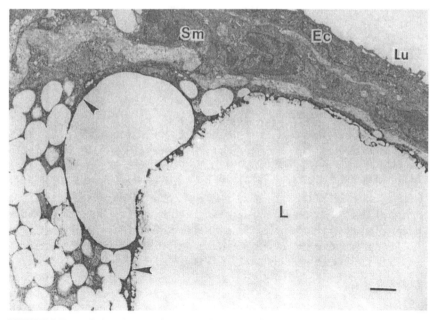

FIGURE 6. Transmission electron micrograph of a portion of an atherosclerotic lesion to demonstrate acid phosphatase as a marker for lysosomes. Dark acid phosphatase reaction product (arrowheads) lines the periphery of a multichambered lysosome in a large macrophage foam cell. The lysosome is filled with lipid (L). Smaller, nonlysosomal lipid deposits (presumably cholesteryl ester) also fill the cytoplasm. The foam cell is separated from the arterial lumen (Lu) by endothelium (Ec) and a layer of smooth muscle cells (Sm). Magnification = × 6950; bar = 1 μm. (Provided by Dr. W. Gray Jerome, Bowman Gray School of Medicine, Winston-Salem, North Carolina.)

and macrophage-derived foam cells differs, as does the intracellular location of the bulk of the stored lipid. In both cell types the predominant lipid is cholesteryl ester, although the level of unesterified cholesterol also is significantly increased (Shio *et al.*, 1978, 1979; Katz *et al.*, 1976). Cholesteryl esters are present in both cytoplasmic lipid droplets and in lysosomes (Shio *et al.*, 1978, 1979; Katz *et al.*, 1976). In the latter compartment, free cholesterol also accumulates (Shio *et al.*, 1978, 1979) and sometimes can be observed as crystalline cholesterol (Shio *et al.*, 1979; Lupu *et al.*, 1987).

It is now generally believed that the first cells that constitute the initial focal lesion in the vessel wall are macrophage-derived foam cells (Stary *et al.*, 1994), with smooth muscle-derived foam cells becoming a more prominent feature as the lesion progresses (Rosenfeld and Ross, 1990; Tsukada *et al.*, 1986; Stary *et al.*, 1994). The temporal sequence of the location of intracellular cholesteryl ester shifts during the development of the lesion. In the early lesion the lipid appears primarily in non-membrane-bound cytoplasmic inclusions (Jerome and Lewis, 1985, 1987; Lewis *et al.*, 1988); however, in more advanced stages of atherosclerosis the deposition of lipid within pleomorphic lysosomal-like organelles becomes a prominent feature of foam cells (Jerome and Lewis, 1985, 1987; Lewis *et al.*, 1988).

An understanding of the metabolic events that result in the production of foam cells and the primary role played by lysosomes in the intracellular metabolism of specific lipoproteins developed from the pioneering studies of Brown and Goldstein (1983, 1986). These investigations demonstrated the presence of receptors on the cell surface that bound and internalized LDL through an interaction of the receptor with the apolipoprotein B component of the LDL (Brown and Goldstein, 1986). Subsequent studies revealed that another apolipoprotein, apo E, also bound to the same receptor (apo B/E receptor) (Wilson *et al.*, 1991). By following the metabolic fate of radiolabeled apo B on LDL, Brown and Goldstein were able to establish the now well-documented pathway in which receptor-bound ligands are internalized via an endocytotic event, followed by the delivery of the ligand, in this case the LDL, to lysosomes following the dissociation of the lipoprotein from the receptor in endocytotic vesicles (Brown and Goldstein, 1983, 1986) (see Chapter 3). The lipid components of the lipoprotein are processed within the lysosomes, and since the predominant lipid in LDL is esterified cholesterol, the hydrolysis of this component has been investigated in detail. Under normal conditions the unesterified cholesterol, either that delivered directly as free cholesterol on the lipoprotein or that generated by hydrolysis of cholesteryl ester by the aCEH (see above), is transported out of the lysosome by a pathway that remains to be elucidated.

The free cholesterol produced by the lipoprotein internalization/lysosome degradation pathway becomes available within the cells for new membrane synthesis, efflux to appropriate extracellular acceptors, or accumulation as cholesteryl ester. This latter process is accomplished through the action of ACAT with the resulting cholesteryl esters stored in cells as cytoplasmic lipid droplets or inclusions (Brown and Goldstein, 1983, 1986). These cholesteryl ester-rich inclusions are similar to the cytoplasmic lipid inclusions that are characteristic of both macrophage- and smooth muscle-derived foam cells of the atherosclerotic plaque (Brown *et al.*, 1980b; Rothblat *et al.*, 1977; Snow *et al.*, 1988). The esterified cholesterol stored in cells in this compartment has been shown to be in a cholesteryl ester cycle, continually being hydrolyzed to free cholesterol by a neutral cholesteryl ester hydrolase (nCEH), and resynthesized by ACAT (Brown *et al.*, 1980b; Glick *et al.*, 1987). The cholesteryl ester cycle can be interrupted by the diversion of the free cholesterol out of the cell by the presence of efficient extracellular cholesterol acceptors (Brown *et al.*, 1980b; Glick *et al.*, 1987).

A key factor in understanding the role of the apo B/E receptor and lysosome lipoprotein processing in the development of the foam cell was the very early observation that the presence of the apo B/E receptor on cells is a highly regulated process, closely linked to the level of cellular cholesterol (Brown and Goldstein, 1983, 1986). Thus, this regulated lipoprotein internalization will limit the amount of excess cholesterol that will accumulate in cells (Brown and Goldstein, 1983, 1986) and should not participate in the processes leading to cholesterol deposition in either lysosome or cytoplasmic compartments within foam cells. The mechanism for macrophage-derived foam cell formation became evident with the demonstration that a number of receptors are present on macrophages that recognize and internalize a wide variety of ligands, including chemically modified or abnormal lipoproteins (Brown and Goldstein, 1983; Goldstein *et al.*, 1979, 1983). These receptors, in contrast to the apo B/E receptor, are not effectively downregulated in response to elevated intracellular cholesterol levels and thus continue to deliver lipid into the cell, even as excess cholesterol is deposited (Brown and Goldstein, 1983; Goldstein *et al.*, 1979, 1983). There is an ever-increasing array of such modified lipoproteins (Aviram, 1993), including acetylated LDL (Goldstein *et al.*, 1979), aggregated LDL (Khoo *et al.*, 1988), oxidized LDL (Sparrow *et al.*, 1989), and LDL associated with mast cell granules (Kovanen, 1991) or extracellular matrix materials (Falcone *et al.*, 1984; Vijayagopal *et al.*, 1985). In addition, other types of lipoproteins serve as ligands for unregulated receptors such as the cholesterol-rich very low density lipoprotein (βVLDL) which is present in cholesterol-fed animals (Van Lenten *et al.*, 1983; Pitas *et al.*, 1983; Wang-Iverson *et al.*, 1985; Gianturco *et al.*, 1988). Although all of the lipoprotein particles that are internalized by receptor-mediated processes eventually appear to undergo processing in lysosomes, the delivery to different intracellular vesicles may differ, depending on the lipoprotein. For example, Tabas *et al.* (1990) have demonstrated that the time course of uptake and the subsequent intracellular distribution of LDL or βVLDL differs in macrophages. The degradation of βVLDL protein and the hydrolysis of βVLDL cholesteryl ester were markedly slower than that of LDL (Tabas *et al.*,1990). In addition, the distribution of fluorescently labeled lipoproteins was different between the two lipoproteins, although both eventually underwent degradation within lysosomes (Tabas *et al.*, 1990). An explanation for this phenomenon may be related to the observation that acetyl LDL and micro-crystalline cholesterol are initially sequestered in a surface-connected compart-ment, prior to uptake and delivery to lysosomes (Kruth *et al.*, 1995). Exposure of macrophages to βVLDL also appears to induce morphological changes associated with membrane invaginations and vesiculation (Jones *et al.*, 1991). It is possible that this network of surface-connected channels forms in macrophages in response to particulate material such as aggregated lipoproteins or crystalline cholesterol. Movement of lipoproteins from this surface-connected compartment to lysosomes appears to be much slower than the transport of lipoproteins via a receptor-mediated endocytotic event (Kruth *et al.*, 1995).

Even with extensive knowledge about receptor-mediated uptake of lipopro-teins and their subsequent metabolism within lysosomes, many aspects of foam cell development remain obscure. The role of the lysosome in the development of the lipid-laden foam cell and atherosclerotic plaque has been approached from two general theoretical points of view: as the site of the primary defect leading to foam cell formation and as an intracellular compartment participating in the normal pro-cessing of lipoproteins and their associated lipids.

In 1974, de Duve and colleagues (de Duve, 1974; Peters and de Duve, 1974) proposed that the deposition of cholesterol in foam cells was a reflection of a rel-ative depletion of aCEH in the target cell within the arterial wall. In this model, the rate of hydrolysis of cholesteryl ester delivered to lysosomes in smooth mus-cle cells or macrophages would be reduced because of a partial deficiency in the activity of aCEH, thus resulting in the accumulation of the lipoprotein-derived lipid within the lysosomal compartment. Variations among individuals in the ex-tent of the aCEH deficiency could, in part, explain the differences in susceptibil-ity to atherosclerosis. There was considerable justification for this hypothesis, particularly based on well-established studies documenting that a complete defi-ciency in aCEH in Wolman disease and cholesteryl ester storage disease (see above) led to massive deposition of esterified cholesterol in tissues. However, fur-ther studies indicated that the basic tenet of this hypothesis appeared unlikely for two reasons. First, an analysis of the lipid composition of lysosomes isolated from foam cells demonstrated that the primary form of cholesterol within the lysosomes was unesterified (Shio *et al.*, 1979) rather than cholesteryl ester, as would be pre-dicted by the enzyme-deficit hypothesis. Second, a careful quantitation of enzyme levels in lipid-laden atheromatous cells revealed a positive correlation between the activity of aCEH and the extent of lipid accumulation (Haley *et al.*, 1980). Thus, although lysosomes play a key role in the normal processing of lipoproteins and lipids and, as discussed below, are important components in the processes leading to the development of foam cells, there is currently no convincing evidence that atherosclerosis can be viewed as a classical lysosomal storage disease, as first sug-gested by de Duve (1974). There have been somewhat contradictory reports that acid lipase activity is lower in mononuclear cells obtained from individuals with hypercholesterolemia (Hagemenas *et al.,* 1984). However, other investigators have not observed a difference between control and hypercholesterolemic patients, but have obtained data indicating reduced acid lipase activity in cells from patients with premature cardiovascular disease (Coates *et al.,* 1986).

Once it was established that there were lipoprotein receptors on macrophages that did not downregulate in response to increasing intracellular cholesterol levels, it was easily demonstrated that exposure of macrophages to modified lipoproteins that were recognized by these "scavenger receptors" resulted in extensive accumulation of cholesteryl ester in cytoplasmic lipid droplets and the acquisition of foam cell morphology (Brown and Goldstein, 1983). However, although the unregulated scavenger receptor model can explain the formation of macrophage-derived foam cells containing cytoplasmic cholesteryl ester inclusions, two major questions persist. First, if smooth muscle cells have only the well-regulated apo B/E receptor, how do they acquire sufficient cholesterol to assume the foam cell phenotype? Second, by what mechanism do macrophage- and smooth muscle-derived foam cells accumulate free cholesterol within lysosomes?

Unregulated uptake of cholesterol by smooth muscle cells would be required for these cells to assume the foam cell phenotype. It has been proposed that this influx of cholesterol could proceed by either phagocytosis of extracellular debris, or through the action of lipoprotein receptors that do not downregulate in response to excess cell cholesterol. Although it was generally believed that smooth muscle cells did not express scavenger receptors, studies by Pitas (1990) demonstrated the presence of such receptors on both rabbit fibroblasts and smooth muscle cells after treatment with phorbol esters; however, exposure of these cells to modified lipoproteins failed to provoke extensive cellular cholesterol deposition and foam cell morphology (Pitas *et al.,* 1992). The resistance of smooth muscle cells to free and esterified cholesterol accumulation may be a reflection of relatively low numbers of scavenger receptors (Pitas *et al.,* 1992), or alternatively, to sluggish activity of ACAT, which would result in an inability to synthesize large amounts of cholesteryl ester.

Another hypothesis to explain the formation of smooth muscle-derived foam cells proposes that the phagocytic uptake of extracellular debris could result in the loading of lysosomes with cholesterol (Wolfbauer *et al.,* 1986). Experimental efforts have been focused on determining the effect of exposing cells to small lipid droplets of the type formed by ACAT in macrophage-derived foam cells (Wolfbauer *et al.,* 1986). Such droplets, which are quite stable (Adelman *et al.,* 1984), could be liberated from dying macrophage foam cells and recycled, through a phagocytic process, into smooth muscle cells. It was established that exposure of a variety of cell types in culture to native cytoplasmic inclusions or artificial lipid droplets resulted in the rapid incorporation of the droplets, producing extensive deposition of cholesteryl ester in the lysosomes (Fig. 7). The lipid incorporated by this phagocytic process rapidly undergoes hydrolysis by aCEH (Minor *et al.,* 1989, 1991). Interestingly, if the droplets are composed of equal amounts of triglycerides and cholesteryl ester, triglyceride hydrolysis is much faster than that of esterified

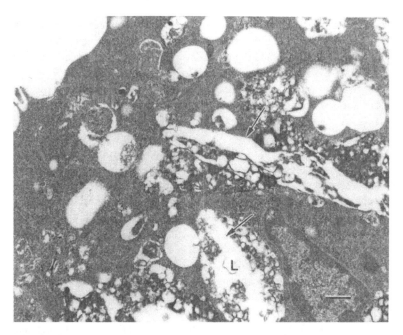

FIGURE 7. Transmission electron micrograph of cultured J774 macrophage incubated for 24 hr in the presence of cholesteryl ester-phospholipid droplets followed by an additional 48 hr incubation without lipid droplets. Lipid accumulation is seen within large lysosomes (L), two of which contain cholesterol crystals (arrows). Magnification = × 15,000; bar = 0.5 μm. (Provided by Dr. W. Gray Jerome, Bowman Gray School of Medicine, Winston-Salem, North Carolina.)

cholesterol (Minor *et al.,* 1989, 1991). As might be anticipated, exposure of macrophages to extracellular cholesteryl ester droplets results in even more rapid uptake and lysosomal hydrolysis than is obtained with smooth muscle cells (Mahlberg *et al.,* 1990). Although some of the free cholesterol liberated by lysosomal hydrolysis of the incorporated cholesteryl ester droplets exited the lysosomal compartment and became available for efflux or re-esterification by ACAT, 20–40% of the free cholesterol appeared to be retained within the lysosomes, leading to a time-dependent accumulation of free cholesterol in lysosomes (Tangirala *et al.,* 1993). This accumulation, when allowed to continue for a number of days, resulted in the formation of intralysosomal cholesterol crystals (Tangirala *et al.,* 1994) of the type characteristic of the atherosclerotic plaque (Small, 1970; Dejager *et al.,* 1993). Not all treated cells exhibited the formation of cholesterol crystals, and the number of crystals within a cell varied from one to four (Tangirala *et al.,* 1994). After the appearance of prominent free cholesterol crystals, the cells appeared to remain viable and retain the ability to translocate large, single crystals (Tangirala *et al.,* 1994). Cells with multiple crystals often appeared damaged, and the presence of extracellular cholesterol crystals was generally associated with cell death (Tangirala *et al.*, 1994). Such a sequence would be consistent with data indicating that the death of macrophage foam cells contributes to the lipid core of the plaque (Ball *et al.*, 1995).

The reason for the accumulation of lysosomal free cholesterol following the phagocytic incorporation of cholesteryl ester droplets may be that the delivery of a large bolus of cholesteryl ester directly into the lysosomal compartment results in the generation of free cholesterol in excess of that which can be removed. The more regulated receptor-mediated uptake of lipoprotein-associated esterified cholesterol probably would distribute the cholesteryl ester among a much larger population of lysosomes, and thus not normally result in lysosomal cholesteryl ester overloading. If the rapid delivery of large amounts of cholesterol to a limited population of lysosomes was responsible for the accumulation and subsequent crystallization of free cholesterol, then a number of extracellular particles can be implicated. For example, aggregates of lipoproteins (Suits *et al.*, 1989), lipoprotein complexes with extracellular matrix material (Falcone *et al.*, 1984; Vijayagopal *et al.*, 1985), or mast cell granules coated with LDL (Wang *et al.*, 1995; Ma and Kovanen, 1995) all can produce the type of cholesteryl ester delivery necessary for the accumulation of free cholesterol in lysosomes and the conversion of smooth muscle cells and macrophages to foam cells. Oxidized LDL presents a particularly interesting candidate lipoprotein for foam cell formation since it has been demonstrated that the degradation of oxidized apolipoprotein B is slower than native apo B, and undegraded apo B accumulates in the lysosomes of macrophages (Lougheed *et al.*, 1991; Roma *et al.*, 1992; Mander *et al.*, 1994). In addition, it has recently been demonstrated that even though esterified cholesterol incorporated as part of oxidized LDL undergoes hydrolysis at rates similar to cholesteryl ester delivered as native LDL, the free cholesterol generated from the oxidized LDL accumulates within lysosomes (Maor and Aviram, 1994). The reason for the lack of transport of free cholesterol out of lysosomes was attributed to the presence of oxysterols in the oxidized LDL, although the mechanism for the free cholesterol trapping remains to be resolved (Maor and Aviram, 1994).

It is clear that lysosomes play an important role in the development of the atherosclerotic lesion and that large quantities of lipid can accumulate in both the cytoplasmic and lysosomal compartments of foam cells. A number of different processes may contribute to the lipid deposition, including unregulated receptor and nonreceptor uptake of lipoproteins and extracellular debris and overloading of lysosomal degradation and transport processes. Figure 8 presents a model for the accumulation of free and esterified cholesterol in the lysosomal and cytoplasmic compartments of both macrophage- and smooth muscle-derived foam cells. In this model the phagocytic uptake of cholesteryl ester inclusions plays an important role in lysosomal cholesterol deposition. It can be anticipated that other particles, particularly aggregated lipoproteins of various types, could produce the same responses as has been demonstrated with lipid droplets.

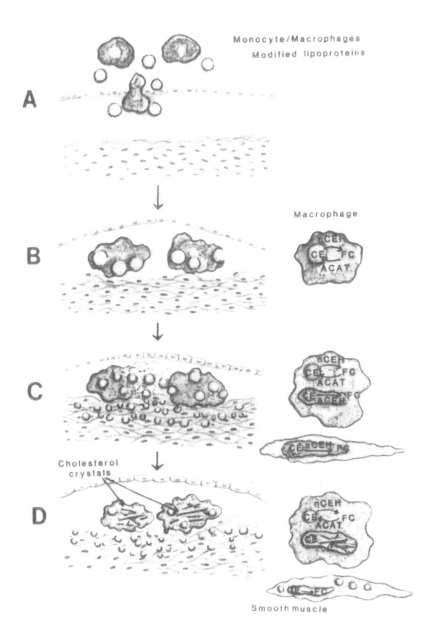

FIGURE 8. Model for the recycling of cholesteryl ester droplets and cholesterol accumulation in foam cells during the progression of atherosclerosis. (A) Infiltration of circulating monocytes into the arterial wall followed by the incorporation of excess cholesterol from modified lipoproteins by the monocyte/macrophages could lead to the formation of earliest macrophage-derived foam cells. (B) In the early lesions, the foam cells are primarily macrophage-derived foam cells with the cholesterol deposited primarily as cytoplasmic inclusions. (C) Progression of the lesion would lead to the lysis of some early foam cells with the liberation of cholesteryl ester (CE) inclusions. The phagocytic incorporation of the liberated inclusions by macrophage leads to lysosomal CE hydrolysis by the action of acid cholesteryl ester hydrolase (aCEH) with the generation of free cholesterol (FC). Rapid CE hydrolysis and FC accumulation in the macrophage foam cells results in FC deposition in the lipid-filled lysosomes. (D) The accumulation of FC in the lipid-filled lysosomes of macrophage foam cells leads to the formation of cholesterol monohydrate crystals. The formation of FC crystals could eventually result in the lysis and death of the cells. From Tangirala *et al.* (1994).

ACKNOWLEDGMENTS. We thank Dr. John R. Wherrett (University of Toronto, Ontario, Canada) for permission to use Fig. 1, and Dr. W. Gray Jerome (Bowman Gray School of Medicine, Winston-Salem, North Carolina) for providing previously unpublished electron micrographs and for critiquing portions of the text. We are grateful to Dr. Richard Anderson (Bowman Gray School of Medicine) for his feedback on some other portions of the text. We also thank Ms. Barbara Engle and Ms. Savoy Moyer for preparation of the manuscript. This research has been supported by NIH Program Project Grant HL22633, a Grant-in-Aid to W.J.J. from the Southeastern Pennsylvania Affiliate of the American Heart Association, NIH Training Grant HL07743, and Intramural Research Grant 95-045-3MCP from the Allegheny-Singer Research Institute.

7. REFERENCES

Adelman, S. J., Glick, J. M., Phillips, M. C., and Rothblat, G. H., 1984, Lipid composition and physical state effects on cellular cholesteryl ester clearance. *J. Biol. Chem.* **259**:13844–13850.

Ahmad, T. Y., Beaudet, A. L., Sparrow, J. T., and Morrisett, J. D., 1986, Human lysosomal sphingomyelinase: Substrate efficacy of apolipoprotein/sphingomyelin complexes, *Biochemistry* **25**:4415–4420.

Aikawa, K., Furuchi, T., Fujimoto, Y., Arai, H., and Inoue, K., 1994, Structure-specific inhibition of lysosomal cholesterol transport in macrophages by various steroids, *Biochim. Biophys. Acta* **1213**:127–134.

Al, B. J., Tiffany, C. W., Gomes de Mesquita, D. S., Moser, H. W., Tager, J. M., and Schram, A. W., 1989, Properties of acid ceramidase from human spleen, *Biochim. Biophys. Acta* **1004**:245–251.

Alpert, A. J., and Beaudet, A. L., 1981, Apolipoprotein C-III-1 activates lysosomal sphingomyelinase in vitro, *J. Clin. Invest.* **68**:1592–1596.

Ameis, D., Brockmann, G., Knoblich, R., Merkel, M., Ostlund Jr., R. E., Yang, J. W., Coates, P. M., Cortner, J. A., Feinman, S. V., and Greten, H., 1995, A 5' splice-region mutation and a dinucleotide deletion in the lysosomal acid lipase gene in two patients with cholesteryl ester storage disease, *J. Lipid Res.* **36**:241–250.

Amidon, B., Schmitt, J. D., Thuren, T., King, L., and Waite, M., 1995, Biosynthetic conversion of phosphatidylglycerol to sn-1:sn-1' bis(monoacylglycerol)phosphate in a macrophage-like cell line, *Biochemistry* **34**:5554–5560.

Anderson, R. A., and Sando, G. N., 1991, Cloning and expression of cDNA encoding human lysosomal acid lipase/cholesteryl ester hydrolase. Similarities to gastric and lingual lipases, *J. Biol. Chem.* **266**:22479–22484.

Anderson, R. A., Byrum, R. S., Coates, P. M., and Sando, G. N., 1994, Mutations at the lysosomal acid cholesteryl ester hydrolase gene locus in Wolman disease, *Proc. Natl. Acad. Sci. USA* **91**:2718–2722.

Anderson, R. G., Brown, M. S., and Goldstein, J. L., 1977, Role of the coated endocytic vesicle in the uptake of receptor-bound low density lipoprotein in human fibroblasts, *Cell* **10**:351–364.

Andrieu, N., Salvayre, R., and Levade, T., 1994, Evidence against involvement of the acid lysosomal sphingomyelinase in the tumour-necrosis-factor- and interleukin-1-induced sphingomyelin cycle and cell proliferation in human fibroblasts, *Biochem. J.* **303**:341–345.

Antonarakis, S. E., Valle, D., Moser, H. W., Moser, A., Qualman, S. J., and Zinkham, W. H., 1984, Phenotypic variability in siblings with Farber disease, *J. Pediatr.* **104**:406–409.

Assmann, G., and Seedorf, U., 1995, Acid lipase deficiency: Wolman disease and cholesteryl ester storage disease, in *The Metabolic and Molecular Bases of Inherited Disease,* 7th Ed. (C. R. Scriver, A. L. Beaudet, W. S. Sly, and D. Valle, eds.), pp. 2563–2587, McGraw-Hill, Inc., New York.

282   William J. Johnson *et al.*

Assmann, G., Fredrickson, D. S., Sloan, H. R., Fales, H. M., and Highet, R. I., 1975, Accumulation of oxygenated steryl esters in Wolman's disease, *Biochemistry* 34:7955–7965.

Aviram, M., 1993, Modified forms of low density lipoprotein and atherosclerosis, *Atherosclerosis* 98:1–9.

Ball, R. Y., Stowers, E. C., Burton, J. H., Cary, N. R. B., Skepper, J. N., and Mitchinson, M. J., 1995, Evidence that the death of macrophage foam cells contributes to the lipid core of atheroma, *Atherosclerosis* 114:45–54.

Bangham, A. D., and Dawson, R. M. C., 1959, The relation between the activity of a lecithinase and the electrophoretic charge of the substrate, *Biochem. J.* 72:486–492.

Bangham, A. D., Standish, M. M., and Watkins, J. C., 1965, Diffusion of univalent ions across lamellae of swollen phospholipids, *J. Mol. Biol.* 13:238–252.

Beckman, J. K., Owens, K., and Weglicki, W. B., 1981, Endogenous lipolytic activities during autolysis of highly enriched hepatic lysosomes, *Lipids* 16:796–799.

Bergstrom, B., and Brockman, H. L., 1984, *Lipases,* Elsevier, New York.

Blanchette-Mackie, E. J., Dwyer, N. K., Amende, L. M., Kruth, H. S., Butler, J. D., Sokol, J., Comly, M. E., Vanier, M. T., August, J. T., Brady, R. O., and Pentchev, P. G., 1988, Type-C Niemann-Pick disease: Low density lipoprotein uptake is associated with premature cholesterol accumulation in the Golgi complex and excessive cholesterol storage in lysosomes, *Proc. Natl. Acad. Sci. USA* 85:8022–8026.

Bleistein, J., Heidrich, H. G., and Debuch, H., 1980, The phospholipids of liver lysosomes from untreated rats, *Hoppe-Seyler's Z. Physiol. Chem.* 361:595–597.

Bradova, V., Smid, F., Ulrich-Bott, B., Roggendorf, W., Paton, B. C., and Harzer, K., 1993, Prosaposin deficiency: Further characterization of the sphingolipid activator protein-deficient sibs, *Hum. Genet.* 92:143–152.

Brown, M. S., and Goldstein, J. L., 1983, Lipoprotein metabolism in the macrophage: Implications for cholesterol deposition in atherosclerosis, *Annu. Rev. Biochem.* 52:223–261.

Brown, M. S., and Goldstein, J. L., 1986, A receptor-mediated pathway for cholesterol homeostasis, *Science* 232:34–47.

Brown, M. S., Goldstein, J. L., Krieger, M., Ho, Y. K., and Anderson, R. G. W., 1979, Reversible accumulation of cholesteryl esters in macrophages incubated with acetylated lipoproteins, *J. Cell Biol.* 82:597–613.

Brown, M. S., Basu, S. K., Falck, J. R., Ho, Y. K., and Goldstein, J. L., 1980a, The scavenger cell pathway for lipoprotein degradation: Specificity of the binding site that mediates the uptake of negatively-charges LDL by macrophages, *J. Supramol. Struct.* 13:67–81.

Brown, M. S., Ho, Y. K., and Goldstein, J. L., 1980b, The cholesteryl ester cycle in macrophage foam cells. Continual hydrolysis and re-esterification of cytoplasmic cholesteryl esters, *J. Biol. Chem.* 255:9344–9352.

Burrier, R. E., and Brecher, P., 1983, Hydrolysis of triolein in phospholipid vesicles and microemulsions by a purified rat liver acid lipase, *J. Biol. Chem.* 258:12043–12050.

Burton, B. K., Emery, D., and Mueller, H. W., 1980, Lysosomal acid lipase in cultivated fibroblasts: Characterization of enzyme activity in normal and enzymatically efficient cell lines, *Clin. Chim. Acta* 101:25–32.

Butler, J. O., Blanchette-Mackie, J., Goldin, E., O'Neill, R. R., Carstea, G., Roff, C. F., Patterson, M. C., Patel, S., Comly, M. E., Cooney, A., Vanier, M. T., Brady, R. O., and Pentchev, P. G., 1992, Progesterone blocks cholesterol translocation from lysosomes, *J. Biol. Chem.* 267:23797–23805.

Callahan, J. W., Jones, C. S., Davidson, D. J., and Shankaran, P., 1983, The active site of lysosomal sphingomyelinase: Evidence for the involvement of hydrophobic and ionic groups, *J. Neurosci. Res.* 10:151–163.

Chen, W. W., and Decker, G. L., 1982, Abnormalities of lysosomes in human diploid fibroblasts from patients with Farber's disease, *Biochim. Biophys. Acta* 718:185–192.

Chen, W. W., and Moser, H. W., 1979, Purification of acid ceramidase from human placenta, *FEBS Proc.* 38:405.

Chen, W. W., Moser, A. B., and Moser, H. W., 1981, Role of lysosomal acid ceramidase in the metabolism of ceramide in human skin fibroblasts, *Arch. Biochem. Biophys.* **208**:444–455.

Coates, P. M., and Cortner, J. A., Hoffman, G. M., and Brown, S. A., 1979, Acid lipase activity of human lymphocytes, *Biochim. Biophys. Acta* **572**:225–230.

Coates, P. M., Langer, T., and Cortner, J. A., 1986, Genetic variation of human mononuclear leukocyte lysosomal acid lipase activity, *Atherosclerosis* **62**:11–20.

Coxey, R. A., Pentchev, P. G., Campbell, G., and Blanchette-Mackie, E. J., 1993, Differential accumulation of cholesterol in Golgi compartments of normal and Niemann-Pick type C fibroblasts incubated with LDL: A cytochemical freeze-fracture study, *J. Lipid Res.* **34**:1165–1176.

Dahl, N. K., Daunais, M. A., and Liscum, L., 1994, A second complementation class of cholesterol transport mutants with a variant Niemann-Pick type C phenotype, *J. Lipid Res.* **35**:1839–1849.

de Duve, C., 1974, The participation of lysosomes in the transformation of smooth muscle cells to foamy cells in the aorta of cholesterol-fed rabbits, *Arterioscler. Thromb.* **12**:1235–1244.

Deems, R. A., Eaton, B. R., and Dennis, E. A., 1975, Kinetic analysis of phospholipase A₂ activity toward mixed micelles and its implications for the study of lipolytic enzymes, *J. Biol. Chem.* **250**:9013–9020.

Dejager, S., Mietus-Snyder, M., and Pitas, R. E., 1993, Oxidized low density lipoproteins bind to the scavenger receptor expressed by rabbit smooth muscle cells and macrophages, *Arterioscler. Thromb.* **13**:371–378.

DeLamatre, J. G., Carter, R. M., and Hornick, C., 1993, Evidence that a neutral cholesteryl ester hydrolase is responsible for the extralysosomal hydrolysis of high-density lipoprotein cholesteryl ester in rat hepatoma cells (Fu5AH), *J. Cell. Physiol.* **157**:164–168.

Dennis, E. A., 1983, Phospholipases, in *The Enzymes,* Vol. XVI (P. Boyer, ed.), pp. 307–353. New York, Academic Press.

Deykin, D., and Goodman, D. S., 1962, The hydrolysis of long-chain fatty acid esters of cholesterol with rat liver enzymes, *J. Biol. Chem.* **237**:3649–3656.

Dinur, T., Schuchman, E. H., Fibach, E., Dagan, A., Suchy, M., Desnick, R. J., and Gatt, S., 1992, Toward gene therapy for Niemann-Pick disease (NPD): Separation of retrovirally corrected and noncorrected NPD fibroblasts using a novel fluorescent sphingomyelin, *Hum. Gene Ther.* **3**:633–639.

Eisen, D., Bartolf, M., and Franson, R. C., 1984, Inhibition of lysosomal phospholipase C and A in rabbit alveolar macrophages, polymorphonuclear leukocytes and rat liver by sodium bisulfite, *Biochim. Biophys. Acta* **793**:10–17.

Erickson, S. K., Lear, S. R., and McCreery, M. J., 1994, Functional sizes of hepatic enzymes of cholesteryl ester metabolism determined by radiation inactivation, *J. Lipid Res.* **35**:763–769.

Falcone, D. J., Mated, N., Shio, H., Minick, C. R., and Fowler, D., 1984, Lipoprotein-heparin-fibronectin-denatured collagen complexes enhance cholesteryl ester accumulation in macrophages, *J. Cell Biol.* **99**:1266–1274.

Farber, S., Cohen, J., and Uzman, L. L., 1957, Lipogranulomatosis: A new lipo-glyco-protein "storage" disease, *J. Mt. Sinai Hospital* **24**:816–837.

Fenson, A. H., Benson, P. F., Neville, B. R., Moser, H. W., Moser, A. E., and Dulaney, J. T., 1979, Prenatal diagnosis of Farber's disease, *Lancet* **2**:990–992.

Fowler, S. D., and Brown, W. J., 1984, Lysosomal acid lipase, in *Lipases,* (B. Bergstrom and H. L. Brockman, eds.), pp. 329–364. Elsevier Science Publishing Company, Inc., New York.

Fowler, S., and de Duve, C., 1969, Digestive activity of lysosomes. III. The digestion of lipids by extracts of rat liver lysosomes, *J. Biol. Chem.* **144**:471–481.

Franson, R. C., and Waite, M., 1973, Lysosomal phospholipases A₁ and A₂ of normal and Bacillus Calmettle Guerin-induced alveolar macrophages, *J. Cell Biol.* **56**:621–627.

Franson, R., Waite, M., and LaVia, M., 1971, Identification of phospholipase A₁ and phospholipase A₂ in the soluble fraction of rat liver lysosomes, *Biochemistry* **10**:1942–1946.

Franson, R., Waite, M., and Weglicki, W. B., 1972, Phospholipase A activity of lysosomes of rat myocardial tissue, *Biochemistry* **11**:472–476.

Freeman, S. J., Davidson, D. J., and Callahan, J. W., 1984, Solid-phase assay for the detection of low-abundance enzymes, and antibodies to enzymes in immune reactions, using acid sphingomyelinase as a model, *Anal. Biochem.* **141**:248–252.

Gatt, S., 1963, Enzymic hydrolysis and synthesis of ceramides, *J. Biol. Chem.* **238**:PC3131–3133.

Gatt, S., 1966, Enzymatic hydrolysis of sphingolipids. I. Hydrolysis and synthesis of ceramides by an enzyme from rat brain, *J. Biol. Chem.* **241**:3724–3730.

Gatt, S., Dinur, T., and Barenholz, Y., 1978, A spectrophotometric method for determination of sphingomyelinase, *Biochim. Biophys. Acta* **530**:503–507.

Gatt, S., Dinur, T., and Barenholz, Y., 1980, A fluorometric determination of sphingomyelinase by use of fluorescent derivatives of sphingomyelin, and its application to diagnosis of Niemann-Pick disease, *Clin. Chem.* **26**:93–96.

Ghosh, P., and Chatterjee, S., 1987, Effects of gentamicin on sphingomyelinase activity in cultured human renal proximal tubular cells, *J. Biol. Chem.* **262**:12550–12556.

Gianturco, S. H., Lin, A. H.-Y., Hwang, S.-L. C., Young, J., Brown, S. A., Via, D. P., and Bradley, W. A., 1988, Distinct murine macrophage receptor pathway for human triglyceride-rich lipoproteins, *J. Clin. Invest.* **82**:1633–1643.

Glaumann, H., and Trump, B. F., 1975, Lysosomal degradation of cell organelles. III. Uptake and disappearance in Kupffer cells in intravenously injected isotope-labeled mitochondria and microsomes in vivo and in vitro, *Lab. Invest.* **33**:262–272.

Glick, J. M., Adelman, S. A., and Rothblat, G. H., 1987, Cholesteryl ester cycle in cultured hepatoma cells, *Atherosclerosis* **64**:223–230.

Glick, J. M., 1990, Intracellular cholesteryl ester hydrolysis and clearance, in *Advances in Cholesterol Research* (M. Esfahani and J. B. Swaney, eds.), pp. 167–197, Telford Press, Caldwell, New Jersey.

Goldberg, D. I., and Khoo, J. C., 1988, Secretion of the lysosomal acid triacylglycerol hydrolase precursor by J774 macrophages, *Biochim. Biophys. Acta* **960**:200–209.

Goldstein, J. L., Dana, S. E., Faust, J. R., Beaudet, A. L., and Brown, M. S., 1975, Role of lysosomal acid lipase in the metabolism of plasma low density lipoprotein, *J. Biol. Chem.* **250**:8487–8495.

Goldstein, J. L., Ho, Y. K., Basu, S. K., and Brown, M. S., 1979, Binding site on macrophages that mediates uptake and degradation of acetylated low density lipoprotein producing massive cholesterol deposition, *Proc. Natl. Acad. Sci. USA* **76**:333–337.

Goldstein, J. L., Kita, T., and Brown, M. S., 1983, Defective lipoprotein receptors and atherosclerosis, *N. Engl. J. Med.* **309**:288–296.

Graber, B., Salvayre, R., and Lavade, T., 1994, Accurate differentiation of neuronopathic and nonneuronopathic forms of Niemann-Pick disease by evaluation of the effective residual lysosomal sphingomyelinase activity in intact cells, *J. Neurochem.* **63**:1060–1068.

Hagemenas, F. C., Manaugh, L. C., Illingworth, D. R., Sundberg, E. E., and Yatsu, F. M., 1984, Cholesteryl ester hydrolase activity in mononuclear cell from patients with type II hypercholesterolemia, *Atherosclerosis* **50**:335–344.

Haley, N. J., Fowler, S., and de Duve, C., 1980, Lysosomal acid cholesteryl esterase activity in normal and lipid-laden aortic cells, *J. Lipid Res.* **21**:961–969.

Hannun, Y. A., 1994, The sphingomyelin cycle and the second messenger function of ceramide, *J. Biol. Chem.* **269**:3125–3128.

Harrison, E. H., 1993, Enzymes catalyzing the hydrolysis of retinyl esters, *Biochim. Biophys. Acta* **1170**:99–108.

Harrison, E. H., Bernard, D. W., Scholm, P., Quinn, D. M., Rothblat, G. H., and Glick, J. M., 1990, Inhibitors of neutral cholesteryl ester hydrolase, *J. Lipid Res.* **31**:2187–2193.

Harrison, E. H., Gad, M. Z., and Ross, A. C., 1995, Hepatic uptake and metabolism of chylomicron retinyl esters: Probable role of plasma membrane/endosomal retinyl ester hydrolase *J. Lipid Res.* **36**:1498–1506.

Harzer, K., Paton, B. C., Poulos, A., Kustermann-Kuhn, B., Roggendorf, W., Grisar, T., and Popp, M., 1989, Sphingolipid activator protein deficiency in a 16-week-old atypical Gaucher disease patient and his fetal sibling: Biochemical signs of combined sphingolipidoses, *Eur. J. Pediatr.* **149**:31–39.

Hassler, D. F., and Bell, R. M., 1993, Ceramidases: Enzymology and metabolic roles, in *Advances in Lipid Research,* Vol. 26 (R. M., Bell, J. A. H. Merrill, and Y. A. Hannun, eds.), pp. 49–57, Academic Press, San Diego, California.

Hayashi, H., Niinobe, S., Matsumoto, Y., and Suga, T., 1981, Effects of Triton WR-1339 on lipoprotein lipolytic activity and lipid content of rat liver lysosomes, *J. Biochem.* **89**:573–579.

Heller, M., and Shapiro, B., 1966, Enzymic hydrolysis of sphingomyelin by rat liver, *Biochem. J.* **98**:763–769.

Henell, F., Ericsson, J. L. E., and Glaumann, H., 1983, Degradation of phagocytosed lysosomes by Kupffer cell lysosomes, *Lab. Invest.* **48**:556–564.

Hilaire, N., Negre-Salvayre, A., and Salvayre, R., 1993, Cytoplasmic triacylglycerols and cholesteryl esters are degraded in two separate catabolic pools in cultured human fibroblasts, *FEBS Lett.* **328**:230–234.

Hoeg, J. M., Demosky, S. J., Pescovitz, O. H., and Brewer, H. B., 1984, Cholesteryl ester storage disease and Wolman's disease: Phenotypic variants of lysosomal acid cholesteryl ester hydrolyase deficiency, *Am. J. Hum. Genet.* **36**:1190–1203.

Holtzman, E., 1989, *Lysosomes,* Plenum Press, New York.

Hornick, C. A., Thouron, C., DeLamatre, J. G., and Huang, J., 1992, Triacylglycerol hydrolysis in isolated endosomes, *J. Biol. Chem.* **267**:3396–3401.

Hostetler, K. Y., and Hall, L. B., 1980, Phospholipase C activity of rat tissues, *Biochem. Biophys. Res. Commun.* **96**:338–393.

Hostetler, K. Y., and Hall, L. B., 1982, Inhibition of kidney lysosomal phospholipases A and C by aminoglycoside antibiotics: Possible mechanism of aminoglycoside toxicity, *Proc. Natl. Acad. Sci. USA* **79**:1663–1667.

Hostetler, K. Y., Yazaki, P. J., and van den Bosch, H., 1982, Purification of lysosomal phospholipase A, *J. Biol. Chem.* **257**:13367–13373.

Hostetler, K. Y., Reasor, M., and Yazaki, P. J., 1985, Chloroquine-induced phospholipid fatty liver: Measurement of drug and lipid concentrations in rat liver lysosomes, *J. Biol. Chem.* **260**:215–219.

Hostetler, K. Y., Huterer, S. J., and Wherrett, J. R., 1992, Biosynthesis of bis(monoacylglycero)phosphate in liver and macrophage lysosomes, in *Methods in Enzymology,* Vol. 209 (E. A. Dennis and D. E. Vance, eds.), pp. 104–110, Academic Press, Inc., San Diego, California.

Hurwitz, R., Ferlinz, K., and Sandhoff, K., 1994, The tricyclic antidepressant desipramine causes proteolytic degradation of lysosomal sphingomyelinase in human fibroblast, *Biol. Chem. Hoppe-Seyler* **375**:447–450.

Huterer, S. J., and Wherrett, J. R., 1989, Formation of bis(monoacylglycero)phosphate by a macrophage transacylase, *Biochim. Biophys. Acta* **1001**:68–75.

Huterer, S. J., Hostetler, K. Y., Gardner, M. F., and Wherrett, J. R., 1993, Lysosomal phosphatidylcholine:bis(monoacylglycero)phosphate acyltransferase:Specificity for the sn-1 fatty acid of the donor and co-purification with phospholipase A1, *Biochim. Biophys. Acta* **1167**:204–210.

Imanaka, T., Amanuma-Muto, K., Ohkuma, S., and Jakano, T., 1984, Characterization of lysosomal acid lipase purified from rabbit liver, *J. Biochem.* **96**:1089–1101.

Irvine, R. F., Hemington, N., and Dawson, R. M. C., 1978, The hydrolysis of phosphatidylinositol by lysosomal enzymes of rat liver and brain, *Biochem. J.* **176**:475–484.

Ishikawa, Y., Nishide, T., Sasaki, N., Shirai, K., Saito, Y., and Yoshida, S., 1988, Hydrolysis of low-density lipoprotein phospholipids in arterial smooth muscle cells, *Biochim. Biophys. Acta* **961**:170–176.

Ishikawa, Y., Nishide, T., Sasaki, N., Shirai, K., Saito, Y., and Yoshida, S., 1989, Effects of chloroquine on the metabolism of phosphatidylcholine associated with low density lipoprotein in arterial smooth muscle cells, *Atherosclerosis* **80**:1–7.

Jerome, W. G., and Lewis, J. C., 1985, Early atherogenesis in White Carneau pigeons. II. ultrastructural and cytochemical observations, *Am. J. Pathol.* **119**:210–222.

Jerome, W. G., and Lewis, J. C., 1987, Early atherogenesis in the white carneau pigeon III. Lipid accumulation in nascent foam cells, *Am. J. Pathol.* **128**:253–264.

Jobb, E. A., and Callahan, J. W., 1989, Biosynthesis of sphingomyelinase in normal and Niemann-Pick fibroblasts, *Biochem. Cell Biol.* **67**:801–807.

Johnson, W. J., Chacko, G. C., Phillips, M. C., and Rothblat, G. H., 1990, The efflux of lysosomal cholesterol from cells, *J. Biol. Chem.* **265**:5546–5553.

Jones, C. S., Shankaran, P., and Callahan, J. W., 1981, Purification of sphingomyelinase to apparent homogeneity by using hydrophobic chromatography, *Biochem. J.* **195**:373–382.

Jones, C. S., Shankaran, P., Davidson, D. J., Poulos, A., and Callahan, J. W., 1983, Studies on the structure of sphingomyelinase. Amino acid composition and heterogeneity on isoelectric focusing, *Biochem. J.* **209**:291–297.

Jones, N. L., Allen, N. S., and Lewis, J. C., 1991, βVLDL uptake by pigeon monocyte-derived macrophages:Correlation of binding dynamics with three-dimensional ultrastructure, *Cell Motil. Cytoskel.* **19**:139–151.

Kamoshita, S., Aron, A. M., Suzuki, K., and Suzuki, K., 1969, Infantile Niemann-Pick disease. A chemical study with isolation and characterization of membranous cytoplasmic bodies and myelin, *Am. J. Dis. Child.* **117**:379–394.

Kamrath, F. J., Dodt, G., Debuch, H., and Uhlenbruck, G., 1984, The isolation of lysosomes from normal rat liver by affinity chromatography, *Hoppe-Seylers's Z, Physiol. Chem.* **365**:539–547.

Kanfer, J. N., Young, O. M., Shapiro, D., and Brady, R. O., 1966, The metabolism of sphingomyelin. I. Purification and properties of sphingomyelin cleaving enzyme from rat liver tissues, *J. Biol. Chem.* **241**:1081–1084.

Katz, S. S., Shipley, G. G., and Small, D. M., 1976, Physical chemistry of the lipids of human atherosclerotic lesions, *J. Clin. Invest.* **58**:200–211.

Khoo, J. C., Miller, E., McLaughlin, P., and Steinberg, D., 1988, Enhanced macrophage uptake of low density lipoprotein after self-aggeration, *Arteriosclerosis* **8**:348–358.

Klein, A., Henseler, M., Klein, C., Suzuki, K., Harzer, K., and Sandhoff, K., 1994, Sphingolipid activator protein D (sap-D) stimulates the lysosomal degradation of ceramide in vivo, *Biochem. Biophys. Res. Commun.* **200**:1440–1448.

Kleinschmidt, T., Christomanou, H., and Braunitzer, G., 1988, Complete amino-acid sequence of the naturally occurring A2 activator protein for enzymic sphingomyelin degradation: Identity to the sulfatide activator protein (SAP-1), *Biol. Chem. Hoppe-Seyler* **369**:1361–1365.

Klements, R., and Lundberg, B., 1984, Purification of lysosomal cholesteryl ester hydrolase from rat liver by preparative isoelectric focusing, *Lipids* **19**:692–698.

Klements, R., and Lundberg, B., 1986, Substrate specificity of lysosomal cholesteryl ester hydrolase isolated from rat liver, *Lipids* **21**:481–485.

Koval, M., and Pagano, R. E., 1990, Sorting of an internalized plasma membrane lipid between recycling and degradative pathways in normal and Niemann-Pick type A fibroblasts, *J. Cell Biol.* **111**:429–442.

Koval, M., and Pagano, R. E., 1991, Intracellular transport and metabolism of sphingomyelin, *Biochim. Biophys. Acta* **1082**:113–125.

Kovanen, P. T., 1991, Mast cell granule-mediated uptake of low density lipoproteins by macrophages: A novel carrier mechanism leading to the formation of foam cells, *Ann. Med.* **23**:551–559.

Krieger, M., Brown, M. S., Faust, J. R., and Goldstein, J. L., 1978, Replacement of endogenous cholesteryl esters of low density lipoprotein with exogenous cholesteryl linoleate, *J. Biol. Chem.* **253**:4093–4101.

Krieger, M., McPhaul, M. J., Goldstein, J. L., and Brown, M. S., 1979, Replacement of neutral lipids of low density lipoprotein with esters of long chain unsaturated fatty acids, *J. Biol. Chem.* **254**:3845–3853.

Kruth, H. S., Skarlatos, S. I., Lilly, K., Chang, J., and Ifrim, I., 1995, Sequestration of acetylated LDL and cholesterol crystals by human monocyte-derived macrophages, *J. Cell Biol.* **129**:133–145.

Kubo, M., and Hostetler, K. Y., 1985, Mechanism of cationic amphiphilic drug inhibition of purified lysosomal phospholipase A₁, *Biochemistry* **24**:6515–6520.

Kudoh, T., and Wenger, D. A., 1982, Diagnosis of metachromatic leukodystrophy, Krabbe disease, and Farber disease after uptake of fatty acid-labeled cerebroside sulfate into cultured skin fibroblasts, *J. Clin. Invest.* **70**:89–97.

Kunze, H., Hesse, B., and Bohn, E., 1982, Hydrolytic degradation of phosphatidylethanolamine and phosphatidylcholine by isolated rat-liver lysosomes, *Biochim. Biophys. Acta* **711**:10–18.

Leighton, F., Poole, B., Beaufay, H., Baudhuin, P., Coffey, J. W., Fowler, S., and de Duve, C., 1968, The large-scale separation of peroxisomes, mitochondria, and lysosomes from the livers of rats injected with Triton WR-1339, *J. Cell Biol.* **37**:482–513.

Levade, T., Gatt, S., Maret, A., and Salvayre, R., 1991a, Different pathways of uptake and degradation of sphingomyelin by lymphoblastoid cells and the potential participation of the neutral sphingomyelinase, *J. Biol. Chem.* **266**:13519–13529.

Levade, T., Gatt, S., land Salvayre, R., 1991b, Uptake and degradation of several pyrenesphingomyelins by skin fibroblasts from control subjects and patients with Niemann-Pick disease, *Biochem. J.* **275**:211–217.

Levade, T., Tempesta, M. C., and Salvayre, R., 1993, The in situ degradation of ceramide, a potential lipid mediator, is not completely impaired in Farber disease, *FEBS Lett.* **329**:306–312.

Levran, O., Desnick, R. J., and Schuchman, E. H., 1991, Niemann-Pick type B. Identification of a single codon deletion in the acid sphingomyelinase gene and genotype/phenotype correlations in type A and B patients, *J. Clin. Invest.* **88**:806–810.

Lewis, J. C., Taylor, R. G., and Ohta, K., 1988, Lysosmal alterations during coronary atherosclerosis in the pigeon: Correlative cytochemical and three-dimensional HVEM/IVEM observations, *Exp. Mol. Pathol.* **48**;103–115.

Liscum, L., and Dahl, N. K., 1992, Intracellular cholesterol transport, *J. Lipid Res.* **33**:1239–1254.

Liscum, L., and Faust, J. R., 1987, Low density lipoprotein (LDL)-mediated suppression of cholesterol synthesis and LDL uptake is defective in Niemann-Pick type C fibroblasts, *J. Biol. Chem.* **262**:17002–17008.

Liscum, L., and Faust, J. R., 1989, The intracellular transport of low density lipoprotein-derived cholesterol is inhibited in Chinese hamster ovary cells cultured with 3-β-[2-(diethylamino)ethoxy] androst-5-en-17-one, *J. Biol. Chem.* **264**:11796–11806.

Liscum, L., and Underwood, K. W., 1995, Intracellular cholesterol transport and compartmentation, *J. Biol. Chem.* **270**:15443–15446.

Liscum, L., Ruggiero, R. M., and Faust, J. R., 1989, The intracellular transport of low density lipoprotein-derived cholesterol is defective in Niemann-Pick type C fibroblasts, *J. Cell Biol.* **108**:1625–1636.

Lougheed, M., Zhang, H. F., and Steinbrecher, U. P., 1991, Oxidized low density lipoprotein is resistant to cathepsins and accumulates within macrophages, *J. Biol. Chem.* **266**:14519–14525.

Lundberg, B. B., Rothblat, G. H., Glick, J. M., and Phillips, M. C., 1990, Effect of substrate physical state on the activity of acid cholesteryl ester hydrolase, *Biochim. Biophys. Acta* **1042**:301–309.

Lupu, F., Danaricu, I., and Simionescu, N., 1987, Development of intracellular lipid deposits in the lipid-laden cells of atherosclerotic lesions, *Atherosclerosis* **67**:127–142.

Ma, H., and Kovanen, P. T., 1995, IgE-dependent generation of foam cells: An immune mechanism involving degranulation of sensitized mast cells with resultant uptake of LDL by macrophages, *Arterioscler. Thromb.* **15**:811–819.

Mahlberg, F. H., Glick, J. M., Jerome, W. G., and Rothblat, G. H., 1990, Metabolism of cholesteryl ester lipid droplets in a J774 macrophage foam cell model, *Biochim. Biophys. Acta* **1045**:291–298.

Mander, E. L., Dean, R. T., Stanley, K. K., and Jessup, W., 1994, Apolipoprotein B of oxidized LDL accumulates in the lysosomes of macrophages, *Biochim. Biophys. Acta* **1212**:80–92.

Maor, I., and Aviram, M., 1994, Oxidized low density lipoprotein leads to macrophage accumulation of unesterified cholesterol as a result of lysosomal trapping of the lipoprotein hydrolyzed cholesteryl ester, *J. Lipid Res.* **35**:803–819.

Marzella, L., Ahlberg, J., and Glaumann, H., 1981, Autophagy, heterophagy, microautophagy and crinophagy as the means for intracellular degradation, *Virchows Arch. Cell. Pathol.* **36**:219–234.

Matsuzawa, Y., and Hostetler, K. Y., 1980, Properties of phospholipase C isolated from rat liver lysosomes, *J. Biol. Chem.* **255**:646–652.

Matsuzawa, Y., Poorthuis, B. J. H. M., and Hostetler, K. Y., 1978, Mechanism of phosphatidylinositol stimulation of lysosomal bis(monoacylglycero)phosphate synthesis, *J. Biol. Chem.* **253**: 6650–6653.

Maziere, J. C., Wolf, C., Maziere, C., Mora, L., Bereziat, G., and Polonovski, J., 1981, Inhibition of human fibroblasts sphingomyelinase by cholesterol and 7-dehydrocholesterol *Biochem. Biophys. Res. Commun.* **100**:1299–1304.

Merrill, A. H., Hannun, Y. A., and Bell, R. M., 1993, Introduction: Sphingolipids and their metabolites in cell regulation, in *Advances in Lipid Research*, Vol. 25 (R. M. Bell, A. H. Merrill, and Y. A. Hannun, eds.), pp. 1–24, Academic Press, Inc., San Diego, California.

Minor, L. K., Rothblat, G. H., and Glick, J. M., 1989, Triglyceride and cholesteryl ester hydrolysis in a cell culture model of smooth muscle foam cells, *J. Lipid Res.* **30**:189–197.

Minor, L. K., Mahlberg, F. H., Jerome, W. G., Lewis, J. C., Rothblat, G. H., and Glick, J. M., 1991, Lysosomal hydrolysis of lipids in a cell culture model of smooth muscle foam cells, *Exp. Mol. Pathol.* **540**:159–171.

Momoi, T., Ben-Yoseph, Y., and Nadler, H. L., 1982, Substrate-specificities of acid and alkaline ceramidases in fibroblasts from patients with Farber disease and controls, *Biochem. J.* **205**:419–425.

Morimoto, S., Martin, B. M., Kishimoto, Y., and O'Brien, J. S., 1988, Saposin D: A sphingomyelinase activator, *Biochem. Biophys. Res. Commun.* **156**:403–410.

Moser, H. W., 1995, Ceramidase deficiency: Farber lipogranulomatosis, in *The Metabolic and Molecular Bases of Inheritied Disease*, 7th Ed. (C. R. Scriver, A. L. Beaudet, W. S. Sly, and D. Valle, eds.), pp. 2589–2599, McGraw-Hill, Inc., New York.

Munford, R. S., Sheppard, P. O., and O'Hara, P. J., 1995, Saposin-like proteins (SAPLIP) carry out diverse functions on a common backbone structure, *J. Lipid Res.* **36**:1653–1663.

Negre, A., Salvayre, R., Rogalle, P., Dang, Q. Q., and Douste-Blazy, L., 1987, Acyl-chain specificity and properties of cholesterol esterases from normal and Wolman lymphoid cell lines, *Biochim. Biophys. Acta* **918**:76–82.

Newrzella, D., and Stoffel, W., 1992, Molecular cloning of the acid sphingomyelinase of the mouse and the organization and complete nucleotide sequence of the gene, *Biol. Chem. Hoppe-Seyler* **373**:1233–1238.

Niemann, A., 1914, Ein unbekanntes Krankheitsbild, *Jarb. Kinderheilk* **79**:1–10.

Nilsson, A., 1969, The presence of spingomyelin- and ceramide-cleaving enzymes in the small intestinal tract, *Biochim. Biophys. Acta* **176**:339–347.

Olsson, J. M., Eriksson, L. C., and Dallner, G., 1991, Lipid compositions of intracellular membranes isolated from rat liver nodules in Wistar rats, *Cancer Res.* **51**:3774–3780.

Osada, J., Aylagas, H., Sanchez-Prieto, J., Sanchez-Vegazo, I., and Palacios-Alaiz, E., 1990, Isolation of rat liver lysosomes by a single two-phase partition on dextran/polyethylene glycol, *Analyt. Biochem.* **185**:249–253.

Paton, B. C., Schmid, B., Kustermann-Kuhn, B., Poulos, A., and Harzer, K., 1992, Additional biochemical findings in a patient and fetal sibilng with a genetic defect in the sphingolipid activator protein (SAP) precursor, prosaposin, *Biochem. J.* **285**:481–488.

Pentchev, P. G., Brady, R. O., Gal, A. E., and Hibbert, S. R., 1977, The isolation and characterization of sphingomyelinase from human placental tissue, *Biochim. Biophys. Acta* **488**:312–321.

Pentchev, P. G., Comly, M. E., Kruth, H. S., Tokoro, T., Butler, J., Sokol, J., Filling-Katz, M., Quirk, J. M., Marshall, D. C., Patel, S., Vanier, M. T., and Brady, R. O., 1987, Group C Niemann-Pick disease: Faulty regulation of low-density lipoprotein uptake and cholesterol storage in cultured fibroblasts, *FASEB J.* 1:40–45.

Pereira, L. V., Desnick, R. J., Adler, D. A., Disteche, C. M., and Schuchman, E. H., 1991, Regional assignment of the human acid sphingomyelinase gene (SMPD1) by PCR analysis of somatic cell hybrids and in situ hybridization to 11p15-11p15.4, *Genomics* 9:229–234.

Peters, T. J., and de Duve, C., 1974, Lysosomes of the arterial wall. II. Subcellular fractionation of aortic cells from rabbits with experimental atheroma, *Exp. Mol. Pathol.* 20:228–256.

Pitas, R. E., 1990, Expression of the acetyl low density lipoprotein receptor by rabbit fibroblasts and smooth muscle cells, *J. Biol. Chem.* 265:12722–12727.

Pitas, R. E., Innerarity, T. L., and Mahley, R. W., 1983, Foam cells in explants of atherosclerotic rabbit aortas have receptors for B-VLDL and modified LDL, *Arteriosclerosis* 3:2–12.

Pitas, R. E., Friera, A., McGuire, J., and Dejager, S., 1992, Further characterization of the acetyl LDL (scavenger) receptor expressed by rabbit smooth muscle cells and fibroblasts, *Arterioscler. Thromb.* 12:1235–1244.

Ponting, C. P., 1994, Acid sphingomyelinase possesses a domain homologous to its activator proteins: Saposins B and D, *Protein Sci.* 3:359–361.

Poulos, A., Ranieri, E., Shankaran, P., and Callahan, J. W., 1984, Studies on the activation of the enzymatic hydrolysis of sphingomyelin liposomes, *Biochim. Biophys. Acta* 793:141–148.

Prensky, A. L., Ferreira, G., Carr, S., and Moser, H. W., 1967, Ceramide and ganglioside accumulation in Farber's lipogranulomatosis, *Proc. Soc. Exp. Biol. Med.* 126:725–728.

Quintern, L. E., Weitz, G., Nehrkorn, H., Tager, J. M., Schram, A. W., and Sandhoff, K., 1987, Acid sphingomyelinase from human urine: Purification and characterization, *Biochim. Biophys. Acta* 922:323–336.

Quintern, L. E., Schuchman, E. H., Levran, O., Suchi, M., Sandhoff, K., and Desnick, R. J., 1989, Isolation of cDNA clones encoding human acid sphingomyelinase. Occurrence of alternatively spliced transcripts, *EMBO J.* 8:2469–2473.

Rao, B. G., and Spence, M. W., 1976, Sphingomyelinase activity at pH 7.4 in human brain and a comparison to activity at pH 5.0, *J. Lipid Res.* 17:506–515.

Rauch, H. J., and Auböck, L., 1983, "Banana bodies" in disseminated lipogranulomatosis (Farber's disease), *Am. J. Dermatopathol.* 5:263–266.

Rhee, S. G., Suh, P. G., Ryu, S.-H., and Lee, S. Y., 1989, Studies of inositol phospholipid-specific phospholipase C, *Science* 244:546–550.

Richards, D. E., Irvine, R. F., and Dawson, R. M. C., 1979, Hydrolysis of membrane phospholipids by phospholipases of rat liver lysosomes, *Biochem. J.* 182:599–606.

Rip, J. W., Blais, M. M., and Jiang, L. W., 1994, Low-density lipoprotein as a transporter of dolichol intermediates in the mammalian circulation, *Biochem. J.* 297:321–325.

Robinson, M., and Waite, M., 1983, Physical-chemical requirements for the catalysis of substrates by lysosomal phospholipase A$_1$, *J. Biol. Chem.* 258:14371–14378.

Rodriguez-Lafrasse, C., Rousson, R., Bonnet, J., Pentchev, P. G., Louisot, P., and Vanier, M. T., 1990, Abnormal cholesterol metabolism in imipramine-treated fibroblast cultures. Similarities with Niemann-Pick type C disease, *Biochim. Biophys. Acta* 1043:123–128.

Roff, C. F., Goldin, E., Comly, M. E., Cooney, A., Brown, A., Vanier, M. T., Miller, S. P. F., Brady, R. O., and Pentchev, P. G., 1991, Type C Niemann-Pick disease: Use of hydrophobic amines to study defective cholesterol transport, *Dev. Neurosci.* 13:315–319.

Roholt, O. A., and Schlamowtitz, M., 1961, Studies of the use of dihexanoyllecithin and other lecithins as substrates for phospholipase A, *Arch. Biochem. Biophys.* 94:364–379.

Roma, P., Bernini, F., Fogliatto, R., Bertulli, S. M., Negri, S., Fumagalli, R., and Catapano, A. L., 1992, Defective catabolism of oxidized LDL by J774 murine macrophages, *J. Lipid Res.* 22:819–829.

Rosenfeld, M. E., and Ross, R., 1990, Macrophages and smooth muscle cell proliferation in atherosclerotic lesions of WHHL and comparably hypercholesterolemic fat-fed rabbits, *Arteriosclerosis* 10:680–687.

Ross, A. C., Go, K., Heider, J., and Rothblat, G. H., 1984, Selective inhibition of acylCoA:cholesterol acyltransferase by compound 58-035, *J. Biol. Chem.* **258**:815–819.

Rothblat, G. H., Rosen Jr., J. M., Insull, W., Yau, A. O., and Small, D. M., 1977, Production of cholesteryl ester-rich anisotropic inclusions by mammalian cells in culture, *Exp. Mol. Pathol.* **26**:318–324.

Rouser, G., Kritchevsky, G., Yamamoto, A., Knudson, A. G. J., and Simon, G., 1968, Accumulation of a glycerolphospholipid in classical Niemann-Pick disease, *Lipids* **3**:287–290.

Rousson, R., Parvaz, P., Bonnet, J., Rodriguez-Lafrasse, C., Louisot, P., and Vanier, M. T., 1993, Preparation of an anti-acid sphingomyelinase monoclonal antibody for the quantitative determination and polypeptide analysis of lysosomal sphingomyelinase in fibroblasts from normal and Niemann-Pick type A patients, *J. Immunol. Methods* **160**:199–206.

Rutsaert, J., Tondeur, M., Vamos-Hurwitz, E., and Dustin, P., 1977, The cellular lesions of Farber's disease and their experimental reproduction in tissue culture, *Lab. Invest.* **36**:474–480.

Sakuragawa, N., 1982, Acid sphingomyelinase of human placenta: Purification, properties, and [125]iodine labeling, *J. Biochem.* **92**:637–646.

Sandhoff, K., and Klein, A., 1994, Intracellular trafficking of glycosphingolipids: Role of sphingolipid activator proteins in the topology of endocytosis and lysosomal digestion, *FEBS Lett.* **346**:103–107.

Sandhoff, K., Harzer, K., and Furst, W., 1995, Sphingolipid activator proteins, in *The Metabolic and Molecular Bases of Inherited Disease*, 7th Ed. (C. R. Scriver, A. L. Beaudet, W. S. Sly, and D. Valle, eds.), pp. 2427–2441, McGraw-Hill, Inc., New York.

Sando, G. N., and Rosenbaum, L. M., 1985, Human lysosomal acid lipase/cholesteryl ester hydrolase: Purification and properties of the form secreted by fibroblasts in microcarrier culture, *J. Biol. Chem.* **260**:15186–15193.

Sawant, P. L., Shibko, S., Kumta, U. S., and Tappal, A. L., 1964, Isolation of rat liver lysosomes and their general properties, *Biochim. Biophys. Acta* **85**:82–92.

Schuchman, E. H., and Desnick, R. J., 1995, Niemann-Pick disease types A and B: Acid sphingomyelinase deficiencies, in *The Metabolic and Molecular Bases of Inherited Disease*, 7th Ed. (C. R. Scriver, A. L. Beaudet, W. S. Sly, and D. Valle, eds.), pp. 2601–2624, McGraw-Hill, Inc., New York.

Schuchman, E. H., Suchi, M., Takahashi, T., Sandhoff, K., and Desnick, R. J., 1991, Human acid sphingomyelinase. Isolation, nucleotide sequence and expression of the full-length and alternatively spliced cDNAs, *J. Biol. Chem.* **266**:8531–8539.

Schuchman, E. H., Levran, O., Periera, L. V., and Desnick, R. J., 1992, Structural organization and complete nucleotide sequence of the gene encoding human acid sphingomyelinase (SMPD1), *Genomics* **12**:197–205.

Scriver, C. R., Beaudet, A. L., Sly, W. S., and Nalle, D. (eds.), 1995, *The Metabolic and Molecular Bases of Inherited Disease*, Vol. II, 7th Ed., McGraw Hill, Inc., New York.

Shio, H., Haley, N. J., and Fowler, S., 1978, Characterization of lipid-laden aortic cells from cholesterol-fed rabbits II. Morphometric analysis of lipid-filled lysosomes and lipid droplets in aortic cell population, *Lab. Invest.* **39**:390–397.

Shio, H., Haley, N. J., and Fowler, S., 1979, Characterization of lipid-laden aortic cells from cholesterol-fed rabbits. III. Intracellular localization of cholesterol and cholsteryl ester, *Lab. Invest.* **41**:160–167.

Small, D. M., 1970, The physical state of lipids of biological importance: cholesteryl esters, cholesterol, triglycerides, in *Surface Chemistry of Biological Systems* (M. Blank, ed.), pp. 55–83, Plenum Press, New York.

Snow, J. W., McCloskey, H. M., Rothblat, G. H., and Phillips, M. C., 1988, Physical state of cholesteryl esters deposited in cultured macrophages, *Biochemistry* **27**:3640–3646.

Sparrow, C. P., Parthasarathy, S., and Steinberg, D., 1989, A macrophage receptor that recognizes oxidized low density lipoprotein but not acetylated low density lipoprotein, *J. Biol. Chem.* **264**:2599–2604.

Spence, M. W., 1993, Sphingomyelinases, in *Advances in Lipid Research*, Vol. 26, (R. M. Bell, A. H. Merrill, and Y. A. Hannun, eds.), pp. 3–23, Academic Press, Inc., San Diego, California.

Spence, M. W., and Callahan, J. W., 1989, Sphingomyelin-cholesterol lipidoses: The Niemann-Pick group of diseases, in *The Metabolic Basis of Inherited Diseases*, 6th Ed. (C. R. Scriver, A. L. Beaudet, W. S. Sly, and D. Valle, eds.), pp. 1655–1676, McGraw-Hill, New York.

Spence, M. W., Clarke, J. T. R., and Cook, H. W., 1983, Pathways of sphingomyelin metabolism in cultured fibroblasts from normal and sphingomyelin lipidosis subjects, *J. Biol. Chem.* **258**:8595–8600.

Spence, M. W., Beed, S., and Cook, H. W., 1986, Acid and alkaline ceramidases of rat tissues, *Biochem. Cell Biol.* **64**:400–404.

Stary, H. C., Chandler, A. B., Glagov, S., Guyton, J. R., Insall, W., Rosenfeld, M. E., Schaffer, S. A., Schwartz, C. J., Wagner, W. D., and Wissler, R. W., 1994, A definition of initial, fatty streak, and intermediate lesions of atherosclerosis, *Arterioscler. Thromb.* **14**:840–856.

Steinberg, S. J., Ward, C. P., and Fensom, A. H., 1994, Complementation studies in Niemann-Pick disease type C indicate the existence of a second group, *J. Med. Genet.* **31**:317–320.

Stoffel, W., and Melzner, I., 1980, Studies in vitro on the biosynthesis of ceramide and sphingomyelin. A reevaluation of proposed pathways, *Hoppe-Seylers Z. Physiol. Chem.* **361**:755–771.

Stoffel, W., Kruger, E., and Melzner, I., 1980, Studies on the biosynthesis of ceramide. Does the reversed ceramidase reaction yield ceramides? *Hoppe-Seyler's Z. Physiol. Chem.* **361**:773–779.

Sugita, M., Dulaney, J. T., and Moser, H. W., 1972, Ceramidase deficiency in Farber's disease (lipogranulomatosis), *Science* **178**:1100–1102.

Sugita, M., Iwamori, M., Evans, J., McCluer, R. H., Dulaney, J. T., and Moser, H. W., 1974, High performance liquid chromatography of ceramides: Application to analysis in human tissues and demonstration of ceramide excess in Farber's disease, *J. Lipid Res.* **15**:223–226.

Sugita, M., Williams, M., Dulaney, J. T., and Moser, H. W., 1975, Ceramidase and ceramide synthesis in human kidney and cerebellum. Description of a new alkaline ceramidase, *Biochim. Biophys. Acta* **398**:125–131.

Suits, A. G., Chait, A., Aviram, M., and Heinecke, J. W., 1989, Phagocytosis of aggregated lipoprotein by macrophages: Low density lipoprotein receptor-dependent foam-cell formation, *Proc. Natl. Acad. Sci. USA* **86**:2713–2717.

Sutrina, S. L., and Chen, W. W., 1982, Metabolism of ceramide-containing endocytotic vesicles in human diploid fibroblasts, *J. Biol. Chem.* **257**:3039–3044.

Suzuki, K., 1987, Enzymatic diagnosis of sphingolipidoses, in *Methods in Enzymology—Complex Carbohydrates (Part E)*, Vol. 138, pp. 727–739, Academic Press, Inc., Orlando, Florida.

Tabas, I., Lim, S., Xu, X.-X., and Maxfield, F. R., 1990, Endocytosed B-VLDL and LDL are delivered to different intracellular vesicles in mouse peritoneal macrophages, *J. Cell Biol.* **1411**:929–940.

Tangirala, R. K., Mahlberg, F. H., Glick, J. M., Jerome, W. G., and Rothblat, G. H., 1993, Lysosomal accumulation of unesterified cholesterol in model macrophage foam cells, *J. Biol. Chem.* **268**:9653–9660.

Tangirala, R. K., Jerome, W. G., Jones, N. L., Small, D. M., Johnson, W. J., Glick, J. M., Mahlberg, F. H., and Rothblat, G., 1994, Formation of cholesterol monohydrate crystals in macrophage-derived foam cells, *J. Lipid Res.* **35**:93–104.

Tiffany, C. W., Al, E., Tager, J. M., Moser, H. W., and Kishimoto, Y., 1987, The use of antibody to characterize control and Farber disease ceramidase, *J. Neurochem.* **48**:S35.

Toppet, M., Vamos-Hurwitz, E., Jonniaux, G., Cremer, N., Tondeur, M., and Pelc, S., 1978, Farber's disease as a ceramidosis: Clinical, radiological and biochemical aspects, *Acta Paediatr. Scand.* **67**:113–119.

Traynor, J. R., and Kalwant, S. A., 1981, Phospholipase A_2 activity of lysosomal origin secreted by polymorphonuclear leucocytes during phagocytosis or on treatment with calcium, *Biochim. Biophys. Acta* **665**:571–577.

Trouet, A., 1974, Isolation of modified liver lysosomes, in *Methods in Enzymology Biomembranes (Part A)*, Vol. 31 (S. Fleischer and L. Packer, eds.), pp. 323–329, Academic Press, Inc., London.

Tsukada, T., Rosenfeld, M. E., Ross, R., and Gown, A. M., 1986, Immunocytochemical analysis of cellular components in atherosclerotic lesions, *Arteriosclerosis* **6**:601–613.

Ullman, M. D., and Radin, N. S., 1972, Enzymatic formation of hydroxy ceramides and comparison with enzymes forming nonhydroxy ceramides, *Arch. Biochem. Biophys.* **152**:767–777.

Van Berkel, T. J. C., Vaandrager, H., Kruijt, J. K., and Koster, J. F., 1980, Characteristics of acid lipase and acid cholesteryl esterase activity in parenchymal and non-parenchymal rat liver cells, *Biochim. Biophys. Acta* **617**:446–457.

Van den Bosh, H., and Aarsman, A. J., 1979, A review on methods of phospholipase A determination, *Agents Actions* **9**:382–389.

Van Lenten, B. J., Fogelman, A. M., Hokom, M. M., Benson, L., Haberland, M. E., and Edwards, P. E., 1983, Regulation of the uptake and degradation of B-very low density lipoprotein in human mono-cyte macrophages, *J. Biol. Chem.* **258**:5151–5157.

Vanier, M. T., 1983, Biochemical studies in Niemann-Pick disease. I. Major sphingolipids of liver and spleen, *Biochim. Biophys. Acta* **750**:178–184.

Verheij, H. M., Slotboom, A. J., and DeHaas, G. H., 1981, Structure and function of phospholipase A$_2$, *Rev. Physiol. Biochem. Pharmacol.* **91**:91–203.

Vijayagopal, P., Srinivasan, S. R., Jones, K. M., Radhakrishnamurthy, B., and Berenson, G. S., 1985, Complexes of low-density lipoproteins and arterial proteoglycan aggregates promote cholesteryl ester accumulation in mouse macrophages, *Biochim. Biophys. Acta* **837**:251–261.

Waite, M., 1987, The phospholipases, in *Handbook of Lipid Research,* Vol. 5 (D. J. Hanahan, ed.) Plenum Press, New York.

Waite, M., 1991, Phospholipases, enzymes that share a substrate class, in *Advances in Experimental Medicine and Biology,* Vol. 279 (A. B. Mukherjee, ed.), pp. 1–22, Plenum Press, New York.

Waite, M., King, L., Thornburg, T., Osthoff, G., and Thuren, T. Y., 1990, Metabolism of phos-phatidylglycerol and bis(monoacylglycero)phosphate in macrophage subcellular fractions, *J. Biol. Chem.* **265**:21720–21726.

Wang, Y., Lindstedt, K. A., and Kovanen, P. T., 1995, Mast cell granule remnants carry LDL into smooth muscle cells of the synthetic phenotype and induce their conversion into foam cells, *Ar-terioscler. Thromb.* **15**:801–810.

Wang-Iverson, P., Ginsberg, H. N., Peteanu, L. A., Le, N., and Brown, W. V., 1985, Apo E-mediated up-take and degradation of normal very low density lipoproteins by human monocyte/macrophages. A saturable pathway distinct from the LDL receptor, *Biochem. Biophys. Res. Commun.* **126**:578–586.

Warner, G. J., Stoudt, G., Bamberger, M., Johnson, W. J., and Rothblat, G. H., 1995, Cell toxicity in-duced by inhibition of acyl coenzyme A:cholesterol acyltransferase and accumulation of unester-ified cholesterol, *J. Biol. Chem.* **270**:5772–5778.

Watanabe, K., Sakuragawa, N., Arima, M., and Satoyoshi, E., 1983, Partial purification and properties of acid sphingomyelinase from rat liver, *J. Lipid Res.* **24**:596–603.

Weglicki, W. B., Ruth, R. C., Owens, K., Griffen, H. D., and Waite, B. M., 1974, Changes in lipid com-position of triton-filled lysosomes during lysis, *Biochim. Biophys. Acta* **337**:145–152.

Weglicki, W. B., Ruth, R. C., Gottwik, M. G., Owens, K., Griffin, H. D., and Waite, B. M., 1975, Phos-pholipase A and acid lipase activity during release of lysosomal hydrolases, *Recent Adv. Stud. Card. Struct. Metab.* **7**:49–59.

Weitz, S., Driessen, J., Brouwer-Kelder, E. M., Sandhoff, K., Barranger, J. A., Tager, J. M., and Schram, A. W., 1985, Soluble sphingomyelinase from human urine as antigen for obtaining anti-sphingomyelinase antibodies, *Biochim. Biophys. Acta* **838**:92–97.

Wightman, P. D., Humes, J. L., Davies, P., and Bonney, R. J., 1981, Identification and characterization of two phospholipase A$_2$ activities in resident mouse peritoneal macrophages, *Biochem. J.* **195**:427–433.

Wilson, C., Wardell, M. R., Weisgraber, K. H., Mahley, R. W., and Agard, D. A., 1991, Three-dimensional structure of the LDL receptor-binding domain of human apolipoprotein E, *Science* **252**:1817–1822.

Wolfbauer, G., Glick, J. M., Minor, L. K., and Rothblat, G. H., 1986, Development of the smooth mus-cle foam cell: Uptake of macrophage lipid inclusions, *Proc. Natl. Acad. Sci. USA* **83**:7760–7664.

Wong, T. K., and Lennarz, W. J., 1982, The site of biosynthesis and intracellular deposition of dolichol in rat liver, *J. Biol. Chem.* **257**:6619–6624.

Wong, T. K., Decker, G. L., and Lennarz, W. J., 1982, Localization of dolichol in the lysosomal fraction of rat liver, *J. Biol. Chem.* **257**:6614–6618.

Yamanaka, T., and Suzuki, K., 1982, Acid sphingomyelinase of human brain: Purification to homogeneity, *J. Neurochem.* **38**:1753–1764.

Yavin, E., and Gatt, S., 1969, Enzymatic hydrolysis of sphingolipids. VIII. Further purification and properties of rat brain ceramidase, *Biochemistry* **8**:1692–1698.

Yedgar, S., and Gatt, S., 1976, Effect of Triton X-100 on the hydrolysis of sphingomyelin by sphingomyelinase of rat brain, *Biochemistry* **15**:2570–2573.

Yedgar, S., and Gatt, S., 1980, Enzymic hydrolysis of sphingomyelin in the presence of bile salts, *Biochem. J.* **185**:749–754.

Yoshida, Y., Arimoto, K., Sato, M., Sakuragawa, N., Arima, M., and Satoyoshi, E., 1985, Reduction of acid sphingomyelinase activity in human fibroblasts induced by AY-9944 and other cationic amphiphilic drugs, *J. Biochem.* **98**:1669–1679.

Chapter 9

Lysosomal Nucleic Acid and Phosphate Metabolism and Related Metabolic Reactions

Ronald L. Pisoni

1. INTRODUCTION

Lysosomes are a major intracellular site for the degradation of nucleic acids. Under conditions of nutritional deprivation, for example, approximately 65% of total rat liver cytoplasmic RNA is degraded per day, with 70–85% of this turnover occurring within the lysosomal compartment (Lardeux and Mortimore, 1987; Lardeux *et al.,* 1987, 1988; Heydrick *et al.,* 1991). Lysosomes contain an active repertoire of hydrolytic enzymes capable of completely degrading large amounts of nucleic acids to their basic constituents. In this chapter, the properties of the enzymes involved in lysosomal metabolism of nucleic acids and phosphates are described. The reader is also referred to earlier reviews by Barrett (1972) and Vaes (1973). In addition to the enzymes involved in lysosomal nucleic acid and phosphate metabolism, the role of lysosomes in the metabolism of cobalamin, folic acid polyglutamate, and coenzyme A are also discussed.

Ronald L. Pisoni   Department of Internal Medicine, Division of Nephrology, The University of Michigan Medical Center, Ann Arbor, MI 48109.

Subcellular Biochemistry, Volume 27: Biology of the Lysosome, edited by Lloyd and Mason. Plenum Press, New York, 1996.

2. NUCLEIC ACID DEGRADATION

Lysosomes contain enzymes capable of degrading DNA and RNA completely to nucleosides and phosphate. Although lysosomal nucleic acid degradative enzymes have been characterized in various tissues, the overall rate and amount of lysosomal nucleic acid degradation have been quantified only for cytoplasmic RNA turnover (Lardeux and Mortimore, 1987; Lardeux et al., 1987, 1988; Heydrick et al., 1991). It can be anticipated that the contribution of the lysosomal nucleic acid degradative pathway to the function of a given cell type will vary depending on the specialized functions of different cells. Thus in hepatocytes, lysosomal nucleic acid degradation may serve largely to support autophagic degradation of endogenous nucleic acids, whereas in macrophages, in which phagocytosis is a major cellular activity, lysosomal acid degradation may serve largely to support the degradation of phagocytosed exogenous nucleic acids.

Early work from de Duve's laboratory provided some of the first descriptions of lysosomes containing enzymes capable of degrading nucleic acids (de Duve et al.,1955; Bowers and de Duve, 1967). During these early years of characterizing the lysosome, electron microscopy served as an important experimental tool for visualizing the role of lysosomes and phagocytosis in degrading nucleic acids. Bensch et al., (1964), in applying electron microscopy to study the process by which mammalian cells phagocytose DNA–protein coacervates, found that particulate DNA–protein coacervates were immediately phagocytosed and completely digested within 30 min when incubated with L-strain fibroblasts in suspension culture. The vacuoles in which the DNA-protein coacervates were hydrolyzed were shown by cytochemical staining reactions to contain acid phosphatase and nucleotidase activities, both of which were strongly inhibited by 10 mM sodium fluoride. In 1970, Arsenis et al. (1970) demonstrated that extracts from Ehrlich ascites tumor lysosomes and rat liver tritosomes (lysosomes made buoyant by the uptake of the detergent Triton WR1339) were able to hydrolyze various DNA and RNA preparations completely to free nucleosides and inorganic phosphate. The pH optimum for overall DNA degradation was 4.3 to 4.8, and for overall RNA degradation 5.0 to 5.5. The release of phosphate from DNA or RNA was dependent on the prior action of nucleases and diesterases in the lysosomal extract, and the phosphatase reaction was suggested as the possible rate-limiting step in overall nucleic acid degradation. These studies along with many others helped demonstrate that nucleic acid degradation within lysosomes is catalyzed by the successive action of several different lysosomal enzymes, as depicted in Fig. 1 for lysosomal RNA degradation. In general, nucleic acids initially are degraded to oligonucleotides either by an acid ribonuclease in the case of RNA or by an acid deoxyribonuclease when DNA is the substrate nucleic acid. Oligonucleotides are further degraded to nucleoside 3′ -phosphates by acid exonuclease, and these monophosphates are then converted to nucleosides and orthophosphate

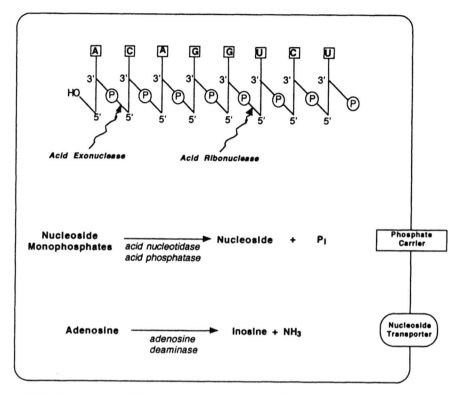

FIGURE 1. Lysosomal RNA degradation: Summary of the different lysosomal enzymes participating in lysosomal RNA metabolism.

either by lysosomal acid nucleotidase or acid phosphatase activities. The nucleoside adenosine can be further metabolized to inosine through the action of an adenosine deaminase, thus far shown to be present in human fibroblast lysosomes (Lindley and Pisoni, 1993). Lysosomal phosphate and nucleoside transport systems then provide routes for carrier-mediated release of orthophosphate and nucleosides from the lysosomal compartment (see Chapter 11).

As shown by Bensch *et al.,* (1964) and others, phagocytosis and pinocytosis are able to deliver large amounts of exogenous nucleic acids to the lysosomal compartment, where they are quickly degraded. Similarly, the processes of autophagy as shown by Lardeux *et al.* (1987) serve to deliver endogenous nucleic acids to lysosomes. These pathways for the delivery of macromolecules to lysosomes are described in detail in Chapters 3–5; they constitute an important part of the overall process of lysosomal nucleic acid degradation. The next few sections of this chapter focus on the characteristics of the individual enzymes involved in lysosomal nucleic acid and phosphate metabolism.

2.1. Acid Ribonuclease

Acid ribonuclease activities have been reported in calf spleen (Maver and Greco, 1949b), human spleen (Dulaney and Touster, 1972), hog spleen (Bernardi and Bernardi, 1966), rat liver (Futai *et al.,* 1969), and HeLa cells (Saha *et al.,* 1979). Futai *et al.* (1969) found that the acid ribonuclease activity from rat liver lysosomes exhibits maximal activity from pH 4.8 to 5.8; displays a molecular weight of 24–28 kDa by gel filtration; and cleaves single-stranded RNA to form products with 3′ -phosphate termini. Similarly, the acid RNase purified from HeLa cell lysosomes has an acid pH optimum, but displays a molecular weight of ~40 kDa by gel filtration, has an acidic isoelectric point of pI 3.0–3.5, and binds tightly to concanavalin A-agarose (Saha *et al.,* 1979). Monovalent cations such as Na^+, and K^+, and NH_4^+ can stimulate acid RNase as much as five-fold depending on the concentration used. In sharp contrast, acid RNase is inhibited $> 90\%$ by 1 mM Cu_2^+, Hg_2^+ and Zn_2^+ and inhibited 75–85% by 1 mM Mg_2^+ or Ca_2^+. In addition, 1 mM *p*-chloromercuribenzoate and 10 mM N-ethylmaleimide also strongly inhibit acid RNase (Saha *et al.,* 1979).

The HeLa cell acid RNase degrades many natural RNAs, including rRNA, tRNA, and T4 mRNA. When synthetic homopolymers were tested as substrates, however, acid RNase could degrade only poly(U), whereas poly(C) and poly(A) were resistant to degradation. Saha *et al.* (1979) also found the acid RNase activity in L cells grown in serum-free media displays properties very similar to those of the HeLa cell acid RNase.

Saha (1982) determined the frequency at which acid RNase hydrolyzes different potential cleavage sites in 16S ribosomal RNA. Acid RNase was found to primarily cleave single-stranded RNA at sites 5′ to uridine residues, yielding oligonucleotides bearing 3′-P and 5′-OH termini. Acid RNase cleavage at uridine residues occurs much more readily when the nucleotide preceding uridine is a purine rather than a pyrimidine, although cleavage at Pyr-U is occasionally observed. In addition to cleavage at uridine residues, the HeLa cell acid RNase may hydrolyze RNA at linkage sites 5′ to guanine residues but with an efficiency much lower than that seen at uridine-containing cleavage sites.

In order for RNA to be degraded within lysosomes, it must be in a single-stranded form. This requirement was shown by Silverstein and Dales (1968), who observed that the reovirus protein coat is quickly stripped away within lysosomes when L cells are infected with reovirus, but the double-stranded reovirus RNA genome remains resistant to degradation. They proposed that this protein uncoating is an obligatory step in the sequence of infection and that the RNA genome is resistant because of its double-stranded character. The viral RNA, however, was easily degraded *in vitro* if it previously had been denatured.

Levels of different lysosomal enzymes, including acid RNase, have been shown to change substantially as cultured cells reach confluence. Acid RNase levels increased six-fold in WI-38 cells and approximately 2-fold in HeLa cells upon reaching confluence (Saha *et al.,* 1981). When examined in several cell lines, acid

RNase levels were found to exceed substantially the amount of enzyme necessary to maintain normal rates of RNA turnover. Varying acid RNase levels as much as 10-fold did not limit the rate of normal RNA turnover. In addition, fibroblasts from patients with I-cell disease contain one third the normal level of acid RNase but show normal rates of RNA turnover (Saha *et al.,* 1981).

2.2. Acid Deoxyribonuclease (DNase II)

Acid deoxyribonuclease, also referred to as DNase II, is a lysosomal enzyme that hydrolyzes the phosphodiester bonds of DNA to form oligonucleotides bearing 3'-phosphate termini. The existence of an acid DNase activity was first reported by Catchside and Holmes (1947) and by Maver and Greco (1949a, b). The name DNase II was suggested to distinguish it from pancreatic DNase, which was called DNase I (Cunningham and Laskowsi, 1953). De Duve *et al.* (1955), using differential centrifugation, first demonstrated that DNase II cosediments with lysosomal fractions, and later Dulaney and Touster (1972) directly isolated DNase II from purified rat liver lysosomes. DNase II has been shown to be present in a wide variety of tissues from various species (Liao *et al.,* 1989; Bernardi, 1968; Cordonnier and Bernardi, 1968), and the gene for human DNase II has been mapped to chromosome 19p (Lord *et al.,* 1990).

In some species and tissues such as porcine spleen, liver, and gastric mucosa, DNase II exists as a heterodimer consisting of a 35-kDa α-chain and a 10-kDa β-chain where as in bovine liver, DNase II is present as a single 26-kDa polypeptide (Liao *et al.,* 1989). Since no free sulfhydryls are detected in DNase II, its eight half-cystine residues are believed to form four disulfide bridges. DNase II displays multiple ionic forms by ion exchange chromatography and thin layer isoelectric focusing with isoelectric points ranging from pI 7.3–10.2 (Bernardi, 1968; Cordonnier and Bernardi, 1968; Liao *et al.,* 1989).

The DNase II enzymes from different tissues all display an acid pH optimum near pH 5.0 and are dependent on monovalent cations at a concentration of ~0.2–0.3 M. In addition, cysteine has been reported to activate acid DNase under some conditions (Maver and Greco, 1949b; Bernardi and Griffe, 1964). When DNA is extensively digested with purified DNase II, oligonucleotides are obtained that have an average length of 10–12 bases. DNase II has the special property of breaking both DNA strands at the same point of scission during hydrolysis. Tsubota *et al.* (1974) have shown that human gastric mucosal DNase II produces double-stranded breaks in a two-step process: the first single-strand break is followed rapidly by the second single-strand break occurring near the same point on the complementary strand. The bond most favored for hydrolysis by DNase II is either d (Gp-Gp) or d(Gp-Cp); d(Cp-Tp) bonds are cut least frequently.

Several reports indicate the presence of an essential histidine residue for DNase II activity (Oshima and Price, 1973; Liao, 1985; Liao *et al.,* 1989). Porcine spleen DNase II is completely inactivated by iodoacetate at pH 4.6 (Oshima and Price, 1973). Inactivation by iodoacetate was found to be a result of modifying a single histidine residue (Oshima and Price, 1973), and a peptide containing this histidine residue was isolated and sequenced from the α-chain of porcine spleen DNase II (Liao, 1985). Protection against iodoacetate inactivation was observed in the presence of substrate DNA, suggesting that the essential histidine residue may be located in the substrate binding site (Oshima and Price, 1973). DNase II is also strongly inhibited by 10 mM sulfate at pH 5 and moderately inhibited by 10 mM Mg_2^+, whereas 10 mM phosphate is only weakly inhibitory (Maver and Greco, 1949b; Bernardi and Griffe, 1964; Oshima and Price, 1974). However, 0.5 mM sulfate produces a 20% increase in DNase activity with a 15–19 fold increase in single strand DNA scission (Oshima and Price, 1974). Bernardi (1968) postulated that DNase II has two active sites that participate sequentially in the double-strand breaks. Oshima and Price (1974) further proposed an allosteric interaction between the two sites to explain the different effects of high and low [SO$_4^{2-}$] on DNase activity. At low sulfate concentrations, sulfate may bind to one site which activates DNase activity, whereas binding of sulfate to the second site at high [SO$_4^{2-}$] results in strong inhibition of DNase activity.

2.3. Acid Exonuclease

Lysosomal acid exonuclease activity has been observed in rat liver lysosomes by Van Dyck and Wattiaux (1968). Lysosomal acid DNase and acid exonuclease were separated by hydroxyapatite chromatography, and the complementary action of the two enzymes resulted in digestion of DNA to nucleoside 3'-monophosphates. Acid exonuclease cleaves off 3'-phosphomononucleotides in a sequential manner from the 5'-hydroxy end of either DNA or RNA oligonucleotides. Bernardi and Bernardi (1968) found spleen acid exonuclease to display maximal activity at pH 5.5 using deoxyribonucleotide as a substrate. Native DNA was found to be a poor substrate for spleen acid exonuclease, being degraded at less than 4% of the rate at which alkali-denatured DNA was hydrolyzed. Furthermore, poly(A), poly(U), and poly(I) could all be degraded by spleen acid exonuclease, whereas poly(C) was resistant to degradation. Acid exonuclease activity does not appear to have any strict metal cation requirement and is not inhibited by 10 mM EDTA. Incubation at 55 °C for 20 min results in a 50% loss in exonuclease activity.

2.4. Adenosine Deaminase

As mentioned earlier, nucleic acids are degraded completely to nucleosides and inorganic phosphate within the lysosomal compartment. The lysosomal enzymes responsible for degrading nucleic acids to oligonucleotides and nucleotide monophosphates have been described in this chapter. Before continuing with the description of the lysosomal hydrolysis of nucleotides to nucleoside and inorganic phosphate, the following describes lysosomal metabolism of adenosine, which is the only nucleoside currently known to be further processed within the lysosomal compartment.

Lysosomal adenosine deaminase activity has recently been described in human fibroblast lysosomes by Lindley and Pisoni (1993). The existence of this lysosomal activity was first suggested by the observation that radiolabeled adenosine taken up by human fibroblast lysosomes was quickly deaminated to inosine (Pisoni and Thoene, 1989). After a 2.5 min uptake of 1 mM [^{3}H]adenosine, $\sim$15% of the total radioactivity recovered from human fibroblast lysosomes was found to have been converted to inosine, whereas after a 15–20 min [^{3}H]adenosine uptake, inosine accounted for the great majority of the radioactivity recovered from fibroblast lysosomes. To further characterize adenosine deaminase activity (ADA) in human fibroblast lysosomes, Lindley and Pisoni (1993) prepared a granular fraction which contained 13% of the total fibroblast adenosine deaminase activity and analyzed the distribution of ADA activity when the granular fraction was separated on a 31% Percoll density gradient. Approximately 60–70% of the total granular fraction ADA activity was found to migrate with lysosomes in these Percoll density gradients. An adenosine deaminase-deficient cell line, GM469, which contains approximately 2% of the normal level of total ADA activity, was found to have very low levels of lysosomal ADA compared to lysosomes from normal human fibroblasts. These results suggested that the catalytic portion of lysosomal ADA is derived from the same gene known to be defective in adenosine deaminase deficiency and which codes for the 38-kDa catalytic subunit of ADA.

Many of the properties displayed by lysosomal ADA are very similar to those reported for other ADA activities, including (a) a K_m of 37 μM for adenosine at pH 5.5, (b) maximum activity in the neutral pH range, but yet displaying 35–40% of the maximal rate at pH 5.3 and greater than 97% inhibition by deoxycoformycin (0.02 mM), coformycin (2.5 mM), adenosine (2.5 mM), 2'-deoxyadenosine (2.5 mM), 6-methylaminopurine riboside (2.5 mM), 2', 3'-isopropylidene adenosine (2.5 mM), erythro-9-(2-hydroxy-3-nonyl) adenine (EHNA, 0.2 mM), Cu^{2+} (2.5 mM), and Hg^{2+}. Furthermore, lysosomal ADA displays a mass > 200 kDa protein by gel filtration, and is quite heat-stable, retaining 70% of its activity when incubated at 65 °C for 80 min.

The presence of ADA activity within human fibroblast lysosomes raises questions as to how this activity is delivered to lysosomes, what the stability of ADA is within the lysosomal compartment, and what role lysosomal deamination of adenosine plays in overall cellular metabolism. Adenosine deaminase activity mainly exists in mammalian cells in two different molecular forms, a low-molecular-weight 38-kDa form and a high-molecular-weight ADA protein complex composed of two 38-kDa ADA catalytic subunits bound to a dimeric 220-kDa ADA complexing protein. The proportion of these two forms of ADA vary in different cell types. Although in many cell types the majority of ADA is cytosolic, some cellular ADA has been found to be associated with the plasma membrane.

The similarities in physical and kinetic properties of lysosomal ADA to those of high-molecular-weight form of ADA which predominates within human fibroblasts suggests that the high-molecular-weight form of ADA may be delivered to the lysosomal compartment and remain active in that location. The ADA-complexing protein may be important for maintaining the stability of ADA activity delivered to the lysosomal compartment by providing enhanced resistance to lysosomal degradation. The ability of lysosomal ADA to maintain 70% of its activity after incubation at 65 °C for 80 min suggests that the active structural form of lysosomal ADA may be strongly held together. However, the mechanism by which ADA is delivered to lysosomes is not known at the present time.

As mentioned earlier, Lardeux and Mortimore (1987) and Heydrick *et al.* (1991) have shown that in perfused rat liver under conditions of nutritional deprivation, a large proportion of cytoplasmic RNA is degraded within lysosomes. In this context, lysosomes clearly play a central role in generating large amounts of nucleosides which can be reused for cell survival. Inosine can serve as a fuel source when converted by nucleoside phosphorylase to hypoxanthine and ribose 1-phosphate, with this latter product contributing to energy production via the pentose pathway. In addition, hypoxanthine can be recycled to purine nucleotide by the salvage pathway. Hence, the role of ADA within lysosomes may be another means by which lysosomes supply nutrients within the cell under certain states of metabolism or nutrition. Compartmentalization of ADA may provide a pathway of metabolism that is advantageous under certain conditions. At low adenosine concentrations, adenosine is primarily phosphorylated in the cytosol by adenosine kinase, which has a much lower K_m for adenosine than does ADA (Kredich and Hershfield, 1989; Fox and Kelley, 1978). In contrast, at high adenosine concentrations, adenosine inhibits adenosine kinase activity, allowing deamination by ADA to be the major route of adenosine metabolism. A consequence of compartmentalizing some ADA activity within lysosomes is that deamination of adenosine to inosine can occur without competition by the cytosolic kinase reaction.

3. PHOSPHATE METABOLISM

3.1. Acid Nucleotidase and Acid Phosphatase

Acid phosphatase denotes the activity of a number of enzymes that can be differentiated according to structural, catalytic, and immunological properties, tissue distribution, and subcellular location. The majority of acid phosphatase activity in rat liver, approximately 70–80% of the total tissue activity, is associated with lysosomes, with the remainder confined to the cytosol. In many tissues, most of the lysosomal acid phosphatase activity is strongly inhibited by L-(+)-tartrate, although a lysosome-associated tartrate-resistant acid phosphatase is present at substantial levels in osteoclasts and other cells of macrophage/monocyte lineage.

Purification of tartrate-inhibitable acid phosphatase from different tissues and various lysosomal preparations has indicated the presence of more than one species (Arsenis and Touster, 1967, 1968; Brightwell and Tappel, 1968b; Igarashi and Hollander, 1968; Heinrikson, 1969; DiPietro and Zengerle, 1967; Saini and Van Etten, 1978b; Gieselmann et al., 1984; Rehkop and Van Etten, 1975). Some of these phosphatase activities are now known to correspond to different glycosylated, processed, or posttranslationally modified forms of the same gene product, while other tartrate-inhibitable phosphatases are the products of different genes. Arsenis and Touster (1967) partially resolved the phosphatase activity of rat liver lysosomes into a nucleotidase and a sugar phosphate phosphohydrolase. As shown in Table I, rat liver acid nucleotidase dephosphorylates a wide variety of ribonucleotides and deoxyribonucleotides but exhibits very little activity with sugar phosphate substrates. In contrast, the sugar phosphate phosphohydrolase is of broad specificity and does not discriminate between sugar phosphates and nucleotides as substrates. Acid nucleotidase displays a pH optimum of 5.0 whereas the broader-specificity lysosomal acid phosphatase has a pH optimum near 4.0.

Acid nucleotidase comprises 5–15% of the total rat liver lysosomal acid phosphatase activity. Its substrate specificity encompasses nucleotides regardless of the linkage position of the phosphate group, as seen by similar activity with 2' -, 3' - and 5' -AMP, or various other 2' - and 3'-monophosphate nucleotides. However, acid nucleotidase poorly recognizes 5' -IMP and 5' -XMP, while cyclic 3', 5' - AMP, sodium pyrophosphate, UDP-glucose, phosphorylcholine, NADP, and NADPH are not substrates. Acid nucleotidase exhibits a K_m of 0.58 mM for 2' - AMP, 0.33 mM for 3' -AMP, and 0.125 mM for 5' -dAMP at pH 4.8 and 37 °C. Substrates, when present near saturating concentrations, exert a substrate inhibition upon acid nucleotidase. In addition, nearly complete inhibition is observed in the presence of 1 mM tartrate, 1 mM p-hydroxymercuribenzoate, or 10 mM EDTA, citrate, or malate, whereas 5 mM N-ethylmaleimide has no effect. Inorganic phosphate (0.02 M) acts as a noncompetitive inhibitor of rat liver acid nucleotidase, and the enzyme is completely inactivated within 10 min at 50 °C. The rat liver acid nucleotidase studied by Arsenis and Touster (1968) has a molecular weight of 80 kDA by sedimentation density gradient centrifugation.

3.2. Physical and Enzymatic Properties of Lysosomal Acid Phosphatase (LAP)

The best characterized lysosomal acid phosphatase (LAP), which appears to be the major form of acid phosphatase in lysosomes from a wide variety of different tissues, is the acid phosphatase purified from human liver and human placenta. Saini and Van Etten (1978b) purified acid phosphatase from human liver 4560-fold. This enzyme appeared to exist as a homodimer, displaying a m.w. of 90–93 kDa by gel filtration, but under denaturing conditions revealed a subunit size of 50–52 kDA by SDS-PAGE. Human liver LAP was also shown to be a

Table I

Relative Rate of Hydrolysis of Phosphate Esters by Rat Liver Acid Nucleotidase and Sugar Phosphate Phosphohydrolase[a,b]

Substrate	Acid nucleotidase	Acid phosphatase
5′-AMP	67	85
3′-AMP	85	83
2′-AMP	55	70
5′-GMP	40	50
5′CMP	72	55
5-UMP	45	50
5′-TMP	64	78
5′dAMP	100	100
5′dGMP	74	76
5′-dCMP	74	60
α-D-GlcUA-1-P	5	115
α-D-Gal-1-P	4	95
α-D-Glc-1-P	3	90
α-D-Man-1-P	3	74
D-Gal-6-P	1	53
D-Glc-6-P	2	80
D-Man-6-P	1	75
α-D-Xyl-1-P	4	68
D-Fru-6-P	5	30
D-Rib-5-P	3	78
β-Glycerol-P	18	73
p-Nitrophenylphosphate	72	200
NaPPi	—	—
Serine-P	—	—
Phosphorylcholine	—	—
6-Phosphogluconate	—	—
3-Phosphoglycerate	—	—
NADP	—	—
UDP-glucose	—	—
Cyclic 3′,5′-AMP	—	—

[a]Phosphatase activity was determined with 6 mM nucleotide or 10 mM sugar phosphate (these being saturating concentrations) in 50 mM sodium malonate buffer at pH 5.0 for acid nucleotidase and at pH 4.0 for sugar phosphate phosphohydrolase. Enzyme activity is displayed relative to the hydrolysis of 5′-dAMP assigned as 100, which corresponds to 0.465 µmol of Pi released per 30 min by 1 µg/ml of acid nucleotidase or 0.81 µmol of Pi released per 30 min by 5 µg/ml of sugar phosphate phosphohydrolase.
[b]Taken from Arsenis and Touster (1967, 1968).

glycoprotein containing 2.85 wt% mannose and 1.09 wt% N-acetylglucosamine. L-(+)-Tartrate is a strong competitive inhibitor ($K_i = 0.43$ µM) of the human liver acid phosphatase, and its substrate specificity is similar to that of other lysosomal acid phosphatases such as the acid phosphatase shown in Table I. The pH dependence of p-nitrophenylphosphate hydrolysis is consistent with the enzyme recognizing the monoanionic form of the substrate.

To test the possible involvement of an active site carboxyl group in the catalytic function of human liver LAP, the effect of Woodward's reagent K on the activity of the enzyme has been examined, since this reagent has high specificity for modifying carboxyl groups. Human liver acid phosphatase is rapidly inactivated by Woodward's reagent K but inactivation is prevented by the presence of the competitive inhibitors phosphate or L-(+)-tartrate (Saini and Van Etten, 1978b). This competitive inhibitor protection suggests that a carboxyl group important for enzymatic activity lies within the substrate binding site of human liver acid phosphatase.

In 1984, Gieselmann et al. purified a tartrate-inhibitable acid phosphatase from human placenta approximately 19,670-fold to apparent homogeneity. Specific antibodies to this enzyme precipitated 85–88% of the tartrate-inhibitable acid phosphatase activity associated with either human fibroblast heavy or light lysosomal fractions obtained from Percoll density centrifugation. Furthermore, the antiserum against the placental LAP precipitated 35–75% of the tartrate-inhibitable acid phosphatase in tissue extracts of human liver, placenta, spleen, kidney, and brain.

The molecular and kinetic parameters of the purified human placental enzyme closely resemble those reported for the acid phosphatase purified from human liver by Saini and van Etten (1978b). The human placental acid phosphatase enzyme elutes as a homodimer with an apparent m.w. of 90 kDa by gel filtration, with each subunit migrating as a 48-kDa polypeptide by SDS-PAGE (Gieselmann et al., 1984). Human placental acid phosphatase binds strongly to concanavalin A Sepharose and exhibits an isoelectric point at pI 6.2.

Purified placental LAP hydrolyzes phosphomonoester bonds in p-nitrophenylphosphate, glucose-6-phosphate, and AMP but does not cleave the phosphate ester bonds in pyrophosphate, ATP, UTP, or NADP (Gieselmann et al., 1984). A K_m of 0.2 mM and a V_{max} of 246 μmol/min/mg is observed for the placental LAP using p-nitrophenylphosphate as substrate. Human placental LAP is strongly inhibited by tartrate (K_i = 0.51 μM), whereas phosphate is a moderate competitive inhibitor with a K_i of 0.8 mM. The purified enzyme exhibits a broad pH optimum between 3.5 and 5.0 with a maximum at pH 4.0. At 50 °C, LAP displays a 74% loss of activity after a 90 min incubation, with more than 95% of activity lost during a 10 min incubation at 70 °C.

3.3. Biosynthesis and Internalization of LAP

Soon after the discovery that many lysosomal enzymes are delivered to the lysosomal compartment by a phosphomannosyl-dependent pathway, it became clear that the delivery of lysosomal acid phosphatase to lysosomes occurs by an alternative mannose 6-phosphate-independent pathway. The work of von Figura and colleagues elegantly delineated the biosynthesis and delivery of acid phosphatase to the lysosomal compartment. Braun et al. (1989) showed that LAP is transported from the *trans*-Golgi network to the cell surface with a half-time of under 10 min. Cell surface LAP then is rapidly internalized, but most of the internalized LAP is transported back to the cell surface. On average, each LAP molecule cycles at least 15 times between the cell surface and endosomes before it is transferred to lysosomes. At equilibrium, some four-fold more LAP precursor is present in endosomes than at the cell surface. The LAP is transferred with a half-time of 5–6 hr from the plasma membrane/endosome pool to dense lysosomes, where it is proteolytically cleaved to its soluble form and then no longer recycled.

Waheed *et al.* (1988) used pulse-chase labeling and immunoprecipitation to demonstrate that human LAP, stably expressed in BHK cells, is first detected in the endoplasmic reticulum as a 61-kDa polypeptide. Endoglucosaminidase H digestion of this protein yields a ladder of seven lower-molecular-weight intermediates, suggesting that all eight potential N-glycosylation sites in LAP are glycosylated. From the endoplasmic reticulum, LAP enters the Golgi apparatus, where its mass increases within about 30 min to 67 kDa and becomes partially endoglucosaminidase H-resistant. Finally, within the next 14 hr the 67-kDa protein is converted to a 63-kDa, membrane-associated protein and a 52-kDa soluble protein. The 52-kDa polypeptide accounts for two thirds of the total LAP and is highly enriched within dense lysosomes. When stably expressing BHK cells were grown for three days in the presence of 10 mM NH_4Cl, the 52-kDa protein was not observed, whereas the 63-kDa, membrane-associated form of LAP accumulated within the dense lysosomes (Fig. 2). This indicated that an acidic pH is not required for transport of LAP to lysosomes.

The internalization of LAP from the cell surface occurs via clathrin-coated pits (Hille *et al.*, 1992; Lehmann *et al.*, 1992). Sorting of membrane proteins and receptors into clathrin-coated transport vesicles is thought to require recognition of their cytoplasmic domains by complexes of special proteins called adaptors. Binding of adaptors to the cytoplasmic domain then mediates assembly of the clathrin coat, which is required for formation of transport vesicles. Binding of HA-2 adaptors to the cytoplasmic domain of the receptors requires an intact internalization signal. The signal for internalization of LAP is a tyrosine residue exposed in a β-turn conformation within the hexapeptide PGYRHV, which lies within the N-terminal 12 amino acids of LAP's cytoplasmic tail (Fig. 2) (Peters *et al.*, 1990; Lehmann *et al.*, 1992). The importance of the cytoplasmic tail Tyr for internalization is supported by site-directed mutagenesis studies showing that LAP mutants, in which this essential Tyr is replaced by Ala, display a 20-fold lower internalization rate, and the Tyr to Phe mutants are internalized 3.5 times more slowly than wild-type LAP (Lehmann *et al.*, 1992). Furthermore, internalization of LAP can be inhibited by microinjection of antibodies against clathrin or a portion of the HA-2 adaptors. In addition, double immunogold labeling at various stages of invaginating coated pits has shown colocalization of LAP with clathrin and plasma membrane HA-2 adaptors. Sosa *et al.* (1993) have also provided di-

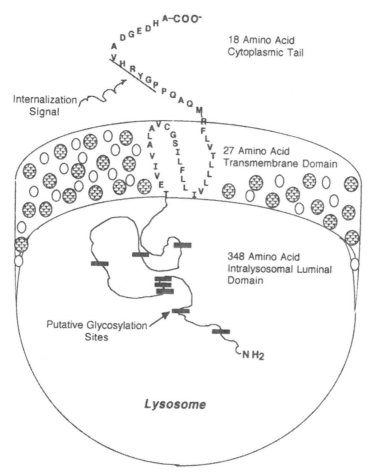

FIGURE 2. Structural diagram of the lysosomal membrane-associated form of lysosomal acid phosphatase (LAP).

rect evidence for high-affinity binding of coated vesicle HA-2 adaptors to a peptide corresponding to the cytoplasmic domain of LAP.

3.4. Proteolytic Processing of LAP in Lysosomes

Lysosomal acid phosphatase initially becomes associated with lysosomes as a membrane protein oriented such that approximately the first 348 N-terminal amino acids reside in the lysosomal lumen, followed by a 27-amino acid transmembrane domain incorporated into the lysosomal membrane and ending with the

C-terminal 18 amino acids forming the cytoplasmic tail (Fig. 2) (Waheed *et al.*, 1988; Pohlmann *et al.*, 1988). This membrane-associated form of LAP is proteolytically processed within lysosomes to a soluble form which remains in the lysosomal compartment (Gottschalk *et al.*, 1989). The proteolytic processing involves at least two sequential cleavages. The first is catalyzed by a thiol proteinase outside of the lysosomal membrane that removes the bulk of the cytoplasmic tail, and the second by an aspartyl proteinase located within lysosomes which releases the luminal part of LAP from the membrane-spanning domain. The first cleavage at the cytoplasmic side of the lysosomal membrane depends on acidification of lysosomes, and the second cleavage inside the lysosomes depends on prior processing of the cytoplasmic tail. Only the membrane-bound LAP intermediate produced by the first proteolytic cleavage reaction, but not the intact precursor, can serve as a substrate for the aspartyl proteinase-catalyzed cleavage inside the lysosomes. This suggests that the conformation or folding of the luminal portion of LAP is controlled in part by its cytoplasmic tail. The soluble mature enzyme is isolated as a ~92-kDa homodimer from tissues.

Lysosomal acid phosphatase is synthesized as a precursor with an apparent m.w. of 69 kDa and processed to a family of 43–47 kDa polypeptides which then dimerize (Waheed and Van Etten, 1985; Waheed *et al.*, 1988). Besides its soluble form in lysosomes, LAP is also present in nonlysosomal organelles, as a membrane-associated form which accounts for ~40% of all LAP.

3.5.   Structure, Expression, and Chromosomal Location of LAP Gene

The amino acid sequence and topography of human LAP is based on the nucleotide sequence of the lysosomal acid phosphatase cDNA and gene isolated from human placenta. Pohlmann *et al.* (1988) used antibodies against purified human placental LAP to screen a λgt11 cDNA library to eventually select a cDNA clone having a 2112-bp insert consisting of a 12-bp 5′ noncoding region, a 1269-bp open reading frame, and an 831-bp 3′ untranslated region with a putative polyadenylation signal upstream of a 3′ polyA tract. The deduced amino acid sequence contains a putative signal sequence of 30 amino acids followed by a sequence of 393 amino acids that contains eight potential glycosylation sites. The hydrophobicity profile reveals a 27-amino acid hydrophobic region situated approximately 18 amino acids before the C-terminus. This hydrophobic sequence corresponds well with the expected position of the transmembrane-spanning domain. Expression of the human placental LAP cDNA in monkey COS cells resulted in the production of enzymatically active acid phosphatase. The 393 amino acids following the putative signal peptide predict a size of 43 kDa and begin with a sequence identical to the previously determined N-terminal amino acid sequence of placental LAP. The predicted size is in good agreement with the 44 kDa protein that corresponds to the deglycosylated precursor of the human fibroblast membrane-associated LAP.

The gene for human LAP is 9 kb in length, comprising 11 exons, and is located on human chromosome 11 (Geier *et al.*, 1989). The high GC content (63.5%), two GC boxes, and a region complying with the properties of a CpG island are consistent with LAP functioning as a housekeeping gene.

Himeno *et al.* (1989) isolated and sequenced a cDNA encoding the acid phosphatase from rat liver lysosomes in which 89% of the amino acids and 67% of the nucleotides are similar to those of human placental LAP cDNA. The amino acid sequence of the hydrophobic domain is identical to that of human placental LAP, and the 18 amino acids comprising the cytoplasmic tail are 94% identical between rat liver LAP and human placental LAP. Northern blot analysis identified a single mRNA species in rat liver which was 2.2 kb in length.

The mouse LAP cDNA has also been isolated and shown by Geier *et al.* (1992) to have a high degree of homology with the human and rat LAP cDNAs. Geier *et al.* (1992) examined the expression of LAP in different mouse tissues, using a portion of the mouse LAP cDNA as a probe. Three different LAP mRNA species of 2.3 kb, 3.2 kb, and 5.2 kb were observed to be differentially expressed in various mouse tissues. It remains to be established whether these three mRNA result from the use of different transcription initiation and termination sites or only from different termination sites. In brain, lung, skeletal muscle, and heart, the 5.2-kb LAP mRNA appeared to predominate, whereas the 3.2-kb mRNA was the major species in kidney; in testis, the 2.3-kb and 3.2-kb mRNAs are expressed at comparable levels. In agreement with its presumed function as a housekeeping gene, a ubiquitous and fairly homogenous expression of the LAP gene was observed in most tissues by *in situ* hybridization. Only in testis and brain was a cell type-dependent expression of LAP found. In the adult mouse brain, LAP mRNA is expressed at high levels in pyramidal cells of the hippocampus, granular cells of the dentate gyrus, Purkinje cells of the cerebellum, and epithelial cells of the choroid plexi. In addition, the neurons in other brain areas express higher amounts of LAP mRNA per cell area than do glial cells, which are only slightly labeled. The biological meaning of these observed differences in LAP mRNA expression and the extent to which these different mRNAs are translated into functional acid phosphatase proteins remains to be elucidated.

Although the substrate specificity of the cloned LAP is now known in terms of the small-molecular-weight substrates which it can act upon, the actual contribution of this particular LAP in a biological process has not been directly demonstrated. In an investigation to determine if LAP can dephosphorylate the mannose 6-phosphate residues in lysosomal proteins and arylsulfatase A, Bresciani *et al.* (1992) found that nontransfected L cells dephosphorylate endogenous lysosomal proteins or internalized arylsulfatase A slowly ($t_{1/2} = 13$ hr). A more than 100-fold overexpression of LAP in these cells, however, did not change the dephosphorylation rate. Although LAP can hydrolyze the sugar phosphate, mannose 6-phosphate, it was unable to dephosphorylate the mannose 6-phosphate residues when they were still bound to the oligosaccharide chain attached to arylsulfatase

A. Furthermore, in BHK cells, which dephosphorylate lysosomal enzymes rapidly, and in mouse Ltk-cells, which dephosphorylate lysosomal enzymes rather slowly, comparable LAP levels are observed. Bresciani *et al.* (1992) concluded from these observations that LAP is not the mannose 6-phosphatase that dephosphorylates lysosomal proteins after their delivery to lysosomes. However, it is possible that removal of mannose 6-phosphate from lysosomal proteins is an ordered, sequential process in which a yet to be established rate-limiting enzyme(s) may be required to act upon the lysosomal proteins or part of the oligosaccharide chain before lysosomal acid phosphatase can gain access to the susceptible mannose 6-phosphate phosphomonoester linkage.

3.6. Sorting of LAP in Polarized Epithelial Cells

Prill *et al.* (1993) have found that expression of LAP in Madin–Darby canine kidney cells results in direct sorting of LAP to the basolateral membrane, rapid endocytosis, and delivery to lysosomes. In contrast, a LAP deletion mutant lacking the cytoplasmic tail is delivered to the apical membrane and internalized slowly. A series of truncation and substitution mutants in the cytoplasmic tail was constructed, and comparison of their polarized sorting and internalization revealed that the determinants for basolateral sorting and rapid internalization reside in the same segment of the cytoplasmic tail.

3.7. Similarities Between LAP and Prostatic Acid Phosphatase

Although prostatic acid phosphatase (PAP) is not associated with lysosomes and is immunologically unrelated to LAP, it has many features similar to those of LAP and may have evolved from a common ancestral gene (Geier *et al.,* 1989). Prostatic acid phosphatase is a soluble, tartrate-inhibitable acid phosphatase that displays a substrate specificity similar to LAP, but it also can hydrolyze choline phosphate (Saini and Van Etten, 1978a). It is secreted in large amounts by prostatic tissue into the seminal fluid, with much smaller amounts expressed in several other tissues including spleen, placenta, and kidney. Similar to LAP, PAP exists as a 90-kDA dimer of two identical ~48-kDa subunits (Derechin *et al.,* 1971; Saini and Van Etten, 1978a; Luchter-Wasyl and Ostrowski, 1974). The deduced amino acid sequences of LAP and PAP show 49% similarity within the sequences that encode the mature forms of the two polypeptides. All six cysteines in the mature LAP protein are conserved in PAP, suggesting they may be important for the tertiary structure of these two acid phosphatases (Geier *et al.,* 1989). Similar to LAP, PAP is inactivated by Woodward's reagent K, (Saini and Van Etten, 1978b) suggesting a stoichiometry of inactivation correlating with one essential active site carboxyl group per subunit. The major nucleotide sequence differences between PAP and LAP are in the 5' and 3' untranslated regions, the N-terminal signal sequence, and the portion corresponding to the transmembrane and cytoplasmic domains of LAP.

Sequences corresponding to these latter two domains of LAP, which are encoded by exon 11, are lacking in the PAP cDNA (Geier *et al.,* 1989; Yeh *et al.,* 1987). Geier *et al.* (1989) have hypothesized that the LAP and PAP genes possibly could have arisen from duplication of a common ancestral gene and that exon 11 has been skipped during evolution of the PAP gene. In the human genome, the PAP gene is located on chromosome 3.

3.8. Tartrate-Resistant Acid Phosphatase (TRAP)

Another acid phosphatase associated with lysosomes is the metalloprotein, tartrate-resistant acid phosphatase (TRAP). In contrast to the major lysosomal acid phosphatases which are strongly inhibited by L-(+)-tartrate, TRAP enzymatic activity is not inhibited by tartrate. In many tissues, TRAP is a minor component of the total acid phosphatase activity. However, in osteoclasts and cells of monocyte/macrophage lineage, TRAP plays a major or substantial role as an acid phosphatase. In addition, TRAP is present at high levels in patients with Gaucher's disease and is a major acid phosphatase serving as a marker for hairy cells in hairy cell leukemia (Li *et al.,* 1970a,b; Yam *et al.,* 1971; Janckila *et al.,* 1992a; Drexler and Gignac, 1994). Tartrate-resistant acid phosphatase has also been referred to as the type 5 acid phosphatase, based on the observation that human nonerythrocytic acid phosphatases separate into seven distinct activity bands when subjected to nondenaturing polyacrylamide gel electrophoresis. The separated phosphatases are designated as type 0, 1, 2, 3, 3b, 4, and 5 according to their increasing mobility toward the cathode under acidic electrophoretic conditions (Janckila *et al.,* 1992a). Only type 5, which is the most cationic, is tartrate-resistant, and thus has been named the type 5 tartrate-resistant acid phosphatase or TRAP.

Subcellular fractionation studies, as well as ultrastructural immunocytochemistry studies, have demonstrated that TRAP is predominantly located within lysosomes (Andersson *et al.,* 1986; Clark *et al.,* 1989; Reinholt *et al.,* 1990). Tartrate-resistant acid phosphatase has been detected in osteoclasts, macrophages, monocytes, bone, spleen, lung, placenta, epidermis, uterine epithelium during pregnancy, and at high levels in Gaucher's disease and hair cell leukemia (Ek-Rylander *et al.,* 1991a; Janckila *et al.,* 1992a, 1995; Drexler and Gignac, 1994). In osteoclasts, lysosomal hydrolases play an important role in the selective degradative action of these cells. Osteoclasts attach to the bone surface so as to form a sealed extracellular pocket, which is acidified by the action of a plasma membrane proton pump, thereby forming an acidic extracellular space into which lysosomal enzymes are secreted.

Tartrate-resistant acid phosphatase hydrolyzes nucleoside triphosphates and disphosphates, pyrophosphate, aryl phosphates, *p*-nitrophenylphosphate, thiamine pyrophosphate, and α-napthylphosphate, but shows little activity with most aliphatic phosphates (Allen *et al.,* 1989; Fukushima *et al.,* 1991; Drexler and Gignac, 1994). In addition, TRAP has been shown to dephosphorylate several different phosphoproteins. Janckila *et al.* (1992b) first demonstrated that the phosphotyrosine-containing peptide Raytide could be dephosphorylated by TRAP. These authors suggested that the TRAP present at high levels in hairy cell leukemia may function as a protein tyrosine phosphatase. High activity of TRAP as a protein tyrosine phosphatase could regulate the activities of protein tyrosine kinases, thereby influencing the growth and differentiation of hairy cells. Recently, Ek-Rylander *et al.* (1994) demonstrated that TRAP of skeletal osteoclasts partially dephosphorylates the bone matrix phosphoproteins, osteopontin and bone sialoprotein. Flores *et al.* (1992) have previously shown that osteoclasts bind to osteopontin and bone sialoprotein coated onto glass, but once these proteins are dephosphorylated, osteoclasts no longer bind to them (Ek-Rylander *et al.*, 1994). These results suggest that secretion of TRAP from osteoclasts into the resorption area could modulate attachment or motility of osteoclasts on the bone surface and affect the development of ruffled borders in bone tissue.

Tartrate-resistant acid phosphatase displays maxmial activity between pH 5 and 6. It is activated by mild reducing agents but is inhibited by molybdate, fluoride, arsenate, phosphate, and dithionite (Allen *et al.*, 1989; Fukushima *et al.*, 1991). A unique feature of TRAP is the presence of iron in its active site which is required for enzymatic activity. Histidine and tyrosine residues are thought to be involved in binding of the iron atoms in TRAP, and in the case of the purified enzyme from bone tissue, 1.7 mol Fe/mol enzyme has been found (Ek-Rylander *et al.*, 1991b; Lord *et al.*, 1991). Iron exists in TRAP as an anti-ferromagnetically coupled binuclear core (Davis and Averill, 1982; Lord *et al.*, 1990) that gives these metalloproteins a characteristic EPR signal and purple color for which they have been referred to as purple acid phosphatases (Antanaitis and Aisen, 1982; Vincent and Averill, 1990). Some of the differences in enzymatic properties between TRAP isolated from different tissues may be due to differences in the redox state of the iron cluster, loss of iron from the cluster, or possible replacement of one of the two irons by a second metal ion (Ling and Roberts, 1993).

Tartrate-resistant acid phosphatase has been purified from a number of different tissues including spleen, human placenta, rat calvaria, pig uterus, osteoclastoma, macrophages, and hairy cells (Ketcham *et al.*, 1989; Ek-Rylander *et al.*, 1991b; Ling and Roberts, 1993). In these tissues, TRAP is found to exist as a basic glycoprotein with a pI of 8.5–9.0, with a strong affinity for concanavalin A, and consists of either one subunit of M_r 30–40 kDa or two dissimilar chains of 16 and 23 kDa (Ek-Rylander *et al.*, 1991a,b; Drexler and Gignac, 1994). Hayman *et al.* (1991) demonstrated that both dissimilar chains of TRAP are transcribed from the same mRNA species, which is translated to yield a single polypeptide that is subsequently cleaved to yield two chains joined by disulfide bond linkage.

Full-length cDNAs have been isolated for TRAPs from rat calvaria, beef spleen, human placenta, and human macrophages (Ketcham *et al.*, 1989; Lord *et al.*, 1990; Ek-Rylander *et al.*, 1991a; Ling and Roberts, 1993). These cDNAs typically contain a 969–981-bp open reading frame that codes for a deduced protein sequence of 323–327 amino acids and displays a putative signal sequence of some 20 amino acids. The mature polypeptide of 306 amino acids has a calculated molecular weight of 34,350 Da and contains two potential N-glycosylation sites. The deduced amino acid sequences of TRAP enzymes from human placenta, beef spleen, and macrophages are 85–94% homologous. The N-terminal amino acid segment of TRAP is 83% homologous to the corresponding region in LAP and PAP. However, a comparison of the entire amino acid sequence of TRAP with that of human prostatic and lysosomal acid phosphatases displays only a 41–44% similarity (Ek-Rylander *et al.*, 1991a; Lord *et al.*, 1990; Yeh *et al.*, 1987; Pohlmann *et al.*, 1988).

Alcantara *et al.* (1994) have prepared constructs containing various segments of the murine TRAP 5' promoter region coupled to a luciferase reporter gene to demonstrate that the 5' flanking region of the TRAP gene contains an iron responsive element. Transfected cells were exposed to 10 µg/ml iron saturated human transferrin to assess iron responsiveness of the constructs. Constructs containing a full-length TRAP promoter responded with a four- to fivefold increase in luciferase activity, whereas constructs containing only nucleotides -363 to $+2$ of the TRAP promoter did not respond when exposed to the iron complex. These data indicate that expression of TRAP is regulated by iron at the level of gene transcription. Additional experiments localized the iron regulatory element within the TRAP promoter region to the 5' flanking sequence between nucleotides -1846 and -1240.

Tartrate-resistant acid phosphatase has been mapped to the short arm of human chromosome 19 (Lord *et al.*, 1990). The TRAP mRNA is expressed at high levels in osteoclasts and in cells of monocyte/macrophage lineage. In most cell types, TRAP mRNA is observed as a single 1.5-kb mRNA species (Lord *et al.*, 1990; Ek-Rylander *et al.*, 1991a,b; Lacey *et al.*, 1994). However, in addition to the 1.5-kb TRAP mRNA, Northern blots have revealed the presence of 2.5-kb and 5-kb TRAP-related mRNA transcripts at low levels in osteoclasts (Lacey *et al.*, 1994). These larger species may represent distinct, yet highly related mRNAs or may result from alternative splicing of the same gene. In neonatal rats, TRAP mRNA is highly expressed in skeletal tissues, with much lower levels ($< 10\%$) detected in spleen, thymus, liver, skin, brain, kidney, lung, and heart (Ek-Rylander *et al.*, 1991).

Levels of TRAP mRNA increase during *in vitro* differentiation and is sensitive to macrophage-activating treatments. During *in vivo* osteoclast formation, it appears that TRAP expression first occurs in mononucuclear cells near the bone surface prior to the appearance of osteoclasts. These mononuclear, TRAP-positive cells are considered to be osteoclast precursors, and such cells increase in parallel with osteoclast-like cells in an *in vitro* osteoclast generating culture system. Furthermore, bone marrow macrophages grown in the presence of macrophage colony stimulatory factor express low levels of the 1.5-kb transcript, which increases upon exposure to interleukin 4 (Lacey *et al.*, 1994). Similarly, Lord *et al.* (1990) have observed that TRAP mRNA levels increase more than 20-fold upon transforming normal human monocytes to macrophages by culture in serum-supplemented medium. In addition, culture of K562 cells in the presence of 10 nM phorbol 12-myristate-13-acetate ester for 48–72 hr enhances cellular TRAP activity approximately 30-fold and leads to a corresponding increase in TRAP mRNA levels (Ketcham et al., 1989; Bevilacqua et al., 1991).

Elucidating osteoclast cell biology and function is important for understanding the mechanisms involved in regulating bone tissue growth and turnover. As part of this effort, a number of factors affecting TRAP levels in skeletal osteoclasts have been highlighted. Moonga et al. (1993) have found that the G protein stimulators, tetrafluoraluminate and cholera toxin, inhibit bone resorption and lead to a dramatic increase in TRAP. In contrast, pertussis toxin, a G_i inhibitor, inhibits bone resorption but slightly decreases TRAP release. These results suggest an involvement of Gs-like G protein in osteoclast TRAP secretion. Zheng et al. (1994) have found that at low concentrations, the peptide osteostatin, which is derived from parathyroid hormone-related protein, is able to inhibit osteoclast TRAP by 50%. It appears that osteostatin has a biphasic effect on TRAP activity, inhibiting its secretion and either suppressing its synthesis or increasing its degradation. Loss of estrogen at menopause can cause rapid loss in bone volume that can be prevented by estrogen therapy. The mechanism by which estrogen modulates resorption activity remains unknown. Recently, Kremer et al. (1995) have shown that estrogen treatment causes a decrease in the steady-state levels of the mRNAs of a variety of lysosomal enzymes in both avian osteoclasts and human osteoclast-like multinucleated cells from giant tumors of the bone. In addition, Zheng et al. (1995) found that administration of 17β-estradiol to ovariectomized rats reduced TRAP and calcitonin gene expression. Overall, these results suggest that one of the mechanisms by which estrogen may decrease bone loss is through its ability to lower mRNA levels of osteoclast bone-resorbing enzymes in bone tissue.

3.9. Relationship of TRAP to Uteroferrin

Uteroferrin is a purple, iron-binding acid phosphatase abundant in the uterine secretions of pregnant sows. Physiological studies in vivo have suggested that uteroferrin, under the influence of progesterone, serves in the delivery of maternal iron to the developing fetal piglet during pregnancy (Ling and Roberts, 1993). The TRAPs share 85% amino acid homology with uteroferrin and are very similar in many other properties. However, TRAP is located intracellularly within lysosomes, whereas uteroferrin is a secreted protein. Recent studies by Ling and Roberts (1993) indicate that uteroferrin and TRAP are derived from the same gene. These authors concluded that the difference in trafficking between uteroferrin and TRAP could be the consequence of minor posttranslational modifications, or that the very high synthetic rates achieved in the uterus in response to progesterone

Table II
Substrate Specificity of Acid Pyrophosphatase[a]

Substrate	Concentration (mM)	Activity (%)	Inhibition by tartrate (%)[b]
ATP	1.5	100	9
CTP	1.5	87	8
UTP	1.5	71	—
GTP	1.5	35	—
TTP	1.5	78	—
PPi	1.0	46	—
Thiamine diphosphate	1.0	15	14
NAD+	—	0	—
NADP+	—	0	—
UDP-glucose	—	0	—
Phosphocreatine	—	0	—

[a]Rat liver acid pyrophosphatase, purified by CM-cellulose chromatography, was assayed for its ability to hydrolyze the indicated substrates at 37 °C in 0.1 M sodium acetate pH 5.2 buffer. Pi released during the incubations was measured by the Fiske–Subbarow method.
[b]Inhibition by 1.4 mM L-(+)-tartrate was for a 20-min incubation.
[c]Taken from Brightwell and Tappel (1968).

could overwhelm the processing enzymes and receptors responsible for lysosomal targeting of TRAP within the Golgi apparatus.

3.10. Acid Pyrophosphatase

Brightwell and Tappel 1967a,b) and Ragab et al. (1968) characterized the acid pyrophosphatase of rat liver lysosomes. It displayed maximal activity against pyrophosphate at pH 5.0–5.5. The substrate specificity of this enzyme includes the nucleoside triphosphates, PPi, FAD and to a lesser extent, thiamine diphosphate (Table II). Apparent K_m values of 0.19 mM for ATP and 0.07 mM for PPi were observed for rat liver acid pyrophosphatase at pH 5.2 and 37 °C. Acid pyrophosphatase fails to hydrolyze NAD, ADP, NADP, UDP-glucose, or phosphocreatine. In contrast to the strong inhibition of acid phosphatase and acid nucleotidase by 1.4 mM L-(+)-tartrate, rat liver acid pyrophosphatase is only weakly inhibited by L-(+)-tartrate. Furthermore, acid pyrophosphatase is not affected by 0.02 mM ouabain, 0.02 mM dinitrophenol, 5 mM GSH, 1 mM N-ethylmaleimide, 1 mM $ZnCl_2$, or 1 mM EDTA, whereas 0.7 mM NaF results in a 50% inhibition.

3.11. Acid Trimetaphosphatase

Acid trimetaphosphatase is a lysosomal enzyme capable of hydrolyzing the phosphodiester bonds in the cyclic phosphate, trimetaphosphate. Acid trimetaphosphatase (TMPase) was first noted by Berg (1960) following histo-chemical staining of various tissues. Berg and Gordon (1960) observed maximal TMPase activity near pH 4.0; this activity was completely inhibited by 10 mM Cu^{2+} or F^-, and to a lesser extent Fe^{2+} or Be^{2+}. Doty and Schofield (1972) ap-plied Berg's histochemical staining method in electron microscopic studies to demonstrate in skeletal muscle that acid trimetaphosphatase was confined largely to lysosomes. Further studies by Doty et al. (1977) showed that tissue staining with acid TMPase serves as a useful histochemical marker for lysosomes in a wide variety of tissues. With most tissues, acid TMPase appears to be more spe-cific than acid phosphatase for staining lysosomal structures. For some cell types there is a good correspondence among the lysosomes stained with either acid TMPase or acid phosphatase. However, Oliver (1980) noted that despite the sim-ilarities in the lysosomal staining pattern of acid TMPase and acid phosphatase, there sometimes are also differences in the lysosome populations that are histo-chemically stained with these two enzymes. In rat exocrine acinar cells, Oliver (1980) observed acid TMPase staining of basal elongated lysosomes which were not stained with acid phosphatase. Similarly, Zhang et al. (1991) have noted a reciprocal relationship between two kinds of lysosomes containing either acid phosphatase or tri-metaphosphatase activity in the transitional epithelium of the rat urinary bladder. However, Petty et al. (1985) have found that the lysosomal staining pattern obtained with acid TMPase was very similar to that of acid phos-phatase in macrophages and monocytes. Furthermore, acid TMPase was present within phagolysosomal structures following macrophage phagocytosis of either latex beads, Corynebacterium parvum, or Saccharomyces cerevisiae. The cell types in which acid TMPase has been shown to be lysosomally associated in-clude hepatocytes, cells of the lacrimal gland, osteoclasts, exocrine acinar cells, thyroid follicular cells, natural killer cells, monocytes, and macrophages (Petty et al., 1985), among others. Localization of acid TMPase in thyroid follicular cells was found to increase after TSH stimulation (Payer et al., 1980). Despite the ever-growing base of knowledge regarding the presence of trimetaphosphatase activity within lysosomes, very little is known about the lysosomal formation of trimetaphosphate. Long chains of inorganic polyphosphate are known to break down very slowly to form trimetaphosphatase at neutral pH (Kulaev, 1979), and the existence of inorganic polyphosphates within human fibroblast lysosomes has been demonstrated. However, the detailed mechanism(s) responsible for lysoso-mal trimetaphosphate formation and the functional significance of acid TMPase remain to be elucidated.

3.12. Polyphosphate Synthesis

During the course of characterizing lysosomal phosphate transport, [^{32}P]or-thophosphate was observed to be rapidly incorporated into trichloroacetic acid-soluble and acid-insoluble products following phosphate uptake into human fibroblast lysosomes (Pisoni, 1991). Pisoni and Lindley (1992) characterized the nature of the ^{32}P-labeled TCA-insoluble product formed in human fibroblast lysosomes and found it to consist of long chains of inorganic polyphosphate ranging in length from 100 to at least 600 phosphate residues (Fig. 3). Thus, although lysosomes generally degrade many kinds of macromolecules, these results provided evidence for at least partial synthesis of inorganic polyphosphates within the lysosomal compartment.

The biological role and enzymatic pathway responsible for lysosomal polyphosphate synthesis are not known at the present time. However, there are several areas in which polyphosphates could play an important role in lysosomal

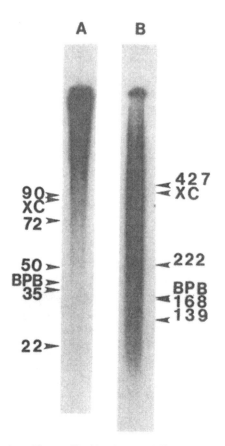

FIGURE 3. Size distribution of human fibroblast lysosomal ^{32}P-polyphosphate on polyacrylamide gels. Human fibroblast lysosomes were incubated with ^{32}P-orthophosphate, and the resulting polyphosphates that form were electrophoresed on (A) a 12% polyacrylamide gel, or (B) 4% polyacrylamide gel along with polyphosphate standards (indicated by arrowheads) and the marker dyes bromphenol blue (BPB) or xylene cyanolle (XC).

function. Inorganic polyphosphates could serve to buffer the acidic intralysosomal pH near 5.0, thereby minimizing effects of fluctuations in lysosomal proton pump activity. The transmembrane potential of lysosomes has been measured to be negative on the inside (Smith *et al.,* 1987). Large amounts of highly negatively charged inorganic polyphosphate species within lysosomes could contribute significantly to the negative lysosomal transmembrane potential. Polyphosphates could also contribute substantially to the osmotic pressure of lysosomes and form complexes with cationic solutes such as arginine which have been observed in yeast vacuoles (Kulaev and Vagabov, 1983; Cramer *et al.,* 1980; Cramer and Davis, 1984). Some lysosomal enzymes are delivered to lysosomes as proenzymes which autoactivate upon binding to negatively charged polymers (Mason and Massey, 1992). Lysosomal polyphosphates could function as the natural substratum for autoactivation of these lysosomal proenzymes. Finally, orthophosphate has been shown to inhibit some lysosomal enzymes. As examples, Pi at concentrations as low as 20–50 μM can inhibit arylsulfatase by 50%, and a K_i of 0.8 mM has been determined for Pi as an inhibitor of LAP activity. Thus, it would appear that maintaining intralysosomal orthophosphate levels at a sufficiently low level would be important to ensure that various lysosomal enzymes are not inhibited. Incorporation of orthophosphate into long chains of inorganic polyphosphate could serve as a mechanism to regulate intralysosomal orthophosphate levels.

A variety of different enzymes involved in the synthesis of inorganic polyphosphate have been isolated from miocroorganisms. For many of these enzymes, a small-molecular-weight phosphorylated substrate serves as the phosphate donor. During lysosomal [^{32}P]phosphate uptake, approximately 50% of the radiolabeled material recovered from lysosomes consists of small-molecular-weight acid-soluble products which have not yet been identified. It is possible that some of these products serve as a substrate or precursor for the formation of inorganic polyphosphate within lysosomes.

Inorganic polyphosphates have been found in various eubacteria, fungi, algae, protozoa, and insects (Kulaev, 1979; Kulaev and Vagabov, 1983). In addition, there have been several reports of the existence of polyphosphate in some mammalian tissues (Cowling and Birnboim, 1994; Kulaev, 1979; Griffin and Penniall, 1966; Offenbacher and Kline, 1984; Gabel and Thomas, 1971; Penniall and Griffin, 1984). Under certain nutritional conditions, inorganic polyphosphates can comprise up to 20% of the cell dry weight of yeast, and they have been found at high concentrations in yeast acidic vacuoles which are analogous to lysosomes in mammalian cells. Despite extensive research with bacteria and fungi, a clear determination of the function of polyphosphates in these microorganisms has not been established.

Inorganic polyphosphates are not likely to be substrates for LAP, as indicated by the inability of acid phosphatase to degrade pyrophosphate and adenosine triphosphate (Table 1). However, the lysosomal enzymes trimetaphosphatase and acid pyrophosphatase can degrade trimetaphosphate and pyrophosphate, respectively. In addition, other lysosomal activities may exist that facilitate inorganic polyphosphate degradation.

The amount of total inorganic polyphosphate in human fibroblast lysosomes has not been quantified. Schneider (1983) noted that rat liver lysosomes contain 0.13 μmol of phosphate/mg protein but the nature of the phosphate species accounting for this total phosphate has not been determined. Kopf-Maier (1990) used electron spectroscopic imaging to analyze the spatial distribution of carbon, nitrogen, oxygen, and phosphorus in mouse liver lysosomes. All lysosomes that were analyzed showed a diffuse and fine granular content of C, N, and O. In contrast, all Kupffer cell lysosomes and lysosomes of hepatocytes located in peribiliary bodies displayed a very high density of phosphorus. The lysosomes of hepatocytes not associated with the biliary canaliculi showed marked heterogeneity in phosphorus content, varying from high concentrations to negligible amounts. The identity of the molecules accounting for the high phosphorus levels in the lysosomes studied by Kopf-Maier remains to be established.

4. OTHER METABOLIC REACTIONS

4.1. Metabolism of Poly (ADP-Ribosylated) Proteins

Some proteins in eukaryotic cells display a posttranslational modification involving the covalent attachment of poly (ADP-ribose) which has been shown to function in regulation of enzyme function, gene expression, and DNA repair (Boulikas, 1992). In a rare genetic disorder, Williams *et al.* (1984) observed the lysosomal accumulation of glutamyl ribose-5-phosphate in an 8-year-old child and proposed that glutamyl ribose-5-phosphate is formed in lysosomes through metabolism of poly (ADP-ribosylated) proteins and that in this particular patient, lysosomal accumulation was the result of a deficiency of ADP-ribose protein hydrolase. The suggested pathway for degradation of poly (ADP-ribosylated) proteins shown in Fig. 4 primarily involves the action of four types of enzymes: a phosphodiesterase, poly (ADP-ribose) glycohydrolase, proteases, and ADP-ribose protein hydrolase. The clinical expression of this lysosomal storage disease predominantly affects brain and kidney function, leading to progressive neurologic deterioration and renal failure.

4.2. Flavin Adenine Dinucleotide Phosphohydrolase

Shin and Mego (1988) have purified and characterized a major rat liver lysosomal membrane glycoprotein capable of hydrolyzing flavin adenine dinucleotide (FAD) to flavin mononucleotide and AMP. This membrane-associated enzyme migrates as a 70-kDa protein by SDS-PAGE and exhibits an apparent K_m of 0.125 mM for FAD at pH 8.5 and 37 °C. The lysosomal membrane FADase displays maximum

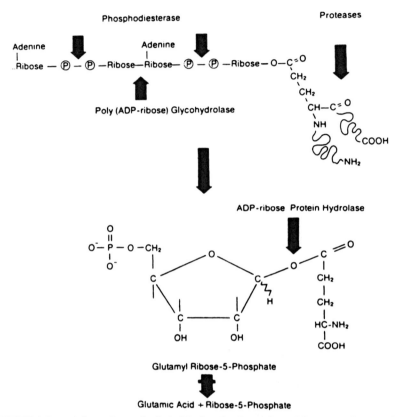

FIGURE 4. Degradative pathway of poly (ADP-ribosylated) proteins and the proposed enzymatic deficiency resulting in the storage of glutamyl ribose-5-phosphate. Reprinted with permission from Williams *et al.* (1984).

activity in the pH range 8.5–9 with very little to no detectable activity at pH < 6.0. Thus, it is difficult to anticipate this enzyme, as described, significantly functioning within the acidic environment of the lysosome. However, the functional orientation of the lysosomal membrane FADase is not known, so it may be involved in reactions on the cytoplasmic side of the membrane, or its pH dependence could be modulated by other cellular factors. In addition to FAD, the lysosomal membrane FADase was also shown to be active with NAD and coenzyme A as substrates, whereas ATP, ADP, AMP, TTP, and pyrophosphate were recognized poorly as substrates. This substrate specificity significantly differs from that of acid pyrophosphatase, which is able to hydrolyze FAD, ATP, and pyrophosphate. Furthermore, antiserum raised against the rat liver lysosomal membrane FADase inhibited the membrane FADase but had no effect on the soluble rat liver acid pyrophosphatase. Lysosomal membrane FADase was activated by Zn^{2+}, inhibited 70–80% by 1 mM EDTA and DTT, and unaffected by NaF or L-(+)-tartrate.

4.3. Coenzyme A Degradation

Coenzyme A is a cofactor important for activity of various enzymes and is primarily located in mitochondria and the cytosol. Several lysosomal enzymes are believed to be involved in coenzyme A degradation and since lysosomes are known to be involved in autophagy of mitochondria, they should be equipped with a repertoire of hydrolytic activities capable of completely degrading coenzyme A. Experimental evidence obtained through a number of studies suggests that coenzyme A is degraded via the series of reactions shown in Fig. 5 (Robishaw and Neely, 1985). In the first step, CoA is dephosphorylated at the 3' position of ribose to form dephospho-CoA. This step is catalyzed by a tartrate-inhibitable, fluoride-sensitive enzyme thought to be LAP. In the second step, a pyrophosphatase cleaves dephospho-CoA to form 4'-phosphopantetheine and 5'-AMP. Skrede (1973) identified a rat liver plasma membrane enzyme catalyzing this reaction that is ten times more active with dephospho-CoA as substrate then CoA. The next step in the degradative pathway is dephosphorylation of 4'-phosphopantetheine by acid phosphatase, which is then followed by pantethinase hydrolyzing pantetheine to

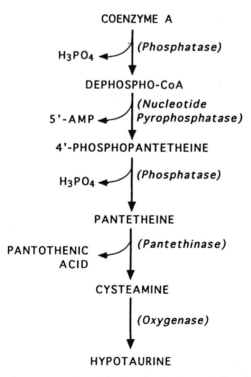

FIGURE 5. Pathway of coenzyme A degradation. Reprinted with permission of Robishaw and Neely (1985).

form pantothenic acid and cysteamine. The subcellular location of pantethinase has not been clearly identified. However, horse kidney pantethinase displays an acidic pH optimum (Cavallini *et al.*, 1968), and rat liver pantethinase appears to be located exclusively in a lysosomal/microsomal fraction (Dupre *et al.*, 1970, 1973; Orloff *et al.*, 1981). In addition , lysosomes are now known to contain a cysteamine transport activity that may be responsible for the lysosomal release of cysteamine generated from CoA degradation (Pisoni *et al.*, 1995). The fate of pantothenic acid resulting from CoA degradation is not known.

As pointed out by Robishaw and Neely (1985), there are a number of logistical problems with the proposed pathway which need to be addressed. The majority of CoA in heart and liver tissue is mitochondrial, whereas many, but not all, of the proposed degradative enzymes are lysosomal. Although CoA certainly could be delivered to lysosomes during autophagy of mitochondria, Smith *et al.* (1989) have observed that liver CoA levels can decrease by 50% within 2 hr of refeeding starved rats. If this decrease represents degradation of CoA, it would appear to occur much too rapidly to be accounted for by lysosomal degradation of whole mitochondria since mitochondria have been estimated to turn over with a half-time of 5–7 days in heart and 5–9 days in liver tissue (Robishaw and Neely, 1985). Mitochondria have been shown to contain a CoA transport system which could regulate mitochondrial CoA efflux (Tahiliani and Neely, 1987; Tahiliani, 1989). However, neither lysosomal CoA transport systems nor other pathways for CoA delivery to lysosomes have been described. Lysosomes would appear to be involved to some extent in coenzyme A turnover, but further investigations are necessary in order to gain a firm understanding of the major pathways determining CoA turnover.

4.4. Folylpolyglutamate Metabolism

Barrueco *et al.* (1992a,b) have studied the metabolism of methotrexate polyglutamates (MTXPG) and have shown the importance of lysosomal folylpolyglutamate (FPG) hydrolase in the degradation of different MTXPGs. Methotrexate, an analog of the folic acid vitamins, is used for cancer treatment since it is a potent inhibitor of dihydrofolate reductase. Methotrexate, taken up by mammalian cells, is converted to methotrexate polyglutamates in the cytosol, a process that involves the sequential addition of glutamyl residues by γ-linkage to the terminal glutamyl moiety of methotrexate. This reaction is catalyzed by the cytosolic enzyme folylpolyglutamate synthetase, which normally converts folic acids to their corresponding polyglutamates. Barrueco *et al.* (1992a) constructed a model for the metabolic turnover of methotrexate polyglutamates (Fig. 6) in which MTXPGs synthesized in the cytosol are first transported into lysosomes by a transport system recognizing the polyglutamates (Barrueco and Sirotnak, 1991). Glutamyl residues are then hydrolyzed from methotrexate polyglutamates by folylpolyglutamate hydrolase. This lysosomal enzyme exhibits maximal activity near pH 6.0

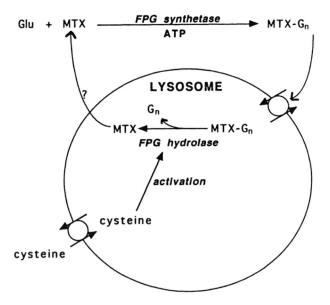

FIGURE 6. Schematic diagram depicting the role of lysosomal transport systems and folypolygluta-mate hydrolase in supporting the lysosomal degradation of methotrexate polyglutamates. Modified from Barrueco, *et al.* (1992).

and is greatly activated by the presence of reduced sulfhydryls. Cysteine was shown to best fulfill the requirements of FPG hydrolase for reduced sulfhydryls and was suggested as the likely sulfhydryl reductant *in vivo*. Comparison of the rates of MTXPG hydrolysis by FPG hydrolase indicated a similar V_{max} for the hydrolysis of MTXPG containing up to four glutamates, although the K_m of FPG hydrolase for the different species of MTXPG is ~13-fold lower for methotrexate tetraglutamate (0.24 μM) compared to methotrexate monoglutamate (3.07 μM).

4.5. Vitamin B₁₂: Endocytosis and Lysosomal Degradation

Vitamin B_{12} , also known as cobalamin, is an essential cofactor for the activity of two mammalian enzymes: methylmalonyl CoA mutase and N_5-methyltetrahydrofolate: homocysteine methyltransferase. Entry of cobalamin into mammalian cells from the plasma depends on cobalamin first combining with the circulating transport protein, transcobalamin (TC II). The transcobalamin II–cobalamin complex then binds to a specific cell surface receptor which internalizes and delivers the complexed cobalamin to lysosomes by receptor-mediated endocytosis (Youngdahl-Turner *et al.*, 1978; Rosenblatt *et al.*, 1984). Within the lysosomal compartment, the transcobalamin II transport protein is degraded by proteolysis, whereas a specific lysosomal cobalamin carrier serves to transport cobalamin from lysosomes to the cytosol where it can be incorporated as an enzymatic cofactor. The work of Rosenblatt *et al.* (1985) was instrumental in determining that the defect in a rare type of methylmalonic aciduria was due to a defect in lysosomal cobalamin transport. The characteristics of this transport system were recently revealed by Idriss and Jonas (1991), who demonstrated saturable cobalamin transport by rat liver lysosomal membrane vesicles with an apparent K_m of 3.5 µM for cobalamin at pH 5.0 and 30 °C. Lysosomal membrane vesicle transport of cobalamin showed maximal transport near pH 5.0 and was trans-stimulated two-to five-fold by cyanocobalamin, adenosylcobalamin, and methylcobalamin. The divalent cations Mg^{2+} and Ca^{2+} stimulated lysosomal cobalamin transport activity, whereas 2 mM EDTA inhibited transport, suggesting that transport was dependent on divalent cations.

5. REFERENCES

Alcantara, O., Reddy, S. V., Roodman, G. D., and Boldt, D. H., 1994, Transcriptional regulation of the tartrate-resistant acid phosphatase (TRAP) gene by iron, *Biochem. J.* **298:**421–425.

Allen, S. H., Nuttleman, P. R., Ketcham, C. M., and Roberts, R. M., 1989, Purification and characterization of human bone tartrate-resistant acid phosphatase, *J. Bone Miner. Res.* **4:**47–55.

Andersson, G. N., Ek-Rylander, B., Hammarstrom, L. E., Lindskog, S., and Toverud, S. U., 1986, Immunocytochemical localization of a tartrate-resistant and vanadate-sensitive acid nucleotide tri- and diphosphatase, *J. Histochem. Cytochem.* **34:**293–298.

Antanaitis, B. S., and Aisen, P., 1982, Detection of a g′ = 1.74 EPR signal in bovine spleen purple acid phosphatase, *J. Biol. Chem.* **257:**5330–5332.

Arsenis, C., and Touster, O., 1967, The partial resolution of acid phosphatase of rat liver lysosomes into a nucleotidase and a sugar phosphate phosphohydrolase, *J. Biol. Chem.* **242:**3399–3401.

Arsenis, C., and Touster, O., 1968, Purification and properties of an acid nucleotidase from rat liver lysosomes, *J. Biol. Chem.* **243:**5702–5708.

Arsenis, C., Gordon, J. S., and Touster, O., 1970, Degradation of nucleic acids by lysosomal extracts of rat liver and Ehrlich ascites tumor cells, *J. Biol. Chem.* **245:**205–211.

Barrett, A. J., 1972, Lysosomal enzymes in *Lysosomes—A Laboratory Handbook* (J. T. Dingle, ed.), pp. 46–135, North-Holland Publishing Co., Amsterdam.

Barrueco, J. R., and Sirotnak, F. M., 1991, Evidence for the facilitated transport of methotrexate polyglutamates into lysosomes derived from S180 cells. Basic properties and specificity for polyglutamate chain length, *J. Biol. Chem.* **266:**11731–11737.

Barrueco, J. R., O'Leary, D. F., and Sirotnak, F. M., 1992a, Metabolic turnover of methotrexate polyglutamates in lysosomes derived from S180 cells. Definition of a two-step process limited by mediated lysosomal permeation of polyglutamates and activating reduced sulfhydryl compounds, *J. Biol. Chem.* **267:**15356–15361.

Barrueco, J. R., O'Leary, D. F., and Sirotnak, F. M., 1992b, Facilitated transport of methotrexate polyglutamates into lysosomes derived from S180 cells. Further characterization and evidence for a simple mobile carrier system with broad specificity for homo- or heteropeptides bearing a C-terminal glutamyl moiety, *J. Biol. Chem.* **267:**19986–19991.

Bensch, K., Gordon, G., and Miller, L., 1964, The fate of DNA-containing particles phagocytized by mammalian cells, *J. Cell Biol.* **21:**105–114.

Berg, G. G., 1960, Histochemical demonstration of acid trimetaphosphatase and tetrametaphosphatase, *J. Histochem. Cytochem.* **8:**92–101.

Berg, G. G., and Gordon, L. H., 1960, Presence of trimetaphosphatase in the intestinal mucosa and properties of the enzyme, *J. Histochem. Cytochem.* **8**:85–91.

Bernardi, G., 1968, Mechanism of action and structure of acid deoxyribonuclease, *Adv. Enzymol.* **31**:1–49.

Bernardi, A., and Bernardi, G., 1966, Studies on acid hydrolases. 3. Isolation and properties of spleen acid ribonuclease, *Biochim. Biophys. Acta.* **129**:23–31.

Bernardi, A., and Bernardi, G., 1968, Studies on acid hydrolases IV. Isolation and characterization of spleen exonuclease, *Biochim. Biophys. Acta* **155**:360–370.

Bernardi, G., and Griffe, M., 1964, Studies on acid deoxyribonuclease. II. Isolation and characterization of spleen-acid deoxyribonuclease, *Biochemistry* **3**:1419–1426.

Bevilacqua, M. A., Lord, D K., Cross, N. C., Whitaker, K. B., Moss, D. W., and Cox, T. M., 1991, Regulation and expression of type V (tartrate-resistant) acid phosphatase in human mononuclear phagocytes, *Mol. Biol. Med.* **8**:135–140.

Boulikas, T., 1992, Poly(ADP-ribose) synthesis and degradation in mammalian nuclei, *Anal. Biochem.* **203**:252–258.

Bowers, W. E., and de Duve, C., 1967, Lysosomes in lymphoid tissue II. Intracelluar distribution of acid hydrolases, *J. Cell Biol.* **32**:339–348.

Braun, M., Waheed, A., and von Figura, K., 1989, Lysosomal acid phosphatase is transported to lysosomes via the cell surface, *EMBO J.* **8**:3633–3640.

Bresciani, R., Peters, C., and von Figura, K., 1992, Lysosomal acid phosphatase is not involved in the dephosphorylation of mannose-6-phosphate containing lysosomal proteins, *Eur. J. Cell Biol.* **58**:57–61.

Brightwell, R., and Tappel, A. L., 1968a, Subcellular distributions and properties of rat liver phosphodiesterases, *Arch. Biochem. Biophys.* **124**:325–332.

Brightwell, R., and Tappel, A. L., 1968b, Lysosomal acid pyrophosphatase and acid phosphatase, *Arch. Biochem. Biophys.* **124**:333–343.

Catchside, D. G., and Holmes, B., 1947, The action of enzymes on chromosomes, *Soc. Exp. Biol. Symp.* **1**:225–231.

Cavallini, D., Dupre, S., Graziani, M. T., and Tinti, M. G., 1968, Identification of pantethinase in horse kidney extract, *FEBS Lett.* **1**:119–121.

Clark, S. A., Ambrose, W. W., Anderson, T. R., Terrell, R. S., and Toverud, S. U., 1989, Ultrastructural localization of tartrate-resistant, purple acid phosphatase in rat osteoclasts by histochemistry and immunocytochemistry, *J. Bone Miner. Res.* **4**:399–405.

Cordonnier, C., and Bernardi, G., 1968, A comparative study of acid deoxyribonucleases extracted from different tissues and species, *Can. J. Biochem.* **46**:989–995.

Cowling, R. T., and Birnboim, H. C., 1994, Incorporation of [^{32}P]orthophosphate into inorganic polyphosphates by human granulocytes and other human cell types, *J. Biol. Chem.* **269**:9480–9485.

Cramer, C. L., and Davis, R. H., 1984, Polyphosphate-cation interaction in the amino acid-containing vacuole of *Neurospora crassa, J. Biol. Chem.* **59**:5152–5157.

Cramer, C. L., Vaughn, L. E., and Davis, R. H., 1980, Basic amino acids and inorganic polyphosphates in *Neurospora crassa:* Independent regulation of vacuolar pools, *J. Bacteriol.* **142**:945–952.

Cunningham, L., and Laskowski, M., 1953, Presence of two different desoxyribonucledepolymerases in veal kidney, *Biochim. Biophys. Acta.* **11**:590–591.

Davis, J. C., and Averill, B. A., 1982, Evidence for a spin-coupled binuclear iron unit at the active site of the purple acid phosphatase from bovine spleen, *Proc. Natl. Acad. Sci. USA* **79**:4623–4627.

de Duve, C., and Berthet, J., 1954, The use of differential centrifugation in the study of tissue enzymes, *Int. Rev. Cytol.* **3**:225–275.

de Duve, C., Pressman, B. C., Gianetto, R., Wattiaux, R., and Appelmans, F., 1955, Tissue fractionation studies. 6. Intracellular distribution patterns of enzymes in rat-liver tissue, *Biochem. J.* **60**:604–619.

Derechin, M., Ostrowski, W., Galka, M., and Barnard, E. A., 1971, Acid phosphomonoesterase of human prostate. Molecular weight, dissociation and chemical composition, *Biochim. Biophys. Acta* **250**:143–154.

DiPietro, D. L., and Zengerle, F. S., 1967, Separation and properties of three acid phosphatases from human placenta, *J. Biol. Chem.* **242**:3391–3396.

Doty, S. B., and Schofield, B. H., 1972, Electron microscopic localization of hydrolytic enzymes in osteoclasts, *Histochem. J.* **4**:245–258.

Doty, S. B., Smith, C. E., Hand, A. R., and Oliver, C., 1977, Inorganic trimetaphosphatase as a histochemical marker for lysosomes in light and electron microscopy, *J. Histochem. Cytochem.* **25**: 1381–1384.

Drexler, H. G., and Gignac, S. M., 1994, Characterization and expression of tartrate-resistant acid phosphatase (TRAP) in hematopoietic cells, *Leukemia* **8**:359–368.

Dulaney, J. T., and Touster, O., 1972, Isolation of deoxyribonuclease II of rat liver lysosomes, *J. Biol. Chem.* **247**:1424–1432.

Dupre, S., Graziani, M. T., Rosie, M. A., Fabi, A., and Del Grosso, E., 1970, The enzymatic breakdown of pantethine to pantothenic acid and cystamine, *Eur. J. Biochem.* **16**:571–578.

Dupre, S., Rosei, M. A., Bellussi, L., Del Grosso, E., and Cavallini, D., 1973, The substrate specificity of pantethinase, *Eur. J. Biochem.* **40**:103–107.

Ek-Rylander, B., Bill, P., Norgard, M., Nilsson, S., and Andersson, G., 1991a, Cloning, sequence, and developmental expression of a type 5, tartrate-resistant, acid phosphatase of rat bone, *J. Biol. Chem.* **266**:24684–24689.

Ek-Rylander, B., Bergman, T., and Andersson, G., 1991b, Characterization of a tartrate-resistant acid phosphatase (ATPase) from rat bone: Hydrodyamnic properties and N-terminal amino acid sequence, *J. Bone Miner. Res.* **6**:365–373.

Ek-Rylander, B., Flores, M., Wendel, M., Heinegard, D., and Andersson, G., 1994, Dephosphorylation of osteopontin and bone sialoprotein by osteoclastic tartrate-resistant acid phosphatase. Modulation of osetoclast adhesion in vitro, *J. Biol. Chem.* **269**:14853–14856.

Flores, M., Norgard, M., Heinegard, D., Reinholt. F. P., and Andersson, G., 1992, RGD-directed attachment of isolated rat osteoclasts to osteopontin, bone sialoprotein, and fibronectin, *Exp. Cell Res.* **201**:526–530.

Fox, I. H., and Kelley, W. N., 1978, The role of adenosine and 2′-deoxyadenosine in mammalian cells, *Annu. Rev. Biochem.* **47**:655–686.

Fukushima, O., Bekker, P. J., and Gay, C. V., 1991, Ultrastructural localization of tartrate-resistant acid phosphatase (purple acid phosphatase) activity in chicken cartilage and bone, *Am. J. Anat.* **191**:228–236.

Futai, M., Miyata, S., and Mizuno, D., 1969, Acid ribonucleases of lysosomal and soluble fractions from rat liver, *J. Biol. Chem.* **244**:4951–4960.

Gabel, N. W., and Thomas, V., 1971, Evidence for the occurrence and distribution of inorganic polyphosphates in vertebrate tissues, *J. Neurochem.* **18**:1229–1242.

Geier, C., von Figura, K., and Pohlmann, R., 1989, Structure of the human lysosomal acid phosphatase gene, *Eur. J. Biochem.* **183**:611–616.

Geier, C., Kreysing, J., Boettcher, H., Pohlmann, R., and von Figura, K., 1992, Localization of lysosomal acid phosphatase mRNA in mouse tissues, *J. Histochem. Cytochem.* **40**: 1275–1282.

Gieselmann, V., Hasilik, A., and von Figura, K., 1984, Tartrate-inhibitable acid phosphatase. Purification from placenta, characterization and subcellular distribution in fibroblasts, *Hoppe-Seyler's Z. Physiol. Chem.* **365**:651–660.

Gottschalk, S., Waheed, A., Schmidt, B., Laidler, P., and von Figura, K., 1989, Sequential processing of lysosomal acid phosphatase by a cytoplasmic thiol proteinase and a lysosomal aspartyl proteinase, *EMBO J.* **8**:3215–3219.

Griffin, J. B., and Penniall, R., 1966, Studies of phosphorus metabolism by isolated nuclei. VI. Labeled components of the acid-insoluble fraction, *Arch. Biochem. Biophys.* **114**:67–75.

Griffin, J. B., Davidian, N. M., and Penniall, R., 1965, Studies of phosphorus metabolism by isolated nuclei. VII. Identification of polyphosphate as a product, *J. Biol. Chem.* **240**:4427–4434.

Hayman, A. R., Dryden, A. J., Cambers, T. J., and Warburton, M. J., 1991, Tartrate-resistant acid phosphatase from human osteoclastomas is translated as a single polypeptide, *Biochem. J.* **277**:631–634.

Heinrikson, R. L., 1969, Purification and characterization of a low molecular weight acid phosphatase from bovine liver, *J. Biol. Chem.* **244**:299–307.

Heydrick, S. J., Lardeux, B. R., and Mortimore, G. E., 1991, Uptake and degradation of cytoplasmic RNA by hepatic lysosomes. Quantitative relationship to RNA turnover, *J. Biol. Chem.* **266**:8790–8796.

Hille, A., Klumperman, J., Geuze, H. J., Peters, C., Brodsky, F. M., and von Figura, K., 1992, Lysosomal acid phosphatase is internalized via clathrin-coated pits, *Eur. J. Cell Biol.* **59**:106–115.

Himeno, M., Fujita, H., Noguchi, Y., Kono, A., and Kato, K., 1989, Isolation and sequencing of a cDNA clone encoding acid phosphatase in rat liver lysosomes, 1989, *Biochem. Biophys. Res. Commun.* **162**:1044–1053.

Idriss, J. -M., and Jonas, A. J., 1991, Vitamin B_{12} transport by rat liver lysosomal membrane vesicles, *J. Biol. Chem.* **266**:9438–9441.

Igarashi, M., and Hollander, V. P., 1968, Acid phosphatase from rat liver, *J. Biol. Chem.* **243**:6084–6089.

Janckila, A. L., Latham, M. D., Lam, K. W., Li, C. Y., and Yam, L. T., 1992a, Heterogeneity of hairy cell tartrate-resistant acid phosphatase, *Clin. Biochem.* **25**:437–443.

Janckila, A. L., Woodford, T. A., Lam, K. W., Li, C. Y., and Yam, L. T., 1992b, Protein-tyrosine phosphatase activity of hairy cell tartrate-resistant acid phosphatase, *Leukemia* **6**:199–203.

Janckila, A. L., Cardwell, E. M., Yam, L. T., and Li, C. Y., 1995, Hairy cell identification by immunohistochemistry of tartrate-resistant acid phosphatase, *Blood* **85**:2839–2844.

Ketcham, C. M., Roberts, R. M., Simmen, R. C., and Nick, H. S., 1989, Molecular cloning of the type 5, iron-containing, tartrate-resistant acid phosphatase from human placenta, *J. Biol. Chem.* **264**:557–563.

Kopf-Maier, P., 1990, The phosphorus content of lysosomes in hepatocytes and Kupffer cells. A study using electon-spectroscopic imaging. *Acta Anat.* **139**:164–172.

Kredich, N. M., and Hershfield, M. S., 1989, Immunodeficiency diseases caused by adenosine deaminase deficiency and purine nucleoside phosphorylase deficiency, in *The Metabolic Basis of Inherited Disease*, 6th Ed. (C. R. Scriver, A. L. Beaudet, W. S. Sly, and D. Valle, eds.), pp. 1045–1075, McGraw-Hill, New York.

Kremer, M., Judd, J., Rifkin, B., Auszmann, J., and Oursler, M. J., 1995, Estrogen modulation of osteoclast lysosomal enzyme secretion, *J. Cell. Biochem.* **57**:271–279.

Kulaev, I. S., 1979, in *The Biochemistry of Inorganic Polyphosphates*, pp. 17–35, John Wiley and Sons, New York.

Kulaev, I. S., and Vagabov, V. M., 1983, Polyphosphate metabolism in micro-organisms, *Adv. Microb. Physiol.* **24**:84–171.

Kumble, K. D., and Kornberg, A., 1995, Inorganic polyphosphate in mammalian cells and tissues, *J. Biol. Chem.* **270**:5818–5822.

Lacey, D. L., Erdmann, J. M., and Tan, H. -L., 1994, Interleukin 4 increases type 5 acid phosphatase mRNA expression in murine bone marrow macrophages, *J. Cell. Biochem.* **54**:365–371.

Lardeux, B. R., and Mortimore, G. E., 1987, Amino acid and hormonal control of macromolecular turnover in perfused rat liver. Evidence for selective autophagy, *J. Biol. Chem.* **262**:14514–14519.

Lardeux, B. R., Heydrick, S. J., and Mortimore, G. E., 1987, RNA degradation in perfused rat liver as determined from the release of [^{14}C]cytidine, *J. Biol. Chem.* **262**:14507–14513.

Lardeux, B. R., Heydrick, S. J., and Mortimore, G. E., 1988, Rates of rat liver RNA degradation in vivo as determined from cytidine release during brief cyclic perfusion in situ, *Biochem. J.* **252**:363–367.

Lehmann, L. E., Eberle, W., Krull, S., Prill, V., Schmidt, B., Sander, C., von Figura, K., and Peters, C., 1992, The internalization signal in the cytoplasmic tail of lysosomal acid phosphatase consists of the hexapeptide PGYRHV, *EMBO J.* **11**:4391–4399.

Li, C. Y., Yam, L. T., and Lam. K. W., 1970a, Acid phosphatase isoenzyme in human leukocytes in normal and pathologic conditions, *J. Histochem. Cytochem.* **18**:473–481.

Li, C. Y., Yam, L. T., and Lam, K. W., 1970b, Studies of acid phosphatase isoenzymes in human leukocytes: Demonstration of isoenzyme cell specificity, *J. Histochem. Cytochem.* **18**:901–910.

Liao, T. -H., 1985, The subunit structure and active site sequence of porcine spleen deoxyribonuclease, *J. Biol. Chem.* **260**:10708–10713.

Liao, T. -H., Liao, W. -C., Chang, H. -C., and Lu, K. -S., 1989, Deoxyribonuclease II purified from the
isolated lysosomes of porcine spleen and from porcine liver homogenates. Comparison with deoxyribonuclease II purified from porcine spleen homogenates, *Biochim. Biophys. Acta* **1007**:15–22.

Lindley, E. R., and Pisoni, R. L., 1993, Demonstration of adenosine deaminase activity in human fibroblast lysosomes, *Biochem. J.* **290**:457–462.

Ling, P., and Roberts, R. M., 1993, Uteroferrin and intracellular tartrate-resistant acid phosphatases are
the products of the same gene, *J. Biol. Chem.* **268**:6896–6902.

Lord, D. K., Cross, N. C. P., Bevilacqua, M. A., Rider, S. H., Gorman, P. A., Groves, A. V., Moss,
D. W., Sheer, D., and Cox, T. M., 1990, Type 5 acid phosphatase. Sequence, expression and chro-
mosomal localization of a differentiation-associated protein of human macrophage, *Eur. J.
Biochem.* **189**:287–293.

Lucther-Wasyl, E., and Ostrowski, W., 1974, Subunit structure of human prostatic acid phosphatase,
Biochim. Biophys. Acta **365**:349–359.

Mason, R. W., and Massey, S. D., 1992, Surface activation of pro-cathepsin L., *Biochem. Biophys. Res.
Commun.* **189**:1659–1666.

Maver, M. E., and Greco, A. E., 1949a, The hydrolysis of nucleoproteins by cathepsins from calf
thymus, *J. Biol. Chem.* **181**:853–860.

Maver, M. E., and Greco, A. E., 1949b, The nuclease activities of cathepsin preparations from calf
spleen and thymus, *J. Biol. Chem.* **181**:861–870.

Moonga, B. S., Pazianas, M., Alam, A. S., Shankar, V. S., Huang, C. L., and Zaidi, M., 1993, Stimulation of a Gs-like G protein in the osteoclast inhibits bone resorption but enhances tartrate-
resistant acid phosphate secretion, *Biochem. Biophys. Res. Commun.* **190**:496–501.

Offenbacher, S., and Kline, E. S., 1984, Evidence for polyphosphate in phosphyorylated nonhistone
nuclear proteins, *Arch. Biochem. Biophys.* **231**:114–123.

Oliver, C., 1980, Cytochemical localization of acid phosphatase and trimetaphosphatase activities in
exocrine acinar cells, *J. Histochem. Cytochem.* **28**:78–81.

Orloff, S., Butler, J. D., Towne, D., Mukherjee, A. B., and Schulman, J. D., 1981, Pantetheinase activity and cysteamine content in cystinotic and normal fibroblasts and leukocytes, *Pediatr. Res.*
15:1063–1067.

Oshima, R. G., and Price, P. A., 1973, Alkylation of an essential histidine residue in porcine spleen deoxyribonuclease, *J. Biol. Chem.* **248**:7522–7526.

Oshima, R. G., and Price, P. A., 1974, Effect of sulfate on the activity and the kinetics of deoxyri-
bonucleic acid degradation by porcine spleen deoxyribonuclease, *J. Biol. Chem.* **249**:4435–4438.

Payer, A. F., Battle, C. L., and Peake, R. L., 1980, Use of osmium-ferrocyanide treatment for improved
lysosomal acid trimetaphosphatase reaction and subcellular detail in thyroid follicular cells, *J.
Histochem. Cytochem.* **28**:183–186.

Penniall, R., and Griffin, J. B., 1984, Studies of phosphorus metabolism by isolated nuclei. XII. Some
fundamental properties of the incorporation of ^{32}P into polyphosphate by rat liver nuclei, *Biosci.
Rep.* **4**:957–962.

Peters, C., Braun, M., Weber, B., Wendland, M., Schmidt, B., Pohlmann, R., Waheed, A., and von
Figura, K., 1990, Targeting of a lysosomal membrane protein: A tyrosine-containing endocytosis
signal in the cytoplasmic tail of lysosomal acid phosphatase is necessary and sufficient for targeting to lysosomes, *EMBO J.* **9**:3497–3506.

Petty, H. R., Hermann, W., and McConnell, H. M., 1985, Cytochemical study of macrophage lysosomal inorganic trimetaphosphatase and acid phosphatase, *J. Ultrastruct. Res.* **90**:80–88.

Pisoni, R. L., 1991, Characterization of a phosphate transport system in human fibroblast lysosomes,
J. Biol. Chem. **266**:979–985.

Pisoni, R. L., and Lindley, E. R., 1992, Incorporation of [^{32}P]orthophosphate into long chains of inorganic polyphosphate within lysosomes of human fibroblasts, *J. Biol. Chem.* **267**:3626–3631.

Pisoni, R. L., and Thoene, J. G., 1989, Detection and characterization of a nucleoside transport system in human fibroblast lysosomes, *J. Biol. Chem.* **264**:4850–4856.

Pisoni, R. L., Park, G. Y., Velilla, V. Q., and Thoene, J. G., 1995, Detection and characterization of a transport system mediating cysteamine entry into human fibroblast lysosomes. Specificity for aminoethylthiol and aminoethylsulfide derivatives, *J. Biol. Chem.* **270**:1179–1184.

Pohlmann, R., Krentler, C., Schmidt, B., Schroder, W., Lorkowski, G., Culley, J., Mersmann, G., Geier, C., Waheed, A., Gottschalk, S., Grsechik, K. -H., Hasilik, A., and von Figura, K., 1988, Human lysosomal acid phosphatase: Cloning, expression and chromosomal assignment, *EMBO J.* **7**:2343–2350.

Prill, V., Lehmann, L., von Figura, K., and Peters, C., 1993, The cytoplasmic tail of lysosomal acid phosphatase contains overlapping but distinct signals for basolateral sorting and rapid internalization in polarized MDCK cells, *EMBO J.* **12**:2181–2193.

Ragab, M. H., Brightwell, R., and Tappel, A. L., 1968, Hydrolysis of flavin-adenine dinucleotide by rat liver lysosomes, *Arch. Biochem. Biophys.* **123**:179–185.

Rehkop, D. M., and Van Etten, R. L., 1975, Human liver acid phosphatases, *Hoppe-Seyler's Z. Physiol. Chem.* **356**:1775–1782.

Reinholt, F. P., Widholm, S. M., Ek-Rylander, B., and Andersson, G., 1990, Ultrastructural localization of a tartrate-resistant acid ATPase in bone, *J. Bone Miner. Res.* **5**:1055–1061.

Robishaw, J. D., and Neely, J. R., 1985, Coenzyme A metabolism, *Am. J. Physiol.* **248**:E1–E9.

Rosenblatt, D. S., Cooper, B. A., Pottier, A., Lue-Shing, H., Matiaszuk, N., and Grauer, K., 1984, Altered vitamin B$_{12}$ metabolism in fibroblasts from a patient with megaloblastic anemia and homocystinuria due to a new defect in methionine biosynthesis, *J. Clin. Invest.* **74**:2149–2156.

Rosenblatt, D. S., Hosack, A., Matiaszuk, N., Cooper, B., and Laframboise, R. L., 1985, Defect in vitamin B$_{12}$ release from lysosomes: Newly described inborn eror of vitamin B$_{12}$ metabolism, *Science* **228**:1319–1321.

Saha, B. K., 1982, Specificity of acid RNase from HeLa cell lysosomes, *Nucleic Acids Res.* **10**:645–652.

Saha, B. K., Graham, M. Y., and Schlessinger, D., 1979, Acid ribonuclease from HeLa cell lysosomes, *J. Biol. Chem.* **254**:5951–5957.

Saha, B. K., Sameshima, M., Sameshima, F., and Schlessinger, D., 1981, Lysosomal enzyme activities and RNA turnover rates in growing and nongrowing WI-38 and HeLa cells, *In Vitro* **17**:816–824.

Saini, M. S., and Van Etten, R. L., 1978a, Dimeric nature and amino acid compositions of homogeneous canine prostatic, human liver and rat liver acid phosphatase isoenzymes. Specificity and pH-dependence of the canine enzyme, *Biochim. Biophys. Acta* **526**:468–478.

Saini, M. S., and Van Etten, R. L., 1978b, A homogeneous isoenzyme of human liver acid phosphatase, *Arch. Biochem. Biophys.* **191**:613–624.

Schneider, D. L., 1983, ATP-dependent acidification of membrane vesicles isolated from purified rat liver lysosomes. Acidification activity requires phosphate, *J. Biol. Chem.* **258**:1833–1838.

Shin, H. J., and Mego, J. L., 1988, A rat liver lysosomal membrane flavin-adenine dinucleotide phosphohydrolase: Purification and characterization, *Arch. Biochem. Biophys.* **267**:95–103.

Silverstein, S. C., and Dales, S., 1968, The penetration of reovirus RNA and initiation of its genetic function in L-strain fibroblasts, *J. Cell Biol.* **36**:197–230.

Skrede, S., 1973, The degradation of CoA: Subcellular localization and kinetic properties of CoA- and dephospho-CoA pyrophosphatase, *Eur. J. Biochem.* **38**:401–407.

Smith, C., Cano, M., and Potyriaj, J., 1978, The relationship between metabolic state and total CoA content of rat liver and heart, *J. Nutr.* **108**:854–862.

Smith, M. L., Greene, A. A., Potashnik, R., Mendoza, S. A., and Schneider, J. A., 1987, Lysosomal cystine transport. Effect of intralysosomal pH and membrane potential, *J. Biol. Chem.* **262**:1244–1253.

Sosa, M. A., Schmidt, B., von Figura, K., and Hille-Rehfeld, A., 1993, In vitro binding of plasma membrane-coated vesicle adaptors to the cytoplasmic domain of lysosomal acid phosphatase, *J. Biol. Chem.* **268**:12537–12543.

Tahiliani, A. G., 1989, Dependence of mitochondrial coenzyme A uptake on the membrane electrical gradient, *J. Biol. Chem.* **264**:18426–18432.

Tahiliani, A. G., and Neely, J. R., 1987, A transport system for coenzyme A in isolated rat heart mitochondria, *J. Biol. Chem.* **262**:11607–11610.

Tsubota, Y., Yamanaka, M., and Takagi, Y., 1974, Mode of action of acid deoxyribonucleases from human gastric mucosa and cervix uteri, *J. Biol. Chem.* **249**:3890–3894.

Vaes, G., 1973, Digestion of nucleic acids and nucleotides, in *Lysosomes and Storage Diseases* (H. G. Hers, and F. Van Hoof, eds.), Academic Press, New York.

Van Dyck, J. M., and Wattiaux, R., 1968, Distribution intracellulaire de l'exonuclease acide dans le foie de rat, *Eur. J. Biochem.* **7**:15–20.

Vincent, J. B., and Averill, B. A., 1990, An enzyme with a double identity: Purple acid phosphatase and tartrate-resistant acid phosphatase, *FASEB J.* **4**:3009–3014.

Waheed, A., and Van Etten, R. L., 1985, Biosynthesis and processing of lysosomal acid phosphatase in cultured human cells, *Arch. Biochem. Biophys.* **243**:274–283.

Waheed, A., Gottschalk, S., Hille, A., Krentler, C., Pohlmann, R., Braulke, T., Hauser, H., Geuze, H., and von Figura, K., 1988, Human lysosomal acid phosphatase is transported as a transmembrane protein to lysosomes in transfected baby hamster kidney cells, *EMBO J.* **7**:2351–2358.

Williams, J. C., Chambers, J. P., and Liehr, J. G., 1984, Glutamyl ribose 5-phosphate storage disease. A hereditary defect in the degradation of poly(ADP-ribosylated) proteins, *J. Biol. Chem.* **259**:1037–1042.

Yam, L. T., Li, C. Y., and Lam, K. W., 1971, Tartrate-resistant acid phosphatase isoenzyme in the reticulum cells of leukemic reticuloendotheliosis, *N. Engl. J. Med.* **284**:357–360.

Yeh, L. C., Lee, A. J., Lee, N. E., Lam, K. W. and Lee, C., 1987, Molecular cloning of cDNA for human prostatic acid phosphatase, *Gene* **60**:191–196.

Youngdahl-Turner, P., Rosenberg, L. E., and Allen, R. H., 1978, Binding and uptake of transcobalamin II by human fibroblasts, *J. Clin. Invest.* **61**:133–141.

Zhang, S. X., Okada, K. T., Garcia del Saz, E., and Seguchi, H., 1991, Ultracytochemical localization of acid phosphatase and trimetaphosphatase activities in the transitional epithelium of the rat urinary bladder, *J. Submicrosc. Cytol. Pathol.* **23**:431–437.

Zheng, M. H., McCaughan, H. B., Papadimitriou, J. M., Nicholson, G. C., and Wood, D. J., 1994, Tartrate resistant acid phosphatase activity in rat cultured osteoclasts is inhibited by a carboxyl terminal peptide (osetostatin) from parathyroid hormone-related protein, *J. Cell. Biochem.* **54**: 145–153.

Zheng, M. H., Lau, T. T., Prince, R., Criddle, A., Wysocki, S., Beilharz, M., Papadimitriou, J. M., and Wood, D. J., 1995, 17-Beta-estradiol suppresses gene expression of tartrate-resistant acid phosphatase and carbonic anhydrase II in ovariectomized rats, *Calcif. Tissue Int.* **56**:166–169.

Chapter 10

Acidification of Lysosomes and Endosomes

Rebecca W. Van Dyke

1. INTRODUCTION: CONCEPT OF ACID INTERIOR

In eukaryotic cells the interior space of endosomes, lysosomes, and a variety of other intracellular organelles is maintained at a pH considerably more acidic than either cytosolic pH or extracellular pH. This acidic vesicle interior can be demonstrated by staining of living cells with pH-sensitive dyes such as the fluorescent tertiary amine acridine orange, which is protonated and trapped in low pH compartments. Cells such as hepatocytes that are endocytically active exhibit literally hundreds of fluorescent acid vesicles (Lake *et al.,* 1987). Concomitant exposure of cells to agents that eliminate pH gradients, such as proton ionophores, or high concentrations of other tertiary amines abolishes this fluorescence, confirming the acid nature of these intracellular compartments and illustrating that vesicle acidification is the product of the active transport of protons rather than simply a Donnan-type equilibrium. Considerable experimental work over the past 15 years has elucidated the mechanisms for acidification of intracellular organelles, including lysosomes and endosomes, and current investigative efforts are focused at understanding regulation of the acidification mechanism.

Rebecca W. Van Dyke   Department of Internal Medicine, University of Michigan Medical School, Ann Arbor, Michigan 48109.

Subcellular Biochemistry, Volume 27: Biology of the Lysosome, edited by Lloyd and Mason. Plenum Press, New York, 1996.

The acid interior of lysosomes and endosomes appears to subserve a variety of important cell functions including receptor-mediated endocytosis and lysosomal degradation of macromolecules, processing and secretion of polypeptide hormones, and storage and excretion of a variety of drugs. In addition, vesicle acidification establishes an electrochemical proton gradient across the membrane of intracellular organelles, and this electrochemical gradient can be exploited as the "driving force" for uptake or release of a variety of organic and inorganic solutes by acidified vesicles. The acid pH inside these organelles also may act as a signal to initiate or terminate changes in vesicle composition and/or function, analogous to more classic second messengers. Sequestration of hydrogen ions in acidified endosomes, lysosomes, and other organelles also may contribute to cell pH regulation. Inhibition of the vacuolar proton pump in intact hepatocytes and opossum kidney cells leads not only to endosome and lysosome alkalinization, but also to rapid cytoplasmic acidification (Bronk and Gores, 1991; Gekle and Silbernagl, 1995; Wadsworth and van Rossum, 1994).

The same acidification mechanism is present on the plasma membrane of certain specialized eukaryotic cells, where it participates in transepithelial proton transport and/or bone reabsorption. Finally, this acid vesicle system has been exploited by a number of viruses and bacterial toxins (Lencer *et al.*, 1992) to facilitate infection and disease. This chapter reviews the acidification mechanisms of lysosomes and endosomes, their regulation, and some aspects of the role of the pH gradient in driving transport of other solutes and in mediating certain drug-induced diseases.

2. VACUOLAR PROTON PUMP

Acidification of the intracellular organelles of eukaryotes is mediated by a unique, electrogenic proton pump called the vacuolar proton pump, or V-type H^+-ATPase. Evidence for such a proton pump includes the observations that organellar acidification is dependent on hydrolysis of ATP in the presence of magnesium and that acidification proceeds in the absence of other ions and in the presence of inhibitors of other known ion pumps (Forgac, 1992; Gluck, 1992, 1993). A separate proton transporting pyrophosphatase has been identified in plant vacuoles, but it appears to play little, if any, role in acidification of eukaryotic intracellular organelles (Coyaud *et al.*, 1987; Hedrich *et al.*, 1989; Sze *et al.*, 1992; Wada and Anraku, 1994).

2.1. Structure

The vacuolar H^+-ATPase is a large, multisubunit structure at least 750 kDa in size. It has been purified from several tissues, including bovine brain clathrin-coated vesicles, bovine kidney, chromaffin granules from bovine adrenal medulla cells, rat liver lysosomes, osteoclasts, and vacuoles from fungi and higher plants

(Arai *et al.*, 1993; Chatterjee *et al.*, 1992b; Floor *et al.*, 1990; Forgac, 1992; Moriyama and Nelson, 1989; Wang and Gluck, 1990). In all of these cell types, the proton pump consists of at least nine different subunits, some in multiple copies. Analogous to the F_1-F_0 H^+-ATPase of mitochondria, which functions as an ATP synthase, the vacuolar H^+-ATPase is thought to consist of a transmembrane portion containing the proton channel (termed the V_0 complex) and a large ball-like head group on the cytosolic side (termed the V_1 complex) (Fig. 1). The V_1 complex contains three 70-kDa catalytic subunits (the A subunit), three 56–60-kDa regulatory B subunits, and one copy each of a 40-kDa C subunit, a 34-kDa D subunit, and a 33-kDa E subunit (Myers and Forgac, 1993b). The V_0 complex consists of six identical 17-kDa subunits (also called the proteolipid subunits) that form a transmembrane proton channel and one copy each of a 19-kDa subunit, a 38-kDa accessory subunit, and a 100–116-kDa transmembrane glycoprotein (Perin *et al.*, 1991; J. Zhang *et al.*, 1992). In addition, 57-and 45-kDa proteins of unknown significance also have been immunoprecipitated with the vacuolar H^+-ATPase (Supek *et al.*, 1994; Xie *et al.*, 1994).

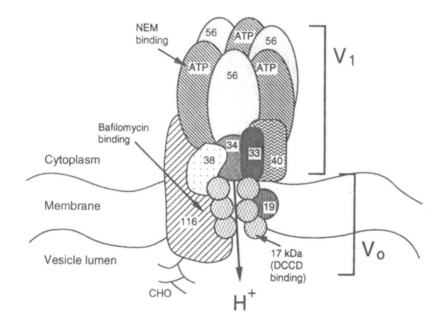

FIGURE 1. A model of the vacuolar H^+-ATPase (based on data and model from Forgac, 1992). Numbers on subunits represent approximate molecular weights. The catalytic subunit (designated by the ATP binding site) is about 70 kDa. The V_1 complex can be released from the membrane and contains the 70-, 56-, 40-, 34-, and 33-kDa subunits, whereas the V_0 complex remains membrane-bound and contains the 116-, 38-, 19- and 17-kDa subunits. Binding sites for three vacuolar H^+-ATPase inhibitors are indicated and CHO indicates glycosylation of the luminal side (which is topographically equivalent to the extracellular space) of the 116-kDa subunit.

The genes for the A, B, and C subunits of the V_1 complex and the proteolipid and 100–116-kDa subunit from the V_0 complex have been cloned (Bernasconi *et al.*, 1990; Lai *et al.*, 1991; Marushack *et al.*, 1992; Nelson *et al.*, 1990; Nezu *et al.*, 1992; Peng *et al.*, 1994a; Puopolo *et al.*, 1991; Wang *et al.*, 1988). Vacuolar H^+-ATPases from different tissues and/or organelles exhibit differences in the molecular weight of some of these subunits on SDS-polyacrylamide gels, and genetic isoforms have been identified. The A subunit is coded by a single gene in most tissues, however an isoform has been identified in osteoclasts (Chatterjee *et al.*, 1992ab; Puopolo *et al.*, 1991). Two or more genes encode the B subunit in eukaryotes, with one isoform predominantly found in kidney cells and the other isoform ubiquitously distributed (Bernasconi *et al.*, 1990; Nelson *et al.*, 1992; Puopolo *et al.*, 1992). Heterogeneity of the B subunit protein has been noted on two-dimensional gel electrophoresis in several tissues (Nelson *et al.*, 1992), and more genetic isoforms may remain to be identified. Similarly, several genetic isoforms of the proteolipid have been identified, and alternative splicing has been shown to result in tissue-specific isoforms of the 100–116-kDa subunit (Peng *et al.*, 1994a).

The A subunit is considered to be the catalytic subunit based on the presence of a high-affinity ATP binding site located within a Walker consensus A sequence and demonstration of binding of various inhibitors to this subunit in an ATP-protectable manner (Forgac, 1992; Gluck, 1992, 1993). The B subunit has been termed a regulatory subunit by analogy to a similar subunit in the F_1-F_0 mitochondrial ATP synthase. The B subunit is highly homologous to the A subunit but does not contain the high-affinity Walker consensus nucleotide binding site. Its actual function is unknown.

The V_1 and V_0 complexes can be separated by treatment with ATP and chaotropic ions at cold temperatures (Fig. 1). The V_1 complex is hydrophilic and soluble in aqueous solutions. This complex of the A, B, C, D, and E subunits exhibits a partial reaction of ATP hydrolysis in the presence of calcium (Peng *et al.*, 1994b). The V_0 complex is hydrophobic and remains embedded in membranes. It exhibits no function on its own, and, in contrast to the analogous F_0 subunit of the mitochondrial proton pump, it does not normally conduct protons when separated from the V_1 complex (Beltrán and Nelson, 1992; Crider *et al.*, 1994; J. Zhang *et al.*, 1992, 1994). Although the V_1 complex plus transmembrane proteolipids reconstitute to form a functional proton pump, full activity requires all of the subunits including the accessory 38-kDa and 100–116-kDa subunits (Peng *et al.*, 1994b).

Although the entire complement of subunits has been identified in the vacuolar proton pump from brain clathrin-coated vesicles, proton pumps from some cell types appear to lack some of the subunits, including the 100–116-kDa accessory protein. This may, however, reflect the extremely labile nature of this protein rather that its actual absence. Further work will be necessary to determine whether, in fact, functional V-type H^+-ATPases from different organelles or different tissues actually exhibit consistent presence or absence of certain subunits.

2.2. Evolution

The vacuolar proton pump is a phylogenetically ancient structure with close ties to the F_1-F_0 H^+-ATPase of eukaryotic mitochondria, the F_1-F_0 H^+-ATPase of chloroplasts, and the cell membrane proton pump of prokaryotes (Kibak et al., 1992; Nelson, 1992). During the evolution of eukaryotes, the mitochondrial and chloroplast F_1-F_0 H^+-ATPases probably evolved from the cell membrane proton pump of those prokaryotes that were endocytosed to become mitochondria and chloroplasts. Early eukaryotes presumably also expressed a proton pump on their plasma membranes, representing the cell membrane proton pump of the prokaryotic cell that became the eukaryotic cell. The vacuolar H^+-ATPase likely derives from endocytosis of this plasma membrane proton pump. Certain specialized eukaryotic cells, such as renal tubular cells, osteoclasts, and macrophages, have, in fact, retained a "vacuolar"-type H^+-ATPase on their plasma membranes.

Based on sequence data, it is likely that the B subunits of all of these pumps arose by duplication of the A subunit gene prior to the development of eukaryotes. During the evolution of eukaryotes, the gene for the proteolipid of the vacuolar proton pump duplicated and fused such that the vacuolar proton pump proteolipid subunit (~ 17 kDa) is twice the size of the proteolipid subunit (~ 8 kDa) in mitochondria, chloroplasts, and bacteria.

2.3. Isoforms

Genetic isoforms of the A, B, E, and 100–116-kDa subunits and the proteolipid have all been identified (Bernasconi et al., 1990; Marushack et al., 1992). In addition, multiple forms of both the B and E subunits have been identified in kidney cells using two-dimensional gel electrophoresis (Hemken et al., 1992). It is quite likely that other isoforms will be identified. Certain isoforms may be tissue-specific and/or organelle-specific, perhaps due to differences in promoter regions and/or different sorting sequences on the translated protein (Hemken et al., 1992; Lee et al., 1995; Nelson et al., 1992).

Isoforms might be directed to different membrane locations by other accessory or associated proteins. A 50-kDa subunit that can be immunoprecipitated with the vacuolar H^+-ATPase from bovine brain clathrin-coated vesicles appears to be the AP-50 subunit of the AP-2 adaptin (Forgac, 1992; Gluck, 1992, 1993; Liu et al., 1994; Myers and Forgac, 1993a). This protein subunit is known to couple receptor tails to clathrin on the plasma membranes of cells. It is possible that this 50-kDa subunit may sort specific proton pump isoform(s) to the plasma membrane/endosome pool. A second adaptin protein, AP-1, is known to be associated with Golgi vesicles and primary lysosomes. A different accessory protein, part of the AP-1 complex, might sort a different proton pump isoform to the Golgi/lysosome circuit (Forgac, 1992).

Table I
Comparison of ATP-Dependent Acidification of
Rat Liver Endosomes and Lysosomes[a]

Parameter	Endosomes	Lysosomes
Maximum rate of acidification in 140 mM KCl	1.01 ΔpH/min (260% of rate in Kgluconate)	0.57 ΔpH/min (190% of rate in Kgluconate)
Steady-state ATP-dependent pH_i in 140 mM KCl	6.08	5.36
Proton leak rate from fully acidified state	0.06 ΔpH/min	0.11 ΔpH/min
Normalized proton leak rate	0.09 ΔpH/min/μM H^+	0.03 ΔpH/min/μM H^+
EC_{50} for MgATP	70 μM	140 μM
EC_{50} for Cl^-	2.8 mM	25 mM
Percent dissipation of $\Delta\Psi$ by 140 mM Cl^-	100%[b]	~50%
Substrate specificity	ATP >> GTP, UTP	ATP > GTP > UTP
Cation preferences	K^+ > Na^+, Li^+ > $NMDG^+$	K^+ > Na^+ > Li^+, $NMDG^+$
Anion preferences	Br^- > Cl^- > gluconate > NO_3^-, PO_4^{2-}, SO_4^{2-}	Cl^-, Br^- > NO_3^-, PO_4^{2-} > gluconate > SO_4^{2-}
Na^+/H^+ exchange	0 to + +	0
Cation channel	+ to + +	0 to +
Inhibitor IC_{50}		
N-ethylmaleimide	~10 μM	~10μM
Dicyclocarbodiimide	~100 μM	~100 μM
Bafilomycin	~5–10 nM	~1 nM

[a]All data except % dissipation of $\Delta\Psi$ data is from Van Dyke, 1993. Endosomes consist of a mixed population of all vesicles that endocytose the fluid-phase marker FITC-dextran during a 10 min period. Lysosomes are purified secondary lysosomes.
[b]Data from Van Dyke, 1988, and Van Dyke and Belcher, 1994. Endosomes consist of purified multivesicular bodies and CURL vesicles.

Other observations suggesting functionally different isoforms of the vacuolar H^+-ATPase include modest differences in the IC_{50} for a variety of inhibitors (Chatterjee *et al.,* 1992b), differences in relative preferences for the substrate ATP over GTP (Van Dyke, 1993; Wang and Gluck, 1990), differences in sensitivity to inhibition by copper (Wang and Gluck, 1990), and small differences in the K_m values for ATP and magnesium (Table I) (David and Baron, 1994; Van Dyke, 1993). Whether these functional isoforms reflect differences in amino acid sequence, subunit composition, and/or effects of membrane lipid composition and associated proteins is not known. It is also not known whether these functional variants exhibit important differences in proton transport in the intact cell.

2.4. Function

The vacuolar H^+-ATPase is an electrogenic pump that uses the energy of ATP hydrolysis to move protons, usually into the interior of a vesicular space (Fig. 2)

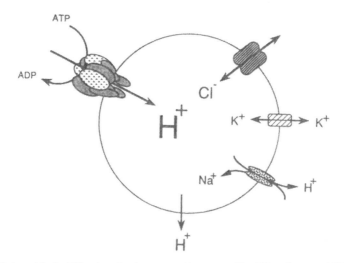

FIGURE 2. A model of acidification of endosomes and lysosomes. Na^+/H^+ exchange and K^+ (cation) channels have been demonstrated in some, but not all endocytic vesicles. (Based on data from Van Dyke, 1988, 1993, 1995; Van Dyke and Belcher, 1994.)

(Anderson *et al.*, 1985; Cidon and Sihra, 1989; Cidon *et al.*, 1983; Floor *et al.*, 1990; Forgac, 1992; Forgac *et al.*, 1983; Fuchs *et al.*, 1989a; Galloway *et al.*, 1983; Hilden *et al.*, 1988; Ohkuma *et al.*, 1982; Sabolic and Burckhardt, 1986; Van Dyke, 1988, 1993; Van Dyke and Belcher, 1994; Van Dyke *et al.*, 1985; Xie *et al.*, 1983). The electrogenic nature of this transport has been demonstrated by the use of membrane potential-sensitive dyes. In the absence of ATP, many vesicular structures exhibit a negative intravesicular membrane potential, possibly a Donnan potential related to fixed negative charges, estimated at between -20 to -70 mV (Harikumar and Reeves, 1983; Henning, 1975; Holz, 1979; Moriyama *et al.*, 1992; Van Dyke *et al.*, 1985). Addition of MgATP results in depolarization of this membrane potential by 40 to 80 mV, resulting in an interior positive membrane potential (Bennett and Spanswick, 1983; Cuppoletti *et al.*, 1987; Dell'Antone, 1984; Harikumar and Reeves, 1983; Holz, 1979; Ohkuma *et al.*, 1983; Van Dyke, 1988, 1993; Van Dyke *et al.*, 1985). The electrical current generated by this proton pump has been measured directly in the apical membrane of turtle bladder epithelial cells (Anderson *et al.*, 1985) and in plant vacuoles (Hedrich *et al.*, 1989).

Ion substitution studies indicate that the vacuolar H^+-ATPase pumps only protons. As expected for an electrogenic ion pump, the H^+-ATPase establishes a transmembrane proton electrochemical gradient consisting of interconvertible chemical and electrical components (Anderson *et al.*, 1985; Van Dyke, 1988). The maximum proton electrochemical gradient developed by this proton pump, when measured across a tight epithelium such as turtle bladder, approximates 180 mV (Anderson *et al.*, 1985). It is not known whether the same proton pump, when functioning in a closed intracellular organelle, develops the same maximum gradient.

The vacuolar H^+-ATPase requires MgATP for maximal proton transport. Manganese may partially substitute for magnesium, however, little or no proton transfer is observed with other divalent cations. Similarly, ATP is the preferred energy source although some forms of the vacuolar H^+-ATPase function modestly well in the presence of GTP (Table I). The proton-to-ATP ratio for this pump is not known precisely but has been estimated to be greater than 1 and possibly as high as 2 (Davies *et al.*, 1994; Nelson, 1992; Steinmetz *et al.*, 1981).

Given the inerconvertibility of pH and membrane potential gradients generated by the H^+-ATPase, it is expected that movement of permeable anions will alter the relative proportions of these two gradients. Indeed this is the case. Permeable anions such as chloride increase the pH gradient and decrease or abolish the membrane potential established by the H^+-ATPase (Fig. 2), whereas addition of permeable protonatable amines or electroneutral cation exchangers such as monensin or nigericin decrease the pH gradient and increase the interior positive vesicle membrane potential (Ohkuma *et al.*, 1983; Van Dyke *et al.*, 1985).

The vacuolar H^+-ATPase has been functionally and immunologically localized to most intracellular membrane-bound compartments, including endosomes, lysosomes, the Golgi apparatus, endoplasmic reticulum, and secretory granules (Forgac, 1992; Gluck, 1992, 1993; Márquez-Sterling *et al.*, 1991; Rodman *et al.*, 1991). In addition, it has been localized to the plasma membrane of certain specialized cells, including the apical membrane of renal epithelia involved in acid secretion, osteoclasts, macrophages, and monocytes (Brown *et al.*, 1992; Chatterjee *et al.*, 1992b; Gluck, 1992; Nordström *et al.*, 1994). In the latter cell types the vacuolar H^+-ATPase may participate in bone reabsorption and in regulation of cytoplasmic pH. However, not all endocytic or exocytic structures contain functional proton pumps, as immunofluorescent studies fail to detect proton pump subunits or measurable acidification in all of these structures (Fuchs *et al.*, 1994; Hammond and Verroust, 1994; Overly *et al.*, 1995; Márquez-Sterling *et al.*, 1991; Rodman *et al.*, 1991; Shi *et al.*, 1991).

2.5. Inhibitors

The vacuolar H^+-ATPase can be distinguished from other families of ion pumps by a unique pattern of inhibitor sensitivities. This pump is exquisitely sensitive to bafilomycin A_1, a fungal antibiotic, with IC_{50} values in the nanomolar range (Table I). Bafilomycin may be the most specific vacuolar H^+-ATPase inhibitor available and it can be used in intact cells. However, its effects on other ion pumps and transporters have not been fully evaluated. Bafilomycin binds to the V_0 complex, probably in or around the proton channel on the 100–116-kDa subunit (Fig. 1); however, the exact site of binding and the mechanism of inhibition are not known (Crider *et al.*, 1994; Gluck, 1993; J. Zhang *et al.*, 1992, 1994).

N-ethylmaleimide (NEM) and other sulfhydryl-reactive agents are potent inhibitors of the vacuolar H^+-ATPases, and this sensitivity distinguishes them from the mitochondrial F_1-F_0H^+-ATPases. N-ethylmaleimide binds to a cysteine residue at position 254 in the A subunit of the pump, which is located in the Walker consensus sequence near the ATP binding site (Fig. 1) (Feng and Forgac, 1992, 1994). Another inhibitor, 7-chloro-4-nitrobenzo-2-oxa-1,3-diazole (NBD-chloride), binds to the same cysteine residue (Forgac, 1992; Gluck, 1993). Position 254 is occupied by a valine residue in the analogous subunit of the F_1-F_0 H^+-ATPases, thus explaining their NEM insensitivity. N-ethylmaleimide inhibits the vacuolar H^+-ATPase at micromolar concentrations, however it is not a useful agent in intact cells as it inhibits many other enzymes.

Dicyclocarbodiimide (DCCD) inhibits all known types of proton pumps, including the vacuolar H^+-ATPase, the F_1-F_0 H^+-ATPases, and a group of unrelated plasma membrane proton pumps found on fungal and plant cells. It binds to the carboxylic acid group of an aspartate residue in the transmembrane proton channel of these proton pumps, forming a covalent bond, and it causes irreversible inhibition at micromolar concentrations (Fig. 1).

The vacuolar H^+-ATPase is also inhibited by a variety of oxyanions, the most potent being nitrate. Nitrate inhibits both the ATPase activity of this pump as well as its capacity to transport protons and does so at high micromolar to low millimolar concentrations (Arai et al., 1989; Moriyama and Nelson, 1989; Van Dyke, 1986; Wang and Gluck, 1990). The effects of nitrate on the vacuolar H^+-ATPase, however, may be difficult to interpret, as nitrate is also a permeable anion that may transiently increase vesicle acidification and, at high concentrations, may act as a chaotropic anion, causing physical disruption of the proton pump and separating the V_1 complex from the V_0 complex (Arai et al., 1989; Moriyama and Nelson, 1989; Wang and Gluck, 1990). At low concentrations, oxyanions such as nitrate, sulfate, and phosphate probably inhibit the proton pump directly, however the mechanism(s) are unknown (Van Dyke, 1986).

Agents that do not inhibit the vacuolar H^+-ATPase or that do so only at extremely high concentrations include oligomycin, vanadate, and ouabain. Oligomycin is a potent inhibitor of the F_1-F_0 H^+-ATPases, while vanadate is a potent inhibitor of the P-type ATPases, a family of ion pumps (including Na^+, K^+-ATPase, H^+/K^+-ATPase, and the Ca^{2+}-ATPases) that undergo a phosphorylation event.

3. LYSOSOME AND ENDOSOME ACIDIFICATION

3.1. Initial Rates of Proton Transport

An important parameter for understanding and comparing proton pump function is measurement of the initial rate of ATP-dependent acidification. Quantitative data is best obtained using ratiometric pH-sensitive fluorescent dyes loaded into the interior space of vesicular structures. One such dye is fluorescein isothiocyanate-dextran, which can be loaded by fluid-phase endocytosis into endosomes and lysosomes. The fluorescent signal from such probes can be calibrated and changes in the internal pH (pH_i) of vesicles followed in real time. Since pH-sensitive dyes measure the free hydrogen ion concentration inside the vesicle space, vesicle volume, buffering capacity, and surface-to-volume ratio are factors important in relating the rate of change of vesicle pH_i to proton pump-mediated proton influx rates. Initial rates of acidification may also be assessed by examining the rate of accumulation of protonatable permeable amines such as the fluorescent dye acridine orange, a probe that can be used with any purified vesicle population. However, changes in acridine orange fluorescence are less sensitive to changes in pH_i than are changes in fluorescin fluorescence, and they cannot be readily calibrated to quantitate proton flux.

Initial rates of ATP-dependent acidification vary with vesicle type and experimental conditions. Rates vary considerably between early endosomes, later endosomes, and lysosomes even when all are prepared from the same tissue (Fig. 3) (Van Dyke, 1993, 1995; Van Dyke and Belcher, 1994). In addition, rates vary depending on the presence of permeable charge-compensating anions such as chloride, with increasing amounts of buffer chloride resulting in increasing rates of acidification up to a maximum (Fig. 3). Experimental observations from a number of vesicular systems indicate that most intracellular organelles exhibit one or more types of chloride conductance that allow passive, electrically driven influx of chloride to neutralize the positive charge of pumped protons, thereby increas-

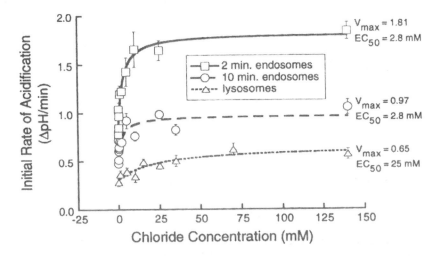

FIGURE 3. Comparison of the initial rates of H^+-ATPase-mediated acidification and effects of chloride for various types of rat liver endocytic vesicles and lysosomes. (Data from Van Dyke, 1993; Anbari *et al.*, 1994.)

ing the rate and extent of accumulation of hydrogen ions while minimizing membrane potential development (Fig. 2) (Arai *et al.*, 1989; Blair *et al.*, 1991; Cuppoletti *et al.*, 1987; Fuchs *et al.*, 1989a; Hilden *et al.*, 1988; Lukacs *et al.*, 1991; Pope *et al.*, 1990; Sabolic and Burckhardt, 1986; Van Dyke, 1988, 1993). Other maneuvers expected to neutralize the interior positive membrane potential of pumped protons also have been found to increase acidification rates. These include efflux of cations such as potassium through either endogenous potassium channels or valinomycin (Fuchs *et al.*, 1989a; Moriyama, 1988; Van Dyke, 1993, 1995; Van Dyke *et al.*, 1994) and influx of other permeable anions such as nitrate. Although nitrate will inhibit the vacuolar H$^+$-ATPase, at least transiently, it can serve as an efficient charge-compensating anion (Van Dyke, 1993; Van Dyke and Belcher, 1994).

Other factors that probably influence the magnitude of the initial rate of ATP-dependent acidification include the number of pumps per vesicle, the density of pumps (in terms of number per surface area or per vesicle volume), the pump rate, and vesicle buffering capacity (Forgac, 1992; Gluck, 1993; Van Dyke, 1993; Van Dyke and Belcher, 1994).

As shown in Fig. 3, the initial rate of acidification is consistently slower for rat liver lysosomes than for other liver vesicles, perhaps due to differences in the surface-to-volume ratio of these structures or to differences in pump density (Van Dyke, 1993; Van Dyke and Belcher, 1994). Overall, initial rates of acidification of a number of acidified organelles have been reported and range from 0.16 ΔpH/min (macrophage phagosomes; Lukacs *et al.*, 1990) to 0.65 ΔpH/min (rat liver lysosomes; Van Dyke, 1993); 0.97–1.81 ΔpH/min (rat liver endosomes; Van Dyke, 1993; Van Dyke and Belcher, 1994); and 4–20 ΔpH/min (rat renal endosomes; Lencer *et al.*, 1990). It is likely that a number of other mechanisms exist for regulating the rates of acidification, as discussed in Section 4.

3.2. Steady-State Parameters

The steady-state pH$_i$ in acidified organelles reflects the algebraic sum of proton influx via the vacuolar H$^+$-ATPase and proton efflux via passive diffusion and any proton-coupled transporters (Fig. 2). Using a variety of techniques and cell types, investigators have measured a wide range of steady-state pH$_i$ values. Estimates for endosomes range from approximately 5.8 to 6.4 (Anbari *et al.*, 1994; Barasch *et al.*, 1991; Lencer *et al.*, 1990; Lukacs *et al.*, 1990; Root *et al.*, 1994; Shi *et al.*, 1991; Van Dyke, 1988, 1993, 1995; Van Dyke and Belcher, 1994). Steady-state pH$_i$ values for lysosomes tend to be considerably lower (Table I). Values measured in isolated lysosomes generally range from approximately 5.2 to 5.5, but for lysosomes in intact cells, values as low as 3.8 to 4.8 have been reported (Moriyama, 1988; Moriyama *et al.*, 1992; Ohkuma and Poole, 1978; Ohkuma *et al.*, 1982; Schneider, 1981; Van Dyke, 1993; Van Dyke *et al.*, 1992b).

Although the initial rate of acidification of lysosomes tends to be slower than that of many other endocytic vesicles, the steady-state pH_i is far more acidic (Fig. 3; Table I). At least two factors may be responsible for this seemingly paradoxical observation. First, endocytic vesicles are an ever-changing population of structures continuously remodeling and evolving toward lysosomes. In the intact cell it is not clear whether a steady-state pH_i is in fact achieved in any endocytic vesicle prior to secondary lysosomes. Indeed, in intact cells the pH_i of endosomes falls with a biphasic rate, rapidly to 6.0–6.2 and more slowly to less than 5.5 (Roederer et al., 1987). However, studies using isolated endosomes and lysosomes also demonstrate a persistent difference in the magnitude of the steady-state acid interior. This may be accounted for at least in part by observed differences in the passive efflux rate, which tends to be greater in endocytic vesicles compared to lysosomes (Table I) (Van Dyke, 1993; Van Dyke and Belcher, 1994).

3.3. Proton Efflux

Protons are not only pumped into, but also leak out of, acidified organelles, including lysosomes. Mechanisms for proton efflux include a diffusional pathway(s) and, in some organelles, proton-coupled antiporters or symporters (see Section 3.5). Passive proton efflux does not appear to occur backwards through the vacuolar H^+-ATPase itself, as, under physiologic circumstances, the proton channel in the V_0 complex does not conduct protons when uncoupled from the V_1 complex (Beltrán and Nelson, 1992; Crider et al., 1994; J. Zhang et al., 1994).

Proton efflux can be quantitated by measuring the rate of realkalinization of acidified vesicles after inhibition of the H^+-ATPase or from pH jump studies in which vesicles are added to acid buffers and the rate of vesicle acidification measured. Proton movement through the diffusional pathway likely is determined by both the transmembrane electrical and chemical gradients, as suggested by the observation that chloride, normally an efficient counter-ion, does not alter proton efflux rates in acidified liver endosomes (Van Dyke, 1993; Van Dyke and Belcher, 1994). Here, the effects of chloride on the transmembrane pH gradient and the membrane potential probably cancel each other out with regard to altering proton efflux rates. By contrast, chloride significantly increases proton leak rates from acidified liver lysosomes, presumably by serving as an efficient counter-ion that alters the transmembrane pH gradient more than the interior positive membrane potential (Van Dyke, 1993).

Proton efflux rates may be limited by these proton diffusional pathways or by counter-ion diffusional pathways. For several types of liver endosomes and macrophage phagosomes, proton permeability appears more rate-limiting than counter-ion availability. For example, valinomycin in potassium buffers increases efflux rates only modestly, whereas the proton ionophore carbonyl cyanide m-chlorophenylhydrazone (CCCP) increases efflux four- to tenfold (Lukacs et al., 1991; Van Dyke, 1993; Van Dyke and Belcher, 1994). In contrast, proton efflux is limited by both proton and counter-ion permeabilities in liver lysosomes, as valinomycin and CCCP both increase proton efflux rates by 200 to 300% (Van Dyke, 1993).

Endosome and lysosome passive proton permeability has been estimated in a few studies. Values obtained in rat liver multivesicular bodies (5×10^{-3} cm/sec) and CURL vesicles (2×10^{-3} cm/sec) are approximately tenfold greater than those measured in rat liver lysosomes (0.3×10^{-3} cm/sec), whereas values estimated for rat renal endosomes (20×10^{-3} cm/sec) and rat renal lysosomes (0.36 cm/sec) are considerably higher (Lencer *et al.*, 1990; Piqueras *et al.*, 1994; Van Dyke, 1993; Van Dyke and Belcher, 1994).

3.4. *In Vivo* versus *in Vitro* Values

The interior pH of lysosomes in intact cells is generally considerably lower (3.8 to 4.8) (Ohkuma and Poole, 1978, Van Dyke *et al.*, 1992b) than that measured in isolated lysosomes (5.0 to 5.5) (Moriyama, 1988; Moriyama *et al.*, 1992; Ohkuma *et al.*, 1982; Schneider, 1981; Van Dyke, 1993). Explanations for these differences include differences in experimental technique and assay temperature, damage to somewhat fragile lysosomes during isolation (Van Dyke, 1993), isolation of a less acidic subset of lysosomes, tissue-specific differences in lysosome steady-state pH_i, and the presence or absence of various activating or inhibiting processes in the *in vivo* versus the *in vitro* situation. Some of these potential regulatory mechanisms are discussed in Section 4.

Since endosome and lysosome acidification is dependent on chloride, the actual ΔpH and membrane potential *in vivo* also may depend on cytoplasmic concentrations of chloride. For example, intracellular chloride in rat hepatocytes is about 25 mM (Fitz and Scharschmidt, 1987), due to a relatively low cell membrane potential and the absence of plasma membrane active chloride transport. This concentration (25mM) of chloride is sufficient to support maximal acidification of rat liver endosomes, however, under these same conditions rat liver lysosomes may not acidify at maximal rates (Fig. 3) (Van Dyke, 1993; Van Dyke and Belcher, 1994). In cells that exhibit a more hyperpolarized cell membrane potential, cytoplasmic chloride concentrations may be much lower, thereby reducing acidification (and increasing the membrane potential) of many intracellular organelles.

3.5. Modification of Acidification by Other Ion Transporters

The vacuolar H^+-ATPase does not function alone in the membrane of intracellular organelles. Transport of other ions and/or protons affects both the rate of vesicle acidification and steady-state pH_i. The transport pathways identified to date are reviewed here. It is quite likely that others remain to be identified.

3.5.1. Chloride Conductance

The best studied ion conductance in intracellular organelles is a chloride conductance (Fig. 2). Interest in the transport pathway of this ion was sparked by the virtually ubiquitous observation that acidification of a variety of intracellular organelles is stimulated by chloride. Extravesicular chloride has been shown to stimulate initial rates of acidification, reduce or eliminate proton pump-generated interior positive membrane potentials, and facilitate development of a more acidic steady-state pH_i (Bennett and Spanswick, 1983; Blair *et al.*, 1991; Fuchs *et al.*, 1989a; Pope *et al.*, 1990; Van Dyke, 1988, 1993; Van Dyke and Belcher, 1994; Wada *et al.*, 1992). *In vitro*, chloride uptake by intracellular organelles can be driven by an interior positive membrane potential, established either by the proton pump or by a potassium gradient with valinomycin (Blair *et al.*, 1991; Hilden *et al.*, 1988; Van Dyke, 1988; Wada *et al.*, 1992), indicating that the pathway is electrogenic.

Availability of chloride appears to increase the initial rate of acidification and steady-state intravesicular hydrogen concentration by 2 to 2.5-fold in a variety of rat liver endosomes and lysosomes (Fig. 3) (Van Dyke, 1993, 1995; Van Dyke and Belcher, 1994). The estimated concentration of chloride for 50% maximal acidification rates (EC_{50}) is different among different vesicles and has been estimated at 2.8 mM for very early rat liver endosomes, 11–16 mM for late endosomes such as multivesicular bodies and compartment for uncoupling of receptor and ligand (CURL) vesicles, and 25 mM for lysosomes (Fig. 3; Table I) (Van Dyke, 1993; Van Dyke and Belcher, 1994). As noted in Section 3.3 above, this chloride conductance also influences proton efflux rates in some, but not all intracellular acidified organelles.

Since many different types of chloride conductances have been identified in plasma membranes from different cells, it is unlikely that endosomes and lysosomes from the same or different cell types exhibit the same chloride conductance. Indeed, the functional characteristics of chloride conductance vary from organelle to organelle, and yeast vacuoles exhibit at least two distinct chloride conductances (Wada and Anraku, 1994). For example, organellar chloride conductances transport nitrate as well as halides, but the order of halide preference varies between different types of vesicles. The preference sequence bromide > chloride > iodide was noted for rat liver CURL vesicles and multivesicular bodies (Van Dyke *et al.*, 1994); the order iodide > bromide > chloride for early rat liver endosomes (Van Dyke *et al.*, 1994); and the order chloride > iodide > bromide for renal endosomes (Reenstra *et al.*, 1992).

At least some of these chloride conductances are blocked by stilbene derivatives (Hedrich and Kurkdjian, 1988; Reenstra *et al.*, 1992; Tilly *et al.*, 1992; Xie *et al.*, 1989). Chloride channels have been partially purified and reconstituted from bovine brain clathrin-coated vesicles (Xie *et al.*, 1989) and from renal membranes (Landry *et al.*, 1993; Reenstra *et al.*, 1992) that may represent some of the vesicular chloride channels identified from acidification studies.

The substantial effect of chloride on rates of vesicle acidification indicates that chloride channels may provide a mechanism for regulating vesicle pH_i. Although some second messengers and protein kinases have been shown to alter chloride transport and/or acidification of some intracellular organelles, these agents appear to have considerable tissue and organellar specificity. The chloride conductances from renal proximal tubule endosomes and bovine brain clathrin-coated vesicles are increased by protein kinase A phosphorylation, associated with an increase in ATP-dependent vesicle acidification (Bae and Verkman, 1990; Mulberg et al., 1991; Reenstra et al., 1992). By contrast, the chloride conductance(s) in liver endosomes and lysosomes are not affected by cAMP (R. W. Van Dyke, unpublished observations). Calcium activates a chloride conductance in secretory vesicles of thyroid parafollicular cells associated with increases in vesicle acidification (Barasch et al., 1988; Tamir et al., 1994). However, in these same cells, lysosome acidification is not altered by calcium. Finally, a nucleotide-activated chloride channel has been identified in rat liver lysosome membranes, however it is not known whether changes in channel opening alter lysosome acidification rates (Tilly et al., 1992).

With cloning of the gene responsible for cystic fibrosis and identification of the cystic fibrosis transmembrane conductance regulator (CFTR) as an ATP-activated, cAMP-regulated chloride channel, a number of investigators have studied whether CFTR is responsible for at least some of the chloride conductance in intracellular organelles. Nasal epithelial cells from patients with cystic fibrosis exhibit defective acidification of endosomes and Golgi apparatus, but not lysosomes (Barasch and Al-Awqati, 1993; Barasch et al., 1991), suggesting that defective chloride conductance may exist in these organelles. However, transfection of normal CFTR into cells that do not express it does not alter rates of acidification or steady-state pH_i of endosomes or lysosomes (Biwersi and Verkman, 1994; Lukacs et al., 1992; Root et al., 1994; Van Dyke et al., 1992b), even though functional CFTR can be demonstrated in the endosomal compartment of these transfected cells by cAMP-mediated increases in passive chloride conductance (Biwersi and Verkman, 1994; Lukacs et al., 1992). Although counter-ion permeability is clearly increased by CFTR in these transfected cells, the lack of change in acidification rates suggests that endogenous counter-ion permeability is not rate-limiting in these organelles. Overall it appears that CFTR is not the principal chloride conductance present in intracellular organelles in most tissues. However, it may play an important role in selected organelles in certain cell types.

3.5.2. Cation Conductance

Ion substitution studies and the use of potassium channel inhibitors suggest the presence of potassium (or cation) conductances in at least some acidified intracellular organelles, including rat liver endosomes and lysosomes (Fuchs et al.,

1989a; Van Dyke, 1993), macrophage phagosomes (Lukacs *et al.,* 1991), and spleen lysosomes (Moriyama, 1988). In these vesicles, potassium efflux increases ATP-dependent vesicle acidification while potassium influx stimulates proton efflux. Electrophysiologic studies have identified a 70-pS cation channel in the membrane of sugar beet vacuoles (Coyaud *et al.,* 1987). Yeast vacuoles have been shown to exhibit a membrane potential-and calcium-gated cation channel permeable to both potassium and sodium (Wada and Anraku, 1994). A combination of electrophysiologic and fluorescent probe studies have demonstrated a cation channel permeable to both potassium and sodium in the membranes of rat liver CURL vesicles that is absent from early liver endosomes (Van Dyke *et al.,* 1994). Since many different types of plasma membrane potassium channels have been identified in different cells, it is quite likely that the cation conductances of endosomes and lysosomes will exhibit considerable tissue and vesicle heterogeneity.

3.5.3. Na$^+$/H$^+$ Exchange

Na$^+$/H$^+$ exchange is virtually ubiquitous on the plasma membrane of cells, where it is thought to function primarily in cell pH regulation. Na$^+$/H$^+$ exchangers have been identified in endosomes from rat kidney and liver as well as adrenal chromaffin granules (Gurich and Warnock, 1986; Haigh and Phillips, 1989; Hilden *et al.,* 1990; Sabolic and Brown, 1990; Van Dyke, 1995). In sodium-containing media this exchanger functions primarily as a proton leak pathway, reducing acidification rates and steady-state pH$_i$ (Van Dyke, 1995). It could also function to load sodium into the interior space of endosomes. However, the cytosolic concentration of Na$^+$ in rat hepatocytes is too low for this exchanger to significantly alter acidification (Van Dyke, 1995). Further, in rat liver, the exchanger functions only in very early endosomes and is absent from later endosomes and lysosomes (Van Dyke, 1995). In these very early endosomes, which are likely to contain extracellular concentrations of sodium, the exchanger may function, albeit transiently, as an acidifying transporter, exchanging intravesicular sodium for cytosolic protons. The absence of Na$^+$/H$^+$ exchange from later endocytic vesicles suggests that this exchanger is either inactivated or rapidly recycled out of early endosomes (Van Dyke, 1995).

3.5.4. Tertiary Amines

A variety of weak bases, especially tertiary amines, can be taken up and concentrated in acidified organelles, especially lysosomes (Cain and Murphy, 1986; MacIntyre and Cutler, 1988; Poole and Ohkuma, 1981; Van Dyke *et al.,* 1992a; Weitering *et al.,* 1977). Some of these organic cations are membrane-permeable in their unprotonated state, however upon entering acidified organelles (via diffusion or a specific transporter), they are protonated and trapped. Examples of such agents include methylamine, ammonium chloride, tubocurarine, and chloroquine.

Uptake of these protonatable organic cations "buffers" intravesicular protons, thereby alkalinizing vesicles and increasing the interior positive vesicle membrane potential. The degree of alkalinization will depend on the extravesicular concentration of organic cation and the rates of continued proton pumping by the vacuolar H^+-ATPase. Thus, at low concentrations, these organic cations may have little effect on vesicle pH_i (Van Dyke *et al.*, 1992a), whereas at higher concentrations considerable accumulation of organic cation occurs, with vacuolation of lysosomes and significant shifts in lysosome pH_i. However, unless high concentrations of these organic cations are administered, lysosome pH_i does not become neutral (Cain and Murphy, 1986; Moriyama *et al.*, 1992; Ohkuma and Poole, 1978; Ohkuma *et al.*, 1983; Tietz *et al.*, 1990). Since these weak bases accumulate in response to a pH gradient, isolated vesicles would be expected to exhibit ATP-dependent uptake of them all (Forgac *et al.*, 1983; Moseley and Van Dyke, 1995; Schneider, 1981; Van Dyke *et al.*, 1992a).

Other "lysosomotropic" amines such as tributylamine appear to be membrane-permeable in both their protonated and unprotonated states and act as ionophores, shuttling H^+ across vesicle membranes (Cain and Murphy, 1986). These agents reduce vesicle acidification and membrane potential but do not cause vacuolation as they cause no net accumulation of osmotically active molecules.

4. REGULATION OF ACIDIFICATION

Many potential mechanisms for regulation of endosome and lysosome acidification exist. These include differences in the number and rate of the vacuolar H^+-ATPase units (Lee *et al.*, 1995), proton leak rates, the presence and absence of proton pump regulatory subunits, biochemical modification of the pump through phosphorylation or sulfhydryl group modification, differences in the size and shape of endocytic vesicles, endogenous activators or inhibitors, and changes in other vesicle transporters. Evidence of genetic and biochemical proton pump isoforms with different inhibitor and high-energy phosphate preferences suggest that regulation may also exist at the level of the proton pump itself. Measurement of acidification in individual vesicles indicates that not all endosomes or lysosomes are equally acidified and that some appear to maintain a neutral pH_i (Hammond and Verroust, 1994; Overly *et al.*, 1995; Shi *et al.*, 1991). Indeed, rat liver endocytic clathrin-coated vesicles do not appear to exhibit ATP-dependent acidification, suggesting that the vacuolar proton pump is either missing from those structures or inactive (Fuchs *et al.*, 1994). Immunolocalization studies using proton pump antibodies show greatly varying densities of proton pumps per membrane surface area, ranging from a high concentration in renal intercalated cells to a low concentration in chromaffin granules and endosomes (Rodman *et al.*, 1991). Further, pools of free V_1 complexes have been found in at least some cell lines (Forgac, 1992; Gluck, 1993; Kane, 1995; Rodman *et al.*, 1994), suggesting that coupling of free V_1 complexes to inactive membrane-bound V_0 complexes could regulate pump function in some membranes or organelles. Finally, plasma membrane insertion and removal of functional proton pumps appear to be important mechanisms for regulation of renal acid or base secretion (Cannon *et al.*, 1985; van Adelsberg and Al-Awqati, 1986), and analogous addition or removal of proton pumps from intracellular organelles might affect vesicle acidification rates.

Another mechanism for regulation of the vacuolar H^+-ATPase is formation of a disulfide bond between two cysteine residues present at the catalytic nucleotide binding site on the A subunit. Cysteine 254, the residue responsible for sensitivity of these proton pumps to sulfhydryl reagents, may be disulfide-bonded to either a proximal cysteine residue or a sulfhydryl agent (Feng and Forgac, 1992, 1994). When cysteine 254 is disulfide-bonded, the vacuolar proton pump is inactive. Preliminary work suggests that approximately 50% of brain clathrin-coated vesicles are in this inactive state, whereas synaptic vesicle proton pumps are fully functional (Rodman *et al.*, 1994).

Potential regulatory mechanisms include cytosolic proteins, isolated by several investigators, that may act as endogenous activators or inhibitors of the vacuolar H^+-ATPase (Xie *et al.*, 1993, K. Zhang *et al.*, 1992a,b) and drugs such as the estrogenic compound ethynyl estradiol that inhibit endosome acidification (Van Dyke and Root, 1993). Second messengers and protein kinases also may directly regulate the vacuolar proton pump, in organellar and tissue-specific ways. For example, protein kinase C increases acidification of Swiss 3T3 fibroblast endosomes and activates an analogous proton pump on the plasma membrane of macrophages and neutrophils (Nanda *et al.*, 1992; Nordstrom *et al.*, 1994; Zen *et al.*, 1992). However, protein kinase A inhibits acidification of fibroblast endosomes, while fibroblast lysosome pH_i is not affected by protein kinase A or C (Zen *et al.*, 1992). In contrast, cAMP and protein kinase C both reduce acidification of Golgi apparatus vesicles from skin fibroblasts, while cAMP treatment increases acidification of rat liver endosomes (Seksek *et al.*, 1995; Van Dyke *et al.*, 1996). How important these regulatory mechanisms are under physiologic circumstances remains to be determined.

Vesicle acidification also may be altered by the other transporters present in vesicle membranes. As noted in Section 3.5.1, cAMP-regulatable chloride conductances may alter acidification of some organelles. As discussed in Section 3.5.3, the presence of Na^+/H^+ exchange might also alter acidification. The presence of another electrogenic ion pump in the same membrane as the vacuolar proton pump would also be expected to limit vesicle acidification due to changes in vesicle membrane potential. In some, but not all cell types, Na^+,K^+-ATPase activity in endosome membranes reduces acidification rates and steady-state intravesicular hydrogen ion concentrations (Cain *et al.*, 1989; Fuchs *et al.*, 1989b; Sipe *et al.*, 1991). Such a role for Na^+,K^+-ATPase is probably limited to early endosomes and only in certain cell types, as this ATPase is not found in liver endosomes or lysosomes (Anbari *et al.*, 1994).

Finally, certain intracellular pathogens such as *Mycobacterium* take up successful residence in the endocytic vesicles of phagocytic macrophages and appear to survive because the invading organism prevents phagosome acidification (Small *et al.,* 1994; Sturgill-Koszycki *et al.,* 1994). The phagosomes from infected macrophages acquire certain lysosomal membrane proteins but not the vacuolar H^+-ATPase, suggesting that *Mycobacterium* (and other successful intracellular pathogens) either prevent fusion of H^+-ATPase-containing vesicles with phagosomes or cause rapid removal of the H^+-ATPase.

5. ROLE OF pH GRADIENT IN DRIVING TRANSPORT OF OTHER SOLUTES

Analogous to the role of plasma membrane Na^+,K^+-ATPase in establishing an electrochemical gradient that drives transport of other solutes via coupled transporters, the vacuolar H^+-ATPase also establishes an electrochemical proton gradient that has, in many systems, been used to drive transmembrane movement of a variety of solutes. These processes have been most fully elucidated for the storage vacuoles of plant cells and yeast, however, identification of proton-coupled transporters in lysosomes and other acidified intracellular organelles in animal cells is proceeding rapidly. Both the pH gradient and membrane potential gradients are potential sources of energy to drive transport. Mechanisms are diverse, including protonation and trapping of diffusible weak bases, proton-coupled antiporters, and proton-coupled symporters.

5.1. Tertiary Amines

As outlined in Section 3.5.4, amphipathic weak bases will be concentrated in any acidified membrane-bound space through inward diffusion of the unprotonated species, followed by protonation and trapping of the charged species. The magnitude of this process will vary from compound to compound depending on relative passive permeability of the unprotonated and protonated species and on the pK of the amine group. In essence, this process resembles an organic solute/H^+ antiporter in that protonation of the organic tertiary amine results in loss of a free intravesicular hydrogen ion. Indeed, if sufficient tertiary amine is available, acidified organelles would be expected to alkalinize significantly, with a concomitant increase in the interior-positive membrane potential, as long as the vacuolar proton pump continues to transport protons. This process also exchanges an osmotically active organic cation for a proton, resulting in osmotic swelling of the acidified organelle. The inorganic cation/H^+ exchangers nigericin and monensin cause similar changes in vesicle pH and membrane potential (Wada and Anraku, 1994; R. W. Van Dyke, unpublished observations).

5.2. Proton-Coupled Antiporters

A variety of proton-coupled antiporters have been identified in the membranes of acidified intracellular organelles. Those best studied include the vacuoles of plants and yeast in which Ca^{2+}/H^+ exchange, Na^+/H^+ exchange, and

K$^+$/H$^+$ exchange have all been identified (Sze *et al.*, 1992; Wada *et al.*, 1992). These exchangers use the electrochemical proton gradient to effect uptake and storage of Ca^{2+}, Na$^+$, and K$^+$ (Sze *et al.*, 1992; Wada *et al.*, 1992). As discussed earlier, Na$^+$/H$^+$ exchange has been identified in rat liver and kidney endosomes but is not present in rat liver lysosomes (Fig. 2). Also, Ca^{2+}/H$^+$ exchange has been identified in acid vesicles from *Dictyostelium* (Rooney and Gross, 1992) and in rabbit renal cortical endosomes (Fig. 4) (Hilden and Madias, 1989). Calcium transport has been identified in human fibroblast lysosomes. However, whether this presents Ca^{2+}/H$^+$ exchange is not known (Lemons and Thoene, 1991).

The best understood proton-coupled antiporters in animal cell acidified organelles are the proton-coupled transporters for biogenic amines and neurotransmitters (Fig. 4). First identified in adrenal chromaffin granules, these transporters couple uptake of catecholamines to efflux of protons (Henry *et al.*, 1994; Johnson *et al.*, 1979). These vesicular monoamine transporters (VMATs) mediate uptake of a variety of biogenic amines, including epinephrine, glutamate, γ-aminobutyric acid, dopamine, and serotonin (Floor *et al.*, 1990; Johnson *et al.*, 1979; Moriyama and Futai, 1990; Tabb *et al.*, 1992; Thomas-Reetz *et al.*, 1993), and are thought to mediate the exchange of a protonated amine for two protons in an electrogenic

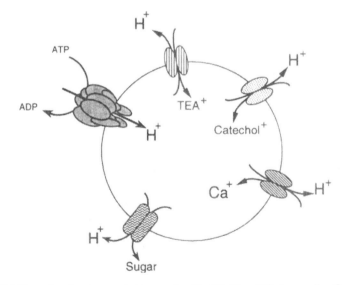

FIGURE 4. Examples of proton-coupled transporters identified in acidified organelles. Catechol/H$^+$ antiport is located in chromaffin granules and neurosecretory granules, while TEA/H$^+$ antiport has been identified in renal endosomes and liver lysosomes. Ca^{+2}/H$^+$ exchange is located in storage vacuoles of yeast and plant cells and it is likely present in renal endosomes. Transport experiments suggest sugar/H$^+$ symport activity in lysosomes. The stoichiometry of these transporters is not necessarily one to one. (Based on data from Moseley and Van Dyke, 1995; Pritchard *et al.*, 1994; Wada and Anraku, 1992; Sze *et al.*, 1992; Hilden and Madias, 1989; Mancini *et al.*, 1989, 1990.)

mechanism. The VMATs are essential for function of variety of neurons as well as adrenal chromaffin cells as they mediate loading of synaptic vesicles and secretory vesicles with high concentrations of neurotransmitters. Genes for two VMATs have been cloned (Henry *et al.,* 1994).

Other organic solute/H^+ antiporters exist in both kidney and liver cells. Using model compounds such as the quaternary amine tetraethylammonium (TEA), investigators have located cation/H^+ antiporters on the apical plasma membrane of renal tubular cells and hepatocytes. Similar proton-coupled organic cation transporters have been identified in endosomes and lysosomes from rat kidney and liver (Fig. 4) (Pritchard *et al.,* 1994; Moseley and Van Dyke, 1995). Uptake of TEA in kidney endosomes and liver lysosomes is dependent on the pH gradient established by the vacuolar proton pump and appears to be independent of organelle membrane potential. The intracellular TEA/H^+ exchanger appears functionally similar to that found on the apical membranes of liver and renal cells, and thus the same transporter may be present on both the apical cell membrane and intracellular organelles.

Since many xenobiotics and drugs are metabolized and excreted by the liver and kidney, the presence of intracellular cation/H^+ antiporters that promote sequestration of cations in membrane-enclosed compartments may help reduce cytoplasmic concentrations of these drugs and promote efficient flux of xenobiotics from blood to cell, thereby rapidly removing them from the systemic circulation (Moseley and Van Dyke, 1995; Pritchard *et al.,* 1994; Van Dyke *et al.,* 1992a). Further, as rat liver lysosomes fuse with the canalicular membrane, discharging their contents by exocytosis into bile, sequestration of xenobiotics in liver lysosomes may provide an additional pathway for removal of potentially toxic cationic xenobiotics.

5.3. Proton-Coupled Symporters

The transmembrane vesicle electrochemical proton gradient also may be used to drive efflux of organic or inorganic solutes out of acidified organelles into cytoplasm through proton-coupled symporters. Proton-coupled symporters for various anions have been characterized in plant and yeast vacuoles (Sze *et al.,* 1992; Wada and Anraku, 1994), and recent data support the existence of proton-coupled symporters in rat liver lysosomal membranes for sugars and vitamin B_{12} (Fig. 4) (Idriss and Jonas, 1991; Mancini *et al.,* 1989, 1990). These transport mechanisms are discussed in detail in Chapter 11.

6.   DISEASES RELATED TO LYSOSOME ACIDIFICATION

A variety of mutations in the acidification process of intracellular organelles have been identified in cultured cells and yeast strains, however all of these are either mild defects in vesicle acidification or are conditionally lethal. Acidification of lysosomes and intracellular organelles is likely so important to normal cell function that any significant mutation in this process is lethal. However, as reviewed in Chapter 11, inherited disorders of specific lysosomal transport mechanisms other than the vacuolar H^+-ATPase are being increasingly recognized. As discussed in Section 3.5.1, mutations in ion transporters such as a chloride channel could modify the degree of acidification of lysosomes or other intracellular organelles, thereby causing cellular malfunction and disease.

6.1. Drug-Induced Phospholipidosis

One well-recognized acquired lysosomal storage disorder is drug-induced lysosomal phospholipidosis (Hostetler, 1984; Kacew, 1987; Reasor, 1989). This topic is covered more fully in Chapter 12. Briefly, a wide variety of cationic amphiphilic drugs accumulate in acidified intracellular organelles, especially lysosomes, and cause accumulation of membranous material called lysosomal lamellar bodies. Drugs known to cause this condition include anti-arrhythmic drugs such as amiodarone and perhexiline, antidepressants, anorectic agents such as chlorphentermine, antimalarials such as chloroquine, and antibiotics such as aminoglycosides (Reasor, 1989). These agents are all organic amines with at least one tertiary nitrogen group, and it is thought that they accumulate in acidified lysosomes and, at high concentrations, inhibit lysosomal phospholipases (Hein, 1990). In humans and animals receiving drugs such as amiodarone, aminoglycosides, and chloroquine, lysosomal phospholipidosis may be seen in variety of tissues, including liver and lung. Pulmonary fibrosis and progressive liver cell damage with cirrhosis are known complications of long-term amiodarone use, however, the relationship between the phospholipidosis observed via electron microscopy, present in all patients who take amiodarone, and the progressive pulmonary and liver fibrosis identified in only some patients is not known (Goldman *et al.*, 1985; Guigui *et al.*, 1988).

7. SUMMARY

Lysosomes, endosomes, and a variety of other intracellular organelles are acidified by a family of unique proton pumps, termed the vacuolar H^+-ATPases, that are evolutionarily related to bacterial membrane proton pumps and the F_1-F_0 H^+-ATPases that catalyze ATP synthesis in mitochondria and chloroplasts. The electrogenic vacuolar H^+-ATPase is responsible for generating electrical and chemical gradients across organelle membranes with the magnitude of these gradients ultimately determined by both proton pump regulatory mechanisms and, more importantly, associated ion and organic solute transporters located in vesicle membranes. Analogous to Na^+,K^+-ATPase on the cell membrane, the vacuolar proton pump not only acidifies the vesicle interior but provides a potential energy source for driving a variety of coupled transporters, many of them unique to specific organelles. Although the basic mechanism for organelle acidification is now well understood, it is already apparent that there are many differences in both the function of the proton pump and the associated transporters in different organelles and different cell types. These differences and their physiologic and pathophysiologic implications are exciting areas for future investigation.

8. REFERENCES

Anbari, M., Root, K. V., and Van Dyke, R. W., 1994, Role of Na,K-ATPase in regulating acidification of early rat liver endocytic vesicles, *Heptaology* **19**:1034–1043.

Anderson, O. S., Silveira, J. E. N., and Steinmetz, P. R., 1985, Intrinsic characteristics of the proton pump in the luminal membrane of a tight urinary epithelium, *J. Gen. Physiol.* **86**:215–234.

Arai, H., Pink, S., and Forgac, M., 1989, Interaction of anions and ATP with the coated vesicle proton pump, *Biochemistry* **28**:3075–3082.

Arai, K., Shimaya, A., Hiratani, N., and Ohkuma, S., 1993, Purification and characterization of lysosomal H+-ATPase, *J. Biol. Chem.* **268**:5649–5660.

Bae, H. R., and Verkman, A. S., 1990, Protein kinase A regulates chloride conductance in endocytic vesicles from proximal tubule, *Nature* **348**:637–639.

Barasch, J., and Al-Awqati, Q., 1993, Defective acidification of the biosynthetic pathway in cystic fibrosis, *J. Cell Science* **17**:229–233.

Barasch, J., Gershon, M. D., Nunez, E. A., Tamir, H., and Al-Awqati, Q., 1988, Thyrotropin induces the acidification of the secretory granules of parafollicular cells by increasing the chloride conductance of the granular membrane, *J. Cell Biol.* **107**:2137–2147.

Barasch, J., Kiss, B., Prince, A., Saiman, L., Gruenert, D., and Al-Awqati, Q., 1991, Defective acidification of intracellular organelles in cystic fibrosis, *Nature* **352**:70–73.

Beltrán, C., and Nelson, N., 1992, The membrane sector of vacuolar H+-ATPases by itself is impermeable to protons, *Acta Physiol. Scand.* **146**:41–47.

Bennett, A. B., and Spanswick, R. M., 1983, Optical measurements of ΔpH and $\Delta\Psi$ in corn root membrane vesicles: Kinetic analysis of Cl$^-$ effects on a proton-translocating ATPase, *J. Membr. Biol.* **71**:95–107.

Bernasconi, P., Rausch, T., Struve, I., Morgan, L., and Taiz, L., 1990, An mRNA from human brain encodes an isoform of the B subunit of the vacuolar H+-ATPase, *J. Biol. Chem.* **265**:17428–17431.

Biwersi, J., and Verkman, A. S., 1994, Functional CFTR in endosomal compartment of CFTR-expressing fibroblasts and T84 cells, *Am. J. Physiol.* **266**:C149–C156.

Blair, H. C., Teitelbaum, S. L., Tan, H. L., Koziol, C. M., and Schlesinger, P. H., 1991, Passive chloride permeability charge coupled to H+-ATPase of avian osteoclast ruffled membrane, *Am. J. Physiol.* **260**:C1315–C1324.

Bronk, S. F., and Gores, G. J., 1991, Efflux of protons from acidic vesicles contributes to cytosolic acidification of hepatocytes during ATP depletion, *Hepatology* **14**:626–633.

Brown, D., Sabolic, I., and Gluck, S., 1992, Polarized targeting of V-ATPase in kidney epithelial cells, *J. Exp. Biol.* **172**:231–243.

Cain, C. C., and Murphy, R. F., 1986, Growth inhibition of 3T3 fibroblasts by lysosomotropic amines: Correlation with effects on intravesicular pH but not vacuolation, *J. Cell. Physiol.* **129**:65–70.

Cain, C. C., Sipe, D. M., and Murphy, R. F., 1989, Regulation of endocytic pH by the Na+,K+-ATPase in living cells, *Proc. Natl. Acad. Sci. USA* **86**:544–548.

Cannon, C., van Adelsberg, J., Kelly, S., and Al-Awqati, Q., 1985, Carbon dioxide–induced exocytotic insertion of H+ pumps in turtle-bladder luminal membrane: Role of cell pH and calcium, *Nature* **314**:443–446.

Chatterjee, D., Chakraborty, M., Leit, M., Neff, L., Jamsa-Kellokumpu, S., Fuchs, R., and Baron, R., 1992a, Sensitivity to vanadate and isoforms of subunits A and B distinguish the osteoclast proton pump from other vacuolar H$^+$ ATPases, *Proc. Natl. Acad. Sci. USA* **89**:6257–6261.

Chatterjee, D., Chakraborty, M., Leit, M., Neff, L., Jamsa-Kellokumpu, S., Fuchs, R., Bartkiewicz, M., Hernando, N., and Baron, R., 1992b, The osteoclast proton pump differs in its pharmacology and catalytic subunits from other vacuolar H$^+$-ATPases, *J. Exp. Biol.* **172**:193–204.

Cidon, S., and Sihra, T. S., 1989, Characterization of a H$^+$-ATPase in rat brain synaptic vesicles, *J. Biol. Chem.* **264**:8281–8288.

Cidon, S., Ben-David, H., and Nelson, N., 1983, ATP-driven proton fluxes across membranes of secretory organelles, *J. Biol. Chem.* **258**:11684–11688.

Coyaud, L., Kurkdjian, A., Kado, R., and Hedrich, R., 1987, Ion channels and ATP-driven pumps involved in ion transport across the tonoplast of sugarbeet vacuoles, *Biochim. Biophys. Acta* **902**:263–268.

Crider, B. P., Xie, X. S., and Stone, D. K., 1994, Bafilomycin inhibits proton flow through the H$^+$ channel of vacuolar proton pumps, *J. Biol. Chem.* **269**:17379–17381.

Cuppoletti, J., Aures-Fischer, D., and Sachs, G., 1987, The lysosomal H$^+$ pump: 8-azido-ATP inhibition and the role of chloride in H$^+$ transport, *Biochim. Biophys. Acta* **899**:276–284.

David, P., and Baron, R., 1994, The catalytic cycle of the vacuolar H$^+$-ATPase, *J. Biol. Chem.* **269**:30158–30163.

Davies, J. M., Hunt, I., and Sanders, D., 1994, Vacuolar H$^+$-pumping ATPase variable transport coupling ratio controlled by pH, *Proc. Natl. Acad. Sci. USA* **91**:8547–8551.

Dell'Antone, P., 1984, Electrogenicity of the lysosomal proton pump, *FEBS Lett.* **168**:15–22.

Feng, Y., and Forgac, M., 1992, A novel mechanism for regulation of vacuolar acidification, *J. Biol. Chem.* **267**:19769–19772.

Feng, Y., and Forgac, M., 1994, Inhibition of vacuolar H$^+$-ATPase by disulfide bond formation between cysteine 254 and cysteine 532 in subunit A, *J. Biol. Chem.* **269**:13224–13230.

Fitz, J. G., and Scharschmidt, B. F., 1987, Intracellular chloride activity in intact rat liver: Relationship to membrane potential and bile flow, *Am. J. Physiol.* **252**:G699–G706.

Floor, E., Leventhal, P. S., and Schaeffer, S. F., 1990, Partial purificaion and characterization of the vacuolar H$^+$-ATPase of mammalian synaptic vesicles, *J. Neurochem.* **55**:1663–1670.

Forgac, M., 1992, Structure and properties of the coated vesicle (H$^+$)-ATPase, *J. Bioenerg. Biomembr.* **24**:341–350.

Forgac, M., Cantley, L., Wiedenmann, B., Altstiel, L., and Branton, D., 1983, Clathrin-coated vesicles contain an ATP-dependent proton pump, *Proc. Natl. Acad. Sci. USA* **80**:1300–1303.

Fuchs, R., Mâle, P., and Mellman, I., 1989a, Acidification and ion permeabilities of highly purified rat liver endosomes, *J. Biol. Chem.* **264**:2212–2220.

Fuchs, R., Schmid, S., and Mellman, I., 1989b, A possible role for Na$^+$,K$^+$-ATPase in regulating ATP-dependent endosome acidification, *Proc. Natl. Acad. Sci. USA* **86**:539–543.

Fuchs, R., Ellinger, A., Pavelka, M., Mellman, I., and Klapper, H., 1994, Rat liver endocytic coated vesicles do not exhibit ATP-dependent acidification in vitro, *Proc. Natl. Acad. Sci. USA* **91**:4811–4815.

Galloway, C. J., Dean, G. E., Marsh, M., Rudnick, G., and Mellman, I., 1983, Acidification of macrophage and fibroblast endocytic vesicles in vitro, *Proc. Natl. Acad. Sci. USA* **80**:3334–3338.

Gekle, M., and Silbernagl, S., 1995, Comparison of the buffer capacity of endocytotic vesicles, lysosomes and cytoplasm in cells derived from the proximal tubule of the kidney (opossum kidney cells), *Pfeügers Arch. Eur. J. Physiol.* **429**:452–454.

Gluck, S. L., 1992, The structure and biochemistry of the vacuolar H$^+$-ATPase in proximal and distal urinary acidification, *J. Bioenerg. Biomembr.* **24**:351–359.

Gluck, S. L., 1993, The vacuolar H$^+$-ATPases: Versatile proton pumps participating in constitutive and specialized functions of eukaryotic cells, *Int. Rev. Cytol.* **137C**:105–137.

Goldman, I. S., Winkler, M. L., Raper, S. E., Barker, M. E., Keung, E., Goldberg, H. I., and Boyer, T. D., 1985, Increased hepatic density and phospholipidosis due to amiodarone, *Am. J. Radiol.* **144**:541–546.

Guigui, B., Perrot, S., Berry, J. P., Fleury-Feith, J., Martin, N., Métreau, J. M., Dhumeaux, D., and Zafrani, E. S., 1988, Amiodarone-induced hepatic phospholipidisis: A morphological alteration independent of pseudoalcoholic liver disease, *Hepatology* **8**:1063–1068.Gurich, R. W., and Warnock, D. G., 1986, Electrically neutral Na^+-H^+ exchange in endosomes obtained from rabbit renal cortex, *Am. J. Physiol.* **251**:F702–F709.

Haigh, J. R., and Phillips, J. H., 1989, A sodium/proton antiporter in chromaffin-granule membranes, *Biochem. J.* **257**:499–507.

Hammond, T. G., and Verroust, P. J., 1994, Heterogeneity of endosomal populations in the rat renal cortex: Light endosomes, *Am. J. Physiol.* **266**:C1783–C1794.

Harikumar, P., and Reeves, J. P., 1983, The lysosomal proton pump is electrogenic, *J. Biol. Chem.* **258**:10403–10410.

Hedrich, R., and Kurkdjian, A., 1988, Characterization of an anion-ermeable channel from sugar beet vacuoles: Effect of inhibitors, *EMBO J.* **7**:3661–3666.

Hedrich, R., Kurkdjian, A., Guern, J., and Flügge, U. I., 1989, Comparative studies on the electrical properties of the H^+ translocating ATPase and pyrophosphatase of the vacuolar-lysosomal compartment, *EMGO J.* **8**:2835–2841.

Hein, L., Lüllmann-Rauch, R., and Mohr, K., 1990, Human accumulation potential of xenobiotics: Potential of catamphiphilic drugs to promote their accumulation via inducing lipidosis or mucopolysaccharidosis, *Xenobiotica* **20**:1259–1267.

Hemkin, P., Guo, X. L., Wang, Z. Q., Zhang, K., and Gluck, S., 1992, Immunologic evidence that vacuolar H^+ ATPases with heterogeneous forms of $M_r = 31{,}000$ subunit have different membrane distributions in mammalian kidney, *J. Biol. Chem.* **267**:9948–9957.

Henning, R., 1975, pH gradient across the lysosomal membrane generated by selective cation permeability and Donnan equilibrium, *Biochim. Biophys. Acta* **401**:307–316.

Henry, J. P., Botton, D., Sagne, C., Isambert, M. F., Desnos, C., Blanchard, V., Raisman-Vozari, R., Krejci, E., Massoulie, J., and Gasnier, B., 1994, Biochemistry and molecular biology of the vesicular monoamine transporter from chromaffin granules, *J. Exp. Biol.* **196**:251–262.

Hilden, S. A., and Madias, N. E., 1989, H^+/Ca^{2+} exchange in rabbit renal cortical endosomes, *J. Membr. Biol.* **112**:131–138.

Hilden, S. A., Johns, C. A., and Madias, N. E., 1988, Cl^--dependent ATP-driven H^+ transport in rabbit renal cortical endosomes, *Am. J. Physiol.* **255**:F885–F897.

Hilden, S A., Ghoshroy, K. B., and Madias, N. E., 1990, Na^+-H^+ exchange, but not Na^+-K^+-ATPase, is present in endosome-enriched microsomes from rabbit renal cortex, *Am. J. Physiol.* **258**:F1311–F1319.

Holz, R. W., 1979, Measurement of membrane potential of chromaffin granules by the accumulation of triphenylmethylphosphonium cation, *J. Biol. Chem.* **254**:6703–6709.

Hostetler, K. Y., 1984, Molecular studies of the induction of cellular phospholipidosis by cationic amphiphilic drugs, *Fed. Proc. Fed. Am. Soc. Exp. Biol.* **43**:2582–2585.

Idriss, J. M., and Jonas, A. J., 1991, Vitamin B_{12} transport by rat liver lysosomal membrane vesicles, *J. Biol. Chem.* **266**:9438–9441.

Johnson, R. G., Pfister, D., Carty, S. E., and Scarpa, A., 1979, Biological amine transport in chromaffin ghosts, *J. Biol. Chem.* **254**:10963–10972.

Kacew, S., 1987, Cationic amphiphilic drug-induced renal cortical lysosomal phospholipidosis: An in vivo comparative study with gentamicin and chlorphentermine, *Toxicol. Appl. Pharmacol.* **91**:469–476.

Kane, P. M., 1995, Disassembly and reassembly of the yeast vacuolar H^+-ATPase *in vivo*, *J. Biol. Chem.* **270**:17025–17032.

Kibak, H., Taiz, L., Starke, T., Bernasconi, P., and Gogarten, J. P., 1992, Evolution of structure and function of V-ATPases, *J. Bioenerg. Biomembr.* **24**:415–424.

Lai, S., Watson, J. C., Hansen, J. N., and Sze, H., 1991, Molecular cloning and sequencing of the cDNAs encoding the proteolipid subunit of the vacuolar H^+-ATPase from a higher plant, *J. Biol. Chem.* **266**:16078–16084.

Lake, J. R., Van Dyke, R. W., and Scharschmidt, B. F., 1987, Acidic vesicles in cultured rat hepato-
    cytes: Identification and characterization of their relationship to lysosomes and other storage vesi-
    cles, *Gastroenterology* **92:**1251–1261.
Landry, D., Sullivan, S., Nicolaides, M., Redhead, C., Edelman, A., Field, M., Al-Awqati, Q., and Ed-
    wards, J., 1993, Molecular cloning and characterization of p64, a chloride channel protein from
    kidney microsomes, *J. Biol. Chem.* **268:**14948–14955.
Lee, B. S., Underhill, D. M., Crane, M. K., and Gluck, S. L., 1995, Transcriptional regulation of the
    vacuolar H$^+$-ATPase B2 subunit gene in differentiating THP-1 cells, *J. Biol. Chem.*
    270:7320–7329.
Lemons, R. M., and Thoene, J. G., 1991, Mediated calcium transport by isolated human fibroblast lyso-
    somes, *J. Biol. Chem.* **266:**14378–14382.
Lencer, W. I., Verkman, A. S., Arnaout, M. A., Ausiello, D. A., and Brown, D., 1990, Endocytic vesi-
    cle from renal papilla which retrieve the vasopressin-sensitive water channel do not contain a
    functional H$^+$ATPase, *J. Cell Biol.* **111:**379–389.
Lencer, W. I., Delp, C., Neutra, M. R., and Madara, J. L., 1992, Mechanism of cholera toxin action on
    a polarized human intestinal epithelial cell line: Role of vesicular traffic, *J. Cell Biol.*
    117:1197–1209.
Liu, Q., Feng, Y., and Forgac, M., 1994, Activity and in vitro reassembly of the coated vesicle (H$^+$)-
    ATPase requires the 50-kDa subunit of the clathrin assembly complex AP-2, *J. Biol. Chem.*
    269:31592–31597.Lukacs, G. L., Rotstein, O. D., and Grinstein, S., 1990, Phagosomal acidification
    is mediated by a vacuolar-type H$^+$-ATPase in murine macrophages, *J. Biol. Chem.* **265:**21099–21107.
Lukacs, G. L., Rotstein, O. D., and Grinstein, S., 1991, Determinants of the phagosomal pH in
    macrophages, *J. Biol. Chem.* **266:**24540–24548.
Lukacs, G. L., Chang, X. B., Kartner, N., Rotstein, O. D., Riodan, J. R., and Grinstein, S., 1992, The
    cystic fibrosis transmembrane regulator is present and functional in endosomes, *J. Biol. Chem.*
    267:14568–14572.
MacIntyre, A. C., and Cutler, D. J., 1988, The potential role of lysosomes in tissue distribution of weak
    bases, *Biopharm. Drug Dispos.* **9:**513–526.
Mancini, G. M. S., de Jonge, H. R., Galjaard, H., and Verheijen, F. W., 1989, Characterization of a pro-
    ton-driven carrier for sialic acid in the lysosomal membrane, *J. Biol. Chem.* **264:**15247–15254.
Mancini, G. M. S., Beerens, C. E. M. T., and Verheijen, F. W., 1990, Glucose transport in lysosomal
    membrane vesicles, *J. Biol. Chem.* **265:**12380–12387.
Márquez-Sterling, N., Herman, I. M., Pesacreta, T., Arai, H., Terres, G., and Forgac, M., 1991, Im-
    munolocalization of the vacuolar-type (H$^+$)-ATPase from clathrin-coated vesicles, *Eur. J. Cell
    Biol.* **56:**19–33.
Marushack, M. M., Lee, B. S., Masood, K., and Gluck, S., 1992, cDNA sequence and tissue expres-
    sion of bovine vacuolar H$^+$-ATPase M$_r$ 70,000 subunit, *Am. J. Physiol.* **263:**F171–F174.
Moriyama, Y., 1988, Potassium ion dependent proton efflux and depolarization from spleen lysosomes,
    Biochem. Biophys. Res. Commun. **156:**211–216.
Moriyama, Y., and Futai, M., 1990, Presence of 5-hydroxytryptamine (serotonin) transport coupled
    with vacuolar-type H$^+$-ATPase in neurosecretory granules from bovine posterior pituitary, *J. Biol.
    Chem.* **265:**9165–9169.
Moriyama, Y., and Nelson, N., 1989, Lysosomal H$^+$-translocating ATPase has a similar subunit struc-
    ture to chromaffin granule H$^+$-ATPase complex, *Biochim. Biophys. Acta* **980:**241–247.
Moriyama, Y., Maeda, M., and Futai, M., 1992, Involvement of a non-proton pump factor (possibly
    Donnan-type equilibrium) in maintenance of an acidic pH in lysosomes, *FEBS Lett.* **1:**18–20.
Moseley, R. H., and Van Dyke, R. W., 1995, Organic cation transport by rat liver lysosomes, *Am. J.
    Physiol.* **268:**G480–G486.
Mulberg, A. E., Tulk, B. M., and Forgac, M., 1991, Modulation of coated vesicle chloride channel ac-
    tivity and acidification by reversible protein kinase A-dependent phosphorylation, *J. Biol. Chem.*
    266:20590–20593.

Myers, M., and Forgac, M., 1993a, The coated vesicle vacuolar (H+)-ATPase associates with and is phosphorylated by the 50-kDa polypeptide of the clathrin assembly protein AP-2, *J. Biol. Chem.* **268:**9184–9186.

Myers, M., and Forgac, M., 1993b, Assembly of the peripheral domain of the bovine vacuolar H+-adenosine triphosphatase, *J. Cell. Physiol.* **156:**35–42.

Nanda, A., Gukovskaya, A., Tseng, J., and Grinstein, S., 1992, Activation of vacuolar-type proton pumps by protein kinase C, *J. Biol. Chem.* **267:**22740–22746.

Nelson, H., Mandiyan, S., Noumi, T., Moriyama, Y., Miedel, M. C., and Nelson, N., 1990, Molecular cloning of cDNA encoding the C subunit of H+-ATPase from bovine chromaffin granules, *J. Biol. Chem.* **265:**20390–20393.

Nelson, N., 1992, Evolution of organellar proton-ATPases, *Biochim. Biophys. Acta* **1100:**109–124.

Nelson, R. D., Guo, X. L., Masood, K., Brown, D., Kalkbrenner, M., and Gluck, S., 1992, Selectively amplified expression of an isoform of the vacuolar H+-ATPase 56-kilodalton subunit in renal intercalated cells, *Proc. Natl. Acad. Sci. USA* **89:**3541–3545.

Nezu, J., Motojima, K., Tamura, H., and Ohkuma, S., 1992, Molecular cloning of a rat liver cDNA encoding the 16 kDa subunit of vacuolar H+-ATPases: Organellar and tissue distribution of 16 kDa proteolipids, *J. Biochem.* **112:**212–219.

Nordström, T., Grinstein, S., Brisseau, G. F., Manolson, M. F., and Rotstein, O. D., 1994, Protein kinase C activation accelerates proton extrusion by vacuolar-type H+-ATPases in murine peritoneal macrophages, *FEBS Lett.* **350:**82–86.

Ohkuma, S., and Poole, B., 1978, Fluorescence probe measurement of the intralysosomal pH in living cells and the perturbation of pH by various agents, *Proc. Natl. Acad. Sci. USA* **75:**3327–3331.

Ohkuma, S., Moriyama, Y., and Takano, T., 1982, Identification and characterization of a proton pump on lysosomes by fluorescein isothiocyanate-dextran fluorescence, *Proc. Natl. Acad. Sci. USA* **79:**2758–2762.

Ohkuma, S., Moriyama, Y., and Takano, T., 1983, Electrogenic nature of lysosomal proton pump as revealed with a cyanine dye, *J. Biochem.* **94:**1935–1943.

Overly, C. C., Lee, K. D., Berthiaume, E., and Hollenbeck, P. J., 1995, Quantitative measurement of intraorganelle pH in the endosomal-lysosomal pathway in neurons by using ratiometric imaging with pyranine, *Proc. Natl. Acad. Sci. USA* **92:**3156–3160.

Peng, S. B., Crider, B. P., Xie, X. S., and Stone, D. K., 1994a, Alternative mRNA splicing generates tissue-specific isoforms of 116-kDa polypeptide of vacuolar proton pump, *J. Biol. Chem.* **269:**17262–17266.

Peng, S. B., Zhang, Y., Crider, B. P., White, A. E., Fried, V. A., Stone, D. K., and Xie, X. S., 1994b, Reconstitution of the recombinant 70-kDa subunit of the clathrin-coated vesicle H+ ATPase, *J. Biol. Chem.* **269:**27778–27782. Perin, M. S., Fried, V. A., Stone, D. K., Xie, X. S., and Sudhof, T. C., 1991, Structure of the 116-kDa polypeptide of the clathrin-coated vesicle/synaptic vesicle proton pump, *J. Biol. Chem.* **266:**3877–3881.

Piqueras, A. I., Somers, M., Hammond, T. G., Strange, K., Harris, H. W., Jr., Gawryl, M., and Zeidel, M. L., 1994, Permeability properties of rat renal lysosomes, *Am. J. Physiol.* **266:**C121–C133.

Poole, B., and Ohkuma, S., 1981, Effect of weak bases on the intralysosomal pH in mouse peritoneal macrophages, *J. Cell Biol.* **90:**665–669.

Pope, A. J., Jennings, I. R., Sanders, D., and Leigh, R. A., 1990, Characterization of Cl−-sensitive fluorescent probe: Reaction kinetic models for voltage- and concentration-dependence of Cl− flux, *J. Membr. Biol.* **116:**129–137.

Pritchard, J. B., Sykes, D. B., Walden, R., and Miller, D. S., 1994, ATP-dependent transport of tetraethylammonium by endosomes isolated from rat renal cortex, *Am. J. Physiol.* **266:**F966–F976.

Puopolo, K., Kumamoto, C., Adachi, I., and Forgac, M., 1991, A single gene encodes the catalytic "A" subunit of the bovine vacuolar H+-ATPase, *J. Biol. Chem.* **266:**24564–24572.

Puopolo, K., Kumamoto, C., Adachi, I., Magner, R., and Forgac, M., 1992, Differential expression of the "B" subunit of the vacuolar H+-ATPase in bovine tissues, *J. Biol. Chem.* **267:**3696–3706.

Reasor, M. J., 1989, A review of the biology and toxicologic implications of the induction of lysoso-
    mal lamellar bodies by drugs, *Toxicol. Appl. Pharmacol.* **97**:47–56.
Reenstra, W. W., Sabolic, I., Bae, H. R., and Verkman, A. S., 1992, Protein kinase A dependent mem-
    brane protein phosphorylation and chloride conductance in endosomal vesicles from kidney cor-
    tex, *Biochemistry* **31**:175–181.
Rodman, J., Feng, Y., Myers, M., Zhand, J., Magner, R., and Forgac, M., 1994, Comparison of the
    coated-vesicle and synaptic-vesicle vacuolar (H^+)-ATPases, *Ann. N. Y. Acad. Sci.* **733**:203–211.
Rodman, J. S., Stahl, P. D., and Gluck, S., 1991, Distribution and structure of the vacuolar H^+ ATPse
    in endosomes and lysosomes from LLC-PK₁ cells, *Exp. Cell Res.* **192**:445–452.
Roederer, M., Bowser, R., and Murphy, R. F., 1987, Kinetics and temperature dependence of exposure
    of endocytosed material to proteolytic enzymes and low pH: Evidence for a maturation model for
    the formation of lysosomes, *J. Cell. Physiol.* **131**:200–209.
Rooney, E. K., and Gross, J. D., 1992, ATP-driven Ca^{2+}/H^+ antiport in acid vesicles from Dic-
    tyostelium, *Proc. Natl. Acad. Sci. USA* **89**:8025–8029.
Root, K. V., Engelhardt, J. F., Post, M., Wilson, J. W., and Van Dyke, R. W., 1994, CFTR does not al-
    ter acidification of L cell endosomes, *Biochem. Biophys. Res. Commun.* **205**:396–401.
Sabolic, I., and Brown, D., 1990, Na^+ (Li^+)-H^+ exchange in rat renal cortical vesicles with endosomal
    characteristics, *Am. J. Physiol.* **258**:F1245–F1253.
Sabolic, I., and Burckhardt, G., 1986, Characteristics of the proton pump in rat renal cortical endocy-
    totic vesicles, *Am. J. Physiol.* **250**:F817–F826.
Schneider, D. L., 1981, ATP-dependent acidification of intact and disrupted lysosomes, *J. Biol. Chem.*
    256:3858–3864.
Seksek, O., Biwersi, J., and Verkman, A. S., 1995, Direct measurement of trans-Golgi pH in living cells
    and regulation by second messengers, *J. Biol. Chem.* **270**:4967–4970.
Shi, L. B., Fushimi, K., Bae, H. R., and Verkman, A. S., 1991, Heterogeneity in ATP-dependent acid-
    ification in endocytic vesicles from kidney proximal tubule, *Biophys. J.* **59**:1208–1217.
Sipe, D. M., Jesurum, A., and Murphy, R. F., 1991, Absence of Na^+,K^+-ATPase regulation of endo-
    somal acidification in K562 erythroleukemia cells, *J. Biol. Chem.* **266**:3469–3474.
Small, P. L. C., Ramakrishnan, L., and Falkow, S., 1994, Remodeling schemes of intracellular
    pathogens, *Science* **263**:637–639.
Steinmetz, P. R., Husted, R. F., Mueller, A., and Beauwens, R., 1981, Coupling between H^+ transport
    and anaerobic glycolysis in turtle urinary bladder: Effects of inhibitors of H^+ ATPase, *J. Membr.
    Biol.* **59**:27–34.
Sturgill-Koszycki, S., Schlesinger, P. H., Chakraborty, P., Haddix, P. L., Collins, H. L., Fok, A. K.,
    Allen, R. D., Gluck, S. L., Heuser, J., and Russell, D. G., 1994, Lack of acidification in *My-
    cobacterium* phagosomes produced by exclusion of the vesicular proton-ATPase, *Science*
    263:678–681.
Supek, F., Supekova, L., Mandiyan, S., Pan, Y. C. E., Nelson, H., and Nelson, N., 1994, A novel acces-
    sory subunit for vacuolar H^+-ATPase from chromaffin granules, *J. Biol. Chem.* **269**:24102–24106.
Sze, H., Ward, J. M., and Lai, S., 1992, Vacuolar H^+-translocating ATPases from plants: Structure,
    function, and isoforms, *J. Bioenerg. Biomembr.* **24**:371–381.
Tabb, J. S., Kish, P. E., Van Dyke, R., and Ueda, T., 1992, Glutamate transport into synaptic vesicles,
    J. Biol. Chem. **267**:15412–15418.
Tamir, H., Piscopo, I., Liu, K. P., Hsiung, S. C., Aldersberg, M., Nicolaides, M., Al-Awqati, Q., Nunez,
    E. A., and Gershon, M. D., 1994, Secretogogue-induced gating of chloride channels in the secre-
    tory vesicles of parafollicular cells, *Endocrinology* **135**:2045–2057.
Thomas-Reetz, A., Hell, J. W., During, M. J., Walch-Solimena, C., Jahn, R., and De Camilli, P., 1993,
    A γ-aminobutyric acid transporter driven by a proton pump is present in synaptic-like microvesi-
    cles of pancreatic β cells, *Proc. Natl. Acad. Sci. USA* **90**:5317–5321.Tietz, P. S., Yamazaki, K.,
    and LaRusso, N. F., 1990, Time-dependent effects of chloroquine on pH of hepatocyte lysosomes,
    Biochem. Pharmacol. **40**:1419–1421.

Tilly, B. C., Mancini, G. M. S., Bijman, J., van Gageldonk, P. G. M., Beerens, C. E. M. T., Bridges, R. J., de Jonge, H. R., and Verheijer, F. W., 1992, Nucleotide-activated chloride channels in lysosomal membranes, *Biochem. Biophys. Res. Commun.* **187**:254–260.

van Adelsberg, J., and Al-Awqati, Q., 1986, Regulation of cell pH by Ca^{+2}-mediated exocytotic insertion of H^+-ATPases, *J. Cell Biol.* **102**:1638–1645.

Van Dyke, R. W., 1986, Anion inhibition of the proton pump in rat liver multivesicular bodies, *J. Biol. Chem.* **261**:15941–15948.

Van Dyke, R. W., 1988, Proton pump-generated electrochemical gradients in rat liver multivesicular bodies, *J. Biol. Chem.* **263**:2603–2611.

Van Dyke, R. W., 1993, Acidification of rat liver lysosomes: Quantitation and comparison with endosomes, *Am. J. Physiol.* **265**:C901–C917.

Van Dyke, R. W., 1995, Na^+/H^+ exchange modulates acidification of early rat liver endocytic vesicles, *Am. J. Physiol.* **269**:C943–C954.

Van Dyke, R. W., and Belcher, J. D., 1994, Acidification of three types of liver endocytic vesicles: Similarities and differences, *Am. J. Physiol.* **266**:C81–C94.

Van Dyke, R. W., and Root, K. V., 1993, Ethinyl estradiol decreases acidification of rat liver endocytic vesicles, *Hepatology* **18**:604–613.

Van Dyke, R. W., Hornick, C. A., Belcher, J., Scharschmidt, B. F., and Havel, R. J., 1985, Identification and characterization of ATP-dependent proton transport by rat liver multivesicular bodies, *J. Biol. Chem.* **260**:11021–11026.

Van Dyke, R. W., Faber, E. D., and Meijer, D. K. F., 1992a, Sequestration of organic cations by acidified hepatic endocytic vesicles and implications for biliary excretion, *J. Pharm. Exp. Ther.* **261**:1–11.

Van Dyke, R. W., Root, K. V., Schreiber, J. H., and Wilson, J. M., 1992b, Role of CFTR in lysosome acidification, *Biochem. Biophys. Res. Commun.* **184**:300–305.

Van Dyke, R. W., Li, C., Gu, H., Lee, D., and Bear, C. E., 1994, Planar lipid bilayer studies of ion channels in liver endocytic vesicles, *Gastroenterology* **106**:A1001.

Van Dyke, R. W., Root, K. V., and Hsi, R. A., 1996, cAMP and protein kinase A stimulate acidification of rat liver endosomes in the absence of chloride, *Biochem. Biophys. Res. Comm.* **222**:312–316.

Wada, Y., and Anraku, Y., 1994, Chemiosmotic coupling of ion transport in the yeast vauole: Its role in acidification inside organelles, *J. Bioenerg. Biomembr.* **26**:631–637.

Wada, Y., Ohsumi, Y., and Anraku, Y., 1992, Chloride transport of yeast vacuolar membrane vesicles: A study of in vitro vacuolar acidification, *Biochim. Biophys. Acta* **1101**:296–302.

Wadsworth, S. J., and van Rossum, G. D. V., 1994, Role of vacuolar adenosine triphosphatase in the regulation of cytosolic pH in hepatocytes, *J. Membr. Biol.* **142**:21–34.

Wang, S. Y., Moriyama, Y., Mandel, M., Hulmes, J. D., Pan, Y. C. E., Danho, W., Nelson, H., and Nelson, N., 1988, Cloning of cDNA encoding a 32-kDa protein, *J. Biol. Chem.* **263**:17638–17642.

Wang, Z. Q., and Gluck, S., 1990, Isolation and properties of bovine kidney brush border vacuolar H^+-ATPase, *J. Biol. Chem.* **265**:21957–21965.

Weitering, J. G., Lammers, W., Meijer, D. K. F., and Mulder, G. J., 1977, Localization of D-tubocurarine in rat liver lysosomes, *Naunyn-Schmiedeberg's Arch. Pharmacol.* **299**:277–281.

Xie, X. S., Stone, D. K., and Racker, E., 1983, Determinants of clathrin-coated vesicle acidification, *J. Biol. Chem.* **258**:14834–14838.

Xie, X. S., Crider, B. P., and Stone, D. K., 1989, Isolation and reconstitution of the chloride transporter of clathrin-coated vesicles, *J. Biol. Chem.* **264**:18870–18873.

Xie, X. S., Crider, B. P., and Stone, D. K., 1993, Isolation of a protein activator of the clathrin-coated vesicle proton pump, *J. Biol. Chem.* **268**:25063–25067.

Xie, X. S., Crider, B. P., Ma, Y. M., and Stone, D. K., 1994, Role of a 50–57-kDa polypeptide heterodimer in the function of the clathrin-coated vesicle proton pump, *J. Biol. Chem.* **269**:25809–25815.

Zen, K., Biwersi, J., Periasamy, N., and Verkman, A. S., 1992, Second messengers regulate endosomal acidification in Swiss 3T3 fibroblasts, *J. Cell Biol.* **119**:99–110.

Zhang, J., Myers, M., and Forgac, M., 1992, Characterization of the V_0 domain of the coated vesicle (H^+)-ATPase, *J. Biol. Chem.* **267**:9773–9778.

Zhang, J., Feng, Y., and Forgac, M., 1994, Proton conduction and bafilomycin binding by the V_0 domain of the coated vesicle V-ATPase, *J. Biol. Chem.* **269**:23518–23523.

Zhang, K., Wang, Z. Q., and Gluck, S., 1992a, Identification and partial purification of a cytosolic activator of vacuolar H^+-ATPases from mammalian kidney, *J. Biol. Chem.* **267**:9701–9705.

Zhang, K., Wang, Z Q., and Gluck, S., 1992b, A cytosolic inhibitor of vacuolar H^+-ATPases from mammalian kidney, *J. Biol. Chem.* **267**:14539–14542.

Chapter 11

Metabolite Efflux and Influx Across the Lysosome Membrane

John B. Lloyd

1. INTRODUCTION

The chapters in Section II of this volume review the catabolic potential of the lysosome, describing how the enzymes of the lysosome degrade the major classes of biopolymers. By a series of sequential hydrolytic steps, most macromolecules are digested to the monomer level. This chapter examines the fate of the end products generated in the lysosome and the role of the lysosome membrane in regulating metabolite flow from lysosome to cytoplasm.

The flow of metabolites across the lysosome membrane is principally in the direction of efflux from the lysosome. However, there are a few substances that are believed to move in the reverse direction. These too will be considered.

The past decade has witnessed the discovery and characterization of numerous metabolite porters in the lysosome membrane. Each of these porters appears to have a definable substrate specificity, and it is not unreasonable to infer a physiologic role in the translocation of certain metabolites across the lysosome membrane. It is salutary to recall that, until the discovery of the cystine porter in 1982,

John B. Lloyd   Department of Pediatrics, Jefferson Medical College, Philadelphia, Pennsylvania 19107, and Division of Developmental Biology, Nemours Research Programs, Wilmington, Delaware 19899.

Subcellular Biochemistry, Volume 27: Biology of the Lysosome, edited by Lloyd and Mason. Plenum Press, New York, 1996.

it was assumed that every substance that could cross the lysosome membrane did so by passive diffusion. As late as 1977 a major review of this field (Reijngoud and Tager, 1977) found "no evidence for transport systems in the lysosome membrane."

In 1996 entirely opposite perceptions are prevalent. So many transport systems have been described that most authors do not even consider the contribution that passive diffusion may make to a solute's translocation. In at least one case (see Section 3.6), this imbalance has caused experimental data to be interpreted as evidence for a carrier, when truly they are more readily explained in terms of passive diffusion.

The aim of this chapter is to consider how each of the metabolic products generated in lysosomes escapes across the membrane into the cytoplasm. An evaluation of each solute's ability to cross by passive diffusion will be followed by a consideration of the properties of any lysosome membrane porter reported to recognize the solute. On the basis of this information, an assessment will be made of the contribution each mechanism makes of the overall efflux process.

2.   MECHANISMS OF METABOLITE EFFLUX AND INFLUX

The permeability properties of the lysosome membrane have been of interest from the earliest days of lysosome research. Indeed the discovery of lysosomes depended in large measure on the impermeability of their membrane to both sucrose and glycerol-2-phosphate. The story is well known (de Duve, 1969) but is worth retelling.

Christian de Duve and his colleagues at the University of Louvain (Belgium) were investigating the properties of the liver enzyme glucose 6-phosphatase. Devising a specific assay was difficult, because liver contains another enzymic activity, acid phosphatase, that also hydrolyzes glucose 6-phosphate. In order to make progress they decided to study the properties of acid phosphatase. In December 1949 they prepared a rat liver homogenate in isotonic sucrose and assayed for acid phosphatase, using glycerol 2-phosphate as substrate. Repeating the assay on the same homogenate a few days later, they found a ninefold increase in activity. Further investigation led to the report published in 1955 (de Duve et al., 1955) that acid phosphatase was one of several enzymes present in an organelle, distinct from the mitochondrion, that had a significant internal osmotic pressure and whose limiting membrane was impermeable to sucrose. The lysosome received its name in this same paper.

Lysosomes are stable when suspended in 250 mM sucrose. This is because there is osmotic balance between the sucrose in the suspending solution and the constituents of the lysosome matrix that give the organelle its osmotic pressure. Lysosomes remain stable in cold 250 mM sucrose for several hours, because the sucrose does not penetrate into the lysosome, nor do the lysosomal constituents escape. Because the membrane is also impermeable to glycerol 2-phosphate, and because the assay conducted by de Duve and colleagues was performed in the presence of 250 mM sucrose, only the acid phosphatase of damaged or broken lysosomes was detected. Even in ice-cold sucrose, lysosomes break after a day or two, and so assays conducted later revealed the total acid phosphatase present.

If lysosomes were impermeable to sucrose (f.w. 342) and glycerol 2-phosphate (f.w. 170), it was considered unlikely that larger molecules could cross. This conclusion was reinforced by the obvious inability of the acid phosphatase (and other lysosomal enzymes) to reach a substrate in the suspending medium, and also by the discovery in the early 1960s of the lysosomal storage diseases. This large group of inborn errors of metabolism arises from the congenital absence from the lysosome of one or other of the enzymes normally present (for recent reviews, see Reuser et al., 1994; Gieselmann, 1995). It became clear that neither the macromolecular substrates of these enzymes nor their partially digested products were able to cross the membrane.

However, the final products of the catabolic pathways must be permeant; otherwise these would accumulate in a similar manner to the biopolymers from which they arose. In a very early paper (Berthet et al., 1951), de Duve showed that lysosomes begin to break when incubated in cold 250 mM glucose or glycerol. They correctly interpreted this result to indicate that glucose, unlike sucrose, "probably enters [the granules] at a slow but significant rate." Although this 1951 paper refers to the acid phosphatase-containing particles as mitochondria, it contains the clear statement that "it is only because they appear largely in a fraction whose content has been identified by several workers with the cell mitochondria that we have used this name to characterize them." There is no hint of this caveat in the preliminary report of the same data (de Duve et al., 1951), evidence that the lysosome concept was probably born during the winter of 1950–51, only a year after the initial experiments of December 1949.

In the 1950s and 1960s these observations on the osmotic properties of lysosomes and on the phenomenon of lysosome storage were extended to generate data that permitted some generalizations to be made about what could and what could not cross the lysosome membrane.

In a review chapter written 25 years ago, Lloyd (1973) discussed the phenomenon of experimental lysosome storage and the information it provided on the permeability of the lysosome membrane. Briefly summarized, the chapter described the ultrastructure of cells exposed to macromolecules such as dextran, polyvinylpyrrolidone, and Triton WR1339. Lysosomes were prominent in these cells and clearly contained deposits of the macromolecule in question. The profiles observed were similar to those seen in the lysosome storage diseases, and it could confidently be inferred that the macromolecules were taken into the cells by endocytosis and, being nondigestible by the lysosomal enzymes, accumulated in the lysosomes.

One of the substances that accumulated in this way into lysosomes was sucrose. This observation provided an independent confirmation of the inference from the osmotic properties of lysosomes that the lysosome membrane is impermeable to sucrose. It also indicated that sucrose must be resistant to hydrolysis by the lysosomal glycosidases. Zanvil Cohn of Rockefeller University was the first to realize that the ability or the failure of an endocytosed substance to change the morphology of lysosomes could be used to indicate its ability to cross the lysosome membrane. His experiments (Cohn and Ehrenreich, 1969) indicated that disaccharides are impermeant but monosaccharides up to molecular weight 220 are permeant. Some disaccharides failed to engorge the lysosomes; these were ones that could be digested to monosaccharides by lysosomal enzymes. Further experiments (Ehrenreich and Cohn, 1969) suggested that amino acids and dipeptides could cross the lysosome membrane regardless of whether they were of the L or D configuration, provided that their formula weight was below 200–220. Large D-dipeptides appeared to be nonpermeant, and the failure of larger L-dipeptides to engorge the lysosomes was attributed to their hydrolysis within the lysosome to the smaller amino acids. These experiments led to the belief that passage of metabolites across the lysosome membrane was governed by molecular weight/volume and was unlikely to involve stereospecific mechanisms such as metabolite porters.

The osmotic properties of lysosomes also suggested an experimental approach to the investigation of the permeability properties of the lysosome membrane. Lysosomes suspended in 250 mM solutions of various solutes remained intact for periods that varied with the nature of the solute. As had been shown by de Duve, sucrose provided prolonged osmotic protection, whereas glucose did not. Investigations of a range of other compounds, some physiological some not (Lloyd, 1969, 1971), suggested an inverse correlation between molecular weight and rate of penetration that was largely consistent with the data from experimental lysosomal storage. Again, no stereospecificity was apparent.

2.1. Passive Diffusion

Throughout the 1970s these data, deriving from two distinct experimental approaches, provided a consistent and credible model for solute penetration across the lysosome membrane. The membrane could be envisioned as imposing a molecular weight ceiling in the region of 200–220. Molecules below this molecular size could cross the membrane; those above could not. By the late 1980s, however, these data appeared increasingly out of line with the emerging picture of the lysosome membrane as replete with a multitude of substrate-specific porters. Was the relationship between apparent penetration rate and formula weight merely fortuitous, or did a second mechanism exist (presumably passive diffusion) operating in tandem with the metabolite transporters?

It was this question that led the present author to revisit the osmotic protection technique that had been used two decades earlier to study the ability of substances to cross the lysosome membrane. This technique has two important advantages. First it does not require the use of highly purified lysosomes, thus avoiding a long preparation procedure that is potentially damaging to the lysosome membrane. Secondly, and crucially for these experiments, it does not require that the substances whose permeation is to be investigated are available in radiolabeled form. Thus structure–activity relationships can be established from data on a large number of compounds.

In the first study (Iveson *et al.*, 1989) 43 aliphatic nonelectrolytes containing only carbon, hydrogen, and oxygen were investigated. The oxygen moieties were present as either C–O–H (hydroxy) or C–O–C (ether, ester, or hemiacetal). Lysosomes were incubated at 25 °C for 60 min in 250 mM solutions, pH 7.0. The rate at which the lysosomes broke, rendering their enzymes available, was taken as an indication of the permeability of the membrane to the solute in question. Many of the compounds used were nonphysiological and unlikely to be recognized by a lysosome-membrane metabolite porter.

The results showed no consistent correlation between molecular weight and rate of lysosome breakage. While D-mannitol (f.w. 182) provided prolonged osmotic protection, tetraethylene glycol (194) was ineffective, indicating very rapid translocation across the membrane. However, there was a clear inverse correlation between a solute's ability to afford osmotic protection and another molecular parameter, the solute's notional hydrogen-bonding capacity. This latter parameter is calculated by summing the hydrogen-bonding capacities of the molecule's individual functional groups. The values used (2.0 for each C–O–H; 0.8 for each C–O–C) were derived from the work of Stein (1967) and Diamond and Wright (1969). For example, mannitol has a hydrogen-bonding capacity of 12.0 (the molecule has six hydroxy groups), while that of tetraethyleneglycol is 6.4 (two alcohol hydroxy groups and three ether linkages).

Table I shows a summary of the data from this study. A simple empirical relationship (see Table II) is apparent between hydrogen-bonding capacity and ability to afford osmotic protection. It is striking that every one of the 43 compounds conforms to this relationship. This observation, and the fact that hydrogen-bonding capacity has been shown to be an excellent predictor of rate of penetration of nonelectrolytes across a variety of biological and artificial membranes (Stein, 1967), is strong evidence that simple diffusion of substances across the lysosome membrane is a reality.

Before considering the physiological relevance of this conclusion, we should ask how it was possible for the investigators of the 1960s and early 1970s to conclude that penetration correlated inversely with a molecule's formula weight. The answer lies partly in the relatively small number of substances tested in these early investigations, but principally in the fact that, within any homologous series, hydrogen-bonding capacity tends to increase with molecular weight. In the more recent investigation the compounds tested were deliberately chosen so that the effects of these two molecular parameters would be distinguished.

Another question relates to the behavior in these experiments of substances known to be recognized by a porter in the rat liver lysosome membrane. D-Glucose and D-mannose are such compounds. It is striking that these compounds behave as if their entry into the lysosomes was solely a correlate of their

Table I
Free Activity of N-Acetyl-β-hexosaminidase in Rat Liver Lysosomes Incubated at 25 °C for 0, 30, or 60 Minutes in 250 mM Solutions of Various Nonelectrolytes

Solute (formula weight)	H-bonding capacity	Free Activity[a]		
		0 min	30 min	60 min
Carbitol (134)	3.6	81	87	92
Ethylene glycol (62)	4.0	90	99	93
Cyclohexane-1,4-diol (116)	4.0	71	92	89
Hexane-1,6-diol (118)	4.0	89	95	100
Cyclohexane-1,4-dimethanol (144)	4.0	92	90	86
Diethylene glycol (106)	4.8	76	94	95
Dipropylene glycol (134)	4.8	81	82	94
Triethylene glycol (150)	5.6	94	98	98
Dimethyl-D-tartrate (178)	5.6	77	84	85
Hexane-1,2,6-triol (134)	6.0	58	81	91
Propane-1,1,1-trimethanol (134)	6.0	68	88	93
Tetraethylene glycol (194)	6.4	82	91	93
Polyethylene glycol 200	6.5	76	92	85
Pentaethylene glycol (238)	7.2	67	92	79
Isopropylidine-D-mannitol (262)	7.2	70	88	88
Erythritol (122)	8.0	9	75	87
Pentaerythritol (136)	8.0	6	74	91
Octane-1,2,7,8-tetrol (178)	8.0	7	87	90
Hexaethylene glycol (282)	8.0	16	84	89
Polyethylene glycol 300	8.4	55	83	82
D-Arabinose (150)	8.8	6	60	81
D-Lyxose (150)	8.8	6	63	81
L-Lyxose (150)	8.8	8	50	69
D-Ribose (150)	8.8	9	73	95
D-Xylose (150)	8.8	5	73	87
2-Deoxy-D-galactose (164)	8.8	10	57	76
D-Fucose (164)	8.8	12	28	51
L-Rhamnose (164)	8.8	7	55	78
3-O-Methylglucose (194)	9.6	6	27	47
L-Arabitol (152)	10.0	10	30	48
Xylitol (152)	10.0	9	27	55
Polyethylene glycol 400	10.3	4	52	62
D-Glucose (180)	10.8	8	39	54
D-Mannose (180)	10.8	9	31	64
D-Sorbose (180)	10.8	7	21	49
Inositol (180)	12.0	10	19	36
Dulcitol (182)	12.0	5	9	26
D-Mannitol (182)	12.0	5	9	12
D-Sorbitol (182)	12.0	6	10	23
Perseitol (212)	14.0	6	5	10
Polyethylene glycol 600	14.0	6	10	17
Sucrose (342)	18.4	5	9	10
Polyethylene glycol 1000	21.5	6	6	6

[a]Free activity is expressed as a percentage of the total activity seen in the presence of Triton X-100 (0.2%). In calculating hydrogen-bonding capacity, oxygen-containing moieties were assigned values of 2.0 (C–O–H), 0.8 (CO$_2$–C or C–O–C) (see the text). Values for free activity are means for at least three experiments. Data from Iveson *et al.* (1989). (Reproduced from Lloyd (1992a), with permission.)

Table II
Relationship between Hydrogen-Bonding Capacity of Nonelectrolytes and Behavior Pattern in Osmotic Protection Experiments[a]

Hydrogen-bonding capacity	Osmotic protection pattern
Above 11.5	Pattern A: sustained protection; free activity remains below 20% for at least 30 min.
7.5–11.5	Pattern B: transient protection; initial free activity below 20%, but rising to above 20% by 30 min.
Below 7.5	Pattern C: no protection; initial free activity above 20%.

[a]Reproduced from Iveson et al. (1989), with permission.

hydrogen-bonding capacity. The probable explanation is that, in the osmotic protection experiments, the solute concentration greatly exceeds the porter's K_m. Under these circumstances the relative contribution of the porter decreases, and hexose translocation is chiefly a function of passive diffusion. At physiological concentrations of glucose or mannose, however, carrier-mediated translocation may be dominant.

The osmotic protection technique has also been used to investigate the permeability of the liver lysosome membrane to some charged molecules. The ω-amino-aliphatic acids carry two charged groups, but at neutral pH have zero net charge. A homologous series of these compounds was studied (Bird and Lloyd, 1990), the results indicating a slow rate of entry into lysosomes. When the quantitative data are compared with those of Iveson et al. (1989), the ω-amino-aliphatic acids appear to behave as if their hydrogen-bonding capacity was above 10.8 but below 12.0. It is noteworthy that Ginsburg and Stein (1987), in discussing the simple aliphatic amino acids, deduced a hydrogen-bonding capacity of 11. Both groups of compounds have identical hydrogen-bonding moieties, a charged primary amine and a charged carboxylate. Bird and Lloyd (1990) also studied the behavior of some dipeptides containing one or both amino acids in the D configuration. Most of these appeared to enter lysosomes even more slowly than the ω-amino-aliphatic acids, from which they differ in also possessing a secondary amide (peptide) group, which should add 2.0 to their hydrogen-bonding capacity. A few D-dipeptides, D-Ala-Gly for example, entered the lysosomes substantially more rapidly, and this was provisionally attributed to their being recognized by a putative L-dipeptide porter (see Section 3.1).

The hydrogen-bonding capacity is a measure of the difficulty a molecule will experience in leaving an aqueous environment for the hydrophobic interior of a biological membrane. Unsurprisingly hydrogen-bonding capacities show a strong inverse correlation with oil–water partition coefficients (Stein, 1967). What then is their value, when the latter can be measured experimentally? The answer is a strictly practical one: Most metabolites are much too hydrophilic for their partition coefficients to be measured. Urea and glycerol are typically the most hydrophilic molecules included in listings of partition coefficients. The hydrogen-bonding capacity is therefore a useful parameter, particularly as it is estimated by simply considering the functional groups possessed by the molecule in question.

Rashid *et al.* (1991) exposed cultured rat fibroblasts to a large number of fluorescent probes and used fluorescence microscopy to examine the subsequent subcellular localization. Recognizing that uptake into lysosomes can occur by endocytosis as well as by permeation across cellular membranes, they reached the conclusion that membrane-permeant probes possess octanol–water partition coefficients between 1 and 10^5.

A study of the permeability of rat kidney lysosomes to seven small nonphysiological nonelectrolytes (Piqueras *et al.*, 1994) found a good correlation with four oil–water (particularly octanol–water) partition coefficients. As just explained, this result is entirely concordant with the reported inverse correlation with notional hydrogen-bonding capacity. Piqueras *et al.* (1994) also found that the permeability of the kidney lysosome membrane correlates poorly with the molecular volume of solutes. While this conclusion is also concordant with the results of the osmotic protection experiments described above, it has to be pointed out that the molecular weight range of the compounds investigated by Piqueras *et al.* (1994) was narrow (from 59 to 92). Also, and necessarily, with the exception of glycerol and urea, the molecules investigated were not of physiological relevance. They would all be predicted to cross the lysosome membrane with ease.

The contribution of carrier-mediated and diffusional mechanisms to the efflux of individual metabolites from the lysosome will be considered in Section 3. We conclude the present section by noting some of the characteristics of passive diffusion. First, it is symmetrically bidirectional: Its rate and direction will be proportional to the concentration gradient across the membrane. Second, it is not stereospecific. Third, unlike carrier-mediated transport and translocation through pores, it does not exhibit cutoff phenomena. There will be no clear dividing line between permeant and nonpermeant molecules. The rather clear divisions seen in Table II can only reflect operationally defined points on a continuum of permeabilities.

2.2. Porters

The methodology necessary to characterize carrier-mediated metabolite uptake into lysosomes differs in one important respect from that used in osmotic-protection studies: It necessitates the separation of lysosomes, uncontaminated by other organelles. This requirement has the disadvantage that the lysosomes whose uptake characteristics are studied comprise a very small and possibly unrepresentative sample of the entire population. The multistage preparation procedure also carries the danger of greater damage to the lysosomes in the course of their isolation.

Most of the work on lysosome membrane porters has used a purified fraction from homogenates of either rat liver or human fibroblasts. In a typical experiment the lysosomes are incubated for periods ranging from a few seconds to a few minutes in the presence of a radiolabeled metabolite. At the end of the incubation the suspension of lysosomes is passed quickly through a filter, which retains the lysosomes. After washing to remove adherent radioactivity, the filter is counted and the kinetics of uptake deduced from the relationship of the radioactivity detected to the duration of incubation. By varying the concentration of the metabolite in the incubation medium, the saturability of the uptake system can be investigated. Likewise, the ability of structural analogues of the metabolite (or of other potential inhibitors of the porter) to alter the kinetics of uptake can be measured.

In other experiments efflux of the labeled metabolite can be measured. This involves a prior incubation to load the lysosomes with the radiolabeled metabolite, followed by a rapid wash at low temperature and reincubation in fresh medium. The rate of loss of radioactivity from the lysosomes, or the rate of appearance of radioactivity in the medium, permits characterization of the translocation of metabolite from out of the lysosome, which is of course the usual physiological direction. In such experiments the ability of substrate analogues to trans-stimulate metabolite egress can be detected. Trans-stimulation is a common characteristic of membrane porters that can translocate their substrates in both inward and outward directions.

The characteristics of the metabolite porters of the lysosome membrane have been reviewed elsewhere (Pisoni and Thoene, 1991; Thoene, 1992; Chou *et al.,* 1992).

3.   EFFLUX OF METABOLIC PRODUCTS FROM THE LYSOSOME

3.1.   Amino Acids and Dipeptides

As explained in Chapter 6, the digestion of polypeptides in lysosomes yields a mixture of amino acids and a few dipeptides. Here, we consider how these products exit the lysosome. The relevant properties of the known amino acid porters in the lysosome membrane will be described, and the possible role that passive diffusion plays in the efflux process will be discussed.

The properties of the lysosomal membrane porters for amino acids were comprehensively reviewed in the three contributions comprising Chapter 3 of Thoene (1992). These reviews focused on each transporter in turn and indicated its probable substrates. In this chapter the converse approach is adopted, asking whether the available data are adequate to explain the efflux of all 20 amino acids.

There appear to be ten distinct lysosome membrane porters for amino acids. Table III shows that, with the apparent exceptions of asparagine and glutamine, all

Table III

Lysosomal Membrane Porters for Amino Acids[a,b]

Amino acid	System c	System d	System e	System f	System h	System l	System p	System t	Cysteine	Cystine
Ala			++	++						−
Arg	++(320 μM)			−						−
Asn										
Asp	−	++(3.4 μM)		−						
Cys (SH)	−								++(53 μM)	
Cys (SS)										++
Gln										
Glu	−	++(9 μM)		−						−
Gly			+	+						
His	+					−				
Ile						+				
Leu		−	+	−	++	++(5–15 μM)				−
Lys	++									
Met	−			−		+				
Phe	−			−	++	++		++		
Pro			−	++(10 μM)			++(70 μM)			
Ser			++	++						
Thr	−	−	++	++						
Trp						+(10–30 μM)		++(10–15 μM)		−
Tyr					++			+		−
Val			+			++				

[a] ++, Demonstrated to be a substrate; values in parentheses indicate K_m; +, A probable substrate—inhibits transport or stimulates countertransport of a known substrate; values in parentheses indicate K_i; −, Unlikely to be a substrate—fails to inhibit transport or stimulate countertransport of a known substrate; No symbol indicates not tested.
[b] Data compiled from Chapter 3 of Thoene (1992) and references therein.

the amino acids arising from the degradation of proteins in lysosomes have been demonstrated to be probable substrates for at least one of these porters.

The presence of these porters in the lysosome membrane offers teleological evidence that they are required for the amino acids to cross. This conclusion is also consistent with the estimated hydrogen-bonding capacity of amino acids (see above, Section 2.1). The zwitterionic form of aliphatic amino acids such as glycine and alanine is estimated (Ginsburg and Stein, 1987) to have a hydrogen-bonding capacity of 11, a value similar to that of the hexoses (10.8). It seems likely that passive diffusion of these neutral amino acids does occur but is augmented by the activity of the porters. Amino acids that possess additional functional groups with hydrogen-bonding capacity, such as hydroxyl, thiol, or amide, or charged moieties (positive or negative) must have total hydrogen-bonding capacity in excess of 12, which seems to be the upper limit for a functional rate of passive diffusion (Table II). They would thus be expected to be wholly dependent on porters for their translocation across the lysosome membrane.

Table III shows the published K_m values for the amino acid porters in the lysosome membrane. They are low, ranging from 10–320 μM. There have been two estimates of the concentrations of free amino acids in purified lysosomes (Harms *et al.*, 1981; Vadgama *et al.*, 1991). Disturbingly these estimates differ substantially, by one or two orders of magnitude, despite being expressed in identical units. Taking the higher of the two sets (Vadgama *et al.*, 1991), the amino acid concentrations lie in the range 15–83 μM, strikingly similar to the range of the porter K_m values. The lower values of Harms *et al.* (1981) indicate intralysosomal amino acid concentrations well below the K_m values of the porters. So in either scenario, flux due to the porters would be responsive to small changes in the concentration of their substrates.

The low K_m values of the amino acid porters probably accounts for the observation that lysosomes are remarkably stable when incubated in 250 mM solutions of amino acids such as alanine and valine (Lloyd, 1971). At such concentrations a bidirectional porter with K_m in the micromolar range would be almost ineffective in achieving the net translocation of amino acid, because it would be rapidly saturated. The rate at which lysosomes break when incubated in 250 mM glycine is comparable to that in the same concentration of an ω-amino-aliphatic acid (Bird and Lloyd, 1990). At these high concentrations translocation into the lysosome is probably principally by passive diffusion.

As indicated in Chapter 6, the products of lysosomal metabolism of polypeptides seem to include some dipeptides. The implication is that not all the 400 possible dipeptides are hydrolyzed by the lysosomal dipeptide(s). Although these enzymes are incompletely characterized, there is good evidence that they have a degree of substrate specificity. Indirect evidence has been presented (Bird and Lloyd, 1990, 1995) for the existence of an L-dipeptide porter (or porters) in the rat liver lysosome membrane, although these have yet to be characterized. A porter would seem necessary for dipeptides to be released intact from the lysosome, since the hydrogen-bonding capacity of the simplest dipeptide must be 2 points higher than that of the simplest amino acid (see section 2.1 above).

3.2. Monosaccharides

The pathways by which the carbohydrate portion of simple and complex polysaccharides, glycoproteins, and glycolipids are degraded in the lysosome are detailed in Chapter 7 of this volume. These pathways yield many other products, in addition to the monosaccharides considered here. The mechanisms by which these other substances leave the lysosome are considered in the relevant sections of the present chapter.

Table IV shows the eleven monosaccharide products of simple and complex polysaccharide degradation in the lysosomes. With one exception all the products are hexoses: six neutral, three acidic, and one amino-sugar. One pentose, D-xylose, is produced. It should be noted that D-ribose and D-deoxyribose are not listed as products of lysosomal metabolism: This is because the degradation of nucleic acids does not proceed beyond the nucleoside level.

Three distinct porters for monosaccharides have been identified in the lysosome membrane (see Tietze, 1992, for an excellent critical review). One porter carries unsubstituted neutral hexoses, one is specific for the acetylhexosamines, and one carries acidic hexoses. Table V summarizes their ability to transport the monosaccharide products of lysosomal catabolism of polysaccharides. Carrier-mediated transport has been demonstrated for all the neutral hexoses, and it is likely that the pentose D-xylose is a substrate for the neutral sugar porter. Of the three acidic hexoses, two are certainly carried by the acidic sugar carrier. L-Iduronic acid has not been tested as a substrate, but is unlikely to be spurned by a porter that recognizes both glucuronic and sialic acid. D-Glucosamine shows only weak trans-stimulation of glucose uptake, and it may not be an effective substrate for the neutral sugar carrier. The existence of a rather specialized mechanism to acetylate D-glucosamine residues intralysosomally (see Section 4.1) probably indicates that this molecule is not able to traverse the lysosome membrane.

As with the amino acids, we now ask how necessary these porters are likely to be in allowing the monosaccharides to escape from lysosome to cytoplasm. As indicated in Section 2.1. above, the neutral hexoses have a notional hydrogen-bonding capacity of 10.8. The same value can be calculated for the acetylhexosamines. The behavior of neutral sugars in osmotic protection experiments is similar to that of nonphysiological molecules of similar hydrogen-bonding capacity, such as polyethylene glycol 400, indicating a significant ability to cross the lysosome membrane by passive diffusion. Why then are there membrane porters for these molecules? Presumably it is because passive diffusion alone cannot achieve the required rate of hexose efflux. It is noteworthy that other authors (Maguire *et al.*, 1983; Jonas *et al.*, 1990) have presented evidence that passive

Table IV
Products of Lysosomal Digestion of Carbohydrates

Monosaccharides	Glycogen	Glycolipids	Glycoproteins	Glycosaminolglycans[a]					
				DS	HS	CS	KS	HA	Hep
Neutral hexoses									
D-Glucose	✓	✓							
D-Galactose		✓	✓			✓	✓		
D-Mannose			✓						
L-Fucose		✓	✓						
N-Acetylglucosamine		✓	✓		✓		✓	✓	
N-Acetylgalactosamine		✓		✓		✓			
Neutral pentoses									
D-Xylose				✓		✓			
Acidic hexoses									
D-Glucuronic acid				✓	✓	✓		✓	✓
L-Iduronic acid				✓	✓				✓
N-Acetylneuraminic acid			✓				✓		
Amino sugars									
d-Glucosamine					✓				

[a] Abbreviations: DS, dermatan sulfate; HS, heparan sulfate; CS, chondroitin sulfates; KS, keratan sulfate; HA hyaluronic acid; Hep, heparin.

Table V
Lysosome Membrane Porters for Monosaccharides[a,b]

Monosaccharide	Acetyl-hexosamine porter	Acidic sugar porter	Neutral sugar porter
Neutral hexoses			
D-Glucose	−	−	++(22–75 mM)
D-Galactose	0	0	++(50–75 mM)
D-Mannose	−	0	++(50–75 mM)
L-Fucose	−	0	++(65 mM)
N-Acetylglucosamine	++(4.4 mM)	−	−
N-Acetylgalactosamine	++(4.4 mM)	−	−
Neutral pentoses			
D-Xylose	0	0	+
Acidic hexoses			
D-Glucuronic acid	−	++(300 μM)	−
L-Iduronic acid	0	0	0
N-Acetylneuraminic acid	−	++(240–700 μM)	−
Aminohexoses			
D-Glucosamine	−	0	0

[a]++, Demonstrated to be a substrate; values in parenthesis indicate K_m; +, A probable substrate—inhibits transport or stimulates countertransport of a known substrate; −, Unlikely to be a substrate—fails to inhibit transport or to stimulate countertransport of a known substrate; 0, Not tested.
[b]Data compiled from Tietze (1992) and references therein.

diffusion and carrier-mediated transport may both contribute to the translocation of neutral sugars across the lysosome membrane.

In striking contrast to the amino acid porters, the neutral sugar carrier in the lysosome membrane has a K_m for glucose in the millimolar range. Three estimates are available: 22, 48, and 75 mM (Jonas *et al.*, 1990; Maguire *et al.*, 1983; Mancini *et al.*, 1990). Because the glucose concentration in the lysosome is likely to be much lower than this, and that in the cytoplasm even lower, the net rate of glucose efflux on the carrier will be proportional to the glucose concentration gradient across the membrane. This is true also of efflux of passive diffusion. It is thus reasonable to postulate two parallel mechanisms for hexose efflux. The relative contribution of these two mechanisms is not known, but it will not vary much with changing intralysosomal glucose concentration.

If the neutral hexoses can cross the lysosome membrane by passive diffusion, pentoses (with one fewer hydroxy group) should cross much more rapidly. It has been estimated (Stein, 1967) that each hydrogen bond decreases permeability by a factor of at least 5, so for example, xylose should be at least 25 times more permeant than glucose. Therefore, although D-xylose is apparently recognized by the neutral sugar carrier (Jonas *et al.*, 1990), and may therefore be a substrate, release by passive diffusion could be the predominant mode of efflux. Certainly in osmotic protection experiments, nonphysiological molecules with similar hydrogen-bonding capacities to the pentoses cross the lysosome membrane rapidly (Table I).

By contrast, monosaccharides that bear a positive or negative charge, in addition to their hemiacetal and multiple alcohol-hydroxy groups, would be poor candidates for passive diffusion. As already noted, hexosamines probably cannot escape from the lysosome until acetylated, and the three acidic hexoses can confidently be predicted to require their membrane porter for efflux from the lysosome.

3.3. Nucleosides

As indicated in Chapter 9, polynucleotides are degraded in lysosomes only as far as nucleosides and phosphate. No nucleotide phosphorylase or hydrolase has been detected in lysosomes, and the only further metabolism of the principal nucleosides found in RNA and DNA is the deamination of some adenosine to inosine (Lindley and Pisoni, 1993).

The lysosomal membrane nucleoside porter has been the subject of only one report (Pisoni and Thoene, 1989). The porter appears to be relatively nonspecific with regard to the purine or pyrimidine base or to the pentose and probably carries all the physiological nucleosides including inosine. However, it does not recognize either nucleotides or free pentose. The carrier demonstrates a K_m for adenosine of 9 mM. Since the hydrogen-bonding capacity of the nucleosides exceeds the value of 11.5 discussed in Section 2.1, it seems likely that passive diffusion plays little if any part in the translocation of nucleosides across the lysosome membrane.

An early osmotic protection study (Burton et al., 1975) showed that the pyrimidine nucleosides can cross the lysosome membrane. The rank order of apparent rate of entry in this experiment correlates with the affinities for the lysosomal nucleoside transporter by Pisoni and Thoene (1989).

3.4. Degradation Products of Lipids

The lipids found in mammalian tissues are chemically diverse. Chapter 8 discusses in detail the catabolic pathways for lipids in lysosomes, and Table VI shows the principal products of these pathways.

The products of lysosomal lipid metabolism include glycerol and hexoses. The efflux of hexoses was considered in Section 3.2. Glycerol has a hydrogen-bonding capacity of only 6, and its behavior in osmotic protection experiments (Table I) leads to a confident prediction that passive diffusion alone would be entirely sufficient to ensure rapid translocation across the lysosome membrane.

Another nonelectrolyte generated in lysosomes is cholesterol, formed by the hydrolysis of cholesteryl esters delivered by receptor-mediated endocytosis of low-density lipoproteins (see Chapter 8). Cholesterol leaves the lysosomes, to be incorporated into other organelles. Niemann-Pick disease type C is a human lysosome storage disease in which export of cholesterol from lysosomes is impaired

Table VI
Products of Lysosomal Digestion of Lipids[a]

	Triglycerides	Cholesterol esters	Sphingophosphatides	Glycerophosphatides
Nonelectrolytes				
Glycerol	√			√
Cholesterol		√		
D-Glucose			√	
D-Galactose			√	
Organic ions				
Sphingosine			√	√
Fatty acid	√	√	√	√
Phosphorylcholine			√	√
Phosphorylethanolamine				√
Phosphorylinositol				√
Phosphorylserine				
Inorganic ions				
Phosphate			?	?

[a]The components of the extended oligosaccharide chains of glycosphingolipids are omitted from this table.

(see Chapter 8), suggesting that a carrier-mediated mechanism is involved. However, the chemical structure of cholesterol is such that efflux by passive diffusion should be adequate, and in their recent review Pentchev *et al.* (1995) express skepticism that Niemann-Pick disease type C is a lysosomal membrane porter defect similar to cystinosis and Salla disease. Nevertheless, recent evidence (Sato *et al.,* 1995) of substrate specificity in sterol efflux from lysosomes adds weight to the argument that a carrier is responsible.

As indicated in Chapter 8, there are still some unanswered questions regarding the degradative pathway of glycerophosphatides in lysosomes. However, it seems reasonable to suppose that the combined effects of phospholipase C and acid lipase will release glycerol and fatty acid from the diglyceride portion of the phosphatide molecule. The fate of lysophospholipids generated by the action of phospholipase A_1 and A_2 is less certain.

It is not clear whether lysosomal enzymes can hydrolyze the phosphate esters of choline, ethanolamine, serine, and inositol. Two acid phosphatases of lysosomes reportedly have no activity toward phosphorylserine or phosphorylcholine (Arsenis and Touster, 1967), but Fowler and de Duve (1969) reported some hydrolysis of the serine, ethanolamine, and choline esters by a lysosomal enzyme preparation. If there is no further catabolism, the phosphate esters listed in Table VI must be regarded as end products.

There are no experimental data concerning the mode of efflux from the lysosome of the singly or doubly charged organic ion products of phospholipid catabolism. Both sphingosine and the long-chain fatty acids are amphipathic, although the hydrophilic moiety bears a positive charge in one case and a negative charge in the other. As explained elsewhere (Lloyd, 1992a), the hydrogen-bonding capacity of the carboxylate anion appears to be about 3, so the anionic form of the long-chain fatty acids can probably cross the lysosome membrane by passive diffusion. This appears to be the case for acetate (Casey *et al.,* 1978). Sphingosine has two aliphatic hydroxy groups in addition to a protonated primary amine moiety, predictive of slow efflux by passive diffusion.

As discussed above (Section 2.1) ω-amino-aliphatic acids cross the lysosome membrane only slowly by passive diffusion. Phosphorylcholine and phosphorylethanolamine are both bipolar molecules, with structural similarities to the ω-amino-aliphatic acids. It may be that they are transported on an as yet unrecognized lysosome membrane porter. The need for a porter would seem still greater in the case of the triply charged phosphorylserine and the singly charged but very hydrophilic phosphorylinositol. Nature has provided the lysosome with porters for serine (see Table III) and for glucose, a molecule with lower notional hydrogen-bonding capacity than inositol. It seems unlikely that the efflux of phosphorylated serine or inositol can adequately occur by passive diffusion.

Despite the lack of evidence, intralysosomal hydrolysis of these phosphate esters cannot be excluded. The putative products, other than serine (Table III) and phosphate (see Section 3.5), are not known to be substrates for any of the known lysosome membrane porters, but some of them may be able to leave the lysosome by passive diffusion, albeit slowly. However, in an early study (Lloyd, 1969), 250 mM inositol was found to afford prolonged osmotic protection to rat liver lysosomes, consistent with its notional hydrogen-bonding capacity of 12 and indicative of poor penetration across the lysosome membrane. Ethanolamine is a structural analog of cysteamine, but is not recognized by the recently reported cysteamine porter (Pisoni et al., 1995); neither is choline. However, choline is a quaternary amine and could be a substrate for the broad-specificity lysosomal membrane amine-proton antiport recently reported by Moseley and Van Dyke (1995) and discussed in Section 3.6.

3.5. Inorganic Anions and Cations

Sulfate and phosphate are the only inorganic ions generated in the lysosome during the catabolism of biopolymers. Sulfate arises from the degradation of glycosaminoglycans and sulfatides. Phosphate is a major product of nucleotide degradation, and small amounts presumably arise from the degradation of phosphorylated proteins and glycoproteins and perhaps also from the degradation of phospholipids (see Section 3.4). Sulfate and phosphate are substrates for specific anion porters in the lysosome membrane.

The sulfate porter (Jonas and Jobe, 1990; Vadgama and Jonas, 1992; Koetters et al., 1995) has a K_m for sulfate of 160 µM and shares many but not all of the properties of the well-characterized band-3 anion porter of the erythrocyte membrane. It recognizes chloride and molybdate, and in vivo sulfate efflux from the lysosome may be accompanied by proton efflux and chloride influx. The porter does not significantly recognize phosphate or bicarbonate. The thyroid hormone status of a rat modifies the characteristics and perhaps the abundance of the lysosomal sulfate porter, although T_3 does not appear to have a direct effect on the porter (Chou et al., 1994).

The phosphate porter in the human fibroblast lysosome membrane (Pisoni, 1991; Pisoni and Thoene, 1992) is distinct from the sulfate porter. This conclusion follows the demonstration that, while arsenate competes strongly, sulfate does not. The porter's K_m for phosphate is 5 µM. Although the porter's physiological role is presumably to mediate the efflux of phosphate from the lysosome, the properties of the porter have been deduced from the results of experiments on the uptake of [^{32}P]orthophosphate. These experiments yielded one unexpected result: Phosphate accumulates in the lysosome in an acid-insoluble form, subsequently shown (Pisoni and Lindley, 1992) to be polyphosphate. In Chapter 9, Pisoni discusses the possible physiological significance of this unique synthetic metabolic pathway in lysosomes.

Pisoni (1991) also showed that ^{32}P radioactivity accumulated in isolated human fibroblast lysosomes when they were incubated with ATP labeled in the terminal phosphate position. Uptake was not seen when the ATP was labeled in the phosphate adjacent to the ribose moiety. Clearly it is not intact ATP that is accumulating, and the simplest interpretation is that phosphate liberated at the cytoplasmic face of the lysosome membrane by the ATP-driven proton pump is being translocated into the lysosome. Whether this observation indicates some degree of coupling between proton and phosphate influx, or whether it is an insignificant epiphenomenon, is not known.

Earlier studies using the osmotic protection methodology, the most systematic of which was by Casey *et al.* (1978), had indicated that the permeability of the lysosome membrane to the inorganic anions decreases according to the order of the so-called lyotropic series: thiocyanate > iodide > chloride > bicarbonate and phosphate > sulfate. As explained in Section 3.1, in the context of amino acid transport, the high concentration of solute required for osmotic protection experiments masks the contribution from high-affinity porters, so that the results indicate the ions' ability to penetrate the lysosome membrane by passive diffusion.

With regard to the inorganic metal cations, only calcium has so far been shown to have a porter in the lysosome membrane (Lemons and Thoene, 1991; Pisoni and Thoene, 1992). The early work of Casey *et al.* (1978) suggests that passive diffusion of the alkali metal cations across the liver lysosome membrane is rather inefficient, with permeabilities decreasing according to the lyotropic series: $Cs^+ > K^+ > Na^+$. Since metal cations are not major products of lysosome metabolism, their transport will not be further considered here.

3.6. Other Metabolites

Cysteamine (β-thiol-ethylamine) is a component of coenzyme A and produced during its degradation, part or all of which may take place in lysosomes (see Chapter 9, Section 4.3). A lysosome membrane porter for cysteamine and some related aliphatic aminothiols has recently been described (Pisoni *et al.,* 1995). Cysteamine transport is not inhibited by cysteine and thus cannot be a manifestation of the previously reported (Pisoni *et al.,* 1990) cysteine porter. By contrast, Pisoni *et al.* (1990) showed that cysteamine does inhibit the cysteine porter and thus may also act as a substrate. Currently, the only putative physiological role for the cysteamine porter is the transport of coenzyme A–derived cysteamine out of the lysosome.

Although cysteamine is a natural metabolite in mammalian cells, its chief interest in the context of lysosomes derives from the use of exogenous cysteamine as a successful treatment for cystinosis, an inherited disease in which the lysosomal cystine porter is defective. The recently discovered cysteamine porter and the cysteine porter are both credible candidates as mediators of the influx into lysosomes of therapeutically administered cysteamine. The products of the intralysosomal reaction of exogenous cysteamine and stored cystine are cysteine and cysteine-cysteamine disulfide. The latter is a close structural analogue of lysine and a substrate for the lysosomal membrane lysine-arginine porter (Pisoni *et al.,* 1987). Thus both products can leave the lysosome on porters that are unaffected in cystinosis.

Pantothenic acid is a putative end product of coenzyme A metabolism in lysosomes (see Chapter 9, Section 4.3). It is an aliphatic carboxylic acid containing an amide moiety and two hydroxy groups. Assuming a hydrogen-bonding capacity of 3 for the carboxylate (see Section 3.4) and 2 each for the hydroxy and amide groups (see Section 2.1), pantothenic acid's notional hydrogen-bonding capacity of 9 is compatible with slow efflux from the lysosome by passive diffusion. If lysosomes are capable of hydrolyzing the amide moiety of panthothenic acid, one of the two products, pantoic acid, should also be capable of diffusion out of the lysosome. β-Alanine, the other product, should have the same diffusional capacity as glycine or alanine, and indeed osmotic protection data (Bird and Lloyd, 1990) suggests that this is so.

Taurine (2-aminoethane-sulfonic acid) is one of the most abundant molecules in the human body: up to 70 g are present in a 70-kg individual (Huxtable, 1992). It is an essential dietary factor for cats and probably also for primates. Other mammalian species need taurine, but have a greater capacity for its biosynthesis from cysteine (Sturman, 1993). The high concentration of taurine in mammalian tissues has led to much speculation as to its major physiological function (Huxtable, 1992; Sturman, 1993). A role in intracellular osmoregulation appears to be the current consensus view, although it is not clear why nature has chosen this compound from so many possible alternatives. Osmotic protection experiments (Bird and Lloyd, 1990) indicate that taurine crosses the liver lysosome membrane quite slowly and at approximately the same rate as its close analogue β-alanine. The doubly charged nature of these two molecules would lead to the expectation of a slow rate of passive diffusion. Subsequently Vadgama *et al.* (1991) reported experiments on the uptake of [³H]taurine by purified rat liver lysosomes, concluding that a high-K_m porter was responsible. It is the view of the present author that the data presented by Vadgama *et al.* (1991) on the relationship of uptake to taurine concentration are more compatible with uptake by passive diffusion.

Another porter deserves brief mention, although it has not yet been shown to have any role in the transport of metabolites across the lysosome membrane. Moseley and Van Dyke (1995) report the presence in rat liver lysosomes of a porter for organic cations. The uptake of quaternary amines such as the tetraethylammonium ion occurs on an antiport carrier that exchanges an organic cation for a proton and so operates in an electroneutral manner. The operation of this porter depletes the intralysosomal proton concentration and so is effectively driven by the ATP-dependent proton pump responsible for maintaining the pH differential between lysosome and cytoplasm. The substrate specificity of this porter appears to include several quaternary amines and probably some tertiary amines, although in the latter case it is difficult to exclude completely amine influx in the unprotonated form. Pritchard *et al.* (1994) have reported what appears to be a very similar antiport in rat kidney cortex endosomes.

3.7.   Nutritional Role of Lysosomal Metabolites

It is axiomatic that, following release across the lysosome membrane into the cytoplasm, the end products of lysosomal metabolism are incorporated into the metabolite pools in the cytoplasm and used in biosynthetic pathways and for further catabolism. It is difficult however to evaluate whether the lysosomes contribute significantly to a cell's nutrient supply. One case in which this is known concerns the yolk sac cells of the early postimplantation rat conceptus. It has been shown (Beckman *et al.*, 1990, 1991) that the endocytic uptake and intralysosomal digestion of proteins by these cells produces virtually all of the amino acids used for the growth of both the yolk sac itself and the embryo proper. Moreover, an interruption of this process, by inhibition of either pinocytosis or lysosomal proteolysis, has serious effects for the embryo, leading to embryonic loss or malformation (Lloyd, 1990).

4.   METABOLITE INFLUX INTO THE LYSOSOME

Although there is every reason to suppose that most of the metabolite porters in the lysosome membrane can convey substrate in either direction, net flux physiologically will be from lysosome to cytoplasm. Biopolymer degradation yields a steady stream of metabolites which must efflux across the lysosome membrane to join the cytoplasmic pools.

Two exceptions have been proposed to this generalization. Acetyl moieties (from acetyl coenzyme A) and cysteine have both been proposed as metabolites whose net flux is from cytoplasm to lysosome.

4.1.   Acetyl Moiety

In two interesting papers (Rome *et al.*, 1983; Bame and Rome, 1985) Rome and colleagues described a novel and at present unique mechanism that achieves the intralysosomal conversion of terminal α-linked hexosamine residues in glycosaminoglycans into their N-acetyl derivatives. Cytoplasmic acetyl coenzyme A donates its acetyl group to an acetyltransferase in the lysosome membrane. The enzyme moves the acetyl moiety across the membrane, donating it to a hexosamine residue in the lysosome matrix. The mechanism of this acetyl carrier is not relevant to the present paper, but has been described in detail (Bame and Rome, 1985). Its physiological significance is that it converts a cationic molecule (hexosamine) into a neutral molecule (acetyl hexosamine). As explained above (Section 3.2) hexosamines are most unlikely to be capable of penetrating the lysosome membrane by passive diffusion, and the lysosome membrane is apparently not equipped with a hexosamine porter. It is thus unsurprising that no free hexosamines are generated in the lysosome from the degradation of heparin and heparan sulfate. Instead the products are ones that can be released from the lysosome on the GlcNAc/GalNAc porter (see Section 3.2). The demonstration that free glucosamine can act as a substrate for the transferase (Bame and Rome, 1985) suggests that, should any free hexosamine be generated in the lysosome, it would be acetylated and released as N-acetylhexosamine.

Chapter 7 discusses this acetylation reaction in the context of the lysosomal degradation of heparan sulfate. Sanfilippo disease type C arises from the absence of a functional acetyltransferase in the lysosome membrane (Bame and Rome, 1986). The consequent failure to acetylate terminal glucosamine moieties blocks the degradation of heparan sulfate, as the N-acetylhexosaminidase that continues the degradative sequence cannot act on a nonacetylated glucosamine.

It is not known whether substrates other than partially degraded glycosaminoglycans can act as substrates for the acetyltransferase.

4.2. Cysteine

In reporting their discovery of the cysteine porter, Pisoni et al. (1990) drew attention to its pH characteristics: The porter's activity falls off sharply below pH 7.2 and above pH 8.0. It would thus be rather ineffective at transporting cysteine out of the lysosome, where the ambient pH is usually less than 6.0. Recently the finding has been confirmed (Pisoni and Velilla, 1995) and extended to indicate that the cysteine porter bears a histidine residue which must be in the unprotonated form for full activity. The pK of the imidazole group of histidine is consistent with a transition to the protonated form at the usual pH of the lysosome.

Many proteins contain both cysteine and cystine residues, and both amino acids might be expected as products of protein degradation within lysosomes. There is much evidence that the cystine residues of disulfide-containing polypeptides are reduced in the lysosome during the course of proteolysis, and Lloyd (1986a) argued that the reducing agent is probably free cysteine. Pisoni et al. (1990) interpreted this proposed mechanism as entailing a net flux of cysteine from cytoplasm to lysosome. However, as pointed out by Lloyd (1992b), this is not usually so: Protein degradation in lysosomes will result in the net generation of both cystine and cysteine. Thus it seems unlikely that cysteine is an exception to the general rule that the flow of amino acids is from lysosome to cytoplasm.

If indeed net cysteine flux is from lysosome to cytoplasm, how can this be harmonized with the decreased activity of the cysteine porter at pH values below 7.2? Although there is no obvious answer to this question, Andersson (1992) points out that it is not unique to the cysteine porter. The system h and system p porters show similar pH characteristics. None of the other amino acid porters in the lysosome membrane nor a recently reported cysteamine porter (Pisoni et al., 1995) appear to have any capacity to carry cysteine. Pisoni and Velilla (1995) propose that the relative ineffectiveness of the cysteine porter at lysosomal pH may serve to maintain a high cysteine concentration in the lysosome. This is an attractive notion, as several lysosomal enzymes are thiol-dependent, and, as noted above, cysteine is probably the agent that reduces the disulfide linkages of proteins during intralysosomal proteolysis.

5. CONCLUSIONS

Like the plasma membrane, the lysosome membrane possesses a capacity for passive diffusion of solutes and is also equipped with a number of substrate-specific metabolite porters. Both mechanisms contribute to the traffic of metabolites between the cytosol and the lysosome matrix. The characteristics of the two mechanisms are now rather well understood, and it is possible to predict with some confidence whether and how any given metabolite will cross the membrane.

This review has focused on the products of intralysosomal metabolism, discussing how they exit the lysosome. It should also be emphasized that the properties of the lysosome membrane make a major contribution to the efficiency of the lysosome as an intracellular digestive system, *by denying passage to the intermediate metabolic products* (Lloyd, 1986b). No mechanism exists for disaccharides, oligopeptides, or nucleotides to cross the lysosome membrane. Thus the intermediate products are retained in the lysosome until digestion to the monomer level is complete.

Finally, although the focus of this chapter has been on metabolites, the characteristics of the lysosome membrane are relevant to the passage of xenobiotics. Because the membrane's metabolite porters are quite substrate-specific, most xenobiotics will cross, or fail to cross, by passive diffusion. Chapter 12 of this volume makes frequent mention of the lysosome membrane as a relevant factor in lysosome pharmacology and pathology.

ACKNOWLEDGMENTS. The author's work is supported by NIH grant HD29902 and by the Nemours Foundation.

6. REFERENCES

Andersson, H. C., 1992, Lysosomal transport of small and large neutral amino acids, in *Pathophysiology of Lysosomal Transport* (J. G. Thoene, ed.), pp. 73–91, CRC Press, Boca Raton, Florida.

Arsenis, C., and Touster, O., 1967, The partial resolution of acid phosphatase of rat liver lysosomes into a nucleotidase and a sugar phosphate phosphohydrolase, *J. Biol. Chem.* 242:3399–3401.

Bame, K. J., and Rome, L. H., 1985, Acetyl coenzyme A: α-Glucosaminide-N-acetyltransferase. Evidence for a transmembrane acetylation mechanism, *J. Biol. Chem.* 260:11293–11299.

Bame, K. J., and Rome, L. H., 1986, Genetic evidence for transmembrane acetylation by lysosomes, *Science* 233:1087–1089.

Beckman, D. A., Pugarelli, J. E., Jensen, M., Koszalka, T. R., Brent, R. L., and Lloyd, J. B., 1990, Sources of amino acids for protein synthesis during early organogenesis in the rat. 1. Relative contributions of free amino acids and proteins, *Placenta* 11:109–121.

Beckman, D. A., Pugarelli, J. E., Koszalka, T. R., Brent, R. L., and Lloyd, J. B., 1991, Sources of amino acids for protein synthesis during early organogenesis in the rat. 2. Exchange with amino acid and protein pools in embryo and yolk sac, *Placenta* **12**:37–46.

Berthet, J., Berthet, L., Appelmans, F., and de Duve, C., 1951, Tissue fractionation studies 2. The nature of the linkage between acid phosphatase and mitochondria in rat-liver tissue, *Biochem. J.* **50**:182–189.

Bird, S. J., and Lloyd, J. B., 1990, Evidence for a dipeptide porter in the lysosome membrane, *Biochim. Biophys. Acta* **1024**:267–270.

Bird, S. J., and Lloyd, J. B., 1995, Mechanism of lysosome rupture by dipeptides, *Cell Biochem. Funct.* **13**:79–83.

Burton, R., Eck, C. D., and Lloyd, J. B., 1975, The permeability properties of rat liver lysosomes to nucleosides, *Biochem. Soc. Trans.* **3**:1251–1253.

Casey, R. P., Hollemans, M., and Tager, J. M., 1978, The permeability of the lysosomal membrane to small ions, *Biochim. Biophys. Acta* **508**:15–26.

Chou, H.-F., Passage, M., and Jonas, A. J., 1994, Regulation of lysosomal sulfate transport by thyroid hormone, *J. Biol. Chem.* **269**:23524–23529.

Cou, H.-F., Vadgama, J., and Jonas, A. J., 1992, Lysosomal transport of small molecules, *Biochem. Med. Metabol. Biol.* **48**:179–193.

Cohn, Z. A., and Ehrenreich, B. A., 1969, The uptake, storage, and intracellular hydrolysis of carbohydrates by macrophages, *J. Exp. Med.* **129**:201–225.

de Duve, C., 1969, The lysosome in retrospect, in *Lysosomes in Biology and Pathology,* Vol. 1 (J. T. Dingle and H. B. Fell, eds.), pp. 3–40, North-Holland, Amsterdam.

de Duve, C., Pressman, B. C., Gianetto, R., Wattiaux, R., and Appelmans, F., 1955, Tissue fractionation studies 6. Intracellular distribution patterns of enzymes in rat-liver tissue, *Biochem. J.* **60**:604–617.

de Duve, C., Berthet, J., Berthet, L., and Appelmans, F., 1951, Permeability of mitochondria, *Nature* **167**:389–390.

Diamond, J. M., and Wright, E. M., 1969, Biological membranes: The physical basis of ion and non-electrolyte selectivity, *Annu. Rev. Physiol.* **31**:581–646.

Ehrenreich, B. A., and Cohn, Z. A., 1969, The fate of peptides pinocytosed by macrophages in vitro, *J. Exp. Med.* **129**:227–245.

Fowler, S., and de Duve, C., 1969, Digestive activity of lysosomes III. The digestion of lipids by extracts of rat liver lysosomes, *J. Biol. Chem.* **244**:471–481.

Gieselmann, V., 1995, Lysosomal storage diseases, *Biochim. Biophys. Acta* **1270**:103–136.

Ginsburg, H., and Stein, W. D., 1987, Biophysical analysis of novel transport pathways induced in red blood cell membranes, *J. Membr. Biol.* **96**:1–10.

Harms, E., Gochman, N., and Schneider, J. A., 1981, Lysosomal pool of free amino acids, *Biochem. Biophys. Res. Commun.* **99**:830–836.

Huxtable, R. J., 1992, Physiological actions of taurine, *Physiol. Rev.* **72**:101–163.

Iveson, G. P., Bird, S. J., and Lloyd, J. B., 1989, Passive diffusion of non-electrolytes across the lysosome membrane, *Biochem. J.* **261**:451–456.

Jonas, A. J., and Jobe, H., 1990, Sulfate transport by rat liver lysosomes, *J. Biol. Chem.* **265**:17545–17549.

Jonas, A. J., Conrad, P., and Jobe, H., 1990, Neutral-sugar transport by rat liver lysosomes, *Biochem. J.* **272**:323–326.

Koetters, P. J., Chou, H.-F., and Jonas, A. J., 1995, Lysosomal sulfate transport inhibitor studies, *Biochim. Biophys. Acta* **1235**:79–84.

Lemons, R. M., and Thoene, J. G., 1991, Mediated calcium transport by isolated human fibroblast lysosomes, *J. Biol. Chem.* **266**:14378–14382.

Lindley, E. R., and Pisoni, R. L., 1993, Demonstration of adenosine deaminase activity in human fibroblast lysosomes, *Biochem. J.* **290**:457–462.

Lloyd, J. B., 1969, Studies on the permeability of rat liver lysosomes to carbohydrates, *Biochem. J.* **115**:703–707.

Lloyd, J. B., 1971, A study of permeability of lysosomes to amino acids and small peptides, *Biochem. J.* **121**:245–248.

Lloyd, J. B., 1973, Experimental support for the concept of lysosomal storage disease, in *Lysosomes and Storage Diseases*, (H. G. Hers and F. van Hoof, eds.), pp. 173–195, Academic Press, New York.

Lloyd, J. B., 1986a, Disulphide reduction in lysosomes, *Biochem. J.* **237**:271–272.

Lloyd, J. B., 1986b, The lysosome membrane, *Trends Biochem. Sci.* **11**:365–368.

Lloyd, J. B., 1990, Cell physiology of the rat visceral yolk sac: A study of pinoytosis and lysosome function, *Teratology* **41**:383–394.

Lloyd, J. B., 1992a, Passive diffusion across the lysosome membrane, in *Pathophysiology of Lysosomal Transport* (J. G. Thoene, ed.), pp. 295–308, CRC Press, Boca Raton, Florida.

Lloyd, J. B., 1992b, Lysosomal handling of cystine residues: Stoichiomentry of cysteine involvement, *Biochem. J.* **286**:979–980.

Maguire, G. A., Docherty, K., and Hales, C. N., 1983, Sugar transport in rat liver lysosomes. Direct demonstration by using labeled sugars, *Biochem. J.* **212**:211–218.

Mancini, G. M. S., Beerens, C. E. M. T., and Verheijen, F. W., 1990, Glucose transport in lysosomal membrane vesicles: Kinetic demonstration of a carrier for neutral hexoses, *J. Biol. Chem.* **265**:12380–12387.

Moseley, R. H., and Van Dyke, R. W., 1995, Organic cation transport by rat liver lysosomes, *Am. J. Physiol.* **268**:G480–G486.

Pentchev, P. G., Vanier, M. T., Suzuki, K., and Patterson, M. C., 1995, Niemann-Pick disease type C: A cellular cholesterol lipidosis, in *The Metabolic and Molecular Bases of Inherited Disease*, Vol. II (C. R. Scriver, A. L., Beaudet, W. S. Sly and D. Valle, eds.), pp. 2625–2639, McGraw-Hill, New York.

Piqueras, A.-I., Somers, M., Hamond, T. G., Strange, K., Harris, H. W., Gawryl, M., and Zeidel, M. L., 1994, Permeability properties of rat renal lysosomes, *Am. J. Physiol.* **266**:C121–C133.

Pisoni, R. L., 1991, Characterization of a phosphate transport system in human fibroblast lysosomes, *J. Biol. Chem.* **266**:979–985.

Pisoni, R. L., and Lindley, E. R., 1992, Incorporation of [^{32}P]orthophosphate into long chains of inorganic polyphosphate within lysosomes of human fibroblasts, *J. Biol. Chem.* **267**:3626–3631.

Pisoni, R. L., and Thoene, J. G., 1989, Detection and characterization of a nucleoside transport system in human fibroblast lysosomes, *J. Biol. Chem.* **264**:4850–4856.

Pisoni, R. L., and Thoene, J. G., 1991, The transport systems of mammalian lysosomes, *Biochim. Biophys. Acta* **1071**:351–373.

Pisoni, R. L., and Thoene, J. G., 1992, Lysosomal phosphate and calcium transport, in *Pathophysiology of Lysosomal Transport* (J. G. Thoene, ed.), pp. 115–131, CRC Press, Boca Raton, Florida.

Pisoni, R. L., Thoene, J. G., and Christensen, H. N., 1985, Detection and characterization of carrier-mediated cationic amino acid transport in lysosomes of normal and cystinotic human fibroblasts. Role in therapeutic cystine removal? *J. Biol. Chem.* **260**:4791–4798.

Pisoni, R. L., and Veilla, Q. V., 1995, Evidence for an essential histidine residue located in the binding site of the cysteine-specific lysosomal transport protein, *Biochim. Biophys. Acta* **1236**:23–30.

Pisoni, R. L., Thoene, J. G., Lemons, R. M., and Christensen, H. N., 1987, Important differences in cationic amino acid transport by lysosomal system c and system y$^+$ of the human fibroblast, *J. Biol. Chem.* **262**:15011–15018.

Pisoni, R. L., Acker, T. L., Lisowski, K. M., Lemons, R. M., and Thoene, J. G., 1990, A cysteine-specific lysosomal transport system provides a major route for the delivery of thiol to human fibroblast lysosomes: Possible role in supporting lysosomal proteolysis, *J. Cell Biol.* **110**:327–335.

Pisoni, R. L., Park, G. Y., Velilla, V. Q., and Thoene, J. G., 1995, Detection and characterization of a transport system mediating cysteamine entry into human fibroblast lysosomes. *J. Biol. Chem.* **270**:1179–1184.

Pritchard, J. B., Sykes, D. B., Walden, R., and Miller, D. S., 1994, ATP-dependent transport of tetraethylammonium by endosomes isolated from rat renal cortex, *Am. J. Physiol.* **266**:F966–F976.

Rashid, F., Horrobin, R. W., and Williams, M. A., 1991, Predicting the behaviour and selectivity of fluorescent probes for lysosomes and related structures by means of structure–activity models, *Histochem. J.* **23:**450–459.

Reijngoud, D.-J., and Tager, J. M., 1977, The permeability properties of the lysosomal membrane, *Biochim. Biophys. Acta* **472:**419–449.

Reuser, A. J. J., Kroos, M. A., Visser, W. J., and Willemsen, R., 1994, Lysosomal storage diseases: Cellular pathology, clinical and genetic heterogeneity, therapy, *Ann. Biol. Clin. (Paris)* **52:**721–728.

Rome, L. H., Hill, D. F., Bame, K. J., and Crain, L. R., 1983, Utilization of exogenously added acetyl coenzyme A by intact isolated lysosomes, *J. Biol. Chem.* **258:**3006–3011.

Sato, Y., Nishikawa, K., Aikawa, K., Mimura, K., Murakami-Murofushi, K., Arai, H., and Inoue, K., 1995, Side-chain structure is critical for the transport of sterols from lysosomes to cytoplasm, *Biochim. Biophys. Acta* **1257:**38–46.

Stein, W. D., 1967, *The Movement of Molecules across Cell Membranes,* Academic Press, New York.

Sturman, J. A., 1993, Taurine in development, *Physiol. Rev.* **73:**119–147.

Thoene, J. G., ed., 1992, *Pathophysiology of Lysosomal Transport,* CRC Press, Boca Raton, Florida.

Tietze, F., 1992, Lysosomal transport of sugars: Normal and pathological, in *Pathophysiology of Lysosome Transport* (J. G. Thoene, ed.), pp. 165–200, CRC Press, Boca Raton, Florida.

Vadgama, J., and Jonas, A., 1992, Lysosomal sulfate transport, in *Pathophysiology of Lysosomal Transport* (J. G. Thoene, ed.), pp. 133–154, CRC Press, Boca Raton, Florida.

Vadgama, J. V., Chang, K., Kopple, J. D., Idriss, J.-M., and Jonas, A. J., 1991, Characteristics of taurine transport in rat liver lysosomes, *J. Cell. Physiol.* **147:**447–454.

Chapter 12

Lysosome Pharmacology and Toxicology

Robert Wattiaux and Simone Wattiaux-De Coninck

1. INTRODUCTION

Soon after the discovery of lysosomes, their role in pharmacology and toxicology was proposed. The nature of the organelles and the fact that all reactions catalyzed by the lysosomal enzymes were hydrolytic permitted one to hypothesize that lysosomes could only play a part in the biotransformation of drugs and toxins. On the other hand, it was not difficult to imagine the necessity of such structures in protecting cells from exogenous material that came into contact with the plasma membrane and was subsequently internalized. In fact, the apparent simplicity of the lysosomes, when compared to other subcellular structures like the mitochondria, makes it easy to conceptualize their functions and malfunctions (de Duve and Wattiaux, 1966). As innumerable articles are published every year on drug–lysosome interaction,* it is evident that this review cannot cover the whole area, but we hope to give a comprehensive account of the fundamental facts pertaining to the subject and illustrate them with examples.

*A Medline search of the last four years has revealed as many as 700 publications on drug–lysosome interaction.

Robert Wattiaux and **Simone Wattiaux-De Coninck**   Laboratoire de Chimie Physiologique, Facultés Universitaires Notre-Dame de la Paix, B5000 Namur, Belgium.

Subcellular Biochemistry, Volume 27: Biology of the Lysosome, edited by Lloyd and Mason. Plenum Press, New York, 1996.

2. EFFECTS OF LYSOSOMES ON DRUGS AND OTHER EXOGENOUS SUBSTANCES

Studies on lysosomes can be carried out with the whole animal, cells in culture, organelles, or purified enzymes. A study involves determining the effect when an exogenous molecule penetrates the lysosomes; this molecule may or may not be modified after its entry into the organelle. The manner of uptake of the molecule into the lysosome and its subsequent fate are relevant factors. Chapter 3 discusses in detail the different modes of entry of molecules into lysosomes, and Chapters 6–9 analyze the metabolic transformations that occur within lysosomes. Here we wish to envisage some of the consequences on the function of the molecules from the transformation they undergo within the lysosome. There are basically two types of consequences: The molecule can be inactivated or it can be activated by being released from a complex used for its transport into the cell and the lysosome. In addition, the intralysosomal pH may affect certain molecules. Modification of exogenous molecules under the influence of the lysosomal enzymes could have an effect on the lysosomes themselves. The last point will be considered in Section 3.

It is appropriate to introduce here the term *lysosomotropic*. As defined by de Duve *et al.* (1974), the term lysosomotropic designates "all substances that are taken up selectively into lysosomes, irrespective of their chemical nature or mechanism of uptake."

2.1. Inactivation of the Molecule

After its uptake by lysosomes, a substance can be degraded and lose its biological activity. A characteristic example is that of polypeptide hormones. It has been shown by a number of researchers that once substances like insulin and growth hormone bind to their receptors on the plasma membrane and exert their specific effect, they are internalized. In most cases, they find themselves in lysosomes where they are degraded by the lysosomal cathepsins (Postel-Vinay *et al.*, 1982). This results in an irreversible loss of biological activity.

An interesting question and one that is probably of great relevance at the moment is the fate of genetic material introduced into the cell by transfection. Most often the vector used is a plasmid into which the gene has been incorporated. It is difficult to imagine the entry of such a molecule into the cell other than by endocytosis. In fact, certain experiments have suggested that such a manner of entry is plausible (Zhou and Huang, 1994; Cappaccioli *et al.*, 1993; Wattiaux *et al.*, 1995). Logically, the endocytosed DNA should reach the lysosomes and be degraded by the nucleases present in these organelles. However, there are many well-documented cases where transfection of DNA has led to its expression, implying that the transfected gene has remained intact and transcription has occurred. In our opinion there is to date no convincing explanation of this phenomenon. It is possible that a small portion of the DNA has escaped degradation by the lysosomal nucleases and is released in the cytosol. If this is the case for nucleic acids, then the same should apply for other endocytosed molecules whose intracellular pathway finally leads to lysosomes where they are degraded. Another possibility is that some nucleic acid molecules leave the vacuolar apparatus (perhaps from the endosomal compartment) before reaching the lysosomes. However, the problem remains of how such large hydrophilic molecules can cross a biological membrane like that of the endosomes.

2.2.   Release from a Transporter

If, by virtue of its size and structure, a drug is captured by a cell through endocytosis, it will eventually find itself in lysosomes. This fact is the basis for the approaches to drug delivery in which pharmacologic molecules are administered as covalent complexes with macromolecules that serve as the vector, the complex being endocytosed by the cell. When the complex reaches the lysosome and if the link to the macromolecule is designed to be degraded by the lysosomal enzymes, the result would be the release of the drug within the lysosome. If the liberated molecule has an appropriate size and hydrophobicity, it could diffuse through the lysosomal membrane and be released into the intracellular medium, where it could exert its pharmacological action. One of the major advantages of such a system from the pharmacologic point of view is that it permits, in principle, specific targeting of the drug. The macromolecule can be so chosen that it is taken up by a specific receptor present on the plasma membrane of the cell to which the drug is to be targeted. In this type of situation, glycoproteins are probably the best vectors, as there exist different types of receptors that recognize the carbohydrate moiety of the protein (e.g., galactose-specific, mannose-specific, etc.). For example, the galactose-specific receptors in the liver are associated mainly with hepatocytes, while those that recognize mannosylated proteins are mostly located on sinusoidal cells (Ashwell and Harford, 1982; Steer and Clarenburg, 1979). Hence, a drug that is bound to a mannosylated protein would be taken up largely by the sinusoidal cells, while that which is complex with a galactoprotein would be targeted to the hepatocytes. The use of such complexes would be especially interesting in antiviral, antiparasitic, and anticancer therapy (Trouet *et al.*, 1982; Monsigny *et al.*, 1988; Trail *et al.*, 1995). An extensive review of polymer conjugates with anticancer activity has been published recently by Putnam and Kopeček (1995).

2.3.   Effect of Intralysosomal pH

The acidic pH of lysosomes could have an effect on the function of a drug localized within it, especially if the drug has acid–base properties. Once such a molecule enters the lysosomes, its overall charge would become more positive or less negative. A most spectacular result is the one observed for weak, relatively hydrophobic bases like aliphatic amines. Such molecules accumulate within lysosomes as a result of two processes. First, as the pH of the lysosome matrix is lower than the pH of the external medium, the dissociation equilibrium of a weak base diffusing into lysosomes is shifted toward the protonated form of the molecule inside the organelles. Second, the diffusion rate of a weak base through a biological membrane is considerably higher for its un-ionized form than for its ionized one. It is easy to see that in such conditions, a weak base diffusing through the lysosomal membrane will be trapped by protonation within the lysosomes. A diffusion equilibrium will be reached, characterized by a higher concentration of the compound inside the lysosomes than outside. In fact, the maximal concentration ratio that will be reached is equal to the ratio of the proton concentration in the lysosomes to that in the external medium (de Duve *et al.*, 1974). This accumulation could have either a favorable pharmacologic effect or a toxic one, depending on whether such an accumulation would lead to the activation or inhibition of one or more of the lysosomal enzymes. It is possible that the therapeutic effect of chloroquine in malaria is exerted to some extent this way, as chloroquine is a dibasic molecule that can accumulate to high concentration in the lysosomes and in the digestive vacuoles of *Plasmodium,* which resemble the lysosomes. Inhibition of protein degradation in these vacuoles due to the high chloroquine concentration could act against the parasite (Krogstad *et al.,* 1985; Krogstad and Schlesinger, 1986). However, other mechanisms are probably also involved in the antimalarial effect of chloroquine (Slater, 1993).

2.4. Sequestration in Lysosomes

Some compounds may accumulate unmodified in lysosomes and are unable to leave. Because of their size and hydrophilicity, such molecules cannot cross the plasma membrane and can gain access into the cell only through endocytosis. If the molecule reaches the lysosomes, but is not degraded by lysosomal enzymes, it remains within these organelles. (As discussed in Section 3.2.2, this may eventually cause problems.) On the other hand, its sequestration within lysosomes prevent it from reaching other intracellular sites where it might have exerted an effect. This is the case with certain antibiotics (Tulkens, 1991). An antibiotic sequestered within lysosomes is unable to exert any effect on bacteria other than those that enter the lysosome. An example is the aminoglycoside antibiotics that enter the cell by fluid-phase endocytosis and find themselves in lysosomes which they are unable to leave. Such an antibiotic can have no effect in the treatment of intracellular bacterial infections except when the microorganism is also present within the lysosome (Tulkens, 1991).

3. EFFECTS OF SUBSTANCES ON LYSOSOMES

The purpose of past and present work on this subject is to determine whether the pharmacologic or toxic effect of molecules is due to effects at the level of the lysosomal system.

3.1. Methodology

All the methods used, whether *in vitro, ex vivo,* or *in vivo,* involve observing the effect of the molecule either on the lysosomal membrane or on hydrolases present in these organelles.

3.1.1. *In Vitro* Methods

3.1.1.a. Lysosomal Membrane. The easiest, oldest, and probably most adequate way of demonstrating the effect of a drug on the lysosomal membrane is to determine whether the integrity of the lysosomal membrane is modified when the drug is incubated with intact lysosomes (de Duve *et al.,* 1962; Wattiaux, 1971). This type of experiment is based on the fact that the latency of the lysosomal enzymes is diminished if the lysosomal membrane is lysed. The principle is illustrated in Fig. 1. The lysosomes are incubated in the presence of a nonpermeant substrate of one of their enzymes. If the membrane is intact, the substrate is unavailable to the enzyme and any activity arises from membrane damage inflicted during preparation. If the substance under investigation is able to lyse the membrane, the enzyme will be able to reach and hydrolyze the substrate. The available activity is normally expressed as a percentage of the total activity of the enzyme, measured when the membrane is fully ruptured, for example, by detergents. The available (or free) activity is generally proportional to the percentage of lysosomes whose membrane is ruptured or made permeable. Obviously, this is a rough approximation that ignores the heterogeneity of lysosomes.

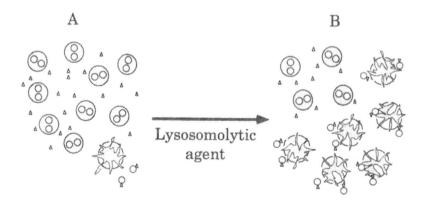

FIGURE 1. Determination of the effect of a lysosomolytic agent: a schematic representation; lysosomal enzyme (○), substrate (△). In the absence of the lytic agent (A), one lysosome out of ten is disrupted; in the presence of the lytic agent (B), six lysosomes are disrupted. As a result, the lysosomal enzymatic activity will be six times higher in the presence of the lysosomolytic agent, provided that the membrane of intact lysosomes is totally impermeable to substrate and that the enzyme concentration is the same within each individual lysosome.

A variation of this procedure is to test the effect of the substance under investigation in a system that destabilizes the lysosomal membrane. Under such conditions, one can establish whether the destabilization of the membrane is affected by the compound or not. Fig. 2 shows the effect of different substsances on the rupture of lysosomal membranes, induced by incubation of the organelles with a permeable substance like glucose. It has been shown (Lloyd, 1969) that glucose and other monosaccharides are unable to afford prolonged osmotic protection to lysosomes. When these organelles are incubated in isoosmotic glucose, a progressive release of lysosomal enzymes takes place, indicating a disruption of the lysosomal membrane. The explanation is that glucose penetrates into lysosomes, induces an osmotic imbalance, and causes an entry of water into the granules, leading to their swelling and finally their rupture. Up to a certain concentration, the compounds used in the experiment illustrated in Fig. 2 inhibit the rupture of the lysosomal membrane. Either they inhibit the diffusion of glucose across the lysosomal membrane or they make the membrane more resistant to swelling (Jadot et al., 1989).

Morphological examination, as illustrated in Fig. 3, may help to determine whether the lysosomal structure is affected or not. It shows the deterioration lysosomes undergo when they are attached by free radicals and the ability of catechin to oppose such a phenomenon (Decharneux et al., 1992).

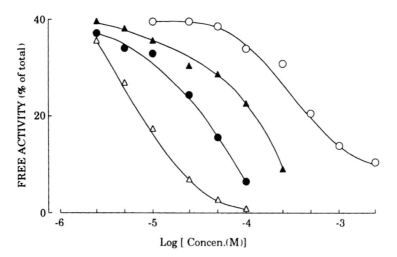

FIGURE 2. Effect of diethylstilbesterol (△), phloretin (●), cytochalasin (▲), and phlorrhizin (○) on the lysis of lysosomes induced by incubation in 0.25 M glucose. The experiment was performed with a rat liver total mitochondrial fraction that contains the majority of the lysosomes. The granules were incubated for 10 min at 25 °C in 0.25 M glucose at pH 7 in the presence of various concentrations of the compounds. Subsequently, the released N-acetylglucosaminidase (free activity) is measured. Free activity, which is proportional to the amount of disrupted lysosomes, is expressed in percent of total activity obtained by disrupting all the lysosomes with 0.1% Triton X-100. (Data from Jadot et al., 1989.)

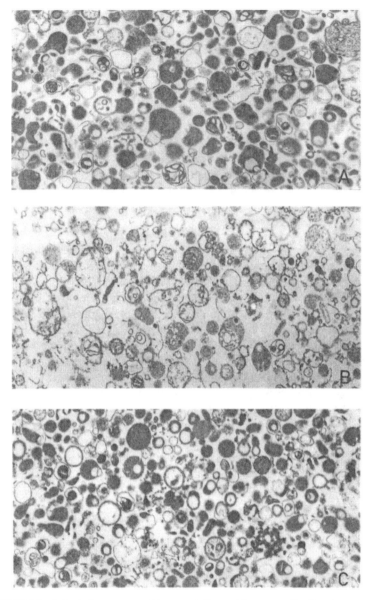

FIGURE 3. Morphological appearance of intact and disrupted lysosomes. Rat liver lysosomes were purified in a Nycodenz gradient and examined under an electron microscope (Wattiaux *et al.*, 1978). (A) Normal preparation; (B) After incubation with a system generating oxygen free radicals (xanthine, xanthine oxidase), many lysosomes are disrupted and empty; (C) Same as in (B) except that incubation was performed in presence of 0.05 mM D-catechin; lysosomes keep their normal configuration. (Data from Decharneux *et al.*, 1992.)

It should be noted that, except for a morphological examination, it is unnecessary to use purified lysosomes in the methodology described above, as the lysosomes are identified by the action of an enzyme that is located only within these organelles; the presence of other subcellular structures does not pose any problem.

3.1.1b. Lysosomal Enzymes.   Classically, the methods used are those that determine the inhibiting or activating effect of substances on the enzymes. It may be more meaningful to work on lysosomal extracts, rather than on purified enzymes, as this ensures one is as close to *in vivo* conditions as possible. For example, in investigating intralysosomal proteolysis, it may be preferable to test the effect of the drug or toxin on the whole proteolytic system obtained with a purified lysosomal extract, rather than one of the purified cathepsins.

3.1.2.   *Ex Vivo* and *In Vivo*

The measurement of free activity of the lysosomal enzymes can also be applied up to a certain point to primary cultures (e.g., hepatocytes), secondary cultures, or tissues. We can presume that if a drug is capable of accessing the lysosomal membrane within the cell, it could induce the release of the lysosomal enzymes into the cytosol. This could be identified by measuring the free activity of one of these enzymes in the homogenate of treated cells. The interpretation of such results is sometimes very difficult. It is possible that the free activity of the enzyme in the homogenate may not arise as a consequence of the action of the drug on the lysosome membrane, it could be due to the sensitivity of the lysosome to the homogenization technique, which could lead to an abnormal lysis of the organelles during the preparation of the homogenate. For example, if for some reason the drug causes an increase in the size of the lysosome (maybe due to the intralysosomal accumulation of the drug), the membrane would be more easily damaged by shearing forces when the tissue is homogenized using a coaxial homogenizer. In such a case the effect of the drug on the stability of the lysosomes within the cell would be purely an artifact.

3.2.   Effects on Physicochemical Properties of Lysosomes

3.2.1.   Lysosomal Labilizers and Stabilizers

The lysis of the lysosomal membrane by pharmacologic or toxic substances has been a subject of interest for many years. If lysosomes were lysed within the cell, this would lead to a considerable perturbation of cellular functions, as the lysosomal hydrolases have the capacity to destroy many cellular constituents, and eventually to cell death. This brings us back to the very old concept that lysosomes are "suicide bags," that is, organelles able to kill the cell in which they are located by allowing an uncontrolled autolytic breakdown when their membrane is disrupted (Wattiaux *et al.,* 1992). Under normal physiological conditions, lysosomes probably do not function as "suicide bags," and their digestion of intracellular components occurs by the controlled process of autophagy. However, it is possible that accidentally, in pathological situations, intracellular release of lysosomal enzymes could take place, leading to cell death.

Injury to the lysosomal membrane caused by a toxic compound may not bring about a true disruption of the membrane, but rather a gradual permeabilization whose extent reflects the severity of the injury. An interesting method to determine whether the lysosomal membrane is ruptured or only permeabilized involves measuring the free and total activities of acid hydrolases as a function of the substrate concentration (de Duve, 1965). If enhancement of free activity due to lysosome injury originates from a disruption of lysosomal membrane, the ratio of free to total activity does not change on raising the substrate concentration in the medium. If the enhancement results from an increase of lysosomal membrane permeability to substrate molecules, the ratio of free to total activity will decrease on raising substrate concentration. In general, when such an approach is used, the results indicate that the loss of latency of injured lysosome enyzmes arises from a true disruption of lysosomal membrane and not from a gradual permeabilization (Burton and Lloyd, 1976; Baccino and Zuretti, 1976).

3.2.1a. Detergents. Many detergents are capable of lysing lysosomes. Of particular interest are the lysosomotropic detergents. These compounds consist of a long aliphatic chain and an amine that becomes protonated at lysosomal pH, giving the compounds detergent properties. Firestone *et al.* (1979) synthesized a series of such compounds; their amine group endows them with lysosomotropic properties and allows them to accumulate in the lysosomes. An example is N-dodecylimidazole. These molecules are highly cytotoxic, and there is a strong correlation between their cytotoxicity and their lysosomotropic nature (Miller *et al.*, 1983). Experimental results indicate that lysosomotropic detergents kill cells by rupturing lysosomes and releasing the lysosomal hydrolases into the intracellular environment. Among the hydrolases, the cysteine proteases play an important role, since specific inhibition of these enzymes makes cells resistant to the toxic action of the lysosomotropic detergents (Wilson *et al.*, 1987).

Jadot *et al.* (1990) have shown that glycyl-D-phenylalanine naphthylamide is cytotoxic to Vero cells. Results suggest that the cytotoxicity of this compound results from its lytic effect on lysosomes. Both phemomena occur in the same concentration range. Moreover, among the subcellular structures investigated, only lysosomes are disrupted by the naphthylamide *in vitro* mitochondria, peroxisome, endoplasmic reticulum, and endosome membranes are not affected. In addition, cytotoxicity of glycyl-D-phenylalanine naphthylamide and its lytic effect on lysosomes are inhibited by nigericin, a substance that is known to increase the intralysosomal pH. Glycyl-D-phenylalanine naphthylamide is a relatively hydrophobic weak base that can cross the lysosomal membrane. Therefore, it is probable that it can accumulate in lysosomes. However, as it cannot be hydrolyzed by cathepsin C since it is a D derivative, lysis of lysosomes that compound cannot be explained by the same mechanism as for glycyl-L-phenylalanine naphthylamide (see Section 3.2.1); in addition, taking into account the concentration required to break lysosomal membranes, an osmotic disruption of lysosomes resulting only from the accumulation of the substance inside the organelles is improbable. The authors have to conclude that glycyl-D-phenylalanine naphthylamide has a direct lytic effect on the lysosomal membrane.

Recently, we have found that cationic lipids used as vectors in DNA transfection are lysosomolytic *in vitro* even more than Triton X-100, a detergent widely used to rupture the lysosomal membrane. Figure 4 shows the loss of latency of β-galactosidase, a lysosomal hydrolase, caused by incubating rat liver lysosomes with lipofectamine, a cationic lipid frequently used in transfection experiments, or with equal concentrations of Triton X-100. Similar results were obtained with related compounds used for transfection. It is thus possible that some of the endocytosed DNA reaching the lysosomes could escape from these organelles before it is degraded.

3.2.1b. Lipid-Soluble Biological Substances. A number of compounds of biological and pharmaceutical interest of capable of inducing the lysis of lysosomes *in vitro*. Diverse steroids, vitamin A, vitamin E, phytol, ubiquinone, and

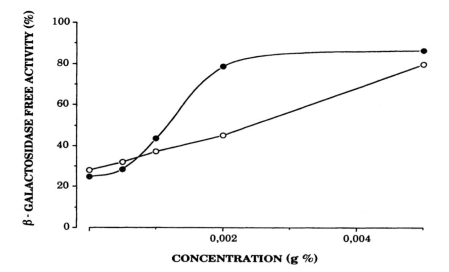

FIGURE 4. Effect of Triton X-100 (○) and lipofectamine (●), a cationic lipid used for transfection on the latency of β-galactosidase. A rat liver light mitochondrial fraction was incubated for 20 min at 37 °C in 0.25 M sucrose, at pH 7.4 in presence of increasing concentrations of lipofectamine or Triton X-100. Then the released β-galactosidase free activity was measured.

others accelerate the lysis of lysosomes when incubated with these organelles at pH 5 (de Duve *et al.*, 1962). In contrast, cholesterol and cortisone protect lysosomes under the same conditions. The protective effect is considerably increased when the lysosomes are isolated from animals treated with Triton WR1339, a non-ionic detergent that accumulates in lysosomes (Wattiaux *et al.*, 1963). Various flavonoids and related substances are potent inhibitors of lysosomolysis provoked by free radicals (Decharneux *et al.*, 1992). Their action is probably not entirely due to their ability to act as scavengers of free radicals; they also have a direct effect on the lysosomal membrane, as suggested by the observation that they are also able to protect lysosomes against an osmotic stress (Decharneux *et al.*, 1992). It is necessary to note that these substances may have biphasic effects that depend on their concentration in the medium. This is well illustrated by the case of diethylstilbestrol. At low concentration, this steroid analogue prevents lysosomal lysis from osmotic stress; however, higher concentrations produce the opposite effect, that is, diethylstilbestrol becomes lysosomolytic (Jadot *et al.*, 1989).

3.2.1c. Dipeptides and Esters of Amino Acids and Peptides.

Incubation of lysosomes in the presence of certain L-dipeptides provokes the lysis of these organelles (Goldman, 1973; Bird and Lloyd, 1995). These dipeptides enter the lysosomes where they are quickly hydrolyzed to amino acids; efflux of these molecules is hypothesized to be slower than dipeptide entry, with a resulting osmotic imbalance leading to swelling of the lysosomes and rupture of their membrane. Rupture of lysosomes can also be caused by incubatiom with esters of certain amino acids and peptides (Goldman and Kaplan, 1973). The proposed mechanism is similar to that advanced to explain the lytic properties of dipeptides, with the additional fact that these esters behave like lysosomotropic weak bases and therefore accumulate in lysosomes.

Glycyl-L-phenylalanine naphthylamide, for which an intralysosomal hydrolysis to dipeptide and naphthylamine by cathepsin C has been demonstrated (Jadot *et al.*, 1984), has the same effect. The lysosomolytic action of glycyl-L-phenylalanine naphthylamide can be used to study the localization of compounds in lysosomes. Since the lysis of an organelle indicates that it contains the lysosomal enzyme finally responsible for the lytic processes, it is a true marker of the lysosomal localization. This can be of particular interest in the study of endocytosis, autophagy, and biogenesis of lysosomal proteins, as the molecules in all these processes find themselves finally in the lysosomes, though their route there may involve a number of other membrane systems (endosomes, Golgi, etc). This would eventually make it difficult to clearly distinguish them from lysosomes by the classical methods of centrifugation. However, we can prove that the compound is in the lysosomes if it is released into the medium when the lysosomes are incubated with glycyl-L-phenylalanine naphthylamide (Jadot and Wattiaux, 1985; Misquith *et al.*, 1988; Berg *et al.*, 1994a,b).

The effect of esters of amino acids and peptides on living cells has been studied extensively. Reeves *et al.* (1981) showed that the methyl ester of leucine can destroy heart lysosomes when the heart is perfused with a medium containing this substance. When mononuclear blood cells are incubated with the methyl ester of L-leucine, the natural killer (NK) cells found in the preparation are selectively killed (Thiele and Lipsky, 1985a,b). Other cells like macrophages and endothelial cells are resistant, even though it has been shown that they accumulate the ester in vacuoles of the cytoplasm. The mechanism by which the NK cells are killed is curious. Apparently, the monocytes in the preparation take up the ester and transform it into a dipeptide ester, Leu-Leu-OMe, by a lysosomal mechanism (transacylation?) that the authors do not describe (Thiele and Lipsky, 1985a). The dipeptide ester is released and taken up by NK cells in which it is converted into a membranolytic product of structure (Leu-Leu)$_n$-OMe in lysosomes, due to a transpeptidation reaction catalyzed by dipeptidylpeptidase I (Thiele and Lipsky, 1990). It is of note that such a reaction can only take place at neutral or slightly alkaline pH; accordingly, the NK cell lysosome pH must first increase by an accumulation of Leu-Leu-OMe in these organelles. This selective destruction of NK cells by (Leu-Leu)$_n$-OMe can be used in the study of graft-versus-host disease, where the NK cells have been directly implicated (Charley *et al.*, 1986; Thiele *et al.*, 1987).

The esters of amino acids and dipeptides have been shown to be potential chemotherapeutic agents, especially in parasitic diseases. In fact, these molecules can be targeted to the parasites, where they accumulate in lysosome-like structures and can lyse cells, as we have seen *in vitro*, and kill the protozoa. Rabinovitch and co-workers (Rabinovitch *et al.*, 1986, 1987; Rabinovitch and Alfieri, 1987) have shown that the esters of amino acids and dipeptides are capable of killing *Leishmania mexicana amazonensis amastigotes* present in cultured macrophages or isolated mouse lesions. Since *Leishmania* possesses organelles that resemble lysosomes (megasomes), the hypothesis is that the toxic effect of the esters on the cells is due to an accumulation of the compound in the subcellular structures of the parasite, leading to a lysis of these structures. A number of experimental arguments favor this hypothesis. The cytotoxicity is pH-dependent and it is reduced by substances such as NH$_4$Cl and monensin that alkalinize the acidic compartments and so reduce the pH gradient required for intralysosomal accumulation of these substances. The esters have to be hydrolyzed after they enter the parasite to exert their lethal effect. An analysis of the relationship between structure and activity shows that the esters of dipeptides are more active than the esters of amino acids. The activity of the heterodipeptide esters depends on the position and nature of the amino acids, the active dipeptides containing at least one hydrophobic amino acid. Hence, if the cytotoxicity of the esters of amino acids and dipeptides toward *Leishmania* depends on their accumulation in megasomes, the lysis of these organelles could be due to osmotic shock. However, we cannot exclude the possibility that this is not the only way lysosomal lysis occurs and that other mechanisms exist that result in a direct effect of the sequestered molecule on the megasomal membrane.

3.2.1d. Photosensitive Substances. Photochemotherapy in the treatment of cancer is still at an experimental stage. However, numerous studies have shown that these cytolytic photosensitive compounds accumulate in lysosomes. From this, it is hypothesized that the lysosomes are the principal target of the photocytotoxic action brought about by these compounds (Lin *et al.*, 1993; Sasaki *et al.*, 1993). It has been found that the accumulation in lysosomes of molecules like the derivatives of Nile Blue can provoke photodestruction of the organelle.

Plasma lipoproteins are the principal transporters of porphyrins and related molecules that might play the role of photosensitive substances in tumors. The plasma lipoproteins are endocytosed and reach the lysosomes, where their degradation could lead to the release of photosensitive substances and exert their phototoxic effect on these organelles. Use of the low density lipoprotein (LDL) receptor pathway could be interesting to exploit. Low density lipoprotein would be an adequate vehicle to direct the photosensitive substance toward the lysosome (Maziere *et al.*, 1991). Another way of targeting photosensitive substances to lysosomes wsa proposed by Bergstrom *et al.* (1994). These authors directed pheophorbide, a cytotoxic agent, by coupling it to a monoclonal antibody against a bladder tumor cell line. After its capture by the cell, the immunotoxin finds itself in lysosomes where it induces a photoactivated cellular lysis.

3.2.1e. Oxygen Free Radical. Free radicals of oxygen seem to play an important role in several physiopathological situations. Intracellular membranes may serve as targets for these radicals, being destroyed by the lipid peroxidation the radicals induce. The lysosomal membrane is sensitive to free radicals. When lysosomes are incubated with systems that generate free radicals, they are ruptured (Decharneux *et al.*, 1992). This phenomenon could explain the cytolytic effect of these radicals when there is a reperfusion in ischemic regions and in aging and other degenerative conditions in which the production of oxygen free radicals is increased. As indicated earlier, certain flavonoids oppose the lysosomolytic effect of free radicals.

3.2.1f. Other Substances. *Trypanosoma brucei* is a protozoan pathogen that infects cattle, causing nagana, but it is incapable of infecting humans because human serum contains a protein factor, part of the high density lipoprotein (HDL) fraction, that has a cytolytic activity against the trypanosome. It has been shown recently (Hager *et al.*, 1994) that this protein is internalized by the trypanosome and reaches the lysosomes where it accumulates, finally leading to their lysis. Hager *et al.* (1994) have proposed that the cytolytic action of the protein on the protozoa arises from the rupture of lysosomes and that the lysosomal hydrolases released ultimately cause the autodigestion of the cells. The mechanism of lysosome disruption is not known. Since cells pretreated with leupeptin are protected against lysis induced by the protein, Hager *et al.* proposed that the lytic effect requires a thiol-proteinase-mediated activation step, perhaps to process the native protein to a toxic form endowed with membranolytic properties.

Recently, Sai *et al.* (1994) observed that lysosomolytic factors were produced in the cytosol of cells treated with GTPγs. They probably comprise one or more proteins of high molecular weight. The effect involves two steps: The first is a modificaiton of the cytosol by GTPγs; the second is the disintegration of lysosomes by the GTPγs-modified cytosol and is an ATP-dependent process. The first step could consist of an activation of a cytosolic phospholipase A_2, which would attack the lysosomal membrane in the second step and cause a rupture of the organelles. The physiological role of this phenomenon as observed *in vitro* is still a hypothetical one. Sai *et al.* suggest that it could be the method of cytolysis induced by agents such as tumor necrosis factor (TNF).

Peroxidized low density lipoproteins are more cytotoxic toward cancer cells than normal cells (Fossel *et al.,* 1994). These lipoproteins are endocytosed via the LDL receptor and reach lysosomes. Apparently, these molecules are capable of labilizing the membrane of lysosomes that contain them, resulting in the rupture of lysosomes in the cell. It is suggested that the cytolytic effect of the peroxidized LDL arises from their ability to lyse lysosomes.

When lysosomes of rat liver are incubated at pH 5 and 37 °C, they are subject to a progressive lysis. We have recently found that this disruption is greatly retarded in very low concentrations of serum. The factor responsible for this action is heat-resistant and can be recovered with high-molecular-weight compounds after gel filtration. The same effect can be achieved with the supernatant obtained after a high-speed centrifugation of the liver and heating for a few minutes at 100 °C.

It has been suggested that urate crystals responsible for gouty inflammation cause intracellular disruption of lysosomes, after phagocytosis. A similar phenomenon could take place after phagocytosis of silica particles, leading to the killing of lung macrophages in silicosis. Although such a hypothesis is plausible, it is not in our opinion sufficiently supported to date by experimental results.

3.2.1g.   Factors Affecting the Stability of Lysosomes.   Many factors are capable of modulating the response of lysosomes to a lysosomolytic agent. For example, the lowering of pH makes lysosomes much more resistant to osmotic stress arising from incubation with isoosmotic glucose (Docherty and Hales, 1979). A possible explanation is that sugar permeation of lysosomes involves a transporter whose efficiency depends on ionization of specific sites. Liver lysosomes loaded with endocytosed Triton WR1339 are much more resistant than nonloaded lysosomes to lysis at pH 5 and 37 °C. The efficacy of cholesterol in protecting these organelles from lysosomolytic agents is also much greater when lysosomes are loaded with Triton WR1339 (Wattiaux *et al.,* 1963). Starvation for a few days also increases the stability of rat liver lysosomes at pH 5 (Fig. 5). The way these fac-

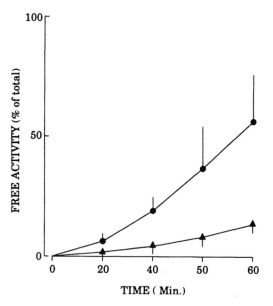

FIGURE 5. Effect of starvation on the stability of lysosomal membrane. Liver homogenates of male Wistar rats were incubated in 0.25 M sucrose at pH 5, 37 °C. Then free N-acetylglucosaminidase was measured. (●) Rats fed *ad libitum*; (▲) rats starved for 4 days. Means with SD of five animals.

tors act is not yet well understood. However, when one compares the response of lysosomes to different lysosomolytic agents, it is important to control the treatment conditions (pH, nutritional state of the animal, etc.).

2.2.2. Modification of the Size and Density of Lysosomes

The entry of extracellular substances into lysosomes can lead to significant modifications in the size and/or density of the lysosomes. Such a situation arises when molecules entering the lysosome accumulate there because they are unable to leave or be degraded. These modifications can be discovered by using analytical centrifugation (differential and isopycnic) and by morphological examination.

3.2.2a. Substances Entering by Endocytosis. A typical example is a natural or artificial polymer which can enter cells only by endocytosis (fluid, absorptive, or receptor-mediated). In such cases, the molecule usually ends up in the lysosome. If it cannot be degraded by the hydrolases, it remains sequestered in these organelles and could provoke an increase in their size and a modification of their density. A particularly striking example is that of Triton WR1339 (Wattiaux *et al.,* 1963). When an animal is injected with this non-ionic detergent, it is endocytosed by the liver, probably bound to plasma lipoproteins. It accumulates in lysosomes, where no lysosomal hydrolase is capable of degrading it. This results in a considerable increase in the size of the organelles, and, as the detergent has a very low density, there is a significant decrease in the density of these subcellular structures. A special case is sucrose. This disaccharide cannot cross the plasma membrane except by endocytosis. Hence, it accumulates in lysosomes, which do not contain invertase. As sucrose is a small molecule, its accumulation leads to an osmotic imbalance that provokes the swelling of the lysosomes (Wattiaux *et al.,* 1964). If before injecting sucrose the lysosomes are loaded with an invertase, accumulation does not take place (Thirion *et al.,* 1983).

When lysosomes are subjected to an increase in size *in vivo,* problems arise, depending on the manner by which the lysosomes recruit supplementary membrane and acquire a greater cellular territory. Swelling of lysosomes *in vitro* rapidly leads to their rupture; the lysosomal membrane is inexpandable. Thus, it is necessary that *in vivo* certain mechanisms are involved that can increase the membrane surface of each lysosome. One possibility is the fusion of a number of lysosomes, which would lead to a decrease in surface/volume ratio of the particle. However, such processes are probably not sufficient to explain, for example, the extent of vacuolation of cells in which chloroquine accumulates.

3.2.2b. Substances Entering by Permeation. Certain low-molecular-weight compounds can diffuse across the lysosomal membrane, at least *in vitro* (Forster and Lloyd, 1988). Increase in the concentration of such substances in the lysosome causes an osmotic imbalance, resulting in a swelling of these organelles, leading to their rupture. Some of these low-molecular-weight compounds are relatively hydrophobic weak bases, for example, alkylamines and chloroquine, which on accumulation in lysosomes attain a concentration much higher than that outside the lysosome. This leads to a significant swelling of the lysosomes, which *in vivo* is seen as a vacuolization of the cell (Ohkuma and Poole, 1981). This swelling induces a decrease in the lysosomal density (Limet *et al.,* 1985).

3.2.3. Modification of the Intralysosomal pH

It is generally accepted that the intralysosomal pH is approximately 5. It is possible that the buffering capacity of the lysosomal matrix and the proton pump in the membrane are capable of maintaining this pH during the digestion of substances within the lysosome. However, accumulation of certain basic compounds in these organelles can lead to an increase in the lysosomal pH. For example, chloroquine can attain a concentration inside the lysosomes considerably higher than in the cytosol (Poole and Ohkuma, 1981). Under these conditions, it is possible that the buffering capacity and the proton pump are unable to counteract such an increase in the lysosomal pH.

3.2.4.   Functional Consequences

What are the functional consequences of a modification in the physicochemical properties of the lysosomes? We see that a drastic modification in the permeability of the lysosomal membrane can have a dramatic effect on the cell. Certainly
experimental observations suggest that lysosomal lysis can induce cell death.
However, it is extremely difficult to establish experimentally that the lysosomes
are really disrupted within the cell and, if this is true, that this rupture is the cause
of cell death. In addition, little is known about the state of the lysosomal hydrolases liberated in the cytosol. The enzymes would be in a pH environment different from the one at which they exhibit maximum activity. In addition, they could
be in the presence of inhibitors, which are absent in the lysosome. Thirdly, the
overall concentration of lysosomal enzymes released in the cytosol is much lower
than when they are confined to lysosomes, since the volume of the lysosome represents a very small proportion of cellular volume. Thus it is necessary to be cau-
tious before attributing cell death to lysosomal lysis. This does not mean that
intracellular lysosomal lysis leaves the cell unaffected. It is evident that any alteration in autophagy or heterophagy would have consequences on the functioning
of the cell.

Obviously, an increase in lysosomal pH due to the effect of certain compounds may have a repercussion on lysosomal functions, leading mainly to an inhibition of acid hydrolases activity, which would finally have an effect on
heterophagy and autophagy. On the other hand, we have no information regarding
the effect of lysosomal swelling on intracellular digestive functions.

3.3.   Effects on Lysosomal Hydrolases

If a compound is an inhibitor of a lysosomal hydrolase and it is capable of
penetrating into lysosomes *in vivo,* it could seriously perturb intralysosomal digestion. Generally this would result in the accumulation of undigested material in
these organelles, which would lead to morphological and biochemical modification. We consider some well-known examples.

3.3.1.   Inhibitors of Proteolytic Activity

When leupeptin, a serine and cysteine protease inhibitor, is administered to
animals, it provokes an augmentation of autophagic vacuoles (Furuno *et al.,*
1982; Ishikawa *et al.,* 1983; Henell and Glaumann, 1984). Lysosomes become
loaded with incompletely degraded cytoplasmic material. Their sedimentation
properties and equilibrium density are altered (Aronson *et al.,* 1981). Intralysosomal sequestration seems to continue, and this leads to the accumulation of soluble material of cytosolic origin in the autophagic vacuoles. The concentration of
lactate dehydrogenase (LDH), for example, is increased in these vacuoles. It is

possible that leupeptin enters lysosomes following its endocytosis. Chloroquine, which can enter lysosomes by permeation, has a similar effect. Microtubule poisons, like vinblastine and colchicine, inhibit protein degradation in isolated cells and cause an accumulation of autophagosomes (Kovacs *et al.,* 1982). It is possible that this phenomenon results from an inhibition of the motility of lysosomes and autophagosomes, preventing their fusion. It should be pointed out that these effects are reversible; the lysosomes become progressively more normal when the administration of the drug is stopped. From a functional point of view, the use of proteolytic inhibitors permits an estimate of the role of autophagy in intracellular proteolysis. It also allows us to study degradation processes that are not exclusively concerned with cellular organelles but play a role in the catabolism of cytosolic proteins (Henell and Glaumann, 1985; Ahlberg *et al.,* 1985).

3.3.2. Cationic Amphiphilic Drugs

A number of cationic amphiphilic drugs are capable of inducing phospholipidosis, an abnormal accumulation of phospholipids in the tissues. They have diverse pharmacologic actions [e.g., antipsychotic (chlorpromazine), antiarrhythmic (propranolol), antidepressive (imipramine)], but their physicochemical properties are similar. Most of them have a hydrophobic cyclic structure and a side chain bearing a charged hydrophilic group. The effect of these drugs on biological membranes has been well studied; the phospholipidosis induced by them is a result of an interaction with membrane phospholipids, which become poorer substrates for phospholipases, or a direct inhibition of phospholipases, or both. The lysosomes could be implicated in phospholipidosis provoked by these drugs, as most of them are lysosomotropic (Lüllmann-Rauch, 1979; Reasor, 1989; Kodavanti and Mehendale, 1990) and can inhibit intralysosomal hydrolysis of phospholipids (Mingeot-Leclercq *et al.,*1988). However, we know that some of these compounds, for example, amiodarone, do not accumulate in lysosomes, but are capable of inducing phospholipidosis (Heath *et al.,* 1985). This would indicate that certain types of phospholipidosis brought on by cationic amphiphilic drugs can be manifested without involving the lysosomal system.

Some dicationic amphiphilic compounds, such as immunomodulatory drugs like tilorone (Burmester *et al.,* 1990; Hein and Lullmann-Rauch, 1989; Lullmann-Rauch, 1994), induce accumulation of sulfated glucosaminoglycans in lysosomes. The storage originates from an impairment of lysosomal degradation of these glycans and leads to a pathological situation reminiscent of inherited mucopolysaccharidoses. The mechanism of inhibition of glycan degradation is not clear. It has been proposed that dicationic compounds cause the formation of insoluble complexes with sulfated glucosaminoglycans and that these would be resistant to lysosomal hydrolases (Lullmann-Rauch and Ziegenhagen, 1991).

3.3.3. Inhibitors of Glycosidases

Swainsonine, an alkaloid from the plant *Swainsona canescens* is a strong inhibitor of lysosomal mannosidase (Dorling *et al.*, 1980); it may also change the stability of the lysosomal membrane (Tulsiani and Touster, 1992). Swainsonine induces a toxicity in grazing livestock after prolonged ingestion of the plant. The disease is similar to the mannosidosis of genetic origin in which there is an absence of lysosomal mannosidase. Hence, swainsonine is an interesting molecule for studying experimentally the pathogenesis of a disease involving lysosomes. Lysosomal storage of glycogen that mimics Pompe disease can be obtained by treating rats with acarbose or castanospermine, inhibitors of α-1,4-glucosidase (Geddes and Taylor, 1985; Konishi *et al.*, 1989; Saul *et al.*, 1985).

4. FINAL COMMENTS

We have tried to show how the lysosomal system may be involved in the pharmacological and toxic effects of biological agents. Although a great deal of work has been devoted to this subject over the years, there has been little in the approach and methodology applied.

A number of substances exist that act on the constituents of the vacuolar system (e.g., Golgi and endosomes) and have a functional relationship with the lysosomes. These compounds would have an indirect effect on the lysosomes by modifying the glycosylation of the lysosomal enzymes or the lipid composition of their membrane.

Finally, lysosomes have a membrane system that is ideal for studying the lytic effect of drugs and toxic substances on biological membranes. When testing the lytic effect of certain pharmacologic or toxic compounds, the red blood cell membrane has been used most frequently as a model for the biological membrane. In this simple model, the lytic effect is manifested as hemolysis. An alternative, which is only slightly more complicated, is the use of lysosomes. Membrane lysis is demonstrated as explained before, by the loss of latency of a lysosomal enzyme. It is not necessary to use a purified preparation of lysosomes; a homogenate is sufficient. The process is quantitative and can be useful as a comparison with hemolysis.

ACKNOWLEDGMENTS.   We thank Dr. Sandra Misquith for help in preparing this manuscript. This work was suported by a grant from the Fonds de la Recherche Scientifique Médicale (contract no. 430395).

5. REFERENCES

Ahlberg, J., Berkenstam, A., Henell, F., and Glaumann, H., 1985, Degradation of short and long lived proteins in isolated rat liver lysosomes, *J. Biol. Chem.* **260**:5847–5854.

Aronson, N. N., Dennis, P. A., and Dunn, W. A., 1981, Metabolism of leupeptin and its effect on autophagy in the perfused rat liver, *Acta Biol. Med. Germ.* **40**:1531–1538.

Ashwell, G., and Harford, J., 1982, Carbohydrate-specific receptors of the liver, *Annu. Rev. Biochem.* **51**:531–554.

Baccino, F. M., and Zuretti, M. F., 1976, Permeability of rat liver lysosome membranes, *Panminerva Med.* **18**:472–491.

Berg, T. O., Stromhaug, P. E., Lovdal, T., Seglen, P. O., and Berg, T., 1994a, Use of glycyl-L-phenylalanine 2-naphthylamide, a lysosome-disrupting cathepsin C substrate, to distinguish between lysosomes and prelysosomal endocytic vacuoles, *Biochem. J.* **300**:229–236.

Berg, T. O., Stromhaug, P. E., Berg, T., and Seglen, P. O., 1994b, Separation of lysosomes and autophagosomes by means of glycyl-phenylalanine-naphthylamide, a lysosome-disrupting cathepsin C substrate, *Eur. J. Biochem.* **221**:595–602.

Bergstrom, L. C., Vucenik, I., Hagen, I. K., Chernomorsky, S. A., and Poretz, R. D., 1994, In vitro photocytotoxicity of lysosomotropic immunoliposomes containing pheophorbide a, with human bladder carcinoma cells, *J. Photochem. Photobiol.* **24**:17–23.

Bird, S. J., and Lloyd, J. B., 1995, Mechanism of lysosome rupture by dipeptides, *Cell Biochem. Funct.* **13**:79–83.

Burmester, J., Handrock, K., and Lüllmann-Rauch, R., 1990, Cultured corneal fibroblasts as a model system for the demonstration of drug-induced mucopolysaccharidosis, *Arch. Toxicol.* **64**:291–298.

Burton, R., and Lloyd, J. B., Latency of some glycosidases of rat liver lysosomes, *Biochem. J.* **160**:631–638.

Cappaccioli, S., Di Pasquale, G., Mini, E., Mazzei, T., and Quattrone, A., 1993, Cationic lipids improve antisense oligonucleotide uptake and prevent degradation in cultured cells and human serum, *Biochem. Biophys. Res. Commun.* **197**:818–825.

Charley, M. R., Thiele, D. L., Bennett, M., and Lipsiky, P. E., 1986, Prevention of lethal murine graft versus host disease by treatment of donor cells with L-leucyl-L-leucine methyl ester, *J. Clin. Invest.* **78**:1415–1420.

Decharneux, Th., Dubois, F., Beauloye, C., Wattiaux-De Coninck, S., and Wattiaux, R., 1992, Effect of various flavonoids on lysosomes subjected to an oxidative and an osmotic stress, *Biochem. Pharmacol.* **44**:1243–1248.

de Duve, C., 1965, The separation and characterization of subcellular particles, *Harvey Lect.* **59**:49–60.

de Duve, C., and Wattiaux, R., 1966, Functions of lysosomes, *Annu. Rev. Physiol.* **28**:435–492.

de Duve, C., De Barsy, T., Poole, B., Trouet, A., Tulkens, P., and Van Hoof, F., 1974, Lysosomotropic agents, *Biochem. Pharamcol.* **223**:2495–2531.

de Duve, C., Wattiaux, R., and Baudhuin, P., 1962, Effect of fat-soluble compounds on lysosomes in vitro, *Biochem. Pharmacol.* **9**:97–116.

Docherty, K., and Hales, C. N., 1979, The effects of ions and pH on the transport of sugars into rat liver lysosomes, *FEBS Lett.* **106**:145–148.

Dorling, P. R., Huxtable, C. R., and Colegate, S. M., 1980, Inhibition of lysosomal α-mannosidase by swainsonine, an indolizidine alkaloid isolated from *Swainsona canescens*, *Biochem. J.* **191**:649–651.

Firestone, R. A., Pisano, J. M., and Bonney, R. J., 1979, Lysosomotropic agents. 1. Synthesis and cytotoxic action of lysosomotropic detergents, *J. Med. Chem.* **22**:1130–1133.

Forster, S., and Lloyd, J. B., 1988, Solute translocation across the mammalian lysosome membrane, *Biochim. Biophys. Acta* **947**:465–491.

Fossel, E. T., Zanella, C. L., Fletcher, J. G., and Hui, K. K., 1994, Cell death induced by peroxidized low-density lipoprotein: Endopepsis, *Cancer Res.* **54**:1240–1248.

Furuno, K., Ishikawa, T., and Kato, K., 1982, Appearance of autolysosomes in rat liver after leupeptin treatment, *J. Biochem.* **91**:1485–1492.

Geddes, R., and Taylor, J. A., 1985, Lysosomal glycogen storage induced by acarbose, a 1,4-α-glucosidase inhibitor, *Biochem. J.* **228**:319–324.

Goldman, R., 1973, Dipeptide hydrolysis within intact lysosomes in vitro, *FEBS Lett.* **33**:208–212.

Goldman, R., and Kaplan, A., 1973, Rupture of rat liver lysosomes mediated by L-amino acid esters, *Biochem. Biophys. Acta* **318**:205–216.

Hager, K. M., Pierce, M. A., Ray Moore, D., Tytler, E. M., Esko, J. D., and Hadjuk, S. L., 1994, Endocytosis of a cytotoxic human high density lipoprotein results in disruption of acidic intracellular vesicles and subsequent killing of African trypanosomes, *J. Cell Biol.* **126**:155–167.

Heath, M. F., Costa-Jussa, F. R., Jacobs, J. M., and Jacobson, W., 1985, The induction of pulmonary phospholipidosis and the inhibition of lysosomal phospholipases by amiodarone, *Br. J. Exp. Pathol.* **66**:391–397.

Hein, L., and Lullmann-Rauch, R., 1989, Mucopolysaccharidosis and lipidosis in rats treated with tilorone analogues, *Toxicology* **58**:145–154.

Hennell, F., and Glaumann, H., 1984, Effect of leupeptin on the autophagic vacuolar system of rat hepatocytes, *Lab. Invest.* **51**:46–56.

Hennell, F., and Glaumann, H., 1985, Participation of lysosomes in basal proteolysis in perfused rat liver. Discrepancy between leupeptin-induced lysosomal enlargement and inhibition of proteolysis, *Exp. Cell Res.* **158**:257–262.

Ishikawa, T., Furuno, K., and Kato, K., 1983, Ultrastructural studies on autolysosomes in rat hepatocytes after leupeptin treatment, *Exp. Cell Res.* **144**:15–21.

Jadot, M., and Wattiaux, R., 1985, Effect of glycyl-L-phenylalanine 2-napththylamide on invertase endocytosed by rat liver, *Biochem. J.* **225**:645–648.

Jadot, M., Colmant, C., Wattiaux-De Coninck, S., and Wattiaux, R., 1984, Intralysosomal hydrolysis of glycyl-L-phenylalanine 2-naphthylamide, *Biochem. J.* **219**:965–970.

Jadot, M., Wattiaux-De Coninck, S., and Wattiaux, R., 1989, The permeability of lysosomes to sugar. Effect of diethylstilbestrol on the osmotic activation of lysosomes induced by glucose, *Biochem. J.* **262**:981–984.

Jadot, M., Biélande, V., Beauloye, V., Wattiaux-De Coninck, S., and Wattiaux, R., 1990, Cytotoxicity and effect of glycyl-D-phenylalanine-2-naphthylamide on lysosomes, *Biochim. Biophys. Acta* **1027**:205–209.

Kodavanti, U. P., and Mehendale, H. M., 1990, Cationic amphiphilic drugs and phospholipid storage disorder, *Pharmacol. Rev.* **42**:327–354.

Konishi, Y., Hata, Y., and Fujimori, K. A., 1989, Formation of glycogenosomes in rat liver induced by injection of acarbose, an α-glucosidase inhibitor, *Acta Histochem. Cytochem.* **22**:227–231.

Kovacs, A. L., Reigh, A., and Seglen, P. O., 1982, Accumulation of autophagosomes after inhibition of hepatocytic protein degradation by vinblastine, leupetin or a lysosomotrophic amine, *Exp. Cell Res.* **137**:191–201.

Krogstad, D. J., and Schlesinger, P. H., 1986, A perspective of antimalarial action: Effects of weak bases on *Plasmodium falciparum*, *Biochem. Pharmacol.* **53**:547–552.

Krogstad, D. J., Schlesinger, P. H., and Gluzman, I. Y., 1985, Antimalarials increase vesicle pH in *Plasmodium falciparum*, *J. Cell Biol.* **101**:2302–2309.

Limet, J. N., Quintart, J., Schneider, Y. J., and Courtoy, P., 1985, Receptor mediated endocytosis of polymeric IgA and galactosylated serum albumin in rat liver, *Eur. J. Biochem.* **146**:539–548.

Lin, C. W., Shulok, J. R., Kirley, S. D., Bachelder, C. M., Flotte, T. J., Sherwood, M. E., Cincotta, L., and Foley, J. W., 1993, Photodynamic destruction of lysosomes mediated by Nile blue photosensitizers, *Photochem. Photobiol.* **58**:81–91.

Lloyd, J. B., 1969, Studies of the permeability of rat liver lysosomes to carbohydrates, *Biochem. J.* **115**:703–707.

Lüllmann-Rauch, R., 1979, Drug-induced lysosomal storage disorder, in *Lysosomes in Applied Biology and Therapeutics* (J. T. Dingle, P. J. Jacques, and I. H. Shaw, eds.), pp. 49–130, North Holland Publishing Co., New York.

Lüllmann-Rauch, R., 1994, Drug-induced intralysosomal storage of sulfated glycosaminoglycans: A methodological pitfall occurring with acridine derivatives, *Exp. Toxicol. Pathol.* **46**:315–322.

Lüllmann-Rauch, R., and Ziegenhagen, M., 1991, Acridine orange, a precipitant for sulfated glycoamino-glycans, causes mucopolysaccharidosis in cultured fibroblasts, *Histochemistry* **95**:263–268.

Maziere, J. C., Morliere, P., and Santus, R., 1991, The role of the low density lipoprotein receptor pathway in the delivery of lipophilic photosensitizers in the photodynamic therapy of tumours, *J. Photochem. Photobiol.* **8**:351–360.

Miller, D. K., Griffiths, E., Lenard, J., and Firestone, R. A., 1983, Cell killing by lysosomotropic detergents, *J. Cell Biol.* **97**:1841–1851.

Mingeot-Leclercq, M. P., Laurent, G., and Tulkens, P. M., 1988, Biochemical mechanism of aminoglycoside-induced inhibition of phosphatidylcholine hydrolysis by lysosomal phospholipases, *Biochem. Pharmacol.* **37**:591–599.

Misquith, S., Wattiaux-De Coninck, S., and Wattiaux, R., 1988, Uptake and intracellular transport in rat liver of formaldehyde-treated bovine serum albumin labelled with ^{125}I-tyramine-cellobiose, *Eur. J. Biochem.* **174**:691–697.

Monsigny, M., Roche, A. C., and Midoux, M., 1988, Endogenous lectins and drug targeting, *Ann. N.Y. Acad. Sci.* **414**:399–414.

Ohkuma, S., and Poole, B., 1981, Cytoplasmic vacuolation of mouse peritoneal macrophages and the uptake into lysosomes of weakly basic substances, *J. Cell Biol.* **90**:656–664.

Pisoni, R. L., and Thoene, J. G., 1991, The transport systems of mammalian lysosomes, *Biochim. Biophys. Acta* **1071**:351–373.

Poole, B., and Ohkuma, S., 1981, Effect of weak bases on the intralysosomal pH in mouse peritoneal macrophages, *J. Cell Biol.* **90**:665–669.

Postel-Vinay, M. C., Kayser, C., and Desbuquois, B., 1982, Fate of injected human growth hormone in the female rat liver "in vivo," *Endocrinology* **111**:244–251.

Putnam, D., and Kopeček, J., 1995, Polymer conjugates with anticancer activity, *Adv. Polym. Sci.* **122**:55–120.

Rabinovich, M., and Alfieri, S. C., 1987, From lysosomes to cells to leishmania: Amino acid esters as potential chemotherapeutic agents, *Braz. J. Med. Biol. Res.* **20**:665–674.

Rabinovitch, M., Zilberfarb, V., and Ramazeilles, C., 1986, Destruction of leishmania mexicana amazonensis amastigotes within macrophages by lysosomotropic amino acid esters, *J. Exp. Med.* **163**:520–536.

Rabinovitch, M., Zilberfarb, V., and Pouchelet, M., 1987, Leishmania mexicana: Destruction of isolated amastigotes by amino acid esters, *Am. J. Trop. Med. Hyg.* **36**:288–293.

Reasor, M. J., 1989, A review of the biology and toxicologic implications of the induction of lysosomal lamellar bodies by drugs, *Toxicol. Appl. Pharmacol.* **97**:47–56.

Reeves, S., Decker, R. S., Crie, J. S., and Wildenthal, K., 1981, Intracellular disruption of rat heart lysosomes by leucine methyl ester: Effects on protein degradation, *Proc. Natl. Acad. Sci. USA* **78**:4426–4429.

Sai, Y., Arai, K., and Ohkuma, S., 1994, Cytosol treated with GTPγS disintegrates lysosomes in vitro, *Biochem. Biophys. Res. Commun.* **198**:869–887.

Sasaki, M., Koyama, S., Tokiwa, K., and Fujita, H., 1993, Intracellular target for alpha-terthienyl photosensitization: Involvement of lysosomal membrane damage, *Photochem. Photobiol.* **57**:796–802.

Saul, R., Ghidoni, J. J., Molyneux, R. J., and Elbein, A. D., 1985, Castanospermine inhibits α-glucosidase activities and alters glycogen distribution in animals, *Proc. Natl. Acad. Sci. USA* **82**:93–97.

Slater, A. F. G., 1993, Chloroquine: Mechanism of drug action and resistance in *Plasmodium falciparum, Pharm. Ther.* **57**:203–235.

Steer, C. J., and Clarenburg, R., 1979, Unique distribution of glycoprotein receptors on parenchymal and sinusoidal cells of rat liver, *J. Biol. Chem.* **254**:4457–4461.

Thiele, D. L., and Lipsky, P. E., 1985a, Regulation of cellular function by products of lysosomal enzyme activity: Elimination of human natural killer cells by a dipeptide methyl ester generated from 1-leucine methyl ester by monocytes or polymorphonuclear leukocytes, *Proc. Natl. Acad. Sci. USA* **82**:2468–2472.

Thiele, D. L., and Lipsky, P. E., 1985b, Modulation of human natural killer cell function by L-leucine methyl ester: Monocyte-dependent depletion from human peripheral blood mononuclear cells, *J. Immunol.* **134**:786–793.

Thiele, D. L., and Lipsky, P. E., 1990, Mechanism of L-leucyl-L-leucine methyl ester mediated killing of cytotoxic lymphocytes: Dependence on a lysosomal thiol protease, dipeptidyl peptidase I, that is enriched in these cells, *Proc. Natl. Acad. Sci. USA* **87**:83–87.

Thiele, D. L., Charley, M. R., Calomeni, J. A., and Lipsky, P. E., 1987, Lethal graft-vs-host disease across major histocompatibility barriers: Requirement for leucyl-leucine methyl ester sensitive cytotoxic T cells, *J. Immunol.* **138**:51–57.

Thirion, J., Thibaut-Vercruyssen, R., Ronveaux-Dupal, M. F., and Wattiaux, R., 1983, Experimental sucrose overloading of rat liver lysosomes: Effect of pretreatment with invertase, *Eur. J. Cell Biol.* **31**:107–113.

Trail, P. A., Willner, D., and Hellström, K. E., 1995, Site-directed delivery of anthracyclines for treatment of cancer, *Drug Dev. Res.* **34**:196–209.

Trouet, A., Masquelier, M., Baurain, R., and Deprez-Campeneere, D., 1982, A covalent linkage between daunorubicin and proteins that is stable in serum and reversible by lysosomal hydrolases, as required for a lysosomotropic drug-carrier conjugate: in vitro and in vivo studies, *Proc. Natl. Acad. Sci. USA* **79**:626–629.

Tulkens, P. M., 1991, Intracellular distribution and activity of antibiotics, *Eur. J. Clin. Microbiol. Infect. Dis.* **10**:100–106.

Tulsiani, D. R. P., and Touster, O., 1992, Evidence that swainsonine pretreatment of rats leads to the formation of autophagic vacuoles and of endosomes with decreased capacity to mature to, or fuse with, active lysosomes, *Arch. Biochem. Biophys.* **296**:556–561.

Wattiaux, R., 1971, Drugs and lysosomes, in *Fundamentals of Biochemical Pharmacology* (Z. M. Bacq, ed.), pp. 176–184, Pergamon Press, London.

Wattiaux, R., Wibo, M., and Baudhuin, P., 1963, Influence of the injection of Triton WR 1339 on the properties of rat liver lysosomes, in *Ciba Foundation Symposium on Lysosomes* (A. V. deReuck and M. P. Cameron, eds.), pp. 176–200, J. & A. Churchill Ltd., London.

Wattiaux, R., Wattiaux-De Coninck, S., Rutgeerts, M. J., and Tulkens, P., 1964, Influence of the injection of a sucrose solution on the properties of rat liver lysosomes, *Nature* **203**:757–758.

Wattiaux, R., Wattiaux-De Coninck, S., Ronveaux-Dupal, M. F., and Dubois, F., 1978, Isolation of rat liver lysosomes by isopycnic centrifugation in a metrizamide gradient, *J. Cell Biol.* **78**:349–368.

Wattiaux, R., Wattiaux-De Coninck, S., Jadot, M., Hamer, I., Bielande, V., and Beauloye, V., 1992, Lysosomes as suicide bags, in *Current Topics in Endocytosis* (P. J. Courtoy, ed.), pp. 433–437, Springer Verlag, Berlin, Heidelberg.

Wattiaux, R., Jadot, M., Dubois, F., Misquith, S., and Wattiaux-De Coninck, S., 1995, Uptake of exogenous DNA by rat liver: Effect of cationic lipids, *Biochem. Biophys. Res. Commun.* **213**:81–87.

Wilson, P. D., Firestone, R. A., and Lenard, J., 1987, The role of lysosomal enzymes in killing of mammalian cells by the lysosomotropic detergent N-dodecylimidazole, *J. Cell Biol.* **104**:1223–1229.

Zhou, Z., and Huang, L., 1994, DNA transfection mediated by cationic liposomes containing lipolysine: Characterization and mechanism of action, *Biochim. Biophys. Acta* **1189**:195–203.

Index